OXFORD MASTER SERIES IN PARTICLE PHYSICS, ASTROPHYSICS, AND COSMOLOGY

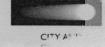

OXFORD MASTER SERIES IN PHYSICS

The Oxford Master Series is designed for final year undergraduate and beginning graduate students in physics and related disciplines. It has been driven by a perceived gap in the literature today. While basic undergraduate physics texts often show little or no connection with the huge explosion of research over the last two decades, more advanced and specialized texts tend to be rather daunting for students. In this series, all topics and their consequences are treated at a simple level, while pointers to recent developments are provided at various stages. The emphasis in on clear physical principles like symmetry, quantum mechanics, and electromagnetism which underlie the whole of physics. At the same time, the subjects are related to real measurements and to the experimental techniques and devices currently used by physicists in academe and industry. Books in this series are written as course books, and include ample tutorial material, examples, illustrations, revision points, and problem sets. They can likewise be used as preparation for students starting a doctorate in physics and related fields, or for recent graduates starting research in one of these fields in industry.

CONDENSED MATTER PHYSICS

1. M.T. Dove: *Structure and dynamics: an atomic view of materials*
2. J. Singleton: *Band theory and electronic properties of solids*
3. A.M. Fox: *Optical properties of solids, second edition*
4. S.J. Blundell: *Magnetism in condensed matter*
5. J.F. Annett: *Superconductivity, superfluids, and condensates*
6. R.A.L. Jones: *Soft condensed matter*
17. S. Tautz: *Surfaces of condensed matter*
18. H. Bruus: *Theoretical microfluidics*
19. C.L. Dennis, J.F. Gregg: *The art of spintronics: an introduction*

ATOMIC, OPTICAL, AND LASER PHYSICS

7. C.J. Foot: *Atomic physics*
8. G.A. Brooker: *Modern classical optics*
9. S.M. Hooker, C.E. Webb: *Laser physics*
15. A.M. Fox: *Quantum optics: an introduction*
16. S.M. Barnett: *Quantum information*

PARTICLE PHYSICS, ASTROPHYSICS, AND COSMOLOGY

10. D.H. Perkins: *Particle astrophysics, second edition*
11. Ta-Pei Cheng: *Relativity, gravitation and cosmology, second edition*

STATISTICAL, COMPUTATIONAL, AND THEORETICAL PHYSICS

12. M. Maggiore: *A modern introduction to quantum field theory*
13. W. Krauth: *Statistical mechanics: algorithms and computations*
14. J.P. Sethna: *Statistical mechanics: entropy, order parameters, and complexity*

Relativity, Gravitation and Cosmology

A Basic Introduction

Second Edition

TA-PEI CHENG

University of Missouri–St. Louis

OXFORD

UNIVERSITY PRESS

OXFORD
UNIVERSITY PRESS

Great Clarendon Street, Oxford ox2 6DP

Oxford University Press is a department of the University of Oxford.
It furthers the University's objective of excellence in research, scholarship,
and education by publishing worldwide in

Oxford New York

Auckland Cape Town Dar es Salaam Hong Kong Karachi
Kuala Lumpur Madrid Melbourne Mexico City Nairobi
New Delhi Shanghai Taipei Toronto

With offices in

Argentina Austria Brazil Chile Czech Republic France Greece
Guatemala Hungary Italy Japan Poland Portugal Singapore
South Korea Switzerland Thailand Turkey Ukraine Vietnam

Oxford is a registered trade mark of Oxford University Press
in the UK and in certain other countries

Published in the United States
by Oxford University Press Inc., New York

First edition 2005
Second edition 2010

British Library Cataloguing in Publication Data
Data available

Library of Congress Cataloging in Publication Data
Data available

Typeset by SPI Publisher Services, Pondicherry, India
Printed in Great Britain
on acid-free paper by
CPI Antony Rowe, Chippenham, Wiltshire

ISBN 978–0–19–957363–9 (Hbk.)
       978–0–19–957364–6 (Pbk.)

1 3 5 7 9 10 8 6 4 2

Contents

Preface

From the preface of the first edition

It seems a reasonable expectation that every student receiving a university degree in physics will have had a course in one of the most important developments in modern physics: Einstein's general theory of relativity. Also, given the exciting discoveries in astrophysics and cosmology of recent years, it is highly desirable to have an introductory course whereby such subjects can be presented in their proper framework. Again, this is general relativity (GR).

Nevertheless, a GR course has not been commonly available to undergraduates or even, for that matter, to graduate students who do not specialize in GR or field theory. One of the reasons, in my view, is the insufficient number of suitable textbooks that introduce the subject with an emphasis on physical examples and simple applications without the full tensor apparatus from the very beginning. There are many excellent graduate GR books; there are equally many excellent "popular" books that describe Einstein's theory of gravitation and cosmology at the qualitative level; and there are not enough books in between. I am hopeful that this book will be a useful addition at this intermediate level. The goal is to provide a textbook that even an instructor who is not a relativist can teach from. It is also intended that other experienced physics readers who have not had a chance to learn GR can use the book to study the subject on their own. As explained below, this book has features that will make such an independent study particularly feasible.

Students should have had the usual math preparation at the calculus level, plus some familiarity with matrices, and the physics preparation of courses on mechanics and on electromagnetism where differential equations of Maxwell's theory are presented. Some exposure to special relativity as part of an introductory modern physics course will also be helpful, even though no prior knowledge of special relativity will be assumed.

The emphasis of the book is pedagogical. The necessary mathematics will be introduced gradually. We first present the metric description of spacetime. Many applications (including cosmology) can already be discussed at this more accessible level; students can reach these interesting results without having to struggle through the full tensor formulation, which is presented in later parts of the book. A few other pedagogical devices are also deployed:

- a **bullet list** of topical headings at the beginning of each chapter serves as the "chapter abstracts," giving the reader a foretaste of upcoming material;

- matter in marked **boxes** are calculation details, peripheral topics, historical tit-bits that can be skipped over depending on the reader's interest;
- **Review questions** at the end of each chapter should help beginning students to formulate questions on the key elements of the chapter; answer keys to these questions are provided at the back of the book; (The practice of frequent quizzes based on these review questions can be an effective means to make sure that each member is keeping up with the progress of the class.)
- **Solutions to more difficult problems** at the end of the book also contains some extra material

. . .

This book is based on the lecture notes of a course I taught for several years at the University of Missouri—St. Louis. Critical reaction from the students has been very helpful. Daisuke Takeshita provided me with detailed comments. My colleague Ricardo Flores has been very generous in answering my questions—be they in cosmology or computer typesetting. The painstaking task of doing all the line-drawing figures was carried out by Cindy Bertram. My editor Sonke Adlung at OUP has given me much support and useful advice. He arranged to have the manuscript reviewed by scholars who provided many suggestions for improvements. To all of them I am much indebted. Additionally, Professor Eric Sheldon was kind enough to read over the entire book and made numerous suggestions for editorial improvements, which were adopted in the later printings of the first edition. Finally, I am grateful to my wife Leslie for her patient understanding during the rather lengthy period that it took me to complete this project.

Preface to the second edition

In this enlarged and updated edition, I have added a chapter on special relativity and another on black holes. With more background on the basics of special relativity, its geometric formulation can be presented more clearly. Black holes being such a fundamental feature of strong gravity, a more thorough exposition is clearly warranted. Astronomy and astrophysics continue their impressive advances in recent years—from the double pulsar as laboratory of GR tests, to new observational evidence for dark matter and dark energy. As a consequence, almost every chapter of this book has been revised and updated.

Given our order of presentation, with the more interesting applications coming before the difficult mathematical formalism, it is hoped that the book can be rather versatile in terms how it can be used. With the additional material in this second edition, we can suggest several possibilities:

1. Parts I, II & III should be suitable for an undergraduate course. The tensor formulation in Part IV can then be used as extracurricular material for instructors to refer to, and for interested students to explore on their own. Much of the intermediate steps being given and more difficult problems

having their solutions provided, this section can in principle be used as self-study material by a particularly motivated undergraduate.

2. The whole book can be used for a senior-undergraduate / beginning-graduate course. For such a slightly more advanced course, Part I can in principle be omitted (or assigned as review reading). To fit into a one-semester course, one may have to leave some applications and illustrative examples to students as self-study topics.

3. The book is also suitable as a supplemental text: for an undergraduate astronomy course on cosmology, to provide a more detailed discussion of general relativity; for a regular advanced GR & cosmology course, to ease the transition for those graduate students not having had a thorough preparation in the relevant area.

4. The book is written having very much in mind readers doing independent study of the subject. The mathematical accessibility, and the various "pedagogical devices" (topic headings, review questions and worked-out solutions, *etc.*) should make it practical for an interested reader to use the book to study GR and cosmology on his or her own.

I am grateful to Oxford University Press for giving me the opportunity to put forward this revised edition. In this effort I have also been helped by Cindy Bertram who did the additional line-drawings and by Jane Dunford-Shore and David Coss, who have read various parts of the new edition and provided helpful comments. As always, I shall be glad to receive, at tpcheng@umsl.edu, reader's comments and spotting of errors. An updated list of corrections will be displayed at the website `http://www.umsl.edu/~tpcheng/grbook.html`.

St. Louis   *T.P.C.*

This book is dedicated to
Professor Ling-Fong Li
for more than 30 years' friendship and enlightenment

PRELIMINARIES

Part
I

Introduction and overview

1

- Relativity means that physically it is impossible to detect absolute motion. This can be stated as a symmetry in physics: physics equations are unchanged under coordinate transformations (i.e. when viewed by different observers).

- Special relativity (SR) is the symmetry with respect to coordinate transformations among inertial frames, general relativity (GR) among more general frames, including accelerating coordinate systems.

- The equivalence between the physics due to acceleration and to gravity means that GR is also the relativistic theory of gravitation, and SR is valid only in the absence of gravity.

- Einstein's motivations to develop GR are reviewed, and his basic idea of curved spacetime as the gravitation field is outlined.

- Relativity represents a new understanding of space and time. In SR we first learn that time is also a frame-dependent coordinate; the arena for physical phenomena is four-dimensional spacetime. GR interprets gravity as the structure of this spacetime. Ultimately, according to Einstein, space and time have no independent existence: they express the relational and causal structure of physical processes in the world.

- GR provides the natural conceptual framework for cosmology. The expanding universe reflects a dynamical spacetime. Basic features of an "exploding space" during the big bang (inflation) and the accelerated expansion during the current epoch (dark energy) can be accommodated simply by a vacuum energy term in the GR field equation, which gives rise to a gravitational repulsive force.

- The experimental foundation of GR will be emphasized in our presentation. The necessary mathematics is introduced as they are needed. After the preliminaries of Part I, we discuss the description of spacetime by the metric function in Part II. From this we can discuss many GR applications, including the study of cosmology, given in Part III. Only in Part IV do we introduce the full tensor formulation of the GR field equations and the ways to solve them.

Einstein's general theory of relativity is a classical field theory of gravitation. It encompasses, and goes beyond, Newton's theory, which is valid only for particles moving with slow velocity (compared to the speed of light) in a weak and static gravitational field. Although the effects of general relativity (GR) are often small in the terrestrial and solar domains, its predictions have been accurately verified whenever high precision observations can be performed. When it comes to situations involving strong gravity, such as compact stellar objects and cosmology, the use of GR is indispensable. Einstein's theory predicted the existence of black holes, where the gravity is so strong that even light cannot escape from them. GR, with its fundamental feature of a dynamical spacetime, offers a natural conceptual framework for cosmology of an expanding universe. Furthermore, GR can simply accommodate the possibility of a constant "vacuum energy density" giving rise to a repulsive gravitational force. Such an agent is the key ingredient of modern cosmological theories of the big bang (the inflationary cosmology) and of the accelerating universe (having a dark energy).

Creating new theories for the phenomena that are not easily observed on earth poses great challenges. We cannot repeat the steps that led to the formulation of Maxwell's theory of electromagnetism, as there are not many experimental results one can use to deduce their theoretical content. What Einstein pioneered was the elegant approach of using physics symmetries as a guide to the new theories that would be relevant to the yet-to-be-explored realms. As we shall explain below, relativity is a coordinate symmetry. Symmetry imposes restrictions on the equations of physics. The condition that the new theory should be reduced to known physics in the appropriate limit often narrows it down further to a very few possibilities. The symmetry Einstein used for this purpose is the coordinate symmetries of relativity, and the guiding principle in the formulation of GR is the "principle of general covariance." In Section 1.1 we shall explain the meaning of a symmetry in physics, as well as present a brief historical account of the formulation of relativity as a coordinate symmetry. In Section 1.2 we discuss the motivations that led Einstein to his geometric view of gravitation that was GR.

Besides being a theory of gravitation, GR, also provides us with a new understanding of space and time. Starting with special relativity (SR), we learnt that time is not absolute. Just like spatial coordinates, it depends on the reference frame as defined by an observer. This leads to the perspective of viewing physical events as taking place in a 4D continuum, called the **spacetime**. Einstein went further in GR by showing that the geometry of this spacetime was just the phenomenon of gravitation and was thus determined by the matter and energy distribution. Ultimately, this solidifies the idea that space and time do not have an independent existence; they are nothing but mirroring the relations among physical events taking place in the world.

General relativity is a classical theory because it does not take into account quantum effects. GR being a theory of space and time means that any viable theory of quantum gravity must also offer a quantum description of space and time. Although quantum gravity[1] is beyond the scope of this book, we should nevertheless mention that current research shows that such a quantum theory has a rich enough structure to be a unified theory of all matter and interactions (gravitation, strong, and electroweak, etc.). Thus the quantum generalization of GR should be **the** fundamental theory in physics.

[1] Currently the most developed study of quantum gravity is string theory. For recent textbook expositions see Zwiebach (2009), Becker, Becker, and Schwarz (2007), and Kiritsis (2007).

In this introductory chapter, we shall put forward several "big motifs" in the theory of relativity without much detailed explanation. Our purpose is to provide the reader with an overview of the subject—a roadmap, so to speak. It is hoped that, proceeding through the subsequent chapters, the reader will have occasion to refer back to this introduction, to see how various themes are substantiated.

1.1 Relativity as a coordinate symmetry

We are all familiar with the experience of sitting in a train, and not able to "feel" the speed of the train when it is moving with a constant velocity, and, when observing a passing train on a nearby track, find it difficult to tell which train is actually in motion. This can be interpreted as saying that no physical measurement can detect the absolute motion of an inertial frame. Thus we have the basic concept of **relativity**, stating that only relative motion is measurable in physics.

In this example, the passenger is an observer who determines a set of coordinates (i.e. rulers and clocks). What this observer measures is the physics with respect to this coordinate frame. The expression "the physics with respect to different coordinate systems" just means "the physics as deduced by different observers." Physics should be independent of coordinates. Such a statement proclaims a **symmetry in physics**: Physics laws remain the same (i.e. physics equations keep the same form) under some **symmetry transformation**, which changes certain conditions, for example, the coordinates. The invariance of physics laws under coordinate transformation is called **symmetry of relativity**. This coordinate symmetry can equivalently be stated as the impossibility of any physical measurement to detect a coordinate change. Namely, if the physics remains the same in all coordinates, then no experiment can reveal which coordinate system one is in, just as the passenger cannot detect the train's constant-velocity motion.

Rotational symmetry is a familiar example of coordinate symmetry. Physics equations are unchanged when written in different coordinate systems that are related to each other by rotations. Rotational symmetry says that it does not matter whether we do an experiment facing north or facing southwest. After discounting any peculiar local conditions, we should discover the same physics laws in both directions. Equivalently, no internal physical measurement can detect the orientation of a laboratory. The orientation of a coordinate frame is not absolute.

1.1.1 From Newtonian relativity to ether

Inertial frames of reference are the coordinate systems in which, according to Newton's first law, a particle will, if no external force acts on it, continue its state of motion with constant velocity (including the state of rest). Galileo and Newton taught us that the physics description would be the simplest when given in these coordinate systems. The first law provides us with the definition of an inertial system (also called Galilean frames of reference). Its implicit message that such coordinate systems exist is its physical content. Nevertheless, the first law does not specify which are the inertial frames in the physical universe. It is an empirical fact[2] that these are the frames moving at constant velocities with respect to the fixed stars—distant galaxies, or, another type of distant matter, the cosmic microwave background (CMB) radiation (see Section 10.5). There are infinite sets of such frames, differing by their relative orientation, displacement, and relative motion with constant velocities. For simplicity we shall ignore the transformations of rotation and displacement of coordinate origin, and concentrate on the relation among rectilinear moving coordinates—frames related by the **boost** transformation.

Physics equations in classical mechanics are unchanged under such boost transformations. That is, no mechanical measurement can detect the moving spatial coordinates. The familiar example of not being able to feel the speed of a moving train cited at the beginning of this section is a simple illustration of this **principle of Newtonian relativity**:[3] "physics laws (classical mechanics) are the same in all inertial frames of reference." In this sense, there is no absolute rest frame in Newtonian mechanics. The situation changed when electromagnetism was included. Maxwell showed a light speed being given by the static parameters of electromagnetism. Apparently there is only one speed of light c regardless of whether the observer is moving or not. Before Einstein, just about everyone took it to mean that Maxwell's equations were valid only in the rest frame of the **ether**, the purported medium for electromagnetic wave propagation. In effect this reintroduced into physics the notion of absolute space (the ether frame).

Also, in Newtonian mechanics the notion of time is taken to be absolute, as the passage of time is perceived to be the same in all coordinates.

1.1.2 Einsteinian relativity

It is in this context that one must appreciate Einstein's revolutionary proposal: All motions are relative and there is no need for concepts such as absolute space. Maxwell's equations are valid in every inertial coordinate system.[4] There is no ether. Light has the peculiar property of propagating with the same speed c in all (moving) coordinate systems—as confirmed by the Michelson–Morley experiment.[5] Furthermore, the constancy of the light speed implies that, as Einstein would show, there is no absolute time.

Einstein generalized Newtonian relativity in two stages:

- **1905** Covariance of physics laws under boost transformations were generalized from Newtonian mechanics to include electromagnetism. Namely, the laws of electricity and magnetism, as well as mechanics, are unchanged under the coordinate transformations that connect different inertial frames of reference. Einstein emphasized that this generalization implied a new kinematics: not only space but also time measurements are coordinate dependent. It is called the principle of **special relativity** because we are still restricted to the special class of coordinates: the inertial frames of reference.

[2]That there should be a physical explanation why distant matter defines inertial frames was first emphasized by Bishop George Berkeley in the eighteenth century, and by Ernst Mach in the nineteenth. A brief discussion of Mach's principle can be found in Box 1.1.

[3]We call it Newtonian relativity as the relativity principle applied to Newtonian physics. It is also known alternatively as **Galilean relativity** as the principle of relativity was first stated by Galileo in *Dialog Concerning the Two Chief World Systems* (1632), and also as it is the symmetry with respect to Galilean transformation, cf. Eq. (2.8).

[4]While emphasizing Einstein's role, we must also point out the important contribution to SR by Henri Poincaré and Hendrik Lorentz. In fact the full Lorentz transformation was originally written down by Poincaré (who named it in Lorentz's honor). Poincaré was the first one to emphasize the view of relativity as a physics symmetry. For accessible accounts of Poincaré's contribution, see Sartori (1996) and Logunov (2001).

[5]Michelson and Morley, using a Michelson interferometer, set out to measure a possible difference in light speeds along and transverse to the orbit motion of the earth around the sun. Their null result confirmed the notion that the light speed was the same in different inertial frames.

- **1915** The generalization is carried out further. **General relativity** is the physics symmetry allowing for more general coordinates, including accelerating frames as well. Based on the empirical observation that the effect of an accelerating frame and gravity is the same (the principle of equivalence), GR is the field theory of gravitation; SR is special because it is valid only in the absence of gravity. GR describes gravity as curved spacetime, which is flat in SR.

To recapitulate, relativity is a coordinate symmetry. It is the statement that physics laws are the same in different coordinate systems. Thus, physically it is impossible to detect absolute motion and orientation because physics laws are unchanged under coordinate transformations. For SR, these are the transformations among Galilean frames of reference (where gravity is absent); for GR, among more general frames, including accelerating coordinate systems.

1.1.3   Coordinate symmetry transformations

Relativity is the symmetry describing the covariance of physics equations (i.e. invariance of the equation form) under coordinate transformations. We need to distinguish among several classes of transformations:

- **Galilean transformation** In classical (nonrelativistic) mechanics, inertial frames are related to each other by this transformation. Thus, by Newtonian relativity, we mean that laws of Newtonian mechanics are covariant under Galilean transformations. From the modern perspective, Galilean transformations such as $t' = t$ are valid only when the relative velocity is negligibly small compared to c.
- **Lorentz transformation** As revealed by SR, the transformation rule connecting all the inertial frames, valid for all relative speed $\leq c$, is the Lorentz transformation. That is, Galilean is the low-speed approximation of the Lorentz transformation. Maxwell's equations were first discovered to possess this symmetry—they are covariant under the Lorentz transformation. It then follows that Newtonian (nonrelativistic) mechanics must be modified so that relativistic mechanics, valid for particles having arbitrary speed up to c, can also have this Lorentz symmetry.
- **General coordinate transformation** The principle that physics equations should be covariant under the general transformations that connect different coordinate frames, including accelerating frames, is GR. Such a symmetry principle is called the **principle of general covariance**. This is the basic guiding principle for the construction of the relativistic theory of gravitation.

Thus, in GR, all sorts of coordinates are allowed—there is a "democracy of coordinate systems." All sorts of coordinate transformations can be used. But the most fruitful way of viewing the transformations in GR is that they are **local** Lorentz transformations (i.e. an independent transformation at every spacetime point), which in the low-velocity limit are Galilean transformations.

1.1.4   New kinematics and dynamics

Einstein's formulation of the relativity principle involves a sweeping change of kinematics: not only space, but also time measurements, may differ in different inertial frames. Space and time are on an equal footing as coordinates of a reference system. We can represent space and time coordinates as the four components of a (spacetime) position vector x^μ ($\mu = 0, 1, 2, 3$), with x^0 being the time component, and the transformation for coordinate differentials is now represented by a 4×4 matrix $\mathbf{A}$,

$$dx^\mu \rightarrow dx'^\mu = \sum_\nu [\mathbf{A}]^\mu_\nu \, dx^\nu, \tag{1.1}$$

just like a rotational coordinate transformation is represented by a 3×3 matrix. The Galilean and Lorentz transformations are **linear** transformations, i.e. the transformation matrix elements do not themselves depend on the coordinates, $[\mathbf{A}] \neq [\mathbf{A}(x)]$. That the transformation matrix is a constant with respect to the coordinates means that one makes the **same** transformation at **every** coordinate point. We call this a **global transformation**. By contrast, general coordinate transformations are **nonlinear** transformations. Recall, for example, the transformation to an accelerating frame, $x \rightarrow x' = x + vt + at^2/2$, is nonlinear in the time coordinate. Here the transformations are coordinate-dependent, $[\mathbf{A}] = [\mathbf{A}(x)]$—a different transformation for each coordinate spacetime point. We call this a **local transformation**, or a **gauge transformation**. Global symmetry leads to kinematic restrictions, while local symmetry is a dynamics principle. As we shall see, the general coordinate symmetry (general relativity) leads to a dynamical theory of gravitation.[6]

1.2   GR as a gravitational field theory

The problem of noninertial frames of reference is intimately tied to the physics of gravity. In fact, the inertial frames of reference should properly be defined as the reference frames having no gravity. GR, which includes the consideration of accelerating coordinate systems, represents a new theory of gravitation.

The development of this new theory is rather unique in the history of physics: it was not prompted by any obvious failure (crisis) of Newton's theory, but resulted from the theoretical research, "pure thought," of one person—Albert Einstein. Someone puts it this way: "Einstein just stared at his own navel, and came up with general relativity."[7]

1.2.1   Einstein's motivations for the general theory

If not prompted by experimental crisis, what were Einstein's motivations in his search for this new theory? From his published papers,[8] one can infer several **interconnected** motivations (Uhlenbeck, 1968):

1. To have a relativistic theory of gravitation. The Newtonian theory of gravitation is not compatible with the principle of (special) relativity as

[6]Following Einstein's seminal work, physicists learned to apply the local symmetry idea also to the internal charge-space coordinates. In this way, electromagnetism as well as other fundamental interactions among elementary particles (strong and weak interactions) can all be understood as manifestation of local gauge symmetries. For respective references of gauge theory in general and GR as a gauge theory in particular, see for example Cheng and Li (1984, 2000).

[7]The reader of course should not take this description to imply that the discovery was in any sense straightforward and logically self-evident. In fact, it took Einstein close to 10 years of difficult research, with many false detours, to arrive at his final formulation. To find the right mathematics of Riemannian geometry, he was helped by his friend and collaborator Marcel Grossmann.

[8]Einstein's classical papers in English translation may be found in the collected work published by Princeton University Press (Einstein, 1989). A less complete, but more readily available, collection may be found in Einstein *et al.* (1952).

it requires the concept of an "action-at-a-distance" force, which implies instantaneous transmission of signals.

2. To have a deeper understanding of the empirically observed equality between inertial mass and gravitational mass.[9]

3. **"Space is not a thing"** Einstein phrased his conviction that physics laws **should not** depend on reference frames, which express the relationship among physical processes in the world and do not have independent existence.

[9]These two types of masses will be discussed in detail in Section 4.2.1.

"Space is not a thing" While the first two of the above-listed motivations will be discussed further in Chapter 4, here we make some comments on the third motivation. Einstein was dissatisfied with the prevailing concept of space. SR confirms the validity of the principle of special relativity: physics is the same in every Galilean frame of reference. But as soon as one attempts to describe physical phenomena from a reference frame in acceleration with respect to an inertial frame, the laws of physics change and become more complicated because of the presence of the fictitious inertial forces. This is particularly troublesome from the viewpoint of relative motion, since one could identify either frame as the accelerating frame. (The example known as Mach's paradox is discussed in Box 1.1.) The presence of the inertial force is associated with the choice of a noninertial coordinate system. Such coordinate-dependent phenomena can be thought of as brought about by space itself. Namely, space behaves as if it is the source of the inertial forces. Newton was thus compelled to postulate the existence of **absolute space,** as the origin of these coordinate-dependent forces. The unsatisfactory feature of such an explanation is that, while absolute space is supposed to have an independent existence, yet no object can act on this entity. Being strongly influenced by the teaching of Ernst Mach, Einstein emphasized that space and time should not be like a stage upon which physical events take place, thus having an existence even in the absence of physical interactions. In Mach and Einstein's view, space and time are nothing but expressing relationships among physical processes in the world—"space is not a thing". Such considerations led Einstein to the belief that the laws of physics should have the same form in all reference frames. Put another way, spacetime is a fixed for pre-GR physics, and in GR it is dynamic, as determined by the matter/energy distribution.

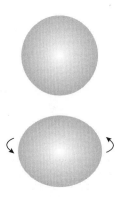

Fig. 1.1 Mach's paradox: Two identical elastic spheres, one at rest, and the other rotating, in an inertial frame of reference. The rotating sphere is observed to bulge out in the equatorial region, taking on an ellipsoidal shape.

Box 1.1 Mach's principle

At the beginning of his 1916 paper on general relativity, Einstein discussed Mach's paradox (Fig. 1.1) to illustrate the unsatisfactory nature of Newton's conception of space as an active agent. Consider two identical elastic spheres separated by a distance much larger than their size. One is at rest, and the other rotating around the axis joining these two spheres in an inertial frame of reference. The rotating body takes on the shape of an ellipsoid. Yet if the spheres are alone in the world, each can be regarded as being

(*cont.*)

[10] An affirmative answer can be argued by invoking the example of "dragging of inertial frames" by a rotating massive source, to be discussed in Section 8.4.1 (see Fig. 8.9).

[11] This is related to the fact that the Einstein theory is a geometric theory restricted to a metric field, as discussed below.

Box 1.1 (*Continued*)

in rotation with respect to the other. Thus there should be no reason for dissimilarity in shapes.

Mach had gone further. He insisted that it is the relative motion of the rotating sphere with respect to the distant masses that was responsible for the observed bulging of the spherical surface. The statement that the "average mass" of the universe gives rise to the inertia of an object has come to be called **Mach's principle**. While in Einstein's theory, the structure of space and time is influenced by the presence of matter in accordance to Mach's idea, the question of whether GR actually incorporates all of Mach's principle is still being debated.[10] For a recent discussion see, for example, Wilczek (2004), who emphasized that even in Einstein's theory not all coordinate systems are on an equal footing.[11] Thus the reader should be aware that there are subtle points with respect to the foundation questions of GR that are still topics in modern theoretical physics research.

1.2.2 Geometry as gravity

Einstein, starting with the equivalence principle (EP)—see Chapter 4—made the bold inference that the proper mathematical representation of the gravitational field is a **curved spacetime** (see Chapter 6). As a result, while spacetime has always played a passive role in our physics description, it has become a dynamic quantity in GR. Recall our experience with electromagnetism; a field theoretical description is a two-step description: the source, e.g. a proton, gives rise to a field everywhere, as described by the **field equations** (i.e. the Maxwell's equations); the field then acts locally on the test particle, e.g. an electron, to determine its motion, as dictated by the **equation of motion** (Lorentz force law).

$$\text{source} \longrightarrow \text{field} \longrightarrow \text{test particle.}$$

GR as a field theory of gravity with curved spacetime as the gravitational field offers the same two-step description. Its essence is nicely captured in an aphorism (by John A. Wheeler):

Spacetime tells matter how to move
Matter tells spacetime how to curve.

Since a test particle's motion in a curved space follows "the shortest possible and the straightest possible trajectory" (called the **geodesic curve**), the GR equation of motion is the **geodesic equation** (see Sections 5.2, 6.2, and 14.1). The GR field equation (the **Einstein equation**) tells us how the source of mass/energy can give rise to a curved space by fixing the curvature of the space (Sections 6.3 and 14.2). This is what we mean by saying that "GR is a geometric theory of gravity," or "gravity is the structure of spacetime."

1.2.3   Mathematical language of relativity

Our presentation will be such that the necessary mathematics is introduced as it is needed. Ultimately what is required for the study of GR is Riemannian geometry.

Tensor formalism   Tensors are mathematical objects having definite transformation properties under coordinate transformations. The simplest examples are scalars and vector components. The principle of relativity says that physics equations should be covariant under coordinate transformation. To ensure that this principle is **automatically** satisfied, all one needs to do is to write physics equations in terms of tensors. Because each term of the equation transforms in the same way, the equation automatically keeps the same form (it is covariant) under coordinate transformations. Let us illustrate this point by the familiar example of $F_i = ma_i$ as a rotational symmetric equation. Because every term of the equation is a vector, under a rotation the same relation $F_i' = ma_i'$ holds in the new coordinate system. The physics is unchanged. We say this physics equation possesses rotational symmetry. (See Section 2.1.1 for more details.) In relativity, we shall work with tensors that have definite transformation properties under ever more general coordinate transformations: the Lorentz transformations and general coordinate transformations (see Chapters 12 and 13). If physics equations are written as tensor equations, then they are automatically relativistic. This is why a tensor formalism is needed for the study of relativity.

Our presentation will be done in the coordinate-based component formalism, although this may lack the deep geometric insight that can be provided by the coordinate-independent formulation of differential geometry. This choice is made so that the reader can study the physics of GR without overcoming the hurdle of another layer of abstraction.

Metric description vs. full tensor formulation   Mathematically understanding the structure of the Einstein field equation is more difficult because it involves the Riemannian curvature tensor. A detailed discussion of the GR field equation and the ways of solving it in several simple situations will be postponed till Part IV. In Part II, our presentation will be restricted mainly to the description of space and time in the form of the metric function, which is a mathematical quantity that describes the shape of space through length measurements. From the metric function one can deduce the corresponding geodesic equation required for various applications. We will demonstrate in Part IV that the metric functions used in Parts II and III are the solutions of the Einstein field equation.

In this introductory chapter, we have emphasized the viewpoint of relativity as a coordinate symmetry. We can ensure that physics equations are covariant under coordinate transformations if they are written as tensor equations. Since the tensor formalism will not be fully explicated until Part IV, this also means that the symmetry approach will not be properly developed until later in the book, in Chapters 12–14.

GR as a geometric theory vs. GR as a theory of a metric field   Instead of emphasizing the geometric language of general relativity, a mathematically equivalent formulation (that's even more like the field theories of other fundamental interactions) is to have GR as a theory of a metric field. The metric function is viewed as the propagating (spin-2) field of gravity, just as electromagnetic potentials is viewed as the propagating (spin-1) field of electromagnetism. This viewpoint also clarifies the origin of GR's nonlinearity. Maxwell's theory is a linear theory because the mediator of the electro magnetic (EM) interaction does not carry EM charge itself—the photon is electrically neutral. Since anything carrying energy and momentum is a source of gravity, the gravitational metric field carries energy and momentum, hence, "gravity charge" also—much in the way the Yang–Mills fields of strong and weak interactions do.[12] In Chapter 15, where we discuss gravitational waves, the metric field viewpoint of GR will be employed. In that discussion we work in the approximation of ignoring the gravity charge of the gravity waves themselves; thus, it's the linearized Einstein theory that we will be working in.

[12]QCD's gluon fields of strong interaction are examples of Yang–Mills fields. These mediating fields among quarks themselves carry, just like quarks, strong interaction charges (called "color"). Such non-Abelian gauge fields are discussed, e.g. in Cheng and Li (1984, 1988).

1.2.4   Observational evidence for GR

Our presentation of general relativity (GR) and cosmology will emphasize heavily the experimental foundation of these subjects. Although the effects GR are small in the terrestrial and solar domains,[13] its predictions have been accurately verified whenever high precision observations can be performed. Notably we have the three classical tests of GR:

[13]Even the largest GR effects in terrestrial and solar domains are limited by small parameters,

$$\frac{G_N M_\oplus}{c^2 R_\oplus} = O\left(10^{-10}\right) \text{ and}$$

$$\frac{G_N M_\odot}{c^2 R_\odot} = O\left(10^{-6}\right),$$

with G_N being Newton's constant, $(M_\oplus, R_\oplus)$ and $(M_\odot, R_\odot)$ being respectively the (mass, radius) values of the earth and the sun.

- the precession of the planet Mercury's perihelion, as discussed in Section 7.3.1;
- the bending of star light by the sun, in Sections 4.3.2 and 7.2.1;
- the redshift of light's frequency in a gravitational field, as in Sections 4.3.1 and 6.2.2.

An electromagnetic signal is delayed while traveling in a warped spacetime; this Shapiro time delay will be studied in Section 7.3.2. We must also use GR for situations involving time-dependent gravitational fields as in emission and propagation of gravitational waves. The existence of gravitational waves predicted by GR has been verified by observing the rate of energy loss, due to the emission of gravitational radiation, in a relativistic binary pulsar systems such as the Hulse–Taylor system (PSR B1913+16), discussed in Section 15.4.3. In recent years a most impressive set of confirmations of GR has been carried out in the newly discovered double pulsar PSR J0737 − 3039A/B (Burgay *et al.*, 2003; Lyne *et al.*, 2004).

The double pulsar system as a unique laboratory for GR tests   A pulsar is a magnetized star whose rapid rotation generates a circulating plasma that serves as a source of beamed radio waves detectable on earth as periodic pulses. PSR J0737 − 3039A/B is a binary system composed of one pulsar with a period of 22 ms (pulsar A) in a 2.4-hour orbit with a younger pulsar with a period of 2.7 s (pular B). Neutron binaries being compact systems in rapid motion exhibit large GR effects. For example, the precession rate of this double pulsar's periastron[14] is $\dot{\omega} \simeq 17$ degrees per year,[15] as compared to planet Mercury's 43 arcseconds per century. Even better, with both neutron stars being pulsars there is an abundant amount of timing data; this has allowed for more than six GR tests in one system. Each GR effect has a unique dependence on the two pulsar masses, M_A and M_B. For example the decay rate of its orbit period $\dot{P}_b (M_A, M_B)$, due to gravitational wave emission, is predicted by GR in Eq (15.71):[16]

$$\dot{P}_{b,GR} = -\frac{192\pi}{5c^5} \left(\frac{2\pi G_N}{P_b}\right)^{5/3} \frac{1 + (73/24)e^2 + (37/96)e^4}{(1 - e^2)^{7/2}} \frac{M_A M_B}{(M_A + M_B)^{1/3}}.$$

As a result, the measured value $\dot{P}_{b,obs} = -1.252(17) \times 10^{-12}$ can then be translated as a (double lined) curve in the "mass–mass diagram" of Fig. 1.2. The other quantities that have been measured in this system are $\dot{\omega}$, the precession of the periastron mentioned above, two parameters, r (range) and s (shape), related to the Shapiro time-delay, and the γ parameter related to special relativistic and gravitational time dilation effects. Evidently, all these curves meet at one point of $M_A = 1.33817(7)$ and $M_B = 1.2487(7)$ in units of the solar mass $M_\odot$. In fact, from this one can infer that GR has been verified at the impressive 0.1% level (Kramer *et al.* 2006). Furthermore, the double pulsar J0737 − 3039A/B has provided us with up-to-now the most precise test of GR's prediction of relativistic (geodetic) spin precession Ω_B with an uncertainty of only 13% (Breton *et al.*, 2008). This phenomenon can be viewed as spin–orbit coupling of the system and is (indirectly) related to the intriguing

[14]This is the point of closest approach between the pulsar and its companion star.

[15]This is more than four times the corresponding value for the Hulse–Taylor binary.

[16]e is the eccentricity of the binary orbit.

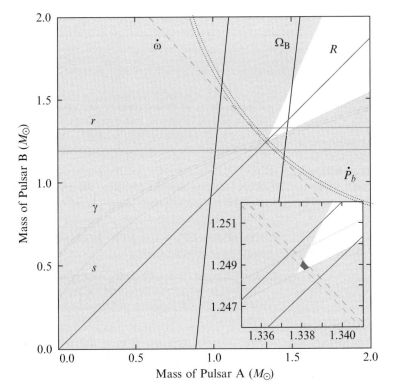

Fig. 1.2 The mass–mass plot that provides a graphical summary of GR parameters $\dot{\omega}$, γ, r, s, $\dot{P}_b$ and Ω_B of the double pulsar J0737 − 3039A/B is from Breton *et al.* (2008); the inset is an expanded view of the region of principal interest. Since both of the projected semimajor axes $x_{A,B} = a_{A,B} (\sin i) /c$ (with i being the inclination between the binary orbit plane and the plane of the sky) have been measured, one can fix the mass ratio $R \equiv M_A/M_B = x_B/x_A$ as the two stars must be orbiting around the system's center of mass (so that $x_A M_A = x_B M_B$). The shaded area is disallowed because mass functions in this region would lead to $\sin i \geq 1$.

GR feature of "dragging of the inertial frame," as discussed in Section 8.4 and in Problem 14.6.

1.2.5 GR as the framework for cosmology

The universe is a huge collection of matter and energy. The study of its structure and evolution, the subject of cosmology, has to be carried out in the framework of GR. The Newtonian theory for a weak and static gravitational field will not be adequate. The large collection of matter and fields means we must deal with strong gravitational effects, and to understand its evolution, the study cannot be carried out in static field theory. In fact, the very basic description of the universe is now couched in the geometric language of general relativity. A "closed universe" is one having positive spatial curvature, an "open universe" is negatively curved, etc. Thus for a proper study of cosmology, we must first learn GR.

Observationally it is clear that GR is needed to provide the conceptual framework for cosmology. The expanding universe reflects a dynamical space-time. Basic features of the big bang (inflation) and the accelerated expansion of the universe (due to dark energy) in the present epoch can be accommodated simply by a vacuum energy term (called the cosmological constant) in the GR field equation, which gives rise to the surprising feature of gravitational repulsion.

Review questions

1. What is relativity? What is the principle of special relativity? What is general relativity?

2. What is a symmetry in physics? Explain how the statement that no physical measurement can detect a particular physical feature (e.g. orientation, or the constant velocity of a lab), is a statement about a symmetry in physics. Illustrate your explanation with the examples of rotation symmetry, and the coordinate symmetry of SR.

3. In general terms, what is a tensor? Explain how a physics equation, when written in terms of tensors, automatically displays the relevant coordinate symmetry.

4. What are inertial frames of reference? Answer this in three ways.

5. The equations of Newtonian physics are unchanged when we change the coordinates from one to another inertial frame. What is this coordinate transformation? The equations of electrodynamics are unchanged under another set of coordinate transformations. How are these two sets of transformations related? (You need only give their names and a qualitative description of their relation.)

6. What is the key difference between the coordinate transformations in special relativity and those in general relativity?

7. What motivated Einstein to pursue the extension of special relativity to general relativity?

8. In the general relativistic theory of gravitation, what is identified as the gravitational field? What is the GR field equation? The GR equation of motion? (Again, only the names.)

9. How does the concept of space differ in Newtonian physics and in Einsteinian (general) relativistic physics?

Special relativity: The basics

2

- Special relativity is introduced as the symmetry of Maxwell's theory of electromagnetism.
- The fact that a signal cannot be transmitted instantaneously led Einstein to propose a new kinematics: the passage of time is different in different inertial frames.
- The operational definition of time, consistent with a universal speed of light, is presented.
- Relativity of simultaneity, time dilation, and length contraction are discussed.
- We present an elementary derivation of the Lorentz transformation and discuss the physical meaning of each term in this transformation.
- We discuss the reciprocity relation in order to show that "relativity is truly relative."
- The pole-and-barn and twin paradoxes are worked out to illustrate some of the counter-intuitive features of relativity.

In this chapter, a brief review of the basic features of special relativity is presented. Its geometric formulation, in terms of flat spacetime, will be introduced in Chapter 3, in preparation for the study of the larger framework of curved spacetime in general relativity.

2.1 Coordinate symmetries

In Chapter 1 we introduced the concept of a symmetry in physics. It is the situation in which physics equations, under some transformation, are unchanged in their form (i.e. they are "covariant"). For a review of the familiar case of rotational symmetry, see Box 2.1. We shall discuss the distinction between the Galilean symmetry of Newtonian mechanics and the Lorentz symmetry of electrodynamics—first their formal aspects in this section, and then their physical basis in Section 2.2. We first introduce the Lorentz symmetry as a mathematical property of the electrodynamics equations. Only afterwards do we discuss the physics—following Einstein's insight—that is implied by such a coordinate symmetry.

Box 2.1 Vector components and rotational symmetry

For a given coordinate system with a set of orthonormal basis vectors $\{\hat{e}_i\}$, a vector—for example, a vector $\vec{V}$—can be represented by its components V_1, V_2, and V_3, with $\{V_i\}$ being the coefficients of expansion of $\vec{V}$ with respect to the basis vectors:

$$\vec{V} = \sum_{i=1}^{3} V_i \hat{e}_i = V_1 \hat{e}_1 + V_2 \hat{e}_2 + V_3 \hat{e}_3. \tag{2.1}$$

With a change of the coordinate basis $\{\hat{e}_i\} \to \{\hat{e}'_i\}$, the same vector would have another set of components (Fig. 2.1):

$$\vec{V} = \sum_{i=1}^{3} V'_i \hat{e}'_i. \tag{2.2}$$

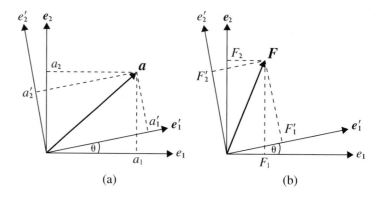

Fig. 2.1 Coordinate change of a vector under rotation. Components of different vectors, whether the acceleration vector as in (a) or the force vector as in (b), all transform in the same way, as in (2.6).

For the example of the coordinate transformation that is a rotation of the coordinate axes by an angle of θ around the z axis, the new position components are related to the original ones by relations that can be worked out geometrically from Fig. 2.1:

$$V'_1 = \cos\theta\, V_1 + \sin\theta\, V_2$$
$$V'_2 = -\sin\theta\, V_1 + \cos\theta\, V_2 \tag{2.3}$$
$$V'_3 = V_3.$$

This set of equations can be written compactly as a matrix (the rotation transformation matrix) acting on the original vector to yield the new position components:

$$\begin{pmatrix} V'_1 \\ V'_2 \\ V'_3 \end{pmatrix} = \begin{pmatrix} \cos\theta & \sin\theta & 0 \\ -\sin\theta & \cos\theta & 0 \\ 0 & 0 & 1 \end{pmatrix} \begin{pmatrix} V_1 \\ V_2 \\ V_3 \end{pmatrix}. \tag{2.4}$$

Expressed in component notation, this equation becomes

$$V_i' = \sum_{j=1}^{3} [\mathbf{R}]_{ij}\, V_j = [\mathbf{R}]_{i1}\, V_1 + [\mathbf{R}]_{i2}\, V_2 + [\mathbf{R}]_{i3}\, V_3, \qquad (2.5)$$

where $[\mathbf{R}]$ is the rotation matrix, and i and j are the row and column indices, respectively. Namely, the components of **any** vector should transform under a rotation of coordinate axes in the manner of (2.5).

Rotational invariance is a prototype of a symmetry in physics. To have rotational symmetry means that the physics is unchanged under a rotation of coordinates (NB: not a rotating coordinate system, in which the coordinate axes have a rotational velocity). Take, for example, the equation of $F_i = ma_i$ ($i = 1, 2, 3$), which is the familiar $\vec{F} = m\vec{a}$ equation written in vector-component form. These components are projections onto the coordinate axes. The equation is said to respect rotational symmetry if the validity of $F_i = ma_i$ in a system O implies the validity of $F_i' = m'a_i'$ in any other system O', which is related to O by a rotation. Since the mass m is a scalar, it, by definition, does not change under a coordinate transformation, whereas the vector components of acceleration and force, a_i and F_i respectively, transform as in (2.5):

$$m' = m, \qquad a_i' = \sum_j [\mathbf{R}]_{ij}\, a_j, \qquad F_i' = \sum_j [\mathbf{R}]_{ij}\, F_j. \qquad (2.6)$$

The validity of $F_i' - m'a_i' = 0$ follows from $F_i - ma_i = 0$ because the transformation matrix $[\mathbf{R}]$ is the **same** for each set of vector components (F_i and a_i):

$$F_i' - m'a_i' = \sum_j [\mathbf{R}]_{ij}\, \left(F_j - ma_j\right) = 0. \qquad (2.7)$$

That each term in this physics equation $F_i = ma_i$ transforms in the same way under the rotational transformation is displayed in Fig. 2.1. Under a transformation, the components of force and acceleration change values. However, since the corresponding components of force and acceleration (e.g. F_1 and a_1) transform in the same way, their relation is not altered and the physics equation keeps the same form. Thus, as $F_i = ma_i$ is a vector equation (or, more generally, a tensor equation), each term of the equation transforms as a vector component under rotation. We see that if a physics equation can be written as a tensor equation, it is covariant (its form is unchanged) and automatically respects rotational symmetry. In plain language, covariance of a physics equation means that the laws of physics are the same in different reference frames. Observers in different frames may disagree about the values of various quantities but, when plugging these values into the physics equations, one finds the equations

(*cont.*)

Box 2.1 (*Continued*)

are satisfied. Namely, the physics relations among these quantities are the same in different reference frames. If this is the case, we then have a physics symmetry.

2.1.1   Newtonian physics and Galilean symmetry

One of the most important lessons Galileo and Newton have taught us is that the simplest way to describe the physical world (hence, the laws of physics) is by using **inertial frames of reference**. The transformation that allows us to go from one inertial frame O with coordinates x_i to another inertial frame O' with coordinates x_i' is the Galilean transformation. If the relative velocity of the two frames is given to be $\vec{v}$ (a constant) and their relative orientations are specified by three angles α, β and γ, the new coordinates are related to the old ones by $x_i \longrightarrow x_i' = [\mathbf{R}]_{ij} x_j - v_i t$, where $[\mathbf{R}] = [\mathbf{R}(\alpha, \beta, \gamma)]$ is the rotation matrix. In Newtonian physics, the time coordinate is assumed to be absolute, i.e. it is the same in every coordinate frame.

For the Galilean transformation, we are mainly interested in coordinate transformations among inertial frames with the same orientation, $[\mathbf{R}(0, 0, 0)]_{ij} = \delta_{ij}$ (see Fig. 2.2). Such a transformation is called a (Galilean) **boost**:

$$x_i \longrightarrow x_i' = x_i - v_i t, \tag{2.8}$$

$$t \longrightarrow t' = t.$$

NB: we are really discussing the relation for the **intervals** between two "events," say, $A = (x_A, t_A)$ and $B = (x_B, t_B)$, and the above relation, simplified to the one spatial dimension case, should be written as $\Delta x' = \Delta x - v \Delta t$, with $\Delta x = x_A - x_B$. Thus, the x's as written in (2.8) are understood to be the interval between position point x and the coordinate origins: $(x_A = x, x_B = 0)$. Similarly, x' is the interval between x' and the origin O'. It is also understood that the coordinate origins of the two systems O and O' coincide at $t = 0$.

Newtonian relativity[1] says that the physics laws (mechanics) are unchanged under the Galilean transformation (2.8). Physically this implies that no mechanical experiment can detect any intrinsic difference between the two

[1] Newtonian relativity being the symmetry under a Galilean transformation is also referred to as Galilean relativity.

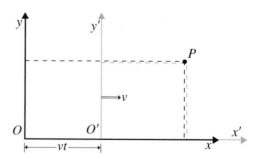

Fig. 2.2 The point P is located at (x, y) in the O system and at (x', y') in the O' system, which is moving with velocity v in the x direction. The coordinate transformations are $x' = x - vt$ and $y' = y$.

inertial frames. It is easy to check that a physics equation such as $G_N m \hat{r} / r^2 = \vec{a}$ does not change its form under the Galilean transformation as the coordinate separation $\vec{r} = \vec{x}_1 - \vec{x}_2$ and the acceleration $\vec{a}$ are unchanged under this transformation.[2] In contrast to these invariant quantities, the velocity vector, $u_i = dx_i / dt$, will change. It obeys the **velocity addition rule**,

$$u_i \longrightarrow u_i' = u_i - v_i, \tag{2.9}$$

which is obtained by differentiation of (2.8).

2.1.2 Electrodynamics and Lorentz symmetry

One can show that Maxwell's equations are not covariant under the Galilean transformation. The easiest way to see this is by recalling the fact that the propagation speed of electromagnetic waves is a constant[3]

$$c = \sqrt{\frac{1}{\mu_0 \epsilon_0}}, \tag{2.10}$$

with ϵ_0 and μ_0, the permittivity and permeability of free space, being the constants appearing in Coulomb's and Ampere's laws. Since ϵ_0 and μ_0 are physical constants having the same values to all observers, c must be the same in all inertial frames. This constancy of the speed of light violates the Galilean velocity addition rule of (2.9). Two alternative conclusions can be drawn from this apparent violation:

- Maxwell's equations are valid only in one inertial frame. Hence the (Newtonian) relativity principle is not applicable. It was thought that, like all mechanical waves, electromagnetic waves must have an elastic medium for their propagation. Maxwell's equations were thought to be valid only in the rest frame of the *ether* medium. The constant c was interpreted to be the wave speed in this ether—the frame of **absolute rest**. This was the interpretation accepted by most nineteenth-century physicists.[4]
- Maxwell's equations do obey the principle of relativity but the relationship between inertial frames is not correctly given by the Galilean transformation. Hence the velocity addition rule of (2.9) is invalid; the correct relation should be such that c can be the same in every inertial frame. The modification of the velocity addition rule must necessarily bring about a change of the time coordinate $t' \neq t$.

The second interpretation turned out to be correct. The measurement made by Michelson and Morley showed that the speed of light is the same in different moving frames.[5] It had been discovered by Poincaré, based on prior work by Lorentz, among others, and independent of Einstein's 1905 work, that Maxwell's equations were covariant under a new boost transformation, "the **Lorentz transformation**" (see Box 2.2). Maxwell's equations keep the same form if one transforms.[6] not only the position variables but also the time variable, from (t, x, y, z) to (t', x', y', z'), representing the coordinates of two

---

[2] Because the $v_i t$ term is the same for both $(x_1)_i$ and $(x_2)_i$, we have an invariant separation: $r_i' = (x_1')_i - (x_2')_i = (x_1)_i - (x_2)_i = r_i$. Time-differentiating the position relation (using $t' = t$), we have $(dx'/dt') = (dx/dt) - v_i$; differentiating one more time, we see that acceleration is unchanged: $(d^2 x'/dt'^2) = (d^2 x/dt^2)$ because the relative velocity is a constant, $dv_i/dt = 0$.

[3] $c = 299\,792\,458$ m/s. In our discussion we shall mostly use the simpler figure of $c = 3 \times 10^8$ m/s.

[4] Many physicists were nonetheless unsettled by the idea of an all-permeating, yet virtually undetectable, medium. The ether must not impede the motion of any object; yet it must be extremely stiff, as the wave speed should be proportional to the stiffness of the medium in which it is propagating.

[5] If there had existed an ether medium, the motion of the earth would have brought about an "ether wind," which would have caused the speed of light to vary with its direction of propagation.

[6] Besides the space and time variables, one must also change electromagnetic quantities $(\vec{E}, \vec{B}, \vec{j}, \rho)$, as detailed in Box 2.2.

frames moving with a relative velocity $\vec{v} = v\hat{x}$:

$$x' = \gamma\,(x - vt)\,, \qquad y' = y, \qquad z' = z, \qquad t' = \gamma\left(t - \frac{v}{c^2}x\right), \quad (2.11)$$

where the parameter γ depends on the relative speed of the two reference frames as

$$\gamma = \frac{1}{\sqrt{1 - \beta^2}}\,, \qquad \beta = \frac{v}{c}. \qquad (2.12)$$

We have, in general, $\beta \le 1$ and $\gamma \ge 1$. We note that the spatial transformation is just the Galilean transformation (2.8) multiplied by the γ factor and is thus reduced to (2.8) in the low velocity limit of $\beta \to 0$, hence $\gamma \to 1$. Please also recall the comments followed Eq. (2.8), about coordinate intervals and origins, which are equally applicable to the Lorentz transformation. We note the presence in both equations of $\gamma = \left(1 - \beta^2\right)^{-1/2}$, which blows up at $\beta = v/c = 1$, indicating that no particle can increase its speed beyond the value c.

Box 2.2 Maxwell's equations and the Lorentz transformation

An electric charge at rest gives rise to an electric, but no magnetic, field. However, the same situation when viewed by a moving observer is seen as a charge in motion, which produces both electric and magnetic fields. Namely, different inertial observers will find different electric and magnetic fields, just as they would measure different position and time intervals. We say that position (x_i) and time (t) coordinates transform into each other in moving frames, $(x_i, t) \to \left(x'_i, t'\right)$, as in (2.11); similarly, this shows that the electric (E_i) and magnetic (B_i) fields can change into each other in moving coordinates, $(E_i, B_i) \to \left(E'_i, B'_i\right)$. When we say that Maxwell's equations are covariant under the Lorentz transformation, we must also specify the Lorentz transformation properties of the fields $\vec{E}$ and $\vec{B}$, as well as the current and charge densities, $\vec{j}$ and ρ. That is, under the Lorentz transformation, not only the space and time coordinates will change, but also the electromagnetic fields and source charge and currents will change. However, the relation among these changed quantities remains the same, as given by Maxwell's equations.

The transformation formulae for these electromagnetic quantities are somewhat simpler when written in the Heaviside–Lorentz system of units,[7] in which the measured parameter[8] is taken to be c, the velocity of the EM wave, instead of (ϵ_0, μ_0). In this system, the Lorentz force law reads

$$\vec{F} = q\left(\vec{E} + \frac{1}{c}\vec{v} \times \vec{B}\right), \qquad (2.13)$$

while Maxwell's equations take on the form

$$\vec{\nabla}\cdot\vec{B} = 0, \qquad \vec{\nabla}\times\vec{E} + \frac{1}{c}\frac{\partial\vec{B}}{\partial t} = 0, \qquad (2.14)$$

$$\vec{\nabla}\cdot\vec{E} = \rho, \qquad \vec{\nabla}\times\vec{B} - \frac{1}{c}\frac{\partial\vec{E}}{\partial t} = \frac{1}{c}\vec{j}. \qquad (2.15)$$

[7] Conversion table from **mks** units to Heaviside–Lorentz units:

mks	Heaviside–Lorentz
$\sqrt{\epsilon_0}\,E_i$ $\longrightarrow$	E_i
$\sqrt{1/\mu_0}\,B_i$ $\longrightarrow$	B_i
$\sqrt{1/\epsilon_0}\,(\rho, j_i)$ $\longrightarrow$	(ρ, j_i)

[8] Recall that in the mks system, while ϵ_0 is a measured quantity, μ_0 is not. μ_0 and the unit of electric current are defined together through Ampere's law.

In this system of units, the Lorentz transformation properties of the electromagnetic fields (going from the O to the O' system) are given by

$$E_1' = E_1, \quad E_2' = \gamma\,(E_2 - \beta B_3), \quad E_3' = \gamma\,(E_3 + \beta B_2)$$
$$B_1' = B_1, \quad B_2' = \gamma\,(B_2 + \beta E_3), \quad B_3' = \gamma\,(B_3 - \beta E_2),$$

$$(2.16)$$

and those of the charge and current densities are given by

$$j_1' = \gamma\,(j_1 - v\rho), \quad j_2' = j_2, \quad j_3' = j_3, \quad \rho' = \gamma\left(\rho - \frac{v}{c^2}j_1\right). \quad (2.17)$$

Using these transformation rules, as well as those for the space and time coordinates (2.11), we can check in a straightforward manner (as in Problem 2.4) that the equations of electromagnetism are unchanged in their form (i.e. they are covariant) under the Lorentz transformation.

2.1.3   Velocity addition rule amended

Maxwell's equations have Lorentz symmetry, and they represent the physics behind the phenomenon of an electromagnetic wave propagating with the same velocity c in all moving frames. The Lorentz transformation must imply a new velocity addition rule which allows for a constant speed of light c in every inertial frame. Writing (2.11) in infinitesimal interval form,

$$dx' = \gamma\,(dx - vdt), \quad dy' = dy, \quad dz' = dz, \quad dt' = \gamma\left(dt - \frac{v}{c^2}dx\right),$$
$$(2.18)$$

we obtain the velocity transformation rule by simply constructing the appropriate quotients:

$$u_x' = \frac{dx'}{dt'} = \frac{dx - vdt}{dt - (v/c^2)\,dx} = \frac{u_x - v}{1 - (vu_x/c^2)}, \qquad (2.19)$$

$$u_y' = \frac{dy'}{dt'} = \frac{dy}{\gamma\,(dt - (v/c^2)\,dx)} = \frac{u_y}{\gamma\,(1 - (vu_x/c^2))}, \qquad (2.20)$$

$$u_z' = \frac{dz'}{dt'} = \frac{u_z}{\gamma\,(1 - (vu_x/c^2))}. \qquad (2.21)$$

For the case in which the velocity we are studying (e.g. the velocity of some particle, or of a light wave) is parallel to the relative velocity of the two frames, and for which both of these are in the x direction ($v = v_x$; we can always pick the axes such that the x axes line up with the relative velocity), we have $u_x = u$ and $u_y = u_z = 0$. The x component of the velocity is also the total velocity, in this case $u = u_x$, and it transforms as,

$$u' = \frac{u - v}{1 - (vu/c^2)}, \qquad (2.22)$$

[9]The Michelson–Morley experiment confirmed the notion that the speed of light c is the same in different inertial frames, and therefore the Galilean velocity addition rule (2.9) is not obeyed. But historically, because Einstein had already been convinced of the validity of a constant speed of light c as a law of physics (hence valid in all inertial frames), this experimental result *per se* did not play a significant role in Einstein's thinking when he developed the theory of special relativity.

[10]Electromagnetism beyond electrostatics and magnetostatics would necessarily involve the description of physics with respect to a moving observer. Yet the resulting electrodynamics equation seem to violate the then-accepted relation between different inertial frames as given by the Galilean transformation.

[11]Kinematics, as opposed to dynamics, is the part of physics that sets up the framework so that a particle's position, velocity, acceleration, etc. can be described, before any consideration of the forces that can act on the particle.

[12]These are the postulates as stated in Einstein's original 1905 paper, *On the electrodynamics of moving bodies*. For a more complete list of SR postulates (for example, the existence of global inertial frames, etc.) see Hamilton (2002).

[13]In this chapter we introduce relativity (following Einstein) as a coordinate symmetry that was first discovered in electromagnetism. This explains the appearance of the speed of electromagnetic waves in the basic postulates of special relativity even though relativity is a concept applicable to all of physics. As our discussion in subsequent sections will show, the relevant issue is the existence of a maximal and absolute speed (c) that any signal can be transmitted. The speed of light has this value, but so does the gravitational wave. In the next chapter we shall discuss this matter further. The fundamental roles played by c in the theory of relativity will be summarized in Section 3.4.

while the y and z components of the velocity remain unchanged ($u'_y = u'_z = 0$). This is just the familiar velocity addition rule (2.9), but with the RHS divided by an extra factor of $(1 - uv/c^2)$. It is easy to check that an input of $u = c$ leads to an output of $u' = c$, showing that the speed of light is the same in every inertial frame of reference.[9]

2.2   The new kinematics

The covariance of the equations of electromagnetism under the Lorentz transformation was independently discovered by Lorentz and Poincaré. But it was Einstein who first emphasized the physical basis of this symmetry. Maxwell's equations for electrodynamics are not compatible with the principle of Newtonian relativity. Most notably, the constancy of the velocity of the electromagnetic wave in every inertial frame violates the familiar velocity addition rule of (2.9). Consequently it is difficult to formulate a consistent electrodynamic theory for a moving observer.[10] Einstein's resolution of these difficulties was stated in an all-embracing new kinematics.[11] An understanding of the physics behind the Lorentz covariance, as first discovered in Maxwell's equations, would involve a revision of our basic notions of space and time. This would have fundamental implications for all aspects of physics, far beyond electromagnetism.

2.2.1   The basic postulates of special relativity

Einstein started with the following basic postulates:[12]

- **Principle of relativity** Physics laws have the same form in every inertial frame of reference. No physical measurement can reveal the absolute motion of an inertial frame of reference.
- **Constancy of the speed of light**[13] This second postulate is certainly consistent with the first one, as the constancy of the speed of light is a feature of electrodynamics (hence a law of physics), and the principle of relativity would lead us to expect the constancy of the speed of light to hold in every frame.

Einstein pointed out that a new kinematics was required in order to fully implement the new symmetry. In particular, different time coordinates were required for different inertial frames when the speed of signal transmission was not infinite. Up until then, no one doubted the absolute nature of time. But Einstein emphasized that the definition of time was ultimately based on the notion of **simultaneity**. However, in order for observers in different frames to agree that two events are simultaneous ($\Delta t = 0$), or to agree on any time interval Δt for that matter, their clocks need to be synchronized. But in order to synchronize their clocks, the observer in one frame must send a signal which is received by the observer in the other frame. If signals could be transmitted instantaneously, then this would be simple. But since all signals travel at a finite speed (with the upper speed limit being the speed of light), simultaneity becomes a relative concept (as will be discussed in Section 2.2.2). As a result, a time interval measured by one inertial observer will differ from that by another who is in relative motion with respect to the first observer.

2.2.2 Relativity of equilocality and simultaneity

To describe a certain quantity as being relative means that it is not invariant under coordinate transformations. In this subsection, we shall consider the various invariants (and noninvariants) under different types of coordinate transformations.

Relativity of spatial equilocality

Two events occurring at the same spatial location in a given inertial frame O' are termed "equilocal." If they do not take place at the same time, we represent their position and time coordinates in the O' system as (x_1', t_1') and (x_2', t_2'), with $\Delta x' = x_2' - x_1' = 0$ and $\Delta t' = t_2' - t_1' \neq 0$. Spatial equilocality for these two events is already a relative property even under a Galilean transformation[14] (2.8),

$$\Delta x = v \Delta t \neq 0, \text{ even though } \Delta x' = 0. \tag{2.23}$$

It is useful to have a specific illustration (Fig. 2.3): A light bulb at a fixed position on a moving train emits two flashes of light. To an observer on the train, these two events are spatially equilocal, although not simultaneous. Clearly this spatial equilocality is a relative concept, because to an observer standing on the rail platform as the trains passes by, the flashes appear to be emitted at two different locations.

Relativity of simultaneity

Einstein pointed out that, when signal transmission cannot be carried out at infinite speed, the simultaneity of two events is a relative concept: two events, that are seen by one observer to be simultaneous are seen by another observer in relative motion to the first to occur at different times.

First we need a commonly agreed upon definition of simultaneity. For example, we can mark off the midpoint between two locations. If two light pulses were sent from this midpoint, their arrival times at these two locations are "simultaneous events" to an observer at rest with respect to these locations. On the other hand, they will be seen as nonsimultaneous by an observer in motion with respect to these locations because one of the locations has moved closer to, and one farther away from, the point of emission as seen by that observer. (For details, see the discussion below.)

[14] Recall our discussion following (2.8) that coordinate transformations always involve intervals: $\Delta x' = \Delta x - v \Delta t$.

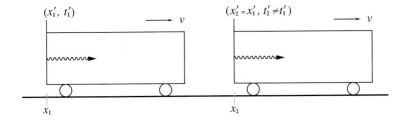

(x_1', t_1') $\longrightarrow v$

$(x_2' = x_1', t_2' \neq t_1')$ $\longrightarrow v$

x_1 x_2

Fig. 2.3 Spatial equilocality of two events is relative if they take place at different times. A light bulb at a fixed position $x_1' = x_2'$ on a moving train flashes at two different times $t_1' \neq t_2'$. To an observer standing on the rail platform, these two events (x_1, t_1) and (x_2, t_2) take place at two different locations $x_1 \neq x_2$.

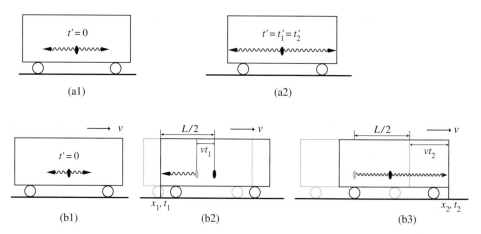

Fig. 2.4 Simultaneity is relative when light is not transmitted instantaneously. Two events (x'_1, t'_1) and (x'_2, t'_2) corresponding to light pulses (wavy lines) arriving at the opposite ends of a moving train, after being emitted from the midpoint (a1), are seen as simultaneous, $t'_1 = t'_2$, by an observer on the train as in (a2). But to another observer standing on the rail platform, these two events (x_1, t_1) and (x_2, t_2) are not simultaneous, $t_1 \neq t_2$, because for this observer the light signals arrive at different times at the two ends of the moving railcar [(b2) and (b3)].

We now illustrate this with a concrete situation (Fig. 2.4). Consider the emission of two light pulses from the midpoint of a railcar towards the front and back ends of the railcar.

- Clearly, to an observer on the railcar these two arrivals will be simultaneous. Let the length of the railcar be L' according to this observer O'. The emission event at the midpoint has the coordinates $(x' = x_0, t' = 0)$ as shown in Fig. 2.4(a1). The arrival events, Fig. 2.4(a2), will be $\left(x'_1 = x_0 - \frac{1}{2}L', t'_1\right)$ and $\left(x'_2 = x_0 + \frac{1}{2}L', t'_2\right)$ at the back and front ends of the railcar respectively. The events are simultaneous because $t'_1 = t'_2 = \frac{1}{2}L'/c$.

- Now consider the observation of the same events by an observer standing outside the railcar as it passes with the speed of v. (One can arrange triggers on the railcar so that when the light pulse hits the ends of the railcar a mark will be left on the ground outside the car.) The ground observer will see the light arrive at the back end first, Fig. 2.4(b2), at $t = t_1$, which is earlier than the time at which the pulse hits the front end of the car, Fig. 2.4(b3), at $t = t_2$. In other words, these two events will be seen as being nonsimultaneous. We calculate these two arrival times in the coordinate frame O of the ground observer by noting that, since the ends of the railcar have traveled the distance vt (towards and away from the emission point) during the light propagation duration t, the distances the light pulses must cover are $\frac{1}{2}L \mp vt$ to reach the back and front ends of the railcar. The propagation times are then

$$t_1 = \frac{\frac{1}{2}L - vt_1}{c}, \qquad t_2 = \frac{\frac{1}{2}L + vt_2}{c}, \qquad (2.24)$$

where L is the length of the railcar as measured by the ground observer. (We are allowing for the possibility that $L \neq L'$.) If we solve Eq. (2.24) for t_1 and t_2, and then express their difference in terms of the shorthand of (2.12) we have

$$\Delta t = t_2 - t_1 = \frac{L}{2c}\left(\frac{1}{1-\beta} - \frac{1}{1+\beta}\right) = \frac{\gamma^2 v L}{c^2}, \qquad (2.25)$$

where Δt is the amount of nonsynchronicity[15] as observed by O.

- Our usual intuitive expectation (that the rail platform observer also sees these as two simultaneous events) turns out to be incorrect because it is based on (nonrelativistic) Newtonian physics. Since the speed of the train is extremely low compared to the speed of light, $v \ll c$, the RHS of Eq. (2.25) is vanishingly small (being v/c times the time for light to traverse the length of the railcar). The nonsimultaneity in the rail platform frame is practically unobservable.

- Spatial equilocality and time simultaneity are obviously similar; yet, while relativity of the first is considered by us to be obvious, the second is rather counter-intuitive. This just reflects the fact that our intuition is based on Newtonian physics, in which space and time are treated very differently: space is relative, time absolute. The symmetry[16] between space and time is one of the key lessons from relativity.

- In the example discussed above, we have chosen the light emitter to be located at the midpoint of the railcar. Had it been located closer to the front end of the car, we would have $t_1' > t_2'$. However, one can easily show that if the train were moving sufficiently fast, we could still obtain $t_1 < t_2$ as in (2.25). That is, even the temporal order of events can be frame-dependent![17]

Frame-dependent time coordinate

The most profound consequence of special relativity is the change it brought about our conception of time. In fact, as we shall see, all major implications of the theory can be traced back to the relativity of time. The above discussion makes it clear that in a world with a finite (and absolute) speed of light, the time interval is a frame-dependent quantity. Thus, any reference system must be specified by four coordinates (x, y, z, t)—that is, by three spatial coordinates and one time coordinate. One can picture a coordinate system being a three-dimensional grid (to determine the position) and a set of clocks (to determine the time of an event), with a clock at every grid point in order to avoid the complication of a time delay between the occurrence of an event and the registration of this event by a clock located a distance away from the event. We require all the clocks to be synchronized (say, against a master clock located at the origin). The synchronization of a clock located at a distance r from the origin can be accomplished by sending out light flashes from the master clock at $t = 0$. When the clock receives the light signal, it should be set to $t = r/c$. Equivalently, synchronization of any two clocks can be checked by sending out light flashes from these two clocks at a given time. If the two flashes arrive at their midpoint at the same time, they are synchronized. The reason that light signals are often used to set and compare time is that this ensures, in the most direct way, that the new kinematical feature of a universal light velocity is properly taken into account.[18]

[15] Further calculations relating to the issue of simultaneity can be found in Section 2.3.1 below, as well as in Problem 2.9.

[16] The statement of "treating space and time symmetrically" means that time is treated as a coordinate, on the same footing as the space coordinates.

[17] For an illustrative example, see the "pole-and-barn paradox" in Box 2.4. If time order can be reversed, one may be concerned about the occurrence of a situation in which effect precedes cause, violating the principle of causality. As we shall discuss later (see Fig. 3.9b), relativity has a structure such that this does not happen.

[18] As Einstein emphasized, any method that is consistent with the principle of relativity is ultimately equivalent to the light signal method.

2.2.3   Time dilation and length contraction

We discuss here two other important physical implications of the SR postulates, **time-dilation** and **length-contraction**:

> A moving clock **appears** to run slow;
> a moving object **appears** to contract.

These physical features underscore the profound change in our conception of space and time brought about by relativity. We must give up our belief that measurements of space and time give the same results for all observers. Special relativity makes the strange claim that observers in relative motion will have different perceptions of distance and time. This means that two identical watches worn by two observers in relative motion will tick at different rates and will not agree on the amount of time that has elapsed between two given events. It is not that these two watches are defective. Rather, it is a fundamental statement about the nature of time.

Time dilation[19]

As a clock ticks away, it marks out a time interval $\Delta t'$ in its own rest frame O' (also called the "comoving frame," with $\Delta t'$ termed the "**proper time**"). In the clock's rest frame, the clock's position is, by definition, unchanged ($\Delta x' = 0$). However, an observer O in motion with respect to this frame sees the clock as a moving clock; the same interval will be viewed by O as ticking away a longer interval:

$$\Delta t = \gamma \Delta t', \qquad \gamma > 1, \qquad (2.26)$$

where γ was defined in (2.12). Thus we say that a moving clock (i.e. moving with respect to the O system) appears to run slow[20] ($\Delta t > \Delta t'$).

There is an easy way to understand the phenomenon of time dilation. Let us consider the most basic of clocks:[21] a light-pulse clock. It ticks away the time by having a light pulse bounce back and forth between two ends separated by a fixed distance d (Fig. 2.5).

For a comoving observer O', one has

$$\Delta t' = \frac{d}{c}. \qquad (2.27)$$

However, for an observer with respect to whom the clock is moving with a velocity v, perpendicular to the light-pulse path,[22] the light pulse will traverse a diagonal distance D in a different time interval Δt

$$\Delta t = \frac{D}{c} = \frac{\sqrt{d^2 + v^2 \Delta t^2}}{c}. \qquad (2.28)$$

[19]Something being "dilated" means it becomes longer. By "time dilation," we mean a time interval becomes longer, and thus the relevant clock is slower.

[20]Time dilation is important in space travel. The (earth) time $\Delta t = v^{-1}L$ required for a spaceship traveling at speed v to reach a distant galaxy (trip-length L) can be extremely long, as the speed is limited by $v < c$. However, for the astronaut in the high-speed spaceship, the (proper) time passage can be much shorter, $\Delta t' = \gamma^{-1}\Delta t$, where $\gamma \gg 1$ for a speed sufficiently close to c.

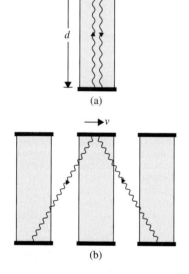

(a)

(b)

Fig. 2.5 Light-pulse clock, with mirrors at the top and bottom of a vertical vacuum chamber, (a) at rest and (b) in motion to the right in the horizontal direction.

[22]In this derivation, we have assumed that the transverse distance d is unchanged in this coordinate transformation. This assumption can be justified from the principle of relativity. For such a discussion see Sartori (1996, p. 89).

[21]A "basic clock" rests on some physical phenomenon that has a direct connection to the underlying laws of physics. Different clocks—mechanical clocks, biological clocks, atomic clocks, or particle decays—simply represent different physical phenomena that can be used to mark time. They are all equivalent if their time intervals are related to the basic-clock intervals by correct physics relations. We note that, while the familiar mechanical pendulum is a convenient basic clock for Newtonian physics, it is no longer so in relativity because the dynamical equation for a pendulum must be modified so that it is compatible with special relativity. In short, a phenomenon such as time dilation holds for any clock; but it is easier to see in the case of a light clock.

Collecting Δt terms, we have

$$\Delta t = \frac{d/c}{\sqrt{1 - v^2/c^2}} = \gamma \Delta t', \qquad (2.29)$$

showing the time dilation result of (2.26).[23]

Length contraction

The length Δx of a moving object, compared to the length $\Delta x'$ of the object as measured in its own rest frame O', appears to be shortened. This phenomenon is often called the **FitzGerald–Lorentz contraction** in the literature:

$$\Delta x = \frac{\Delta x'}{\gamma}, \qquad \gamma > 1. \qquad (2.30)$$

Consider the specific example of length measurement of a railcar.[24] Let there be a clock attached to a fixed marker on the ground. A ground observer O, watching the train moving to the right with speed v, can measure the length L of the car by reading off the times when the front and back ends of the railcar pass this marker on the ground:

$$L = v\,(t_2 - t_1) \equiv v \Delta t. \qquad (2.31)$$

But for an observer O' on the railcar, these two events correspond to the passing of the two ends of the car by the (ground-) marker as the marker is seen moving to the left. O' can similarly deduce the length of the railcar in her reference frame by reading the times from the ground clock:

$$L' = v\,(t'_2 - t'_1) \equiv v \Delta t'. \qquad (2.32)$$

These two unequal time intervals in (2.31) and (2.32) are related by the above-considered time dilation (3.54): $\Delta t' = \gamma \Delta t$, because Δt is the time recorded by a clock at rest, while $\Delta t'$ is the time recorded by a clock in motion (with respect to the observer O'). From this we immediately obtain

$$L = v \Delta t = \frac{v \Delta t'}{\gamma} = \frac{L'}{\gamma}, \qquad (2.33)$$

which is the claimed result (2.30) of length contraction.[25] We also note:

- Length contraction is only in the direction of relative motion of the frames. Thus the volume contraction is the same as the length contraction $V = \gamma^{-1} V'$, and not $\gamma^{-3} V'$.
- We see that the derivation of length contraction invokes relativity of simultaneity (by way of time dilation). Why should length comparison involve time and the concept of simultaneity? This connection becomes more understandable when we consider the simplest way to measure the length of a moving object without the employment of a clock. If you want to measure the length of a moving car, you certainly would want to measure its front and back ends simultaneously.

Example: Pion decay length in the laboratory   Time dilation and length contraction appear counter-intuitive to us because our intuition is based on non-relativistic physics ($v \ll c$); in our everyday lives, we generally deal

[23] Time dilation seems counter-intuitive. This is so because our intuition has been built up from familiar experiences with phenomena having velocities much less than c. Actually, it is easier to understand the physical result of time dilation at an extreme speed of $v \lesssim c$. Let us look at this phenomenon in a situation in which $v = c$. In this case, time is infinitely dilated. Imagine a rocket ship traveling at $v = c$ as it passes a clock tower. The rocket pilot (the observer in the O frame) will see the clock (at rest in the O' frame) as infinitely dilated (i.e. stopped) at the instant when the ship passes the tower. This must be so because the light image of the clock cannot catch up with the rocket ship.

[24] The set-up is similar to that shown in Figs. 2.3 and 2.4.

[25] NB: We generally use the O' system for the rest frame of whatever we are most interested in (be it a clock, or some object whose length we are measuring as in $\Delta t = \gamma \Delta t'$ or $\Delta x = \gamma^{-1} \Delta x'$). When using the results in (2.26) and (2.30), one must be certain which is the rest-frame in the case being discussed and not blindly copy any written equation. For example, in the derivation here we have $\Delta t' = \gamma \Delta t$ rather than the usual $\Delta t = \gamma \Delta t'$.

with velocities that are very low compared to the speed of light, and so the relativistic effects are unobservably small to us, being $O(v/c)$. However, in high-energy particle physics experiments, one uses accelerators to push elementary particles to speed very close to the speed of light. As a result, one routinely encounters relativistic effects in such laboratories. Here we discuss such an example. In high-energy proton–proton collisions a copious number of pions (a specie of subatomic particles) are produced. Even though pions have a half-life of $\tau_0 = 1.77 \times 10^{-8}$ s (they decay into a final state of a muon and a neutrino via the weak interaction), they can be collimated to form pion beams for other high-energy physics experiments. This is possible because the pions produced from proton–proton collisions have high kinetic energy, hence high velocity. Because of this high velocity, there is a time dilation effect under SR, which means that the pions have more time before they decay; therefore, they travel further (their decay length is longer). For example, if a beam of pions has a velocity of $0.99c$, it retains half of its intensity even after traveling a distance close to 38 m. This may be surprising because a naive calculation of $\tau_0 c = (1.77 \times 10^{-8}) \times (3 \times 10^8) = 5.3$ m would lead one to expect a much shorter decay length. We shall discuss this problem from two perspectives:

- **Time dilation in the laboratory frame**   The naive calculation is incorrect because the half-life $\tau_0 = 1.77 \times 10^{-8}$ s is the lifetime measured by a clock at rest with respect to the pion. The speed of $0.99c$ corresponds to a γ of 7.1. In the laboratory, the observer will see the pion decay time dilated to $\tau = \gamma \tau_0 = 7.1 \times 1.77 \times 10^{-8}$ s $= 1.26 \times 10^{-7}$ s. Therefore, the decay length is seven times longer, or close to 38 m, as seen in the laboratory.
- **Length contraction in the pion rest frame**   In the rest frame of the pion, this 38 m in the laboratory is viewed as having a contracted length of 5.3 m, which when divided by the particle speed of $0.99c$ yields its half-life of $\tau_0 = 1.77 \times 10^{-8}$ s.

2.3   Lorentz transformation

There are several ways to obtain the explicit form of Lorentz transformation (2.11). Here we will use the length contraction relation (2.30). In the next chapter (Box 3.1), we will present a derivation based more directly on the universality of the speed of light c in every coordinate frame.

We first consider the measurement of a rod of length L lying on the x' axis of the O' system, which is moving with speed v with respect to the O coordinate system. Let the origins of the two coordinate systems coincide at $t = t' = 0$. After a time interval t, we have the situation as shown in Fig. 2.6(a), with the right-hand end of the rod having the position coordinates with respect to these two systems as $x' = L$ and $x = vt + (L/\gamma)$, where we have used the length contraction result of (2.30) for the rod length in O. Combining these two equations, we get the relation

$$x' = \gamma (x - vt).\tag{2.34}$$

This is the first part of the Lorentz transformation (2.11).

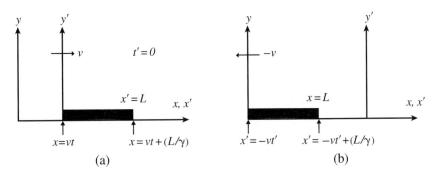

Fig. 2.6 Length contraction as viewed in two reference frames. In (a) a rod (length L) is at rest in the O' system, which moves to the right with respect to the O coordinates. Thus its length is measured in O to be L/γ. In (b) the rod is at rest in the O system, which is viewed by O' as moving to the left. Now its length is measured to be L/γ in O'.

We now repeat the analysis with the rod at rest in the O system and moving with the speed of $-v$ in the O' system. After a time interval t' we have the situation as shown in Fig. 2.6(b). The right-hand end of the rod now has the coordinates $x = L$ and $x' = -vt' + (L/\gamma)$ in the O and O' systems, respectively. This leads to

$$x = \gamma \left(x' + vt'\right).\tag{2.35}$$

We now substitute (2.34) into (2.35), $x = \gamma \left(\gamma x - \gamma vt + vt'\right)$, which can be written as $vt' = \gamma vt - \gamma \left(1 - \gamma^{-2}\right) x$, or

$$t' = \gamma \left(t - \frac{v}{c^2}x\right).\tag{2.36}$$

This is the second part of the Lorentz transformation presented in (2.11). We note that this derivation is based on the physics of length contraction, which in turn relies on time dilation. Thus, the whole derivation of the Lorentz transformation rests on the relativity of time measurement.

Inverse Lorentz transformation In Eqs. (2.34) and (2.36) the coordinates (x, t) are expressed in terms of (x', t'). This is the Lorentz transformation of the O system into the O' system. We can easily use (2.34) and (2.36) to solve, mathematically, for (x, t) in terms of (x', t') with the result:

$$x = \gamma \left(x' + vt'\right)\tag{2.37}$$

$$t = \gamma \left(t' + \frac{v}{c^2}x'\right).\tag{2.38}$$

This is the inverse Lorentz transformation, and it can be easily understood by the fact that, while the O' system is moving with v in the O coordinates, the O system is moving with $-v$ in the O' coordinates. This is how we obtained (2.35) in the above derivation. In summary, to go between the O and O' systems all we need to do is to interchange $(x, t \leftrightarrow x', t')$ and $(-v \leftrightarrow v)$.

2.3.1   Physical meaning of various transformation terms

The Lorentz transformation, Eqs. (2.34) and (2.36), is of fundamental importance in relativity; it will be useful to have a better understanding of its physical significance.

- In the spatial transformation[26]

$$\Delta x' = \gamma \left(\Delta x - v \Delta t \right), \tag{2.39}$$

  the $\Delta x - v \Delta t$ factor is the same one as in the familiar Galilean transformation (2.8). The overall factor γ reflects the physics of length contraction (2.30). In particular for the length measurement of a moving object (for which $\Delta t = 0$, since the length measurements must be made simultaneously in the frame that is moving with respect to the object being measured), the above relation reduces to $\Delta x' = \gamma \Delta x$. This says that the moving frame length Δx is shorter than the rest frame length $\Delta x'$ by a factor of γ.

- The (inverse) time transformation

$$\Delta t = \gamma \left(\Delta t' + \frac{v}{c^2} \Delta x' \right) \tag{2.40}$$

  has two terms.

  - The first term represents the time dilation effect of (2.26) with $\Delta t'$ being the proper time interval if the object (clock) is at rest (i.e. displacement $\Delta x' = 0$) in the O' frame.
  - The second term is the amount of nonsynchronization that has developed in the moving O frame, between two clocks located at different positions $\left(\Delta x' \neq 0 \right)$ in the O' frame. The two clocks are synchronized $\left(\Delta t' = 0 \right)$ in the O' frame. However for the observer in the O system, there will be a lack of synchronization (see Box 2.3 for an example of the role the non-synchronization term plays in a self-consistent description of time evolution by different observer.), according to (2.40), equal to

$$\Delta t_{\text{n-syn}} = \gamma \frac{v}{c^2} \Delta x'. \tag{2.41}$$

Box 2.3 Nonsynchronicity, time dilation, and the resolution of a reciprocity puzzle

In Section 2.2.2, we first illustrated the relativity of simultaneity with an example (see Fig. 2.4). Two events, corresponding to the arrival of the light flashes at the front and rear ends of the railcar, are simultaneous according to the railcar observer; but they are seen by the outside (ground) observer as taking place at two separate instances, as shown in Eq. (2.25). We can rewrite the amount of nonsimultaneity as

$$\Delta t = \gamma \frac{v}{c^2} \Delta x', \tag{2.42}$$

where we have applied the length contraction formula of $\gamma L = L' = \Delta x'$. ($L'$ is the rest-frame length of the railcar.) If we compare (2.42) to the time dilation relation $\Delta t = \gamma \Delta t'$, it suggests a railcar time-separation of $\Delta t' = v \Delta x'/c^2$. Here is the puzzle: this contradicts the condition of $\Delta t' = 0$ which we used to set up the problem to begin with. What went wrong?

The resolution comes with the realization that the clocks in two separate inertial frames (here O = ground, O' = railcar) cannot both be synchronized everywhere (within their respective frames). The nonsynchronicity of clocks within a given frame increases with separation between the clocks in that frame. Let us say that all of the clocks in the O frame are synchronized. In particular, the two clocks located at the rear and front ends of the car are synchronized,[27] $t_2(0) - t_1(0) = 0$. But the two corresponding clocks in the railcar O' frame must be out of sync, by an amount proportional to their distance separation: $t_2'(0) - t_1'(0) = \gamma v \Delta x'/c^2$, as indicated by (2.41). All these points (time-dilation and nonsynchronicity) are nicely summarized by the two terms in the time transformation formula (2.40).

[27](0) indicates at some definite instant, for example, at the instant of light emission at the midpoint of the car.

2.3.2 The relativistic invariant interval

Under the Galilean transformation, while the position interval is relative, $\Delta x \neq \Delta x'$, the time interval $\Delta t = \Delta t'$ is absolute. Under the Lorentz transformation, both are relative. The question arises: is there any quantity that is absolute under the new (Lorentz) transformation among inertial coordinates? There is indeed one. A particular combination of the space and time interval is invariant:

$$\Delta s^2 = \Delta x^2 - c^2 \Delta t^2.$$

This can be easily checked:

$$\Delta s'^2 \equiv \Delta x'^2 - c^2 \Delta t'^2 = \gamma^2 \left[(\Delta x - v\Delta t)^2 - c^2 \left(\Delta t - \frac{v}{c^2} \Delta x \right)^2 \right]$$

$$= \gamma^2 \left[\left(1 - \frac{v^2}{c^2} \right) \Delta x^2 - \left(1 - \frac{v^2}{c^2} \right) c^2 \Delta t^2 \right] = \Delta x^2 - c^2 \Delta t^2 \equiv \Delta s^2.$$

Proper time   What is the physical meaning of this interval? Why should we expect it to be the same in every coordinate frame? This interval is directly related to the time interval $\Delta \tau$ (called the *proper time*), which is the time interval as measured in the rest frame of the clock (where there is no displacement $\Delta x = 0$)

$$\Delta s^2 = -c^2 \Delta \tau^2. \tag{2.43}$$

Since there is only one rest-frame[28] for a clock, its time interval must be unique—all observers should agree on its value. This is the physical basis for the invariance of this quantity. Furthermore, as we shall demonstrate in Box 3.1, Δs is absolute because the light speed c is absolute.

[28] Recall that we are discussing the property of an interval between two events $\Delta x = x_1 - x_2$ and $\Delta t = t_1 - t_2$, etc. There is only one coordinate frame in which $\Delta x = 0$: The object (clock) does not undergo any displacement—hence it is the rest frame of the clock.

2.3.3   Relativity is truly relative

Two observers O and O' are in relative motion. O sees a clock (at rest) in O' as moving and hence it runs slow compared to her own clock (at rest in O), according to the time-dilation relation (3.54). However, just as correctly, observer O' sees the clock in O as moving (with speed $-v$) and concludes it is slow. Who is right? Which clock is slow? Is there a contradiction? According to SR both conclusions are correct; there is full reciprocity, just that we are talking of two different situations. It is helpful to use both the Lorentz transformations, (2.34) and (2.36), as well as their inverse, (2.37) and (2.38).

- For a clock at rest in the O' frame, setting $\Delta x' = 0$ in (2.38), we have $\Delta t = \gamma \Delta t'$. Namely, the observer in O sees the clock in O' as running slow.
- For a clock at rest in the O frame, setting $\Delta x = 0$ in (2.36) we have $\Delta t' = \gamma \Delta t$. Namely, the observer in O' sees the clock in O as running slow.

Thus there is no contradiction because we are talking about two different situations—in one case, $\Delta x' = 0$, and in the other, $\Delta x = 0$.

Perhaps a more concrete way to present this is to reexamine our first derivation of time dilation (2.29) using the "light clock," in which we focused on a clock at rest in the O' frame. This clock was moving in the O frame. We could instead focus on a clock at rest in the O frame, which would be a moving clock with respect to the O' frame. In order to make a change, all we need to do to Eq. (2.29) is to switch $(x, t \leftrightarrow x', t')$ and $(-v \leftrightarrow v)$. Since the velocity appears only as v^2 in (2.29), we would get $\Delta t' = \gamma \Delta t$ instead of $\Delta t = \gamma \Delta t'$.

In short, the key observation is that two different time comparisons are involved, and so there is no contradiction.[29] Thus, we can say that "relativity is truly relative."

[29]Here we have discussed explicitly the reciprocity of time measurements. Clearly, the same reciprocity holds for length measurements as well. In fact, such an example is given by the pole-and-barn paradox discussed in Box 2.4.

2.3.4   Two paradoxes as illustrative SR examples

Here we present two instructive examples, involving relativistic velocities, that shed light on several basic concepts in relativity: the pole-and-barn paradox (Box 2.4) and the twin paradox (Box 2.5). They first appear to be paradoxical; however, through careful analysis in the context of the new SR kinematics, any apparent contradiction is resolved.

Box 2.4 The pole-and-barn paradox

Imagine a runner rushing through a barn carrying a long pole horizontally (see, e.g. Sartori, 1996). Let's say that the rest-frame lengths of the pole and barn are the same (L). From the perspective of the barn/ground observer, Fig. 2.7(a), the length contraction of the moving pole allows it be contained within the barn for a moment—the two ends of the pole, A and B, can be situated inbetween the front (F) and rear (R) doors of the barn. On the other hand, from the perspective of the runner/pole, Fig. 2.7(b), the barn is moving towards the runner and its length contraction would mean that the barn cannot possibly enclose the pole—the two ends of the pole would appear, for a moment, to protrude out from the front and rear doors of the barn. While we might be able to accept the counter-intuitive length contraction of the pole and barn as viewed by different observers, it is still worthwhile to clarify the different containment outcomes.

We need to make a careful analysis of what we mean operationally by the "enclosure" of the pole by the barn, or of the barn by the pole, in different coordinate frames. A reliable way to make this analysis is to carefully list the sequence of events that are involved. The process of the pole rushing past the barn involves four distinct events: the first event (call it AF) is the one in which the front end of the pole first reaches the front door of the barn; the last event (call it BR) is the one in which the rear end of the pole finally leaves the rear door of the barn. Between these two occurrences, there are two other events: AR in which the front end of the pole reaches the rear door, and BF in which the back end of the pole leaves the front door of the barn. The temporal order of these in-between events, AR and BF, is different, depending on whether the barn completely contains the pole, or the pole protrudes out of the barn. In the first sequence, as seen by the ground/barn observer, BF happens before AR, while in the second sequence, as seen by the runner, AR happens before BF. Thus, the resolution of the "pole-and-barn paradox" comes with the realization that the temporal order of events can be changed in different coordinate frames.Let the event time for BF in the ground/barn frame be $t_{BF} = 0$, and let the event AR take place an interval $\Delta t = t_{AR}$ later. Similarly, in the pole/runner's rest frame, $t'_{BF} = 0$, and $t'_{AR} = \Delta t'$. Based on the above discussion, we expect Δt and $\Delta t'$ to have opposite sign. Since the two situations should be completely symmetric, they should have the same magnitude; hence, $\Delta t = -\Delta t'$. A straightforward calculation[30] yields $\Delta t = -\Delta t' = \left(L - \gamma^{-1}L\right)/v$. Indeed, these two time measurements are correctly related by the Lorentz transformation $\Delta t' = \gamma\left(\Delta t - v\Delta x/c^2\right)$ with $\Delta x = L$. (The relevant spacetime diagram will be discussed in Problem 3.13.)

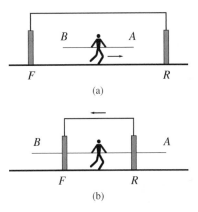

Fig. 2.7 A pole-carrying runner passes through a barn, with a front door F and a rear door R. (a) The ground/barn observer sees the sequence of events as AF, BF, AR, BR. (b) The runner sees the sequence as AF, AR, BF, BR.

[30]In the O frame, where the pole can be completely contained inside the barn, the time interval for A to reach the rear door after the event of B passing the front door is just the time interval for the pole to cover the length-difference of the barn and the (contracted) pole: $\Delta t = t_{AR} - t_{BF} = \left(L - \gamma^{-1}L\right)/v$. We can plug this into the Lorentz transformation in order to check the relation between Δt and $\Delta t'$

$$\Delta t' = \gamma(\Delta t - vL/c^2)$$
$$= (\gamma L/v)(1 - \gamma^{-1}) - v^2/c^2$$
$$= -(L/v)(1 - \gamma^{-1}) = -\Delta t.$$

Box 2.5 The twin paradox

The well-known "twin paradox" is a reciprocity puzzle that sheds light on several basic concepts in relativity. We shall explain its resolution as being

(cont.)

Box 2.5 (*Continued*)

related to the involvement of noninertial frames (see e.g. Ellis and Williams, 1988).

The paradox as a reciprocity puzzle Two siblings, Al and Bill, are born on the same day. At some time during their lifetime, Al goes on a long journey at high speed in a spaceship; Bill stays at home. The biological clock of Al will be measured by the stay-at-home Bill to run slow. When Al returns he should be younger than Bill. Let us consider the case of Al traveling outward at $\beta_1 = 4/5$ for 15 years, and then returning (traveling "inward") at $\beta_2 = -4/5$ for 15 years. Both periods of 15 years are measured in Al's rocket ship with a $\gamma = 5/3$ for both $\beta_1 = 4/5$ and $\beta_2 = -4/5$. When Al and Bill meet again, Al should be younger: while Al has aged 30 years, Bill has aged $5/3 \times 30 = 50$ years according to SR time dilation. Letting the number of years Al has aged be A and the number of years Bill has aged be B, we have

$$A = 30, \quad B = 50 \qquad \text{(Bill's viewpoint).} \qquad (2.44)$$

Of course, this SR prediction of asymmetric aging of the twins, while counter-intuitive based on our low-velocity experience, is not "paradoxical." It is just an example of time dilation—which is counter-intuitive, but real. However, there appears to be a reciprocity puzzle. If relativity is truly relative, we could just as well consider this separation and reunion from the viewpoint of Al, who sees Bill as being in a moving frame. So, when Bill "returns," it is Bill who should have stayed younger. From the viewpoint of Al, the stay-at-home Bill should be younger: while A has aged 30 years, B has aged only $(5/3)^{-1} \times 30 = 18$ years.

$$A = 30, \quad B = 18 \qquad \text{(Al's viewpoint).} \qquad (2.45)$$

Thus, the number of years Bill has aged has been found in one case to be 50 and in another case 18—a full 32 year difference. Unlike the reciprocity examples discussed previously, in this case, these theoretical predictions cannot both be correct—as the twins will meet again to compare their ages directly. Which viewpoint, which theory, is the correct one?

Checking theory by measurements To answer this question, we can carry out the following measurement of Bill's age. Let the stay-at-home Bill celebrate his birthdays by setting off firework displays, which Al, with a powerful telescope onboard his spaceship, can always observe. The theory can be checked by counting the number of firework flashes Al sees during his 30-year-long journey. If he sees 18 flashes, then Al's viewpoint is right; if 50, then Bill is right. **During the outward-bound journey,** Al sees flashes at interval of Δt_A, which differs from the interval $\Delta t_B = 1$ year that Bill sees: first of all, there is the time-dilation effect $\gamma \Delta t_B$, but there is also the fact that between the flashes Al and Bill, because Al is moving away, have increased their distance by an amount $v \Delta t_B$. Therefore, to reach Al in the space ship, the light signals (i.e. the flashes)[31] have to take an extra interval of $v\Delta t_B/c$, which also has to be dilated by a factor of γ:

$$\Delta t_A^{(out)} = \gamma\,(1+\beta_1)\,\Delta t_B = \frac{5}{3}\left(1+\frac{4}{5}\right) = 3 \text{ years}.$$

Namely, during the 15-year outward-bound journey, Al sees Bill's flashes every three years, for a total of five flashes. **During the inward-bound journey,** the relative velocity reverses sign ($\beta_2 = -\beta_1$); Al sees flashes at an interval of

$$\Delta t_A^{(in)} = \frac{5}{3}\left(1-\frac{4}{5}\right) = \frac{1}{3} \text{ years}.$$

Thus, Al sees a total of 45 flashes on his inward-bound journey. Al sees a total of 50 flashes on his entire journey. This proves that Bill's viewpoint is correct: while the traveling twin Al has aged 30 years, the stay-at-home twin Bill has aged 50 years. But there is still the following question: why is Bill's viewpoint correct, and not Al's ?

The reciprocity puzzle resolved The reciprocity situation as discussed in Section 2.3.3 is not applicable here: While Bill remains in an inertial frame of reference throughout the journey, Al does not—because Al must turn around! Thus Al's viewpoint cannot be represented by a single inertial frame of reference. This goes beyond special relativity because noninertial frames are involved. At least two inertial frames must be used to describe Al's round-trip journey. The turn-around of going from one inertial frame to the other must necessarily involve acceleration.[32]

[31] This is a particular realization of the non-synchronicity of clocks as discussed in Eq. (2.40). Keep in mind that the fireworks act as clocks in this situation.

[32] As will be worked out in Problem 3.11 and Problem 3.12, the fact that Al's journey involves a change of the inertial frames, $O' \rightarrow O''$, accounts for "the missing 32 years," discussed after Eq. (2.45).

Review questions

1. Two inertial frames are moving with respect to each other with velocity $\vec{v} = v\hat{x}$. Write out the Lorentz transformation of coordinates $(t, x, y, z) \longrightarrow (t', x', y', z')$? Show that in the low velocity limit it reduces to the Galilean transformation.

2. From the Lorentz transformation, derive the velocity transformation rule in one spatial dimension $(u \rightarrow u')$ that generalizes the familiar low-velocity relation of $u' = u - v$, where v is the relative speed between the two inertial coordinate frames.

3. Under a Lorentz transformation, the electric and magnetic fields transform into each other. Use the familiar laws of electromagnetism to explain why, for example, a static electric field due to an electric charge, when viewed by a moving observer, is seen to have a magnetic field component.

4. The Lorentz transformation was first discovered as some mathematical property of Maxwell's equations. What did Einstein do to provide it with a physical basis?

5. What are the two postulates of special relativity?

6. When the new kinematics of SR are studied (e.g. clock synchronization, time dilation as illustrated by a light clock, etc.), one often employs light signals. Why is this so?

7. What is the meaning of "relativity of simultaneity?" Why is this counter-intuitive?

8. Relativity teaches us to treat the relativity of spatial equi-locality and the relativity of temporal simultaneity on the same footing. What is the general lesson we can learn from this?

9. Use the SR postulates and a simple light clock to derive the physical result of time dilation, $\Delta t = \gamma \Delta t'$. What is the relationship of the clock to each of the two reference frames O and O'?

10. Use the SR postulates and the physics of time dilation to derive the physical result of length contraction, $\Delta x = \gamma^{-1}\Delta x'$, of an object as viewed by two observers O and O' in relative motion. What is the relationship of this object to each of these two reference frames?

11. What are the various terms in the Lorentz transformation that represent the respective physics of length contraction, time dilation and relativity of simultaneity. Present reasoning for your identification.

12. Under the Lorentz transformation, not only the spatial interval Δx, but also the time interval Δt, are relative. What combination of spatial and time intervals remains absolute? What does one mean by "relative" and "absolute" in this context? Give a simple *physical explanation* as to why we expect this combination to be absolute.

Problems

2.1 **Newtonian relativity**   Consider a few definite examples of Newtonian mechanics that are unchanged by the Galilean transformation of (2.8) and (2.9):

(a) Show that force as given by the product of acceleration and mass $\vec{F} = m\vec{a}$, as well as a force law such as Newton's law of gravity $\vec{F} = G_N\left(m_1 m_2/r^2\right)\hat{r}$, remain the same in every inertial frame.

(b) Energy and momentum conservation laws hold in different inertial frames. Consider the one-dimensional collision of two particles having initial (mass, velocity) of $(3, 8)$ and $(5, -4)$ in some appropriate units and final (mass, velocity) of $(3, -7)$ and $(5, 5)$. Check that energy and momentum are conserved when viewed by observers in two reference frames having a relative speed of 2 in some appropriate units. Also work out the same conservation law for the general case of $(m_1, v_1) + (m_2, v_2) \to (m_1, v_1') + (m_2, v_2')$ with a relative velocity u between the observers in the two frames.

2.2 **Inverse Lorentz transformation**   The Lorentz transformation of (2.11) may be written in matrix form as

$$\begin{pmatrix} ct' \\ x' \\ y' \\ z' \end{pmatrix} = \begin{pmatrix} \gamma & -\beta\gamma & 0 & 0 \\ -\beta\gamma & \gamma & 0 & 0 \\ 0 & 0 & 1 & 0 \\ 0 & 0 & 0 & 1 \end{pmatrix} \begin{pmatrix} ct \\ x \\ y \\ z \end{pmatrix}. \qquad (2.46)$$

Find the inverse transformation, i.e. find the coordinates (ct, x, y, z) in terms of $\left(ct', x', y', z'\right)$, by using the physical expectation that the inverse should be given by changing the sign of the relative velocity v. Show that the transformation found in this way is indeed inverse to the matrix in (2.46) by explicit matrix multiplication.

2.3 **Lorentz transformation of derivative operators**   The Lorentz transformation as shown in Eq. (2.46) can be written in component notation as

$$x'^\mu = \sum_\nu [\mathbf{L}]^\mu_\nu x^\nu, \qquad (2.47)$$

where $\mu = 0, 1, 2, 3$ with $x^0 = ct$, $x^1 = x$, $x^2 = y$ and $x^3 = z$. Here we seek the transformation $\bar{\mathbf{L}}$ for the coordinate derivatives

$$\partial'_\mu = \sum_\nu \bar{\mathbf{L}}^\nu_\mu \partial_\nu, \qquad (2.48)$$

where ∂_μ is the shorthand notation for $\partial/\partial x^\mu$

$$\partial_\mu = \left(\frac{\partial}{c\partial t}, \frac{\partial}{\partial x}, \frac{\partial}{\partial y}, \frac{\partial}{\partial z} \right).$$

Following the steps outlined below, show that the coordinate derivative operators $\left(\frac{\partial}{\partial t}, \frac{\partial}{\partial x^i} \right)$ transform oppositely ($v \to -v$) as compared to the manner in which the coordinates $\left(t, x^i\right)$ transform. (2.11):

$$\frac{\partial}{\partial t'} = \gamma\left(\frac{\partial}{\partial t} + v\frac{\partial}{\partial x}\right),$$

$$\frac{\partial}{\partial x'} = \gamma\left(\frac{\partial}{\partial x} + \frac{v}{c^2}\frac{\partial}{\partial t}\right), \qquad (2.49)$$

$$\frac{\partial}{\partial y'} = \frac{\partial}{\partial y} \qquad \frac{\partial}{\partial z'} = \frac{\partial}{\partial z}.$$

That is, the transformation matrices are related as $\bar{\mathbf{L}} = \left[\mathbf{L}^{-1}\right]$.

(a) Obtain this result by the standard chain rule of differentiation: $\partial/\partial x' = \left(\partial x/\partial x'\right)\partial/\partial x + \left(\partial t/\partial x'\right)\partial/\partial t$ and the explicit form of the Lorentz transformation (2.11).

(b) Obtain this result by the observation that $\partial x^\nu/\partial x^\mu = \delta_{\mu\nu}$ is an invariant, hence the transformations of ∂_μ and x^ν must "cancel" each other.

2.4 **Lorentz covariance of Maxwell's equations**   Show that Maxwell's equations and the Lorentz force law as given are covariant under the Lorentz transformation as given in (2.49), (2.16), and (2.17). *Suggestion:* Work with the homogeneous and non-homogeneous parts of Maxwell's equations separately. For example, show that

$$\nabla' \cdot \mathbf{B}' = 0, \qquad \nabla' \times \mathbf{E}' + \frac{1}{c}\frac{\partial \mathbf{B}'}{\partial t'} = 0, \quad (2.50)$$

follows, by Lorentz transformation, from

$$\nabla \cdot \mathbf{B} = 0, \qquad \nabla \times \mathbf{E} + \frac{1}{c}\frac{\partial \mathbf{B}}{\partial t} = 0. \qquad (2.51)$$

2.5 **From Coulomb's to Ampere's law**   We can derive $F_y = ma_y$ and $F_z = ma_z$ from $F_x = ma_x$ by rotation transformations. Show that, by Lorentz transformations, one can similarly derive (a) Ampere's law (with conduction and displacement currents) from Coulomb's/Gauss's law, and vice versa; and (b) the magnetic Gauss's law (absence of magnetic monopole) from Faraday's law of induction, and vice versa.

2.6 **Length contraction and light-pulse clock**   In Section 2.2.3, we used a light-pulse clock to demonstrate the phenomenon of time dilation. This same clock can be used to demonstrate length contraction: the length of the clock L can be measured through the time interval Δt that it takes a light pulse to make the trip across the length of the clock and back: $2L = c\Delta t$. Deduce the length contraction formula (2.30) in this setup. *Suggestion:* In the case of time dilation, we had the clock moving in the

direction perpendicular to the rest-frame light path. Here you want it be parallel. Also, you will need to use the time-dilation formula (2.26) to deduce the final length-contraction result, just as the derivation given in Section 2.2.3.

2.7 **Two spaceships passing one another**   Two spaceships traveling in opposite directions pass one another at a relative speed of 1.25×10^8 m/s. The clock on one spaceship records a time duration of 9.1×10^{-8} s for it to pass from the front end to the tail end of the other ship. What is the length of the second ship as measured in its own rest frame?

2.8 **Invariant spacetime interval and relativity of simultaneity**   Two events are spatially separated $\Delta x \neq 0$, but simultaneous, $\Delta t = 0$, in one frame. When viewed in another inertial frame, they are no longer seen as taking place at the same time, $\Delta t' \neq 0$. Find this time separation $\Delta t'$ in terms of Δx and $\Delta x'$ in two ways: (a) using the invariant spacetime interval, and (b) using the Lorentz transformation.

2.9 **More simultaneity calculations**   Work out the spacetime coordinates (x, t) of the two light pulses emitted from the midpoint of a railcar and arriving at the front and back ends of the railcar as described in Section 2.2.2, cf. Fig. 2.4.

(a) Let the O' coordinates be the railcar observer system, and O the platform observer system. Given $\Delta t' = 0$, use the Lorentz transformation and its inverse to find the relations among Δt, Δx, and $\Delta x'$.

(b) One of the relations obtained in (a) should be $\Delta x = \gamma \Delta x'$. Is this compatible with the derivation of length contraction as done in Section 2.2.3? Explain.

(c) An observer can locate the time the light emission took place by calculating the time it took the light signal to reach the observer. If the interval is t_1 then it must be emitted at time $-t_1$. (Namely, we define the arrival time as being $t = 0$.) By the same token, the emission must have taken place at a location $x_1 = -ct_1$. In this way, verify the relation $\Delta x = \gamma \Delta x'$ discussed above.

2.10 **Reciprocity of twin-paradox measurements**   In Box 2.5, we worked out the following twin-paradox measurement: the traveling Al, during his 30-year journey, sees 50 birthday fireworks set off by the stay-at-home Bill. We now consider the reciprocal measurement of Bill watching the birthday fireworks set off by Al. If reciprocity works, in Bill's 50-year passage, he should observe 30 of Al's annual celebrations. This should be so, even though each observer sees the other's clock to run slow. Work this out explicitly.

2.11 **Velocity addition in the twin paradox**   Consider the twin paradox given in Box 2.5. Just before the traveling twin (Al) turned around his rocketship (after traveling 15 years), his clock told him that the stay-at-home twin (Bill) had aged 9 years since his departure. But immediately after the turn-around, his clock found Bill's elapsed time to be 41 years. Use the velocity addition rule and time dilation to account for this change of 32 years.

RELATIVITY
Metric Description of Spacetime

Special relativity: The geometric formulation

3

- The constancy of light speed c allows us to interpret geometrically the relativistic invariant interval s as a length in the 4D manifold, called Minkowski spacetime, with a metric equal to **diag(-1,1,1,1)**.
- The physics of SR, such as time dilation and length contraction, follows directly from the Lorentz transformation, which is a rotation in Minkowski spacetime.
- We introduce 4-vectors as the simplest example of tensors in Minkowski space, and construct their scalar products. Besides the 4-position vector x^μ, we also introduce the 4-velocity U^μ and 4-momentum p^μ, with components that transform into each other under a Lorentz transformation.
- The principal features of a spacetime diagram and its representation of Lorentz transformations are presented. In particular, the causal structure of SR is clarified in terms of lightcones in a spacetime diagram.

The new kinematics of special relativity discussed in the previous chapter can be expressed elegantly in the geometric formalism of a four-dimensional manifold, known as **spacetime**, as first formulated by Herman Minkowski. The following are the opening words of an address he delivered at the 1908 Assembly of German National Scientists and Physicians held in Cologne, Germany.

The views of space and time which I wish to lay before you have sprung from the soil of experimental physics, and therein lies their strength. They are radical. Henceforth space by itself, and time by itself, are doomed to fade away into mere shadows, and only a kind of union of the two will preserve an independent reality.

In this geometric formulation, the stage on which physics takes place is Minkowski spacetime with the time coordinate being on an equal footing with the spatial coordinates. Here we introduce **flat spacetime**, which is the spacetime that is appropriate for the physics of special relativity. This prepares us for the study of the larger framework of **curved spacetime** in general relativity.

3.1   Minkowski spacetime

The fact that measurement results for time, as well as space, may be different for different observers means that time must be treated as a coordinate in much the same way as the spatial coordinates. The unification of space and time can be made explicit when space and time coordinates appear in the same position vector. A coordinate transformation may be regarded as a rotation in this 4D space, with the possibility of changing space and time coordinates (t, x, y, z) into each other—in much the same way the (x, y, z) coordinates change into each other under an ordinary 3D rotation. The 4D Minkowski spacetime has coordinates $\{x^\mu\}$. Henceforth, all Greek indices[1] such as μ will have the range 0, 1, 2, 3. Namely,

$$x^\mu = \left(x^0, \ x^1, \ x^2, \ x^3\right) = (ct, \ x, \ y, \ z). \tag{3.1}$$

We have already shown in Section 2.4.2 that the interval between the coordinate origin and the position coordinate x^μ,

$$s^2 = -c^2 t^2 + x^2 + y^2 + z^2, \tag{3.2}$$

is a relativistic invariant. Namely, if we were to make a coordinate transformation, with the new coordinates being $x'^\mu = (ct', \ x', \ y', \ z')$, the corresponding spacetime interval $s'^2 = -c^2 t'^2 + x'^2 + y'^2 + z'^2$ would have the same value $s'^2 = s^2$. In Section 2.4.2 we demonstrated this invariance by a calculation using the explicit form of the Lorentz transformation. In Box 3.1, we present another derivation, showing that this result follows directly from the basic postulate that the speed of light, c, is the same in every reference frame: s is absolute because c is absolute.

This interval Δs has the physical significance of being directly related to the proper time: $\Delta s^2 = -c^2 \Delta \tau^2$. Furthermore, we recall that this rest-frame time τ is related to the coordinate time t by the relativistic time-dilation formula of $t = \gamma \tau$. As discussed in Box 3.1, for a light ray, this interval vanishes; therefore, the concept of proper time is not applicable for a light ray. This is entirely consistent with the fact that there does not exist a coordinate frame in which the light velocity vanishes—that is, a frame in which light is at rest.

In 4D Euclidean space with Cartesian coordinates (w, x, y, z), the invariant length[2] is given as $s^2 = w^2 + x^2 + y^2 + z^2$. The minus sign in front of the $c^2 t^2$ term in (3.2) means that if we regard ct as the fourth dimension, the relationship between coordinate and length measurements differs from that in Euclidean space. We say that Minkowski space is pseudo-Euclidean. In Section 3.1.1, we shall introduce generalized coordinates and distance measurements (via the metric) in Minkowski spacetime. In subsequent chapters, the same formalism will be shown to be applicable to coordinates in the warped spacetime of general relativity.

[1] Superscripts rather than subscripts are being used because this is the customary mathematical notation for coordinates in relativity, and it is used because relativity is expressed in tensor notation, which uses superscript indices for "contravariant components" of vectors and subscript indices for "covariant components" of vectors. Tensor notation will be explained in detail in Chapter 12. Until then, we shall always denote components of a 4D vector by using a superscript index. That is, we only need to work with contravariant components.

[2] We are familiar with the idea that in 3D Euclidean space the relation between coordinates and squared length is $l^2 = x^2 + y^2 + z^2$; hence, this straightforward generalization to 4D Euclidean space. However, in the following presentation, we shall refer to any such (quadratic in coordinates) invariant as a squared "length," regardless of whether we can actually visualize it as a quantity that can be measured by a yardstick or not.

Box 3.1 Δs is absolute because c is absolute

In the geometric formulation of special relativity, the spacetime interval

$$\Delta s^2 = \Delta x^2 + \Delta y^2 + \Delta z^2 - c^2 \Delta t^2, \tag{3.3}$$

where $\Delta x = x_2 - x_1$, etc. plays a central role. Here we show that the invariance under a Lorentz transformation of this interval follows directly from the basic postulate of special relativity: the speed of light, c, is the same in every inertial reference frame. The interval Δs is absolute because c is absolute (Landau and Lifshitz, 1975).

First consider the special case in which the two events, $(\vec{x}_1, t_1)$ and $(\vec{x}_2, t_2)$, are connected by a light signal. The interval Δs^2 must vanish because in this case $|\Delta \vec{x}|/\Delta t = c$. When observed in another frame O', this interval also has a vanishing value $\Delta s'^2 = 0$, because the velocity of light remains the same in the new frame O'. From this, we infer that for *any* interval Δs connecting two events (not necessarily by a light signal), Δs and $\Delta s'$ must always be proportional to each other (because, if Δs^2 vanishes, so must $\Delta s'^2$):

$$\Delta s'^2 = F \, \Delta s^2. \tag{3.4}$$

The proportionality factor F can in principle depend on the coordinates and the relative velocity of these two frames: $F = F(\vec{x}, t, \vec{v})$. However the requirement that space and time be homogeneous (i.e. there is no privileged point in space and time) implies that there cannot be any dependence on $\vec{x}$ and t. That space is isotropic means that the proportionality factor cannot depend on the direction of the relative velocity $\vec{v}$ of the two frames. Thus, we can at most have it be dependent on the magnitude of the relative velocity, $F = F(v)$. We are now ready to show that, in fact, $F(v) = 1$.

Besides the system O', which is moving with velocity $\vec{v}$ with respect to system O, let us consider yet another inertial system O'', which is moving with a relative velocity of $-\vec{v}$ with respect to the O' system:

$$O \xrightarrow{\vec{v}} O' \xrightarrow{-\vec{v}} O''. \tag{3.5}$$

From the above consideration, and applying (3.4) to these frames:

$$\Delta s'^2 = F(v) \, \Delta s^2,$$
$$\Delta s''^2 = F(v) \, \Delta s'^2 = [F(v)]^2 \, \Delta s^2. \tag{3.6}$$

However, it is clear that the O'' system is in fact just the O system. This requires that $F(v)^2 = 1$. The solution $F(v) = -1$ being nonsensical, we conclude that this interval Δs is indeed an invariant: $\Delta s'' = \Delta s' = \Delta s$. Every inertial observer will see the same light velocity, and will therefore obtain the same value for this particular combination of space and time intervals.

3.1.1 Basis vectors, the metric and scalar product

To set up a coordinate system for the 4D Minkowski space means to choose a set of four basis vectors $\{\mathbf{e}_\mu\}$, where $\mu = 0, 1, 2, 3$. Each $\mathbf{e}_\mu$, for a definite index value, is a 4D vector. (Figure 3.1 illustrates a case for a 2D space.) In contrast to the Cartesian coordinate system in Euclidean space (see Box 2.1),

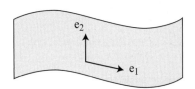

Fig. 3.1 Basis vectors for a 2D surface.

[3]$\delta_{\mu\nu}$ is the Kronecker delta:

$$\delta_{\mu\nu} = \begin{cases} 1; & \mu = \nu \\ 0, & \mu \neq \nu. \end{cases}$$

We can interpret its values as the elements of an identity matrix: $[\mathbf{1}]_{\mu\nu} = \delta_{\mu\nu}$.

this in general is not an orthonormal set, $\mathbf{e}_\mu \cdot \mathbf{e}_\nu \neq \delta_{\mu\nu}$,[3] Nevertheless, we can represent such a collection of scalar products among the basis vectors as a symmetric matrix, called the **metric**, or the **metric tensor**:

$$\mathbf{e}_\mu \cdot \mathbf{e}_\nu \equiv g_{\mu\nu}. \tag{3.7}$$

We can display the metric as a 4×4 matrix with elements being dot products of basis vectors:

$$[\mathbf{g}] = \begin{pmatrix} g_{00} & g_{01} & .. \\ g_{10} & g_{11} & .. \\ & & \end{pmatrix} = \begin{pmatrix} \mathbf{e}_0 \cdot \mathbf{e}_0 & \mathbf{e}_0 \cdot \mathbf{e}_1 & .. \\ \mathbf{e}_1 \cdot \mathbf{e}_0 & \mathbf{e}_1 \cdot \mathbf{e}_1 & .. \\ & & \end{pmatrix}. \tag{3.8}$$

Thus, the diagonal elements are the (squared) lengths of the basis vectors, $|\mathbf{e}_0|^2$, $|\mathbf{e}_1|^2$, etc. while the off-diagonal elements represent their deviations from orthogonality. Any set of mutually perpendicular bases would be represented by a diagonal metric matrix, with diagonal entries of 1 if the basis vectors were of unit length (which is not required). For a Euclidean space with Cartesian coordinates, we have $g_{\mu\nu} = \delta_{\mu\nu}$.

We can expand any 4D vector[4] in terms of the basis vectors,

[4]We denote 4D vectors with boldfaced letters, such as "**A**", and 3D vectors with an arrow on the top, such as "$\vec{A}$".

$$\mathbf{A} = \sum_\mu A^\mu \mathbf{e}_\mu \equiv A^\mu \mathbf{e}_\mu, \tag{3.9}$$

where the coefficients of expansion $\{A^\mu\}$ are labeled with superscript indices. From this point on, we shall adopt the **Einstein summation convention** of omitting the explicit display of the summation sign ($\sum$) whenever we have a pair of repeated indices[5] in one term of an expression, such as μ in (3.9). Consider the scalar product of two vectors and make their respective expansions, $\mathbf{A} \cdot \mathbf{B} = (A^\mu \mathbf{e}_\mu) \cdot (B^\nu \mathbf{e}_\nu) = (\mathbf{e}_\mu \cdot \mathbf{e}_\nu) A^\mu B^\nu$, which can be written as

[5]Such repeated indices are called "dummy indices" and we are free to change their names, for example, $A^\mu \mathbf{e}_\mu = A^\nu \mathbf{e}_\nu = A^\lambda \mathbf{e}_\lambda$, etc. Also, note that although we apply this rule to any pair of repeated indices, strictly speaking it should always be a pair of repeated indices with one superscript index and the other subscript index. See Chapter 12 for further details.

$$\mathbf{A} \cdot \mathbf{B} = g_{\mu\nu} A^\mu B^\nu, \tag{3.10}$$

or displayed in matrix form as

$$\mathbf{A} \cdot \mathbf{B} = \begin{pmatrix} A^0 & A^1 & .. \end{pmatrix} \begin{pmatrix} g_{00} & g_{01} & .. \\ g_{10} & g_{11} & .. \\ & & \end{pmatrix} \begin{pmatrix} B^0 \\ B^1 \\ \\ \end{pmatrix}. \tag{3.11}$$

The metric allows us to express the scalar product in terms of the vector components. A key feature is that all the indices are summed over (said to be "contracted") and the result has no free index left over. A scalar means that it is the same in all coordinates; it is unchanged under a transformation. In particular for the case $\mathbf{A} = \mathbf{B} = \mathbf{x}$, the position vector, then the invariant squared length $s^2 = \mathbf{x} \cdot \mathbf{x}$ can be written by (3.10) as

$$s^2 = g_{\mu\nu} x^\mu x^\nu. \tag{3.12}$$

[6]Here x^μ are understood to be the interval $\triangle x^\mu$ between the position point at x^μ and the origin of the coordinate system.

We refer to this equation as the **metric equation**; it plays a central role in the geometric formulation of relativity. The coordinate-dependent metric turns the coordinate-dependent position components[6] $\{x^\mu\}$ into a coordinate-independent length s^2.

3.1.2 The Minkowski metric and Lorentz transformation

In Minkowski space we have the position 4-vector $x^\mu = (x^0, x^1, x^2, x^3) = (ct, x, y, z)$; the invariant length $s^2 = -c^2 t^2 + x^2 + y^2 + z^2$ can be identified with the scalar product formula of (3.10) and (3.11):

$$s^2 = \mathbf{x} \cdot \mathbf{x} = \begin{pmatrix} x^0 & x^1 & x^2 & x^3 \end{pmatrix} \begin{pmatrix} -1 & & & \\ & 1 & & \\ & & 1 & \\ & & & 1 \end{pmatrix} \begin{pmatrix} x^0 \\ x^1 \\ x^2 \\ x^3 \end{pmatrix} \qquad (3.13)$$

$$= \eta_{\mu\nu} x^\mu x^\nu. \qquad (3.14)$$

Thus, the Minkowski space, with a pseudo-Cartesian coordinate system, has the metric

$$g_{\mu\nu} = \eta_{\mu\nu} = \operatorname{diag}(-1, 1, 1, 1). \qquad (3.15)$$

Because the metric $\eta_{\mu\nu}$ is constant (independent of position and time), Minkowski spacetime is a flat space,[7] as opposed to a curved space. It differs from the familiar Euclidean space by having a negative $\eta_{00} = -1$. We can think of x^0 as having an imaginary length. As we shall discuss in subsequent chapters, the spacetime manifold is warped in the presence of matter and energy. In Einstein's general theory of relativity, curved spacetime is the gravitational field and the metric $g_{\mu\nu}(x)$ for such a warped spacetime is necessarily position-dependent. The pseudo-Euclidean flat spacetime is obtained only in the **absence** of gravity. This is the limit of special relativity.

Lorentz transformations are the coordinate transformations between two frames moving with a constant velocity with respect to each other in Minkowski spacetime.[8] For example a boost with velocity v in the x^1 direction changes the vector components $x^\mu \to x'^\mu$ as shown in (2.11) of the previous chapter. We can write this transformation in matrix-component form as

$$\begin{pmatrix} x'^0 \\ x'^1 \\ x'^2 \\ x'^3 \end{pmatrix} = \begin{pmatrix} \gamma & -\beta\gamma & 0 & 0 \\ -\beta\gamma & \gamma & 0 & 0 \\ 0 & 0 & 1 & 0 \\ 0 & 0 & 0 & 1 \end{pmatrix} \begin{pmatrix} x^0 \\ x^1 \\ x^2 \\ x^3 \end{pmatrix}, \qquad (3.16)$$

where β and γ are defined in (2.12). We shall demonstrate in Box 3.2 that this explicit form of the Lorentz transformation follows directly from the requirement that the transformation leaves the length (in fact, any scalar product) and the metric of the space invariant. The Lorentz coordinate transformation of (3.16) may be written in matrix-component form as

$$x'^\mu = \mathbf{L}^\mu{}_\nu x^\nu. \qquad (3.17)$$

L denotes the 4×4 Lorentz transformation matrix, with the left index (μ) being the row index and the right index (ν) the column index.[9] The components x^μ being those of a prototype 4-vector, any 4-vector **A** must have components A^μ that transform into components A'^μ of the same vector in another coordinate system, under a boost in the x^1 direction, in exactly the same

[7]In Chapter 5 we shall present a brief introduction to the geometric description of a curved space. We start with the familiar 2D space. Thus, say, a horizontal sheet of paper is "flat" and a spherical surface is "curved." This constant metric is a sufficient but not a necessary condition for a flat space. A flat plane can still have a position-dependent metric if we adopt a system such as the polar coordinate system.

[8]In our presentation, we shall restrict ourselves to coordinate transformations under which the coordinate origin $(t, x) = (0, 0)$ is fixed. These are Lorentz transformations. The combined transformations of a Lorentz and a coordinate translation ($x^\mu \to x^\mu + a^\mu$ with a^μ being a constant) is called a *Poincaré transformation*.

[9]We follow this convention regardless of whether the index is a superscript or subscript.

way as given in (3.16):

$$A'^{\mu} = \mathbf{L}^{\mu}_{\nu} A^{\nu}. \tag{3.18}$$

Box 3.2 Lorentz transformation as a rotation in 4D spacetime

The key feature of Minkowski spacetime is that it has the pseudo-Euclidean metric of (3.15). A (squared) length in this space is given by $s^2 = -c^2t^2 + x^2 + y^2 + z^2$. We can make this identification even more obvious by working with an imaginary coordinate $w = ict$ so that $s^2 = w^2 + x^2 + y^2 + z^2$. The coordinate transformation $\mathbf{L}$ can be thought of as a "rotation" of the 4D spacetime coordinates.[10] Any rotational transformation (by definition) preserves the length of vectors. This condition that a rotational transformation be length preserving is enough to fix the explicit form of the Lorentz transformation.[11] Consider the relation between two inertial frames connected by a boost (with velocity v) in the $+x$ direction. Since the (y, z) coordinates are not affected, we have effectively a two-dimensional problem. The rotation relations are just like (2.3):

$$w' = \cos\phi\, w + \sin\phi\, x,$$
$$x' = -\sin\phi\, w + \cos\phi\, x. \tag{3.19}$$

Plugging in $w = ict$, we have

$$ct' = \cos\phi\, ct - i\sin\phi\, x,$$
$$x' = -i\sin\phi\, ct + \cos\phi\, x. \tag{3.20}$$

Reparametrizing the rotation angle as $\phi = i\psi$ and using $-i\sin(i\psi) = \sinh\psi$ and $\cos(i\psi) = \cosh\psi$,[12] we get

$$ct' = \cosh\psi\, ct + \sinh\psi\, x,$$
$$x' = \sinh\psi\, ct + \cosh\psi\, x. \tag{3.21}$$

Thus, in the (ct, x) space a Lorentz boost transformation has the matrix form of

$$[\mathbf{L}(\psi)] = \begin{pmatrix} \cosh\psi & \sinh\psi \\ \sinh\psi & \cosh\psi \end{pmatrix}. \tag{3.22}$$

To relate the parameter ψ, called the **rapidity** parameter,[13] to the boost velocity v, we concentrate on the coordinate origin $x' = 0$ of the O' system. Plugging $x' = 0$ into (3.21):

$$x' = 0 = ct\sinh\psi + x\cosh\psi \quad \text{or} \quad \frac{x}{ct} = -\frac{\sinh\psi}{\cosh\psi}. \tag{3.23}$$

The coordinate origin $x' = 0$ moves with velocity $v = x/t$ along the x axis of the O system:

$$v = \frac{x}{t} = -c\frac{\sinh\psi}{\cosh\psi} \quad \text{or} \quad \frac{\sinh\psi}{\cosh\psi} = -\frac{v}{c} = -\beta. \tag{3.24}$$

---

[10]For an introductory discussion of rotation, see Box 2.1.

[11]A more formal proof can be found in Chapter 12. The components of a vector A^{μ} change under a coordinate transformation as in (3.18) with its length being an invariant, $g'_{\mu\nu} A'^{\mu} A'^{\nu} = g_{\mu\nu} A^{\mu} A^{\nu}$. If $\mathbf{L}$ does not change the metric $\mathbf{g}' = \mathbf{g}$, the corresponding symmetry is called an **isometry**. Such a transformation must satisfy the "generalized orthogonality condition," $\mathbf{L}\mathbf{g}\mathbf{L}^{\mathsf{T}} = \mathbf{L}^{\mathsf{T}}\mathbf{g}\mathbf{L} = \mathbf{g}$. In Euclidean space with $\mathbf{g} = \mathbf{1}$, this reduces to the familiar statement (Problem 3.3) that the transformation matrix must be an orthogonal matrix (i.e. the transposed matrix is the inverse matrix: $\mathbf{L}^{\mathsf{T}} = \mathbf{L}^{-1}$). Further details will be provided in Chapter 12. In Problems 3.4 and Problem 12.7, the reader will be asked to show that the explicit form of $\mathbf{L}$ can be determined from such conditions.

[12]Recall the identities $\sinh\psi = (e^{\psi} - e^{-\psi})/2$ and $\sin\phi = (e^{i\phi} - e^{-i\phi})/2i$; also, $\cosh\psi = (e^{\psi} + e^{-\psi})/2$ and $\cos\phi = (e^{i\phi} + e^{-i\phi})/2$.

[13]The rapidity parameter ψ has the property that, while the addition of relative velocities is complicated as in (2.22), relative rapidity is additive because the hyperbolic tangent, which is the velocity, satisfies the trigonometry identity

$$\tanh(\psi_1 \pm \psi_2) = \frac{\tanh\psi_1 \pm \tanh\psi_2}{1 \pm \tanh\psi_1 \tanh\psi_2}.$$

From the identity $\cosh^2 \psi - \sinh^2 \psi = 1$, which may be written as $\cosh \psi \sqrt{1 - (\sinh^2 \psi / \cosh^2 \psi)} = 1$, we find

$$\cosh \psi = \gamma \quad \text{and} \quad \sinh \psi = -\beta \cosh \psi = -\beta\gamma, \qquad (3.25)$$

where $\beta = v/c$ and $\gamma = (1 - \beta^2)^{-1/2}$. The coordinate transformation in matrix form (3.22) is found to be

$$\begin{pmatrix} ct' \\ x' \end{pmatrix} = \gamma \begin{pmatrix} 1 & -\beta \\ -\beta & 1 \end{pmatrix} \begin{pmatrix} ct \\ x \end{pmatrix}, \qquad (3.26)$$

which is just the Lorentz transformation stated in (3.16).

3.2   Four-vectors for particle dynamics

As a further application of Minkowski 4-vectors, we consider some of the basic quantities involved in the description of particle dynamic: velocity, energy and momentum, as well as acceleration. We shall see that many interesting relativistic features can already be deduced by the proper construction of quantities (vectors, scalars, etc.) that have definite transformation properties under 4D rotation (Lorentz transformation) in Minkowski spacetime.

3.2.1   The velocity 4-vector

We have already shown in Chapter 2 [see Eqs. (2.19)–(2.22)] that the velocity components have rather complicated Lorentz transformation properties. This is because ordinary velocity dx^μ / dt is not a proper 4-vector; namely

$$\frac{dx'^\mu}{dt'} \neq \mathbf{L}^\mu_\nu \frac{dx^\nu}{dt} \quad \text{as} \quad t' \neq t. \qquad (3.27)$$

While dx^μ is a 4-vector $\left(dx'^\mu = \mathbf{L}^\mu_\nu dx^\nu\right)$, the ordinary time coordinate t is not a Lorentz scalar—it is a component of a 4-vector: $x^\mu = \left(ct, x^1, x^2, x^3\right)$. Consequently, the quotient dx^μ / dt cannot be a 4-vector. This suggests that, in order to construct a velocity 4-vector, we should differentiate the displacement with respect to the proper time τ, which is a Lorentz scalar (recall that $s^2 = -c^2\tau^2$ is invariant under a Lorentz transformation):

$$U^\mu = \frac{dx^\mu}{d\tau}, \qquad (3.28)$$

and we have, as in (3.18),

$$U'^\mu = \mathbf{L}^\mu_\nu U^\nu. \qquad (3.29)$$

The relation between the 4-velocity U^μ and dx^μ / dt can be readily deduced, as coordinate time and proper time are related by the time dilation relation of

$t = \gamma \tau$, with

$$\gamma = \left(1 - \frac{v^2}{c^2}\right)^{-\frac{1}{2}} \quad \text{and} \quad v = \left|v^i\right| \quad \text{with} \quad v^i = \frac{dx^i}{dt}. \tag{3.30}$$

We have the components of 4-velocity related to the ordinary velocity components v^i with ($i = 1, 2, 3$) as

$$U^\mu = \frac{dx^\mu}{d\tau} = \gamma \frac{dx^\mu}{dt} = \gamma \left(c, v^1, v^2, v^3\right). \tag{3.31}$$

As an instructive exercise (Problem 3.7), the reader is invited to deduce the velocity transformation rule (2.22) from the fact that U^μ is a 4-vector as in (3.29). It is easy to check that the "4-velocity length squared" $|U|^2 \equiv \eta_{\mu\nu} U^\mu U^\nu$ is a Lorentz scalar, see Eq. (3.10), with the same value in every coordinate frame:

$$|U|^2 = \gamma^2 (-c^2 + v^2) = -c^2, \tag{3.32}$$

where we have used the definition of γ as given in (3.30). For any material particle ($v < c$), we have $|U|^2 = -c^2$. For photons (and any other particles with zero rest mass[14]), which can only travel at $v = c$, Eq. (3.32) would have a vanishing RHS:

$$|U|^2 = \gamma^2 (-c^2 + c^2) = 0. \tag{3.33}$$

Thus U^μ is a null 4-vector. We note that because there is no rest frame for a photon, the concept of "the proper time of the photon" does not exist. In that case, one must replace τ by some curve parameter (say, λ) of the photon's trajectory, $x^\mu(\lambda)$; and the 4-velocity invariant becomes

$$|U|^2 = \eta_{\mu\nu} U^\mu U^\nu = \eta_{\mu\nu} \frac{dx^\mu}{d\lambda} \frac{dx^\nu}{d\lambda}, \tag{3.34}$$

which vanishes because the invariant spacetime separation for light is $\eta_{\mu\nu} dx^\mu dx^\nu = 0$. That is, the 4-velocity for a light ray has zero length in spacetime (a null 4-vector).

3.2.2   Relativistic energy and momentum

For momentum, we naturally consider the product of the invariant mass m with the 4-velocity of (3.31):

$$p^\mu \equiv mU^\mu = \gamma \left(mc, mv^i\right), \tag{3.35}$$

with mv^i being the components of the nonrelativistic 3-momentum $\vec{p} = m\vec{v}$. The spatial components of the relativistic 4-momentum p^μ are the components of the relativistic 3-momentum, $p^i = \gamma mv^i$, which reduces to mv^i in the nonrelativistic limit of $\gamma = 1$. What then is the zeroth component of the 4-momentum? Let's take its nonrelativistic limit ($v \ll c$):

$$p^0 = mc\gamma = mc \left(1 - \frac{v^2}{c^2}\right)^{-\frac{1}{2}}$$

[14] For massless particles in general, see the discussion below as well as that at the end of Section 3.2.2.

$$\xrightarrow{\text{NR}} mc \left(1 + \frac{1}{2} \frac{v^2}{c^2} + \cdots \right) = \frac{1}{c} \left(mc^2 + \frac{1}{2} mv^2 + \cdots \right). \quad (3.36)$$

The presence of the kinetic energy term $\frac{1}{2} mv^2$ in the nonrelativistic limit naturally suggests that we interpret cp^0 as the relativistic energy $E = cp^0 = \gamma mc^2$, which has a nonvanishing value mc^2 even when the particle is at rest

$$p^\mu = \left(E/c, \ p^i \right). \quad (3.37)$$

According to (3.32) and (3.35), the invariant square of the 4-momentum $\eta_{\mu\nu} p^\mu p^\nu$ must be $-(mc)^2$. Plugging in (3.37), we obtain the important relativistic energy–momentum relation:

$$E^2 = \left(mc^2 \right)^2 + (\vec{p}c)^2 = m^2 c^4 + \vec{p}^2 c^2. \quad (3.38)$$

Once again, for a particle with mass,[15] we have the components of the relativistic 3-momentum and the relativistic energy as

$$p^i = \gamma mv^i \quad \text{and} \quad E = \gamma mc^2. \quad (3.39)$$

Thus the ratio of a particle's momentum to its energy can be expressed as that of velocity over c^2

$$\frac{p^i}{E} = \frac{v^i}{c^2}. \quad (3.40)$$

Massless particles always travel at speed c

With $m = 0$, we can no longer define the 4-momentum as $p^\mu = mU^\mu$; nevertheless, since a massless particle has energy and momentum, we can still assign a 4-momentum to such a particle, with components just as in (3.37). When $m = 0$, the relation (3.38) with $p \equiv |\vec{p}|$ becomes

$$E = pc. \quad (3.41)$$

Plugging this into the ratio of (3.40), we obtain the well-known result that massless particles such as photons and gravitons[16] always travel at the speed of $v = c$. Another way of saying the same thing: with this energy momentum relation (3.40), the 4-momentum of a massless particle[17] must have components $p^\mu = \left(p, \ p^i \right)$ which, just like its 4-velocity, is manifestly a null 4-vector.

Box 3.3 The wavevector

Here we discuss the wave 4-vector, which is closely related to the photon 4-momentum. Recall that for a dynamic quantity $A(\vec{x}, t)$ to be a solution to the wave equation, its dependence on the space and time coordinates must be in the combination of $(\vec{x} - \vec{v}t)$, where $\vec{v}$ is the wave velocity. A harmonic electromagnetic wave is then proportional to $\exp i(\vec{k} \cdot \vec{x} - \omega t)$, with $k = |\vec{k}| = 2\pi/\lambda$ being the wavenumber, and $\omega = 2\pi/T$ being the

(cont.)

---

[15] The concept of a velocity-dependent mass $m^* \equiv \gamma m$ is sometimes used in the literature, so that $p^i = m^* v^i$ and $E = m^* c^2$. In our discussion we will avoid such usage and restrict ourselves only to the Lorentz scalar mass m, which equals m^* in the rest frame of the particle $\left(m^*|_{v=0} = m \right)$, hence, m is called the **rest mass**.

[16] Gravitons are the quanta of the gravitational field, just as photons are the quanta of the electromagnetic field.

[17] While we do not have $p^\mu = mU^\mu$, we still have the proportionality of the 4-momentum to its 4-velocity, $p^\mu \propto U^\mu$, with the 4-velocity defined as $U^\mu = dx^\mu/d\lambda$. In fact one can choose the curve parameter λ in such a way that $p^\mu = dx^\mu/d\lambda$.

Box 3.3 (*Continued*)

angular frequency corresponding to a wave period of T, and their being related to the velocity of light by $\omega/k = c$. The phase factor $(\vec{k} \cdot \vec{x} - \omega t)$, which basically counts the number of peaks and troughs of the wave, must be a frame-independent quantity (i.e. a Lorentz scalar). To make this scalar nature explicit, we write this phase in terms of the 4-vector $x^\mu = (ct, \vec{x})$ as

$$\vec{k} \cdot \vec{x} - \omega t = \left(ct, \vec{x} \right) \begin{pmatrix} -1 \\ & 1 \end{pmatrix} \begin{pmatrix} \omega/c \\ \vec{k} \end{pmatrix} \equiv \eta_{\mu\nu} x^\mu k^\nu.$$

From our knowledge that x^μ is a 4-vector and $\eta_{\mu\nu} x^\mu k^\nu$ a scalar, we conclude[18] that ω and $\vec{k}$ must also form a 4-vector, the "wavevector:" $k^\mu = (\omega/c, k^i)$. Under the Lorentz transformation, the components of the wavevector transform, according to (3.18), as

$$k^\mu \longrightarrow k'^\mu = [\mathbf{L}]^\mu_{\ \nu} k^\nu. \tag{3.42}$$

Specifically under a Lorentz boost in the $+x$ direction, we have from (3.16):

$$k'_x = \gamma \left(k_x - \beta \frac{\omega}{c} \right) \tag{3.43}$$

$$\omega' = \gamma \left(\omega - c\beta k_x \right) = \gamma \left(\omega - c\beta k \cos\theta \right) \tag{3.44}$$

where θ is the angle between the boost direction $\hat{x}$ and the direction of wave propagation $\hat{k}$. Since $ck = \omega$, we obtain in this way the **relativistic Doppler formula**,

$$\omega' = \frac{(1 - \beta \cos\theta)}{\sqrt{1 - \beta^2}} \omega \tag{3.45}$$

which is to be compared to the nonrelativistic Doppler relation $\omega' = (1 - \beta \cos\theta) \omega$. We note that in the nonrelativistic limit there is no Doppler shift in the transverse direction, $\theta = \pi/2$, as compared to the relativistic "transverse Doppler effect" of $\omega' = \gamma\omega$. (One can trace the origin of this new effect back to SR time dilation.) In the longitudinal direction, $\theta = 0$, we have the familiar relation of

$$\frac{\omega'}{\omega} = \sqrt{\frac{1 - \beta}{1 + \beta}}, \tag{3.46}$$

which has the low-velocity (v) approximation of ($v \ll c$)

$$\frac{\omega'}{\omega} \approx (1 - \beta) \quad \text{or} \quad \frac{\Delta\omega}{\omega} \approx -\frac{v}{c}. \tag{3.47}$$

Because of the $\omega = ck$ relation, the wave 4-vector $k^\mu = (k, k^i)$ has a null invariant length $\eta_{\mu\nu} k^\mu k^\nu = 0$, compatible[19] with the property of the 4-momentum for a photon as shown in (3.33).

[18] This follows from the quotient theorem, which is described in Problem 12.6.

[19] This is expected from the quantum mechanical relation of $p^\mu = \hbar k^\mu$, which has the zeroth component corresponding to $E = \hbar\omega$ and the spatial components, the de Broglie relation of $p = \hbar k = h/\lambda$.

Box 3.4 Covariant force

Just as the ordinary velocity $\vec{v}$ has a complicated Lorentz property—and so we introduced the object of 4-velocity—it is also not easy to relate different components of the usual force vector $F^i = dp^i/dt$ in different moving frames. However, the notions of 4-velocity and 4-momentum naturally lead us to the definition of 4-force, or the covariant force, as

$$K^\mu \equiv \frac{dp^\mu}{d\tau} = m\frac{dU^\mu}{d\tau},\qquad (3.48)$$

which, using (3.37), has components[20]

$$K^\mu = \frac{dp^\mu}{d\tau} = \gamma\frac{d}{dt}\left(E/c,\ p^i\right) = \gamma\left(\dot{E}/c,\ F^i\right).\qquad (3.49)$$

Next we show that the rate of energy change $\dot{E}$ is given, just as in nonrelativistic physics, by the dot product $\vec{F}\cdot\vec{v}$. Because $|U|^2$ is a constant, its derivative vanishes.

$$0 = m\frac{d}{d\tau}\left(\eta_{\mu\nu}U^\mu U^\nu\right) = 2\eta_{\mu\nu}m\frac{dU^\mu}{d\tau}U^\nu = 2\eta_{\mu\nu}K^\mu U^\nu,\qquad (3.50)$$

where we have used (3.48) to reach the last equality. Substituting in the components of K^μ and U^ν from (3.49) and (3.31), we have

$$0 = \eta_{\mu\nu}K^\mu U^\nu = \gamma^2(-\dot{E} + \vec{F}\cdot\vec{v}),\qquad (3.51)$$

thus $\dot{E} = \vec{F}\cdot\vec{v}$. With this expression for $\dot{E}$, we can display the components of the covariant force (3.49) as

$$K^\mu = \gamma\left(\vec{F}\cdot\vec{v}/c,\ F^i\right).\qquad (3.52)$$

[20]One must keep in mind that the 3-force vector $F^i = dp^i/dt$ here is a differential of the relativistic momentum; hence it is related to the nonrelativistic force by $F^i = \gamma F^i_{\text{NR}}$.

3.3 The spacetime diagram

Space and time coordinates are labels of physical processes taking place, one "event" following another, in the world. Any two given events may or may not be causally connected. Relativity brings about a profound change in this causal structure of space and time, which can be nicely visualized in terms of the **spacetime diagram**.[21] For a given event P at a particular point in space and particular instant of time, all the events that could in principle be reached by a particle starting from P one collectively labeled as the **future of P**, while all the events from which a particle can arrive at P form the **past of point P**. In order to appreciate the nontrivial causal structure brought about by the new relativistic conception of space and time, let us first recall the corresponding structure that one had assumed in pre-relativity physics. Here the notion of simultaneous events is of key importance. Those events that are neither the future nor the past of the event P form a 3D set of events **simultaneous with P**. This notion of simultaneous events allows one to discuss, in pre-relativity

[21]To have the same length dimension for all coordinates, the temporal axis is represented by $x^0 = ct$.

physics, all of space at a given instant of time, and as a corollary, allows one to study space and time separately. In relativistic physics, the events that fail to be causally connected to event P are much larger than a 3D space. As we shall see, all events outside the future and past lightcones are causally disconnected from the event P, which lies at the tip of the lightcones in the spacetime diagram.

3.3.1   Basic features and invariant regions

An event with coordinates (t, x, y, z) is represented by a **worldpoint** in the spacetime diagram. The history of events becomes a line of worldpoints, called a **worldline**. In Fig. 3.2, the 3D space is represented by a 1D x axis. In particular, a light signal $\Delta s^2 = 0$ passing through the origin is represented by a straight worldline at a 45° angle with respect to the axes: $\Delta x^2 - c^2 \Delta t^2 = 0$, thus $c \Delta t = \pm \Delta x$. Any line with constant velocity $v = |\Delta x / \Delta t|$ would be a straightline passing through the origin. We can clearly see that those worldlines with $v < c$, corresponding to $\Delta s^2 < 0$, would make an angle greater than 45° with respect to the spatial axis (i.e. above the worldlines for a light ray). According to relativity, no worldline can have $v > c$. If there had been such a line, it would correspond to $\Delta s^2 > 0$, and would make an angle less than 45° (i.e. below the light worldline). Since $\Delta s^2 = \Delta x^2 + \Delta y^2 + \Delta z^2 - c^2 \Delta t^2$ is invariant, it is meaningful to divide the spacetime diagram into regions, as in Fig. 3.3, corresponding to

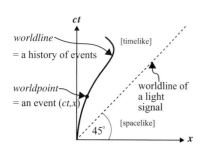

Fig. 3.2 Basic elements of a spacetime diagram, with two spatial coordinates (y, z) suppressed.

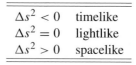

$$
\begin{array}{ll}
\Delta s^2 < 0 & \text{timelike} \\
\Delta s^2 = 0 & \text{lightlike} \\
\Delta s^2 > 0 & \text{spacelike}
\end{array}
$$

where the names of the region are listed on the right-hand column. The coordinate intervals being $c\Delta t = ct_2 - ct_1$, $\Delta x = x_2 - x_1$, etc. consider the separation of two events: one at the origin $(ct_1, \vec{x}_1) = (0, \vec{0})$, the other at a point in one of the regions $(ct_2, \vec{x}_2) = (ct, \vec{x})$:

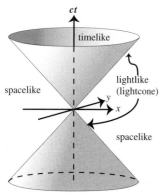

Fig. 3.3 Invariant regions in the spacetime diagram, with two of the spatial coordinates suppressed.

- The light-like region has all the events which are connected to the origin with a separation of $\Delta s^2 = 0$. This corresponds to events that are connected by light signals. The 45° incline in Fig. 3.3, in which two spatial dimensions are displayed, forms a **lightcone**. It has a slope of unity, which reflects the fact that the speed of light is c. A vector that connects an event in this region to the origin, called **a light-like vector, is** a non-zero 4-vector having zero length, a **null vector**. The lightcone surface is a null 3-surface.

- The space-like region has all the events which are connected to the origin with a separation of $\Delta s^2 > 0$. (The 4-vector from the origin in this region is a **space-like vector**, having a positive squared length.) In the spacelike region, it takes a signal traveling at a speed greater than c in order to connect an event to the origin. Thus, an event taking place at any point in this region cannot be influenced causally (in the sense of cause-and-effect) by an event at the origin. We can alternatively explain it by going to another frame O' resulting in different spatial and time intervals $\Delta x' \neq \Delta x$ and $\Delta t' \neq \Delta t$. However the spacetime interval is unchanged, $\Delta s'^2 = \Delta s^2 > 0$. The form of (3.3), with the spatial terms being positive and the time term negative, suggests that we can always find an O' frame such that this event would be viewed as taking place at the same time $\Delta t' = 0$ as the event at the origin but at different locations $\Delta x' \neq 0$. This makes it clear that such a worldpoint (an event) **cannot be causally connected** to an event at the origin because the two events would have to be connected by an instantaneous signal, which is not possible, as no signal can travel faster than c. Thus the causally disconnected 3D space (represented by a horizontal plane) in pre-relativity physics is now enlarged to a much large region—all of the 4D subspace outside the lightcone.

- The time-like region has all the events which are connected to the origin with a separation of $\Delta s^2 < 0$. (The 4-vector from the origin in this region is a **time-like vector**, having a negative squared length.) One can always find a frame O' such that such an event takes place at the same locations, $x' = 0$, but at different time, $t' \neq 0$. This makes it clear that events in this region **can be causally connected** with the origin. In fact, all the world-lines passing through the origin will be confined to this region, inside the lightcone.[22] In Fig. 3.3, we have displayed the lightcone structure with respect to the origin of the spacetime coordinates ($t = 0$, $\vec{x} = 0$). It should be emphasized that each point in a spacetime diagram has a lightcone. The time-like regions with respect to several worldpoints are represented by the lightcones shown in Fig. 3.4. If we consider a series of lightcones having their vertices located along a given worldline, each subsequent segment must lie within the lightcone of that point (at the beginning of that segment). It is clear from Fig. 3.4 that any particle can only proceed in the direction of ever-increasing time. We cannot stop our biological clocks!

[22]The worldline of an inertial observer (i.e. moving with constant velocity) must be a straight line inside the lightcone. This straight line is just the time axis of the coordinate system in which the inertial observer is at rest.

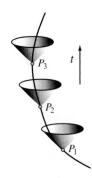

Fig. 3.4 Lightcones with respect to different worldpoints, P_1, P_2, …, etc. along a time-like worldline, which can only proceed in the direction of ever-increasing time as each segment emanating from a given worldpoint must be contained within the lightcone with that point as its vertex.

3.3.2   Lorentz transformation in the spacetime diagram

The nontrivial parts of the Lorentz transformation (3.16) of intervals (taken, for example, with respect to the origin) are

$$\Delta x' = \gamma \left(\Delta x - \beta c \Delta t \right), \quad c \Delta t' = \gamma \left(c \Delta t - \beta \Delta x \right). \tag{3.53}$$

We can represent these transformed axes in the spacetime diagram:

- The x' axis corresponds to the $c\Delta t' = 0$ line. This corresponds, according the second equation above, to a line satisfying the relationship $c\Delta t = \beta \Delta x$. Hence, the x' axis is a straight line in the (x, ct) plane with a slope of $c\Delta t / \Delta x = \beta$.
- The ct' axis corresponds to the $\Delta x' = 0$ line. This corresponds, according the first equation above, to a line satisfying the relationship $\Delta x = \beta c \Delta t$. Hence, the ct' axis is a straight line with a slope of $c\Delta t / \Delta x = 1/\beta$.

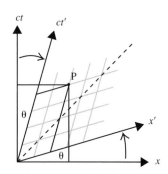

Fig. 3.5 Lorentz rotation in the spacetime diagram. The space and time axes rotate by the same amount but in opposite directions so that the lightcone (the dashed line) remains unchanged. The shaded grid represents lines of fixed x' and t'.

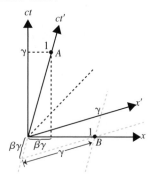

Fig. 3.6 Scale change in a Lorentz rotation. A unit length on the ct' axis has a *longer* projection, γ, onto the ct axis. The event A $(ct' = 1, x' = 0)$ in the O' frame is seen by an observer in the O frame to have coordinates $(ct = \gamma, x = \beta\gamma)$. Similarly, the event B, with coordinates $(ct = 0, x = 1)$ in the O frame, has coordinates $(ct' = \gamma\beta, x' = \gamma)$ in the O' frame. The two sets of dotted lines passing through worldpoints A and B are parallel lines to the axes of (ct, x) and (ct', x'), respectively.

Depending on whether β is positive or negative, the new axes either "close in" or "open up" from the original perpendicular axes. Thus we have the **opposite-angle rule**: the two axes make opposite-signed rotations of $\pm\theta$ (Fig. 3.5). The x axis rotates by $+\theta$ relative to the x' axis; the ct axis, by $-\theta$ relative to the ct' axis. The physical basis for this rule is the need to maintain the same slope ($= 1$; i.e. **equal angles** with respect to the two axes) for the lightcone in every inertial frame so that light speed is the same in every frame. Another important feature of the diagrammatic representation of the Lorentz transformation is that the new axes will have a scale **different** from the original one. Namely, the unit-lengths along the axes of the two systems are different. Let us illustrate this by an example. Consider the separation (from the origin O) of an event A on the ct' axis, which has O' system coordinates $(ct' = 1, x' = 0)$, see Fig. 3.6. What O system coordinates (ct, x) does the worldpoint have?

$$x' = \gamma (x - \beta ct) = 0 \Rightarrow \quad x = \beta ct,$$

$$ct' = \gamma (ct - \beta x) = ct\gamma \left(1 - \beta^2\right) = ct/\gamma = 1.$$

Hence this event has $(ct = \gamma, x = \gamma\beta)$ coordinates in the O system. Evidently, as $\gamma > 1$, a unit vector along the ct' direction has "projection" on the ct axis that is longer than unit length. This is possible only if there is a scale change when transforming from on reference system to another.

Consider another separation of an event B on the x axis, which has O coordinates $(ct = 0, x = 1)$. It is straightforward to check that it has O' system coordinates $(ct' = -\gamma\beta, x' = \gamma)$, again showing a difference in scales of the two systems.

Box 3.5 Time dilation and length contraction in the spacetime diagram

The physics behind the scale changes discussed above is **time dilation** and **length contraction**. While the algebra involved in deriving these results from the Lorentz transformation (3.53) is simple, in order to obtain the correct result, one has to be very clear as to exactly what is being measured (spatial length or time), and which frame is being chosen as the rest frame for the appropriate object (in the derivation of time dilation, the object is a clock; in the derivation of length contraction, the object is any object whose spatial length is being measured). Knowing which frame has been chosen as the rest frame for the appropriate object, one must then have a mathematical way to express the fact that the clock is not moving or the object is not moving; one must have a mathematical condition or input to apply to the Lorentz transformation equations. The conventional way of doing this is to set $\Delta x' = 0$ for time dilation, and $\Delta t = 0$ for length contraction. For time dilation, we pick the O' frame to be the rest frame of the clock; since the clock is at rest in this frame, $\Delta x' = 0$. This is the crucial input into the Lorentz transformation equations in the derivation of time dilation. For length contraction, we also consider the O' frame to be the rest frame, but now it is the rest frame for an object whose length is being measured. The crucial input for this derivation is $\Delta t = 0$, reflecting the fact that the observer in the O frame must measure the ends of the object simultaneously.

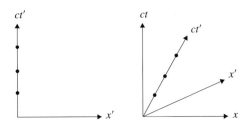

Fig. 3.7 Worldline of a clock, ticking at equal intervals: viewed in the rest frame of the clock, the O' system, and viewed in the moving frame, the O coordinate system.

Time dilation A clock, ticking away in its own rest frame O' (also called the comoving frame), is represented by a series of worldpoints (the ticks of the clock) equally spaced on a vertical worldline ($\Delta x' = 0$) in the (ct', x') spacetime diagram. These same worldpoints when viewed in another inertial frame O, in which the O' system moves with $+v$ along the x axis, will appear as lying on an inclined worldline (Fig. 3.7). From the Lorentz transformation (2.40), as well as our previous discussion of the scale change under Lorentz rotation (also see Fig. 3.8), it is clear that the relationship between the time intervals in the two frames is

$$\Delta t = \gamma \Delta t' \quad \gamma > 1. \tag{3.54}$$

Thus we say that a moving clock (i.e. moving with respect to the O system) appears (Δt) to run slow. NB: keep in mind $\Delta x' = 0$; that is, there is no spatial displacement in the clock's rest frame, the comoving frame. However, there is a spatial displacement in the other, moving frame: $\Delta x \neq 0$.

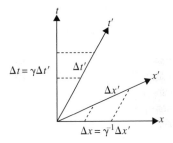

Fig. 3.8 Scale changes associated with the Lorentz rotation, reflecting the physics phenomena of time dilation and length contraction. The clock and object are moving with respect to the O system, but are at rest with respect to the O' system.

Length contraction To obtain a length $\Delta x = x_1 - x_2$ in the O system of an object at rest in the O' system (and, therefore, moving with respect to the O system), we need to measure two events (t_1, x_1) and (t_2, x_2) simultaneously $\Delta t = t_1 - t_2 = 0$. (If you want to measure the length of a moving car, you certainly would not want to measure its front and back locations at different times!) The same two events,[23] when viewed in the rest frame of the object $\Delta x' = x'_1 - x'_2$, will be measured according to (3.53) to have a greater separation (cf. Fig. 3.8):

$$\Delta x' = \gamma \Delta x > \Delta x. \tag{3.55}$$

[23]While we have simultaneous measurements in the moving frame, $\Delta t = 0$, these two events would be viewed as taking place at different times in the rest frame, $\Delta t' = \gamma \left(\Delta t - \frac{v}{c^2} \Delta x \right) \neq 0$. Of course, in the rest frame of the object, there is no need to perform the measurements simultaneously: in order to measure the front and back ends of a parked car, it is perfectly all right to make one measurement, take a lunch break, and then come back to measure the other end.

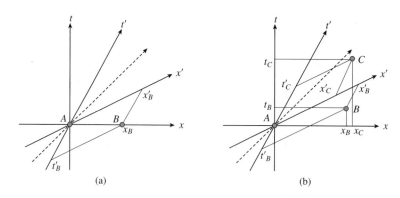

Fig. 3.9 (a) Relativity of simultaneity: $t_A = t_B$ but $t'_A > t'_B$. (b) Relativity of event order: $t_A < t_B$ but $t'_A > t'_B$. However, there is no change of event order with respect to A for all events located above the x' axis, such as event C. This certainly includes the situation in which C is located in the forward lightcone of A (above the dashed line).

Relativity of simultaneity, event-order and causality

It is instructive to use the spacetime diagram to demonstrate some of the physical phenomena we have discussed previously. In Fig. 3.9, we have two events A and B, with A being the origin of the coordinate system O and O': $\left(x_A = t_A = 0, x'_A = t'_A = 0\right)$. In Fig. 3.9(a), the events A and B are simultaneous, $t_A = t_B$, with respect to the O system. But in the O' system, we clearly have $t'_A > t'_B$. This shows the relativity of simultaneity. In Fig. 3.9(b), we have $t_A < t_B$ in the O frame, but we have $t'_A > t'_B$ in the O' frame. Thus, the temporal order of events can be changed by a change of reference frames. However, this change of event order can take place only if event B is located in the region below the x' axis.[24] This means that if we increase the relative speed between these two frames O and O' (with the x' axis moving ever closer to the lightcone) more and more events can have their temporal order (with respect to A at the origin) reversed as seen from the perspective of the moving observer. On the other hand, for all events above the x' axis, the temporal order is preserved. For example, with respect to event C, we have both $t_A < t_C$ and $t'_A < t'_C$. Now, of course, the region above this x' axis includes the forward lightcone of event A. This means that for two events that are causally connected (between A and any worldpoint in its forward lightcone), their temporal order cannot be changed by a Lorentz transformation. The principle of causality is safe under special relativity.

[24]The x' axis having a $1/\beta$ slope means that the region below it corresponds to $(\Delta x/\Delta t) > c/\beta$. This is clearly in agreement with the Lorentz transformation of $\Delta t' = \gamma (\Delta t - \beta \Delta x/c)$ to have opposite sign to Δt.

3.4   The geometric formulation of SR: A summary

Let us summarize the principal lessons we have learnt from this geometric formulation of special relativity:

- The stage on which physics takes place is Minkowski spacetime with the time coordinate being on an equal footing with spatial ones. "Space and time are treated symmetrically." A spacetime diagram is often useful in clarifying ideas in relativity, especially its causal structure.
- Minkowski spacetime has a pseudo-Euclidean length (squared) of $\Delta s^2 = -c^2 \Delta t^2 + \Delta x^2 + \Delta y^2 + \Delta z^2$. This relation between length and coordinate $\left[\Delta s^2 = g_{\mu\nu} \Delta x^\mu \Delta x^\nu\right]$ can be stated by saying that

Minkowski space has a flat-space metric of $g_{\mu\nu} = \text{diag}(-1, 1, 1, 1) \equiv \eta_{\mu\nu}$.

- Δs is invariant under transformations among inertial frames of reference. Such length-preserving transformations in a space with the metric $\eta_{\mu\nu}$ are just the Lorentz transformations, from which we can derive all the physical consequences of time dilation, length contraction, relativity of simultaneity, etc. Thus, in this geometric formulation, we can think of the metric as embodying all of special relativity.

- That one can understand special relativity as a theory of flat geometry in Minkowski spacetime is the crucial step in the progression towards general relativity. In GR, as we shall see, this geometric formulation is generalized into a warped spacetime. The corresponding metric must be position-dependent, $g_{\mu\nu}(x)$, and this metric acts as the generalized gravitational potential.

- In our historical introduction, SR seems to be all about light; the speed c actually plays a much broader role in relativity:

  - c is the universal maximal and absolute speed of signal transmission: massless particles (e.g. photons and gravitons) travel at the speed c, while all other ($m \neq 0$) particles move at a slower speed.

  - c is the universal **conversion factor** between space and time coordinates[25] that allows space and time to be treated symmetrically (i.e. on an equal footing) in relativity.

  - c is absolute and is just the speed so that $\Delta s^2 = -c^2\Delta t^2 + \Delta\vec{x}^2$ is an invariant interval under coordinate transformations. This allows Δs to be viewed as the length in spacetime. Thus, constancy of c underlies the entire geometric formulation of relativity.

[25] Since space and time have different units, we must have a conversion factor connecting the space and time coordinates. We mention two other fundamental physics conversion factors: Newton's constant G_N and Planck's constant h (see Section 6.3.2).

Review questions

1. (a) Give the definition of the metric tensor in terms of the basis vectors.

   (b) What is the invariant interval (the "length") Δs^2 between two neighboring events with coordinate separation $(c\Delta t, \Delta x, \Delta y, \Delta z)$ in Minkowski spacetime?

   (c) When the metric is displayed as a square matrix, what is the interpretation of its respective diagonal and off-diagonal elements?

   (d) What is the metric for an n-dimensional Euclidean space, in Cartesian coordinates, with a (squared) length given by $\Delta s^2 = \Delta x_1^2 + \Delta x_2^2 + \cdots + \Delta x_n^2$?

2. (a) What is the essential input needed for the proof that $\Delta s^2 = -c^2\Delta t^2 + \Delta\vec{x}^2$ has the same value in every inertial frame of reference?

   (b) What is the physical meaning of s that every observer can agree on its value?

3. What are the components of the position 4-vector in Minkowski spacetime?

4. If A^μ is a 4-vector (e.g. it is the 4-velocity or 4-momentum, etc.), how do these components transform under a coordinate transformation $A^\mu \to A'^\mu$ if the position 4-vector changes as $x^\mu \to x'^\mu = [\mathbf{L}]^\mu_\nu x^\nu$?

5. Under a coordinate change $O \to O'$, how is $\eta_{\mu\nu}A^\mu B^\nu$ related to $\eta_{\mu\nu}A'^\mu B'^\nu$?

6. From the condition $\Delta s'^2 = \Delta s^2$, derive the explicit form of the Lorentz transformation for a boost $\vec{v} = +v\hat{x}$.

7. Why is dx^μ/dt not a 4-vector? How is it related to the velocity 4-vector U^μ? The squared length of the 4-velocity $\eta_{\mu\nu} U^\mu U^\nu$ should be a Lorentz invariant; what is this invariant?

8. What are the definitions (in terms of particle mass and velocity) of relativistic energy E and momentum $\vec{p}$ of a particle? Display their non-relativistic limits. What components of momentum 4-vector p^μ are E and $\vec{p}$? How are they related to each other?

9. In the spacetime diagram, display the time-like, space-like, and light-like regions. Also, draw in a worldline for some inertial observer.

10. The coordinate frame O' is moving at a constant velocity v in the $+x$ direction with respect to the coordinate frame O. Display the transformed axes (x', ct') in a two-dimensional spacetime diagram with axes (x, ct). You are not asked to solve the Lorentz transformation equations; but only to justify the directions of the new axes.

11. Two events A and B are simultaneous $(t_A = t_B)$, but not equilocal $(x_A \neq x_B)$, in coordinate frame O. Use a spacetime diagram to show that these same two events are viewed as taking place at different times, $t'_A \neq t'_B$, by an observer in the O' frame (in motion with respect to the O frame).

12. In a spacetime diagram, display two events with a temporal order of $t_A > t_B$ in the O frame such that they can possibly appear to have a reversed order $t'_A < t'_B$ in the O' frame. What is the condition that Δx, Δt and v (relative speed between the O and O' frames) must satisfy in order to have this reversal of temporal order? Explain why event A cannot possibly be caused by event B.

13. **Length contraction** means that the measured length interval of $\Delta x = x_1 - x_2$ is less than the corresponding rest-frame length $\Delta x' = x'_1 - x'_2$. What is the condition on the time coordinates of these two events, (t_1, t_2) and (t'_1, t'_2), and why is this a necessary condition? **Time dilation** means that the measured time interval of $\Delta t = t_1 - t_2$ is longer than the corresponding rest-frame interval $\Delta t' = t'_1 - t'_2$. What is the condition on the spatial coordinates, (x_1, x_2) and (x'_1, x'_2), of these two events? Why is this a necessary condition? Use these conditions and the Lorentz transformation to derive the result of length contraction and time dilation, respectively.

Problems

3.1 **Inverse basis**   We can introduce a set of "inverse basis vectors," denoted by an upper index, $\{e^\mu\}$ which when multiplied with the basis vectors yield the identity: $e^\mu \cdot e_\nu = \delta^\mu_\nu$ and the products among the inverse bases form the "inverse metric" $g^{\mu\nu} = e^\mu \cdot e^\nu$.

   (a) Show that one can use the inverse bases to project out the components A^μ from the vector $\mathbf{A}$.

   (b) Prove that $g^{\mu\nu}$ is the inverse to the metric $g_{\mu\nu}$.

3.2 **Contraction and dummy indices**   A scalar product $\mathbf{A} \cdot \mathbf{B} = g_{\mu\nu} A^\mu B^\nu$ can be written as a "contraction" of two vectors with upper and lower indices $\mathbf{A} \cdot \mathbf{B} = A^\mu B_\mu$ where $B_\mu = g_{\mu\nu} B^\nu$. We have used the Einstein summation convention and summed over repeated indices. As a practice of "contraction over repeated (dummy) indices," prove that the scalar product of symmetric and antisymmetric tensors will always vanish: $T_{\mu\nu} S^{\mu\nu} = 0$ where $T_{\mu\nu} = -T_{\nu\mu}$ and $S^{\mu\nu} = +S^{\nu\mu}$.

3.3 **Rotation matrix is orthogonal**   Explicitly demonstrate that the rotation matrix in Eq. (2.4) satisfies the relation

$$\left[\mathbf{R}^{-1}(\theta) \right] = [\mathbf{R}(-\theta)] = \left[\mathbf{R}^\mathsf{T}(\theta) \right]$$

hence the orthogonality condition $\mathbf{R}\mathbf{R}^\mathsf{T} = \mathbf{1}$.

3.4 **Orthogonality fixes the rotation matrix**   In Problem 3.3, you have been asked to show from the explicit form of a rotational matrix that it is an orthogonal matrix. Here you are asked to prove the converse: the orthogonality condition can fix the rotation matrix explicitly. Consider a rotation around the z axis:

$$\begin{pmatrix} x' \\ y' \end{pmatrix} = \begin{pmatrix} a & b \\ c & d \end{pmatrix} \begin{pmatrix} x \\ y \end{pmatrix},$$

where the effective 2×2 rotation matrix $[\mathbf{R}]$ with real elements (a, b, c, d) must satisfy the orthogonal condition $[\mathbf{R}][\mathbf{R}]^\mathsf{T} = [\mathbf{1}]$ so that the length $x^2 + y^2$ is an invariant. Show that this condition fixes the explicit form of the rotation matrix to be

$$[\mathbf{R}] = \begin{pmatrix} \cos\theta & \sin\theta \\ -\sin\theta & \cos\theta \end{pmatrix}.$$

3.5 Group property of Lorentz transformations Use simple trigonometry to show that the rotation and boost operators given in (2.4) and (3.22) satisfy the group property:

$$\mathbf{R}(\theta_1)\mathbf{R}(\theta_2) = \mathbf{R}(\theta_1 + \theta_2)$$
$$[\mathbf{L}(\psi_1)][\mathbf{L}(\psi_2)] = [\mathbf{L}(\psi_1 + \psi_2)]. \quad (3.57)$$

The expression of "group" used above is in the sense of a mathematical group, which is composed of elements satisfying multiplication rules such as the ones discussed here.

3.6 Group multiplication leads to velocity addition rule Use (3.57) of Problem 3.5 to prove the velocity addition rule (2.22).

3.7 Lorentz transform and velocity addition rule The velocity 4-vector U^μ being a 4-vector has the Lorentz boost transformation as in Eq. (3.16). From this, derive the velocity addition rule (2.22).

3.8 Antiproton production threshold Because of baryon number conservation, the simplest reaction to produce an antiproton $\bar{p}$ in proton–proton scattering is $pp \to ppp\bar{p}$. Knowing that the rest energy of a proton is $m_pc^2 = 0.94\,\text{GeV}$, use the invariant $\eta_{\mu\nu}p^\mu p^\nu$ to find the minimum kinetic energy a (projectile) proton must have in order to produce an antiproton after colliding with another (target) proton at rest.

3.9 A more conventional derivation of the Doppler effect A light signal of frequency ω sent from (x, t) is received at (x', t') with frequency ω'. The receiver is moving in the $+x$ direction with velocity v. One can derive the (longitudinal) Doppler formula (3.46) by the observation that the phase of a light wave (essentially the counting of peaks) remains the same for the sender and receiver: $d\phi = \omega d\tau = \omega' d\tau'$ (where τ and τ' are the proper times of the sender and receiver, respectively). From this, using the time dilation formula, one can relate the ratio ω'/ω to the coordinate time ratio dt'/dt, and finally to the relative velocity $\beta = v/c$.

3.10 Twin paradox measurements and the Doppler effect In presenting the twin paradox (Box 2.5) we used the traveling Al's observation of Bill's annual (birthday) fireworks to determine their respective ages. In the outward bound part of the journey ($\beta = 4/5$) Al sees the firework every three years, and in the inward bound journey, every four months. Discuss these time intervals from the viewpoint of the Doppler effect and show that the results are compatible with formula (3.46).

3.11 Spacetime diagram for the twin paradox Provide a spacetime diagram corresponding to the twin paradox discussed in Box 2.5. Let Bill's rest frame O having coordinates (x, t), the outbound-Al's rest frame O' with (x', t'), and inbound-Al's frame O'' with (x'', t''). Draw your diagram so that the perpendicular lines represent the (x, t) axes.

(a) Mark the event when Al departs in the spaceship by the worldpoint O; the event when Al returns and is reunited with Bill by the worldpoint Q; and the event corresponding to the event when Al turns around (from outward bound to inward bound) by the point P. Thus, the stay-at-home Bill has worldline OQ, and Al's worldline has two segments: OP for the outward bound and PQ for the inward bound parts of his journey.

(b) On the t axis, which should coincide with the worldline OQ, also mark the points M, P' and P'' which should be simultaneous with the turning point P as viewed in the coordinate frames of O, O' and O'', respectively.

(c) Indicate the time values of $t_M, t_{P'}$ and $t_{P''}$ (i.e. the elapsed times since Al's departure at point O in the (x, t) coordinate system. In particular, show how changing the inertial frame from O' and O'' brings about a time change of 32 years in the O frame.

3.12 The twin paradox–the missing 32 years In Problem 3.11, the event P' on Bill's worldline is viewed by the outward-bound Al to be simultaneous to the turning-point P just *before* Al turns around, and the event P'' simultaneous to P just *after*. They are viewed to have different time by the stay-at-home Bill, $t_{P'} \neq t_{P''}$. This just emphasizes again the point that time is just another coordinate label. When we change the frame of reference, all coordinates make their corresponding changes. Two different points P' and P'' in Bill's rest frame (the O system) are simultaneous to P when viewed from two different inertial frames, the O' and O'' systems, respectively. Thus a difference $t_{P'} \neq t_{P''}$ is brought about simply by a change of coordinate: $O' \longrightarrow O''$. Our discussion just below (2.41) and (2.45) suggests that $t_{P''} - t_{P'} = 32$ years. We have already calculated $t_{P'} = 9$ years by the relative motion between the O and O' frames. Let us verify the expected result of $t_{P''} = 41$ years in two ways.

(a) Calculate $t_{P''} = t_Q - t_{QP''}$, where $t_{QP''}$ is the time interval measured in Bill's rest frame of the second 15 year of Al's journey (for the worldline PQ).

(b) The turning point P is seen in the outward-bound O' frame to have the time $t'_{OP} = 15$ year; from the perspective of the inward-bound Al (O'' frame) how long an interval t''_{OP} does this first half of the journey appear to be? $t_{P''}$ can then be obtained by noting that

$t_{P''}$ is the O frame measurement of this t''_{OP} time interval.

3.13 Spacetime diagram for the pole-and-barn paradox
Draw a spacetime diagram for the pole-and-barn paradox as discussed in Box 2.4. Let (x, t) be the coordinates for the ground (barn) observer, with (x', t') the rest frame of the runner (pole). Show the worldlines for the front door (F), rear door (R) of the barn, and front end (A), back end (B) of the pole. Your diagram should display the order reversal phenomenon discussed in Box 2.4: $t_{AR} > t_{BF}$ and $t'_{AR} < t'_{BF}$.

The principle of equivalence

- After a review of the Newtonian theory of gravitation in terms of its potential function, we take the first step in the study of general relativity (GR) with the introduction of the equivalence principle (EP).
- The *weak EP* (the equality of the gravitational and inertial masses) is extended by Einstein to the *strong EP*, the equivalence between inertia and gravitation for all interactions. This implies the existence of "local inertial frames" at every spacetime point. In a sufficiently small region, the "local inertial observer" will not sense any gravity effect.
- The equivalence of acceleration and gravity means that GR (physics laws valid in all coordinate systems, including accelerating frames) must necessarily be a theory of gravitation.
- The strong EP is used to deduce the results of gravitational redshift and time dilation, as well as gravitational bending of a light ray.
- Motivated by EP physics, Einstein proposed the identification of a gravitational field with curved spacetime.

Soon after completing his formulation of special relativity (SR) in 1905, Einstein started working on a relativistic theory of gravitation. In this chapter, we cover the period 1907–1911, when Einstein relied heavily on the equivalence principle (EP) to extract some general relativity (GR) results. Not until the end of 1915 did he work out fully the ideas of GR. By studying the consequences of EP, he concluded that the proper language for GR is Riemannian geometry. The mathematics of curved space will be introduced in Chapter 5 and the geometric representation of the gravitational field in Chapter 6.

4.1 Newtonian gravitation potential—a review

Newton formulated his theory of gravitation through the concept of action-at-a-distance force

$$\vec{F}(\vec{r}) = -G_{\mathrm{N}} \frac{mM}{r^2} \hat{r} \tag{4.1}$$

where G_{N} is **Newton's constant**, the point mass M is located at the origin of the coordinate system, and the test mass m is at position **r**.

Just as in the case of electrostatics $\vec{F}(\vec{r}) = q'\vec{E}(\vec{r})$, we can cast this in the form

$$\vec{F} = m\vec{g}. \tag{4.2}$$

This defines the gravitational field $\vec{g}(\vec{r})$ as the gravitational force per unit mass. Newton's law, in terms of this gravitational field for a point mass M, is

$$\vec{g}(\vec{r}) = -G_{\mathrm{N}}\frac{M}{r^2}\hat{r}. \tag{4.3}$$

Just as Coulomb's law can be equivalently stated as Gauss's law for the electric field, this field, Eq. (4.3), can be expressed, for an arbitrary mass distribution, as Gauss's law for the gravitational field:

$$\oint_S \vec{g} \cdot d\vec{A} = -4\pi G_{\mathrm{N}} M. \tag{4.4}$$

The area integral on the left-hand side (LHS) is the gravitational field flux through any closed surface S, and M on the right-hand side (RHS) is the total mass enclosed inside S. This integral representation of Gauss's law (4.4) can be converted into a differential equation: We will first turn both sides into volume integrals by using the divergence theorem on the LHS (the area integral into the volume integral of the divergence of the field) and by expressing the mass on the RHS in terms of the mass density function ρ:

$$\int \vec{\nabla} \cdot \vec{g} \, dV = -4\pi G_{\mathrm{N}} \int \rho \, dV.$$

Since this relation holds for any volume, the integrands on both sides must also be equal:

$$\vec{\nabla} \cdot \vec{g} = -4\pi G_{\mathrm{N}} \rho. \tag{4.5}$$

This is Newton's field equation in differential form. The gravitational potential[1] $\Phi(\vec{r})$ being defined through the gravitational field $\vec{g} \equiv -\vec{\nabla}\Phi$, the field equation (4.5) becomes

$$\nabla^2 \Phi = 4\pi G_{\mathrm{N}} \rho. \tag{4.6}$$

To obtain the equation of motion of a test particle in a given gravitational field, we insert (4.2) into Newton's second law $\vec{F} = m\vec{a}$ to get

$$\frac{d^2\vec{r}}{dt^2} = \vec{g}, \tag{4.7}$$

which has the outstanding feature of being totally independent of any properties (mass, charge, etc.) of the test particle. Expressed in terms of the gravitational potential, it can now be written as

$$\frac{d^2\vec{r}}{dt^2} = -\vec{\nabla}\Phi. \tag{4.8}$$

We note that the Newtonian field theory of gravitation, as embodied in (4.6) and (4.8), is not compatible with SR as space and time coordinates are not treated on equal footings. In fact Newtonian theory is a **static** field

[1] We have the familiar example of the potential for a spherically symmetric source with total mass M being given by $\Phi = -G_{\mathrm{N}} M/r$.

theory. Stated in another way, these equations are comparable to Coulomb's law in electromagnetism. They are not complete, as the effects of motion (i.e. magnetism) are not included. This "failure" just reflects the underlying physics that only admits an (instantaneous) action-at-a-distance description, which implies an infinite speed of signal transmission, incompatible with the principle of relativity.

4.2   EP introduced

In this section, several properties of gravitation will be presented. They all follow from an empirical principle, called by Einstein the **principle of the equivalence of gravitation and inertia**. The final formulation of Einstein's theory of gravitation, the general theory of relativity, automatically and precisely contains this equivalence principle. Historically, it is the starting point of a series of discoveries that ultimately led Einstein to the geometric theory of gravity, in which the gravitation field is warped spacetime.

4.2.1   Inertial mass vs. gravitational mass

One of the distinctive features of the gravitation field is that its equation of motion (4.8) is totally independent of the test particle's properties. This comes about because of the cancellation of the mass factors in $m\vec{g}$ and $m\vec{a}$. Actually these two masses correspond to very different concepts:

- The **inertial mass**

$$\vec{F} = m_{\mathrm{I}}\vec{a} \qquad (4.9)$$

  enters into the description of the response of a particle to **all** forces.
- The **gravitational mass**

$$\vec{F} = m_{\mathrm{G}}\vec{g} \qquad (4.10)$$

  reflects the response[2] of a particle to a **particular** force: gravity. The gravitational mass m_{G} may be viewed as the "gravitational charge" placed in a given gravitational field $\vec{g}$.

Now consider two objects A and B composed of different material, one of copper and the other of wood. When they are let go in a given gravitational field $\vec{g}$, e.g. "being dropped from the Leaning Tower of Pisa" (see Box 4.1), they will, according to (4.9) and (4.10), obey the equations of motion:

$$(\vec{a})_{\mathrm{A}} = \left(\frac{m_{\mathrm{G}}}{m_{\mathrm{I}}}\right)_{\mathrm{A}} \vec{g}, \quad (\vec{a})_{\mathrm{B}} = \left(\frac{m_{\mathrm{G}}}{m_{\mathrm{I}}}\right)_{\mathrm{B}} \vec{g}. \qquad (4.11)$$

Part of Galileo's great legacy to us is the **experimental observation** that all bodies fall with the same acceleration, $(\vec{a})_{\mathrm{A}} = (\vec{a})_{\mathrm{B}}$, which leads to the equality,

$$\left(\frac{m_{\mathrm{G}}}{m_{\mathrm{I}}}\right)_{\mathrm{A}} = \left(\frac{m_{\mathrm{G}}}{m_{\mathrm{I}}}\right)_{\mathrm{B}}. \qquad (4.12)$$

[2]One should in principle distinguish between two separate gravitational charges: a "passive" gravitational mass as in (4.10) is the response to the gravitational field, and an "active" gravitational mass is the source of the gravitational field $\vec{g} = -G_{\mathrm{N}}m_{\mathrm{G}}\vec{r}/r^3$. These two masses can be equated by way of Newton's third law.

The mass ratio, having been found to be universal for all substances as in (4.12), can then be set, by appropriate choice of units, equal to unity. This way we can simply say

$$m_I = m_G. \tag{4.13}$$

Even at the fundamental particle physics level, where matter is made up of protons, neutrons, and electrons (all having different interactions) bound together with different binding energies, it is difficult to find an a priori reason to expect such a relation (4.12). As we shall see, this is the empirical foundation underlying the geometric formulation of the relativistic theory of gravity that is GR.

Box 4.1 A brief history of the EP: from Galileo and Newton to Eötvös

There is no historical record of Galileo having dropped anything from the Leaning Tower of Pisa. Nevertheless, to refute Aristotle's contention that heavier objects would fall faster than light ones, he did report performing experiments of sliding different objects on an inclined plane, Fig. 4.1(a). (The slower fall allows for more reliable measurements.) More importantly, Galileo provided a theoretical argument, "a thought experiment," in the first chapter of his *Discourse and Mathematical Demonstration of Two New Sciences*, in support of the idea that all substances should fall with the same acceleration. Consider any falling object. Without this universality of free fall, the tendency of different components of the object to fall differently would give rise to internal stress and could cause certain objects to undergo spontaneous disintegration. The nonobservation of this phenomenon could then be taken as evidence for equal accelerations.

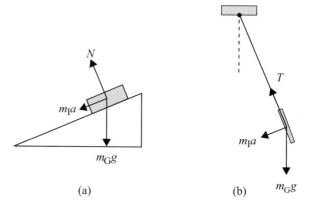

Fig. 4.1 Both the gravitational mass and inertia mass enter in the phenomena: (a) a sliding object on an inclined plane, and (b) oscillations of a pendulum.

(a)   (b)

Newton went further by translating this universality of free fall into the universal proportionality of the inertial and gravitational masses (4.12) and built the equality $m_I = m_G$ right into the foundations of mechanics. Notably, he discussed this equality in the very first paragraph of his *Principia*. Furthermore, he improved upon the empirical check of

Galileo's result (4.12) by experimenting with a pendulum, Fig. 4.1(b), see Problem 4.1,

$$\delta_{AB} \equiv \left| \frac{(m_I/m_G)_A - (m_I/m_G)_B}{(m_I/m_G)_A + (m_I/m_G)_B} \right| \leq 10^{-3}. \tag{4.14}$$

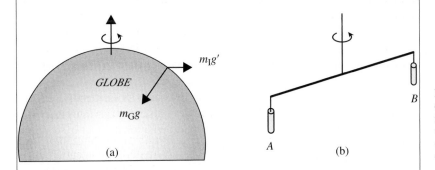

(a)   A   (b)   B

Fig. 4.2 The Eötvös experiment to detect any difference between the ratio of the gravitational to inertial masses of substance A vs. B. The centrifugal acceleration can be decomposed into the vertical and horizontal components, $\vec{g}' = \vec{g}'_v + \vec{g}'_h$.

The Eötvös experiment and modern limits   At the end of the nineteenth century, the Hungarian baron Roland von Eötvös pointed out that any possible nonuniversality of this mass ratio (4.12) would show up as a horizontal twist τ in a torsion balance, Fig. 4.2(b). Two weights composed of different substances A and B are hung at the opposite ends of a rod, which is in turn hung from the ceiling by a fiber at a midpoint having respective distances of l_A and l_B from the two ends. Because of earth's rotation, we are in a noninertial frame of reference. In order to apply Newton's laws, we must include the inertial force, as represented by the centrifugal acceleration $\vec{g}'$, Fig. 4.2(a). In the vertical direction we have the gravitational acceleration g, and the (tiny and, for our simplified calculation, negligible) vertical component g'_v. In the horizontal direction the only nonzero torque is due to the horizontal component g'_h. The equilibrium conditions of a vanishing total torque are:

vertical balance $\left[(m_G)_A\, l_A - (m_G)_B\, l_B \right] g = 0$ (4.15)

horizontal balance $\left[(m_I)_A\, l_A - (m_I)_B\, l_B \right] g'_h = \tau.$ (4.16)

The equality of $(m_G)_A\, l_A = (m_G)_B\, l_B$ from the equilibrium condition of (4.15) means that the twist in (4.16) is related to the sought-after nonuniversality:

$$\tau = m_G l \left[\left(\frac{m_I}{m_G} \right)_A - \left(\frac{m_I}{m_G} \right)_B \right] g'_h. \tag{4.17}$$

In this way Eötvös greatly improved the limit of (4.14) to $\delta_{AB} \leq 10^{-9}$. More recent experiments by others, ultimately involving the comparison of the falling earth and moon in the solar gravitational field, have tightened this limit further to 1.5×10^{-13}.

4.2.2   EP and its significance

In the course of writing a review article on SR in 1907, Einstein came upon what he later termed **"my happiest thought:"** He recalled the fundamental experimental result of Galileo that all objects fall with the same acceleration. "Since all bodies accelerate the same way, an observer in a freely falling laboratory will not be able to detect any gravitational effect (on a point particle) in this frame." Or, "gravity is transformed away in reference frames in free fall."

Principle of equivalence stated   Imagine an astronaut in a freely falling spaceship. Because all objects fall with the same acceleration, a released object in the spaceship will not be seen to fall with respect to its surroundings. Thus, from the viewpoint of the astronaut, gravity is absent; everything becomes weightless. To Einstein, this vanishing of the gravitational effect is so significant that he elevated it (in order to focus on it) to a physical principle: **the equivalence principle.**

$$\left(\begin{array}{c} \text{Physics in a frame freely falling in a gravity field} \\ \text{is equivalent to} \\ \text{physics in an inertial frame without gravity} \end{array} \right).$$

Namely, within a freely falling frame, where the acceleration exactly cancels the uniform gravitational field, no sign of either acceleration or gravitation can be found by any physical means. Correspondingly,

$$\left(\begin{array}{c} \text{Physics in a nonaccelerating frame with gravity } \vec{g} \\ \text{is equivalent to} \\ \text{physics in a frame without gravity but accelerating with } \vec{a} = -\vec{g} \end{array} \right).$$

Absence of gravity in an inertial frame   Thus according to the EP, accelerating frames of reference can be treated in exactly the same way as inertial frames. They are simply frames with gravity. From this we also obtain a **physics definition of an inertial frame**, without reference to any external environment such as fixed stars, as **the frame in which there is no gravity**. Einstein realized the unique position of gravitation in the theory of relativity. Namely, he understood that the question was not how to incorporate gravity into SR but rather how to use gravitation as a means to broaden the principle of relativity from inertial frames to all coordinate systems including accelerating frames.

From EP to gravity as the structure of spacetime   If we confine ourselves to the physics of mechanics, EP is just a restatement of $m_I = m_G$. But once it is highlighted as a principle, it allowed Einstein to extend this equivalence between inertia and gravitation to **all physics**: (not just to mechanics, but also electromagnetism, etc.) This generalized version is sometimes called the **strong equivalence principle**. Thus the "weak EP" is just the statement of $m_I = m_G$, while the "strong EP" is the principle of equivalence applied to all physics. In the following, we shall still refer to, for short, the strong equivalence principle as EP. Because the motion of a test body in a gravitational field is independent of the properties of the body,

Einstein came up with the idea that the effect on the body can be attributed directly to some spacetime feature, and gravity is nothing but the structure of a warped spacetime. This road from EP to GR can also be viewed as follows. The equivalence of an accelerated frame to that with gravity means that we cannot really say that gravity brings about a particle's acceleration. This can be phrased as gravity not being a force; a test body in whatever gravitational field just "moves freely in the spacetime with gravity." Any nontrivial motion is attributed to the structure of spacetime brought about by gravity. Gravity can cause the fabric of spacetime to warp, and the shape of spacetime responds to the matter in the environment.

The strength of a gravitational field   Ordinarily we expect the gravitational effect to be very small as Newton's constant G_N is very small. One way to get an order of magnitude idea is by taking the ratio of the gravitational energy and the (relativistic) rest energy $E_{rel} = m_I c^2$:

$$\varepsilon = \frac{E_{grav}}{E_{rel}} = \frac{m_G \Phi}{m_I c^2} = \frac{\Phi}{c^2} = \frac{G_N M}{c^2 r}, \tag{4.18}$$

where $\Phi = -G_N M / r$ is the gravitational potential for the spherically symmetric case. Near earth's surface, $\Phi = gh$, the product of gravitational acceleration and height, we have $\varepsilon = gh/c^2 = O(10^{-15})$ for a typical laboratory distance range $h = O(10\,\text{m})$ in terrestrial gravity. Such a small value basically reflects the weakness of the gravitational interaction. Only in extraordinary situations of a black hole (extremely compact object) or cosmology (extreme massive system) with huge M to r ratios will the parameter ε approach the order of unity. For further discussion, see the introductory paragraphs of Chapter 9.

4.3   Implications of the strong EP

The strong EP implies, as we shall show in this section, that gravity can bend a light ray, shift the frequency of an electromagnetic wave, and cause clocks to run slow. Ultimately, these results suggested to Einstein that the proper framework to describe the relativistic gravitational effects is a curved spacetime.

To deduce the effect of gravity on certain physical phenomena, we shall use the following general procedure:

1. One first considers the description by an observer inside a spaceship in free fall. According to EP there is no gravitational effect in this inertial frame and SR applies.
2. One then considers the same situation from the viewpoint of an observer watching the spaceship from outside: there is a gravitational field and the first (freely falling) observer is seen to be accelerating in this gravitational field. Namely, this second observer defines an inertial frame, with gravity being treated as one of the forces $\vec{F} = m\vec{g}$.

3. The combined effects of acceleration and gravity, as seen by the second observer, must then reproduce the SR description as recorded by the inertial observer in free fall. "Physics should be independent of coordinate choices."

Bending of a light ray—a qualitative account

Let us first study the effect of gravity on a light ray traveling (horizontally) across a spaceship which is falling in a constant (vertical) gravitational field $\vec{g}$. From the viewpoint of the astronaut in the spaceship, EP informs us there is no detectable effect associated either with gravity or with acceleration: the light travels straight across the spaceship from one side to the other: in this coordinate frame, the light is emitted at a height h, received at the same height h on the opposite side of the spaceship, Fig. 4.3(a). But, to an observer outside the spaceship, there is a gravitational field $\vec{g}$ and the spaceship is accelerating (falling) in this gravitational field. The straight trajectory of the light signal in the freely falling spaceship will appear to bend, Fig. 4.3(b). Thus, to this outside observer, a light ray is seen to bend in the gravitational field.

We do not ordinarily see such "falling of light rays" because, for the gravitational field and distance scale that we are familiar with, this bending effect is unobservably small. Consider a lab with a width of 300 m. The duration for a light ray to travel across the lab would be $1\,\mu s$. During this interval, the distance y that the lab has fallen (amount of the bending) is extremely small:[3] $y = \mathfrak{g}t^2/2 \simeq 5 \times 10^{-12}\,\text{m} = 0.05\,\text{Å}$. This EP consideration suggests that a light ray would be bent by any massive object. The quantitative relation

[3] See also (4.18).

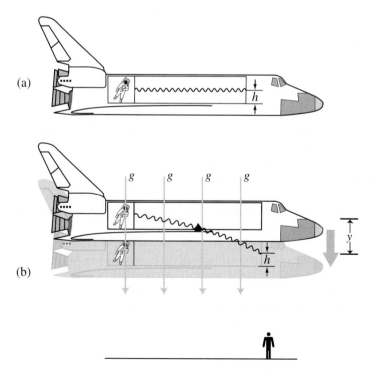

Fig. 4.3 According to the equivalence principle, a light ray will 'fall' in a gravitational field. (a) To the astronaut in the freely falling spaceship (an inertial observer in a gravity-free environment), the light trajectory is straight. (b) To an observer outside the spaceship, the astronaut is accelerating (falling) in a gravitational field. The light ray will be bent so that it reaches the opposite side of the lab at a height $y = gt^2/2$ below the initial point.

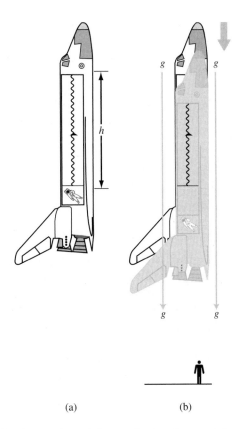

(a)   (b)

Fig. 4.4 According to the equivalence principle, the frequency of a light ray is redshifted when moving up against gravity. (a) To an inertial observer in the freely falling spaceship, there is no frequency shift. (b) To an observer outside the spaceship, this astronaut is accelerating in a gravitation field; the null frequency shift result comes about because of the cancellation between the Doppler blueshift and the gravitational redshift.

between the deflection angle and the gravitational potential will be worked out in Section 4.3.3.

4.3.1 Gravitational redshift and time dilation

Gravitational redshift

In Fig. 4.3, we discussed the effect of a gravitational field on a light ray with its trajectory transverse to the field direction. Now let us consider the situation when the field direction is parallel (or antiparallel) to the ray direction as in Fig. 4.4.

Here we have a receiver placed directly at a distance h above the emitter in a downward-pointing gravitational field $\vec{g}$. Just as the transverse case considered above, we first describe the situation from the viewpoint of the astronaut (in free fall), Fig. 4.4(a). EP informs us that the spaceship in free fall is an inertial frame without gravity. Such an observer will not be able to detect any physical effects associated with gravity or acceleration. In this free-fall situation, the astronaut should not detect any frequency shift: the received light frequency ω_{rec} is the same as the emitted frequency ω_{em}:

$$(\Delta\omega)_{\text{ff}} = (\omega_{\text{rec}} - \omega_{\text{em}})_{\text{ff}} = 0, \tag{4.19}$$

where the subscript ff reminds us that these are the values as seen by an observer in free fall.

From the viewpoint of the observer outside the spaceship, there is gravity and the spaceship is accelerating (falling) in this gravitational field, Fig. 4.4(b). Assume that this spaceship starts to fall at the moment of light emission. Because it takes a finite amount of time $\Delta t = h/c$ for the light signal to reach the receiver on the ceiling, it will be detected by a receiver in motion, with a velocity $\Delta u = \mathfrak{g}\Delta t$. The familiar Doppler formula in the low-velocity approximation (3.48) would lead us to expect a frequency shift of

$$\left(\frac{\Delta\omega}{\omega}\right)_{\text{Doppler}} = \frac{\Delta u}{c}. \tag{4.20}$$

Since the receiver has moved closer to the emitter, the light waves must have been compressed, and this shift must be toward the blue

$$(\Delta\omega)_{\text{Doppler}} = (\omega_{\text{rec}} - \omega_{\text{em}})_{\text{Doppler}} > 0. \tag{4.21}$$

We have already learned in (4.19), as deduced by the observer in free fall, that the received frequency did not deviate from the emitted frequency. This physical result must hold for both observers, so the blueshift in (4.21) must somehow be cancelled. To the observer outside the spaceship, gravity is also present. We can recover the nullshift result if there is another physical effect of **light being redshifted by gravity**, with just the right amount to cancel the Doppler blueshift of (4.20). (See also Problem 4.4.)

$$\left(\frac{\Delta\omega}{\omega}\right)_{\text{gravity}} = -\frac{\Delta u}{c}. \tag{4.22}$$

We now express the relative velocity on the RHS in terms of the gravitational potential difference $\Delta\Phi$ at the two locations:

$$\Delta u = \mathfrak{g}\Delta t = \frac{\mathfrak{g}h}{c} = \frac{\Delta\Phi}{c}. \tag{4.23}$$

When (4.22) and (4.23) are combined, we obtain the formula of **gravitational frequency shift**

$$\frac{\Delta\omega}{\omega} = -\frac{\Delta\Phi}{c^2}, \tag{4.24}$$

namely,[4]

$$\frac{\omega_{\text{rec}} - \omega_{\text{em}}}{\omega_{\text{em}}} = -\frac{\Phi_{\text{rec}} - \Phi_{\text{em}}}{c^2}. \tag{4.25}$$

A light ray emitted at a lower gravitational potential point ($\Phi_{\text{em}} < \Phi_{\text{rec}}$) with a frequency ω_{em} will be received at a higher gravitational field point as a lower frequency ($\omega_{\text{em}} > \omega_{\text{rec}}$) signal, that is, it is redshifted, even though the emitter and the receiver are not in relative motion.

The Pound–Rebka–Snider experiment

In principle, this gravitational redshift can be tested by a careful examination of the spectral emission lines from an astronomical object (hence large gravitational potential difference). For a spherical body (mass M and radius R),

[4]Whether the denominator is ω_{rec} or ω_{em}, the difference is of higher order and can be ignored in these leading order formulas.

the redshift formula of (4.24) takes on the form

$$\frac{\Delta\omega}{\omega} = \frac{G_N M}{c^2 R}. \tag{4.26}$$

We have already commented on the smallness of this ratio in (4.18). Even the solar redshift has only a size $O\,(10^{-6})$, which can easily be masked by the standard Doppler shifts due to thermal motion of the emitting atoms. It was first pointed out by Eddington in the 1920s that the redshift effect would be larger for white dwarf stars; with their masses comparable to the solar mass and much smaller radii, the effect could be 10–100 times larger. But in order to obtain the mass measurement of the star, it would have to be in a binary configuration, for instance the Sirius A and B system. In such cases the light from the white dwarf Sirius B suffers scattering by the atmosphere of Sirius A. Nevertheless, some tentative positive confirmation of the EP prediction had been obtained. However, conclusive data did not exist in the first few decades after Einstein's paper. Surprisingly this EP effect of the gravitational redshift was first verified in a series of terrestrial experiments when Pound and his collaborators (1960 and 1964) succeeded in measuring the truly small frequency shift of radiation traveling up $h = 22.5$ m, the height of an elevator shaft in the building housing the Harvard Physics Department:

$$\left|\frac{\Delta\omega}{\omega}\right| = \left|\frac{gh}{c^2}\right| = O\,(10^{-15}). \tag{4.27}$$

Normally, it is not possible to fix the frequency of an emitter or absorber to a very high accuracy because of the energy shift due to thermal recoils of the atoms. However, with the Mössbauer effect,[5] the emission line width in a rigid crystal is as narrow as possible, limited only by the quantum mechanical uncertainty principle $\Delta t\,\Delta E \geq \hbar$, where Δt is given by the lifetime of the unstable (excited) state. Thus a long-lived state would have a particularly small energy–frequency spread. The emitting atom that Pound and Rebka chose to work with is an excited atom Fe*-57, which can be obtained through the nuclear beta decay of cobalt-57. It makes the transition to the ground state by emitting a gamma ray: Fe* $\to$ Fe $+\gamma$. In the experiment, the γ-ray emitted at the bottom of the elevator shaft, after climbing the 22.5 m, could no longer be resonantly absorbed by a sheet of Fe in the ground state placed at the top of the shaft. To prove that the radiation has been redshifted by just the right amount $O\,(10^{-15})$, Pound and Rebka artificially introduced an (ordinary) Doppler blueshift, by moving the detector slowly towards the emitter, just the right amount to compensate for the gravitational redshift. In this way, the radiation is again resonantly absorbed. What was the speed with which they must move the receiver? From (4.24) and (4.20) we have

$$\underbrace{\frac{gh}{c^2}}_{\text{gravity}} = \underbrace{\frac{\Delta\omega}{\omega}}_{\text{Doppler}} = \frac{u}{c} \tag{4.28}$$

with

$$u = \frac{gh}{c} = \frac{9.8 \times 22.5}{3 \times 10^8} = 7.35 \times 10^{-7}\ \text{m/s}. \tag{4.29}$$

[5] The Mössbauer effect—When emitting light, the recoil atom can reduce the energy of the emitted photon. In reality, since the emitting atom is surrounded by other atoms in thermal motion, this brings about recoil momenta in an uncontrollable way. (We can picture the atom as being part of a vibrating lattice.) As a result, the photon energy in different emission events can vary considerably, resulting in a significant spread of their frequencies. This makes a measurement of the atomic frequency to high enough precision impossible for purposes of testing the gravitational redshift. But in 1958 Mössbauer made a breakthrough when he pointed out, and verified by observation, that crystals with high Debye–Einstein temperature, that is, having a rigid crystalline structure, could pick up the recoil by the entire crystal. Namely, in such a situation, the emitting atom has an effective mass that is huge. Consequently, the atom loses no recoil energy, and the photon can pick up all the energy-change of the emitting atom, and the frequency of the emitted radiation is as precise as it can be.

It is such a small speed that it would take $h/u = c/\mathfrak{g} = O\,(3 \times 10^7\,\text{s}) \simeq 1$ year to cover the same elevator shaft height. Of course this velocity is just the one attained by an object freely falling for a time interval that takes the light to traverse the distance h. This is the compensating effect we invoked in our derivation of the gravitational redshift at the beginning of this section.

Gravitational time dilation

At first sight, this gravitational frequency shift looks absurd. How can an observer, **stationary** with respect to the emitter, receive a different number of wave crests per unit time than the emitted rate? Here is Einstein's radical and yet simple answer: while the number of wave crests does not change, the time unit itself changes in the presence of gravity. The clocks run at different rates when situated at different gravitational field points: there is a **gravitational time dilation** effect.

The frequency being proportional to the inverse of the local proper time rate

$$\omega \sim \frac{1}{d\tau} \tag{4.30}$$

the gravitational frequency shift formula (4.25) can be converted to a time dilation formula

$$\frac{d\tau_1 - d\tau_2}{d\tau_2} = \frac{\Phi_1 - \Phi_2}{c^2}, \tag{4.31}$$

or

$$d\tau_1 = \left(1 + \frac{\Phi_1 - \Phi_2}{c^2}\right) d\tau_2. \tag{4.32}$$

For a static gravitational field, this can be integrated to read

$$\tau_1 = \left(1 + \frac{\Phi_1 - \Phi_2}{c^2}\right) \tau_2, \quad \text{or} \quad \frac{\Delta\tau}{\tau} = \frac{\Delta\Phi}{c^2}. \tag{4.33}$$

Namely, the clock at the higher gravitational potential point will run faster. This is to be contrasted with the special relativistic time dilation effect—clocks in relative motion run at different rates. Here we are saying that two clocks, even at rest with respect to each other, also run at different rates if the gravitational fields at their respective locations are different. Their distinction can be seen in another way: in SR time dilation each observer sees the other's clock run slow (see Section 2.3.3 "relativity is truly relative"), while with gravitational dilation, the observer at a higher gravitational potential point sees the lower clock run slow, and the lower observer sees the higher clock run fast. For two clocks in a gravitational field and also in relative motion, we have to combine the gravitational and relative motion frequency-shift results to obtain

$$\tau_1 = \left(1 + 2\frac{\Delta\Phi}{c^2} - \frac{u^2}{c^2}\right)^{1/2} \tau_2. \tag{4.34}$$

Time dilation—another derivation

As discussed in our SR chapters, the standard method to compare clock readings is through the exchange of light signals. Since the light frequency is directly related to the clock rate, the method to derive the gravitational redshift by an exchange of light signals would involve a comparison of light frequencies. For this we must take into account the effect of the gravitational redshift. Hence, using this standard method of comparing clocks (at two different locations) by light signal exchanges is just the route we have followed in arriving at the gravitational time dilation result of (4.32) via the gravitational redshift relation.

We now present another derivation of (4.32) that will display its relation (and its compatibility) to the familiar SR effect of time dilation. One way two clocks located at different gravitational potential points, clock-1 at Φ_1 and clock-2 at Φ_2, can be compared is to liken each of them to a third clock in such a way as to be free of gravity's effect. Let clock-3 fall freely in this gravitational field; when it passes by clock-1 it has speed u_1 and clock-2 has speed u_2. At the instant when clock-3 passes by clock-1, both clocks are at the **same** gravitational potential point of Φ_1; a comparison of their clock rates involves only the effect of their relative motion (clock-1 at rest while clock-3 in motion) with gravity not being an issue.[6] The comparison when clock-3 passes by clock-2 at Φ_2 with speed u_2 can be similarly carried out. All three clocks being identically constructed, we have the SR effect of

$$t_1^{\mathrm{ff}} = \gamma_1 d\tau_1, \quad \text{and} \quad t_2^{\mathrm{ff}} = \gamma_2 d\tau_2, \tag{4.35}$$

where the time intervals $t_{1,2}^{\mathrm{ff}}$ are those of the freely-falling clock-3, as seen by the stationary clocks-1 and -2 (having their own proper interval $d\tau_{1,2}$) at the respective potential points Φ_1 and Φ_2, with $\gamma_{1,2} = (1 - u_{1,2}^2/c^2)^{-\frac{1}{2}}$. Clock-3 being in free-fall (and gravity transformed away in this reference frame), nothing distinguishes these two position points for this clock; we must have $dt_1^{\mathrm{ff}} = dt_2^{\mathrm{ff}}$. The time-dilation result of (4.32) can then be derived by connecting the two equations in (4.35):

$$\frac{d\tau_1}{d\tau_2} = \frac{\gamma_2}{\gamma_1} = \left(\frac{1 - u_1^2/c^2}{1 - u_2^2/c^2}\right)^{1/2}$$

$$\simeq 1 - \frac{1}{2}\frac{u_1^2 - u_2^2}{c^2} = 1 + \frac{\Phi_1 - \Phi_2}{c^2}, \tag{4.36}$$

where, to reach the second line, we have dropped terms $O\left(u^4/c^4\right)$ in the power series expansions of the denominator and of the square root. In the last equality we have used the low velocity version (consistent with our presentation) of the energy conservation relation for the freely falling clock-3: the change in kinetic energy must be equal to minus potential energy change, $\frac{1}{2}m\Delta u^2 = -m\Delta\Phi$. This derivation of (4.32) shows that gravitational time dilation is entirely compatible with the previously known SR time dilation effect—just as we have shown its compatibility with the Doppler frequency shift in our first derivation (4.22) of the gravitational redshift (4.24).

Time dilation test by an atomic clock The gravitational time dilation effects have been tested directly by comparing the times kept by two cesium atomic clocks: one flown in an airplane at high altitude h (about 10 km)

[6]Clearly, when clock-3 passes clock-1 (or clock-2), these two clocks are at the same location; hence, no light exchanges are required in order to compare the clock rates.

in a holding pattern, for a long time τ, over the ground station where the other clock sits. After correction of the various background effects (mainly SR time dilations), the high altitude clock was found to gain over the ground clock by a time interval of $\Delta\tau = (gh/c^2)\tau$ in agreement with the expectation given in (4.33) (Hafele and Keating, 1972). For an application in the Global Position System, see Problem 4.3.

4.3.2   Light ray deflection calculated

The clocks run at different rates at locations where the gravitational field strengths are different. Since different clock rates will lead to different speed measurements, even the speed of light can be measured to have different values! We are familiar with the light speed in different media being characterized by a varying index of refraction. Gravitational time dilation implies that even in the vacuum there is an effective index of refraction when a gravitational field is present. Since the gravitational field is usually inhomogeneous, this index is generally a position-dependent function.

Gravity-induced index of refraction in free space

At a given position r with gravitational potential $\Phi(r)$ a determination of the light speed involves the measurement of a displacement dr for a time interval $d\tau$ as recorded by a clock at rest at this position. The resultant ratio

$$\frac{dr}{d\tau} = c \tag{4.37}$$

is the light speed according to the local proper time. This speed c is a universal constant. Because of gravitational time dilation, as stated in (4.32), an observer at another position (with a different gravitational potential) would obtain a different value for this speed when using a clock located at the second position. In fact, a common choice of time coordinate is that given by a clock located far away from the gravitational source. For two positions $r_1 = r$ and $r_2 = \infty$, with r_2 being the reference point $\Phi(\infty) = 0$, while $\tau(r)$ is the local proper time, the clock at $r = \infty$ gives the coordinate time $t \equiv \tau(\infty)$. Equation (4.36) then yields the relation between the local time (τ) and the coordinate time (t) as

$$d\tau = \left(1 + \frac{\Phi(r)}{c^2}\right) dt. \tag{4.38}$$

This implies that the speed of light as measured by the remote observer is reduced by gravity as

$$c(r) \equiv \frac{dr}{dt} = \left(1 + \frac{\Phi(r)}{c^2}\right) \frac{dr}{d\tau} = \left(1 + \frac{\Phi(r)}{c^2}\right) c. \tag{4.39}$$

Namely, the speed of light will be seen by an observer (with his coordinate clock) to vary from position to position as the gravitational potential varies from position to position. For such an observer, the effect of the gravitational field can be viewed as introducing an **index of refraction** in the space:

$$n(r) \equiv \frac{c}{c(r)} = \left(1 + \frac{\Phi(r)}{c^2}\right)^{-1} \simeq 1 - \frac{\Phi(r)}{c^2}. \tag{4.40}$$

We will state the key concepts behind this position-dependent speed of light once more: we are not suggesting that the deviation of $c(r)$ from the constant c means that the physical velocity of light has changed, or that the velocity of light is no longer a universal constant in the presence of gravitational fields. Rather, it signifies that the clocks at different gravitational points run at different rates. For an observer with the time t measured by clocks located far from the gravitational source (taken to be the coordinate time), the velocity of the light **appears to this observer** to slow down. A dramatic example is offered by the case of black holes (to be discussed in Chapter 8). There, as a manifestation of an infinite gravitational time dilation, it would take an infinite amount of coordinate time for a light signal to leave a black hole. Thus, to an outside observer, no light can escape from a black hole, even though the corresponding proper time duration is perfectly finite.

Bending of a light ray—the EP expectation

We can use this position-dependent index of refraction to calculate the bending of a light ray by a transverse gravitational field via the Huygen's construction. Consider a plane light wave propagating in the $+x$ direction. At each time interval Δt, a wavefront advances a distance of $c\Delta t$, see Fig. 4.5(a). The existence of a transverse gravitational field (in the y direction) means a nonvanishing derivative of the gravitational potential $d\Phi/dy \neq 0$. A Change of the gravitational potential means a change in $c(r)$ and this leads to tilting of the wavefronts. We can then calculate the amount of the bending of the light ray by using the diagram in Fig. 4.5(b). A small angular deflection can be related to distances as

$$(d\phi) \simeq \frac{(c_1 - c_2)dt}{dy} \simeq \frac{dc(r)(dx/c)}{dy}. \tag{4.41}$$

Working in the limit of weak gravity with small $\Phi(r)/c^2$ (or equivalently $n \simeq 1$), we can relate $dc(r)$ to a change of index of refraction as

$$dc(r) = cdn^{-1} = -cn^{-2}dn \simeq -cdn. \tag{4.42}$$

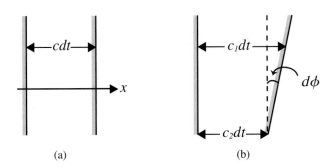

(a)   (b)

Fig. 4.5 Wavefronts of a light trajectory. (a) Wavefronts in the absence of gravity. (b) Tilting of wavefronts in a medium with an index of refraction varying in the vertical direction so that $c_1 > c_2$. The resultant light bending is signified by the small angular deflection $d\phi$.

Namely, Eq. (4.41) becomes

$$(d\phi) \simeq -\frac{\partial n}{\partial y}dx. \tag{4.43}$$

But from (4.40) we have $dn(r) = -d\Phi(r)/c^2$, thus $c^2 d\phi = (\partial\Phi/\partial y)\,dx$. In this way, integrating (4.43), we obtain the total deflection angle

$$\delta\phi = \int d\phi = \frac{1}{c^2}\int_{-\infty}^{\infty}\frac{\partial\Phi}{\partial y}dx = \frac{1}{c^2}\int_{-\infty}^{\infty}(\vec{\nabla}\Phi\cdot\hat{\mathbf{y}})dx. \tag{4.44}$$

The integrand is the gravitational acceleration perpendicular to the light path. We shall apply the above formula to the case of the spherical source $\Phi = -G_N M/r$, and $\vec{\nabla}\Phi = \hat{\mathbf{r}}G_N M/r^2$. Although the gravitational field will no longer be a simple uniform field in the $\hat{\mathbf{y}}$ direction, our approximate result can still be used because the bending takes place mostly in the small region of $r \simeq r_{\min}$. See Fig. 4.6.

$$\delta\phi = \frac{G_N M}{c^2}\int_{-\infty}^{\infty}\frac{\hat{\mathbf{r}}\cdot\hat{\mathbf{y}}}{r^2}dx = \frac{G_N M}{c^2}\int_{-\infty}^{\infty}\frac{y}{r^3}dx, \tag{4.45}$$

where we have used $\hat{\mathbf{r}}\cdot\hat{\mathbf{y}} = \cos\theta = y/r$. An inspection of Fig. 4.6 also shows that, for small deflection, we can approximate $y \simeq r_{\min}$, hence

$$r = (x^2 + y^2)^{1/2} \simeq (x^2 + r_{\min}^2)^{1/2} \tag{4.46}$$

leading to

$$\delta\phi = \frac{G_N M}{c^2}\int_{-\infty}^{\infty}\frac{r_{\min}}{(x^2 + r_{\min}^2)^{3/2}}dx = \frac{2G_N M}{c^2 r_{\min}}. \tag{4.47}$$

With a light ray being deflected by an angle $\delta\phi$ as shown in Fig. 4.6, the light source at S would appear to the observer at O to be located at S'. Since the deflection is inversely proportional to $r_{\min}$, one wants to maximize the amount of bending by having the smallest possible $r_{\min}$. For light grazing the surface of the sun, $r_{\min} = R_\odot$ and $M = M_\odot$, Eq. (4.47) gives an angle of deflection $\delta\phi = 0.875''$. As we shall explain in Section 7.2.1, this is exactly **half** of the correct GR prediction for the solar deflection of light from a distant star.

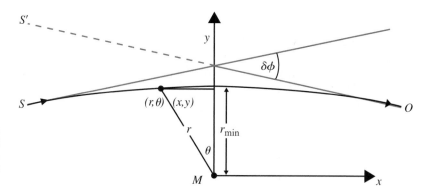

Fig. 4.6 Angle of deflection $\delta\phi$ of light by a mass M. A point on the light trajectory (solid curve) can be labled either as (x, y) or (r, θ). The source at S would appear to the observer at O to be located at a shifted position of S'.

4.3.3 Energy considerations of a gravitating light pulse

Erroneous energy considerations

Because light gravitates (i.e. it bends and redshifts in a gravitational field), it is tempting to imagine that a photon has a (gravitational) mass. One might argue as follows: from the viewpoint of relativity, there is no fundamental difference between mass and energy, $E = m_I c^2$. The equivalence $m_I = m_G$ means that any energy also has a nonzero "gravitational charge"

$$m_G = \frac{E}{c^2}, \qquad (4.48)$$

and hence will gravitate. The gravitational redshift formula (4.24) can be derived by regarding such a light-pulse losing "kinetic energy" when climbing out of a gravitational potential well. One can even derive the light deflection result (4.47) by using the Newtonian mechanics formula[7] of a moving mass (having velocity u) being gravitationally deflected by a spherically symmetric mass M (Fig. 4.6),

$$\delta\phi = \frac{2G_N M}{u^2 r_{\min}}. \qquad (4.49)$$

For the case of the particle being a photon with $u = c$, this just reproduces (4.47). Nevertheless, such an approach to understanding the effect of gravity on a light ray is **conceptually incorrect** because

- A photon is not a massive particle, and it cannot be described as a nonrelativistic massive object having a gravitational potential energy.
- This approach makes no connection to the underlying physics of gravitational time dilation.

The correct energy consideration

The energetics of gravitational redshift should be properly considered as follows (Schwinger, 1986; Okun *et al.*, 2000). Light is emitted and received through atomic transitions between two atomic energy levels of a given atom:[8] $E_1 - E_2 = \hbar\omega$. We can treat the emitting and receiving atoms as nonrelativistic massive objects. Thus when sitting at a higher gravitational potential point, the receiver atom gains energy with respect to the emitter atom,

$$E_{\rm rec} = E_{\rm em} + mgh.$$

We can replace the mass by (4.48) so that, to the leading order, $E_{\rm rec} = (1 + gh/c^2)E_{\rm em}$. This is a multiplicative energy shift of the atomic levels; all the energy levels (and their differences) of the receiving atom are "blueshifted" with respect to those of the emitter atom by

$$(E_1 - E_2)_{\rm rec} = \left(1 + \frac{gh}{c^2}\right)(E_1 - E_2)_{\rm em}, \qquad (4.50)$$

hence a fractional shift of atomic energy

$$\left(\frac{\Delta E}{E}\right)_{\rm atom} = \frac{gh}{c^2} = \frac{\Delta\Phi}{c^2}. \qquad (4.51)$$

[7] Equation (4.49) is quoted in the small angle approximation of a general result that can be found in textbooks on mechanics. See, for example, Eq. (4.37) in Kibble (1985).

[8] We have used the fact that the energy of a light ray is proportional to its frequency. For most of us the quantum relation $E = \hbar\omega$ comes immediately to mind, but this proportionality also holds in classical electromagnetism where the field is pictured as a collection of harmonic oscillators.

On the other hand, the traveling light pulse, neither gaining nor losing energy along its trajectory, has the **same** energy as the emitting atom. But it will be **seen** by the blueshifted receiver atom as redshifted:

$$\left(\frac{\Delta E}{E}\right)_\gamma = -\frac{\Delta\Phi}{c^2} = \frac{\Delta\omega}{\omega}, \tag{4.52}$$

which is the previously obtained result (4.24). This approach is conceptually correct as

- Atoms can be treated as nonrelativistic objects having gravitational potential energy mgh.
- This derivation is entirely consistent with the gravitational time dilation viewpoint: the gravitational frequency shift does not result from any change of the photon properties. It comes about because the standards of frequency (i.e. time) are different at different locations. This approach in fact gives us a physical picture of how clocks can run at different rates at different gravitational field points. An atom is the most basic form of a clock, with time rates being determined by transition frequencies. The fact that atoms have different gravitational potential energies (hence different energy levels) naturally give rise to different transitional frequencies, hence different clock rates.

The above discussion also explains why the usual erroneous derivations of treating photons as nonrelativistic massive particles with gravitational potential energy can lead to the correct EP formulas: the observed change of photon properties is due to the change in the standard clocks (atoms), which can be correctly treated as nonrelativistic masses with gravitational energy.

The various results called "Newtonian" In this connection, we should also clarify the often-encountered practice of calling results such as (4.47) a Newtonian result. By this it is meant that the result can be derived in the pre-Einsteinian-relativity framework where particles can take on **any** speed we wish them to have. There does not exist the notion of a "low-velocity nonrelativistic limit." Consequently, it is entirely correct to use the mechanics formula (4.49) for a light particle which happens to propagate at the speed c. However, one should be aware of the difference between this Newtonian (pre-relativistic) framework and the proper Newtonian limit, which we shall specify in a later discussion, Sections 6.2.1 and 14.2.2, corresponding to the situation of nonrelativistic velocity, and a static weak gravitational field. In this contemporary sense, (4.47) is not a result valid in the Newtonian limit.

4.3.4 Einstein's inference of a curved spacetime

[9]Especially, the gravitational equation of motion (4.8) being totally independent of any property of the test particle suggested to Einstein that the gravitational field, unlike other force fields, is related to some fundamental feature of spacetime.

Aside from the principle of relativity, EP is the most important physical principle underlying Einstein's formulation of a geometric theory of gravity. It not only allows the accelerating frames to be treated on an equal footing the inertial frames and giving these early glimpses of the GR phenomenology, but also the study of EP physics[9] led Einstein to propose that gravity represents the structure of curved spacetime. We shall explain this connection in Chapter 6, after learning some mathematics of curved space in the following chapter.

Review questions

1. Write out, in terms of the gravitational potential $\Phi(x)$, the field equation and the equation of motion for Newton's theory of gravitation. What is the distinctive feature of this equation of motion (as opposed to that for other forces)?

2. What is the inertial mass? What is the gravitational mass? Give the simplest experimental evidence for their ratio being a universal constant (i.e. independent of the material composition of the object).

3. What is the equivalence principle? What is weak EP? Strong EP?

4. Give a qualitative argument showing why EP can lead to the expectation of gravitational bending of a light ray.

5. Provide two derivations of the formula for the gravitational frequency shift:

$$\frac{\Delta\omega}{\omega} = -\frac{\Delta\Phi}{c^2}.$$

   (a) Use the idea that gravity can be transformed away by taking a reference frame in free fall.

   (b) Use the idea that atomic energy levels will be shifted in a gravitational field.

6. Derive the gravitational time dilation formula

$$\frac{\Delta\tau}{\tau} = \frac{\Delta\Phi}{c^2}$$

in two ways:

   (a) from the gravitational frequency-shift formula;

   (b) directly from the considerations of three identically constructed clocks: two stationary at potential points Φ_1 and Φ_2, respectively, and the third one in free fall passing by the first two.

7. Deduce the relation between coordinate time t (defined as the time measured by a clock located far away from any gravitational source) and local proper time $\tau(r)$ at a position with gravitational potential $\Phi(r)$

$$dt = \frac{d\tau}{\left(1 + \Phi/c^2\right)}.$$

8. The presence of a gravitational field implies the presence of an effective index of refraction in free space. How does one arrive at this conclusion? Does this mean that the speed of light is not absolute? Give an example of the physical manifestations of this index of refraction.

Problems

4.1 Inclined plane, pendulum and EP

   (a) **Inclined plane** For the frictionless inclined plane (with angle θ) in Fig. 4.1(a), find the acceleration's dependence on the ratio m_I/m_G. Thus a violation of the equivalence principle would show up as a material-dependence in the time for a material block to slide down the plane.

   (b) **Pendulum** For a simple pendulum on the surface of the earth, cf. Fig. 4.1(b), find its oscillation period's dependence on the ratio m_I/m_G.

4.2 Two EP brain-teasers

   (a) **Forward leaning balloon** Use EP to explain the observation that a helium balloon *leans forward* in a (forward-) accelerating vehicle, see Fig. 4.7(a).

   (b) **A toy for Einstein** On his 76th birthday Einstein received a gift from his Princeton neighbor Eric Rogers. It was a "toy" composed of a ball attached, by a spring, to the inside of a bowl, which was just the right size to hold the ball. The upright bowl is fasten to a broom-stick, see Fig. 4.7(b). What is the

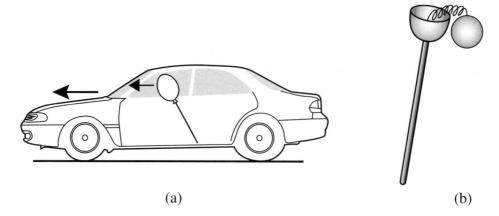

(a)   (b)

Fig. 4.7 Illustrations for the two EP brain-teasers in Problem 4.2.

surefire way, as suggested by EP, to pop the ball back into the bowl each time?

4.3 **The Global Position System**   The signals from the 24 GPS satellites (in six evenly distributed orbit planes) enable us to fix our location on earth to a high degree of accuracy. Each satellite is at such an elevation so as to revolve around the earth every 12 hours. In order to be accurate to within a few meters the satellite clocks must be highly accurate, as 10 nanoseconds translate into a light distance of 3 meters. The atomic clocks on the satellites indeed have the capability of keeping time highly accurately, e.g. to parts in 10^{13} over many days. (To be accurate over a long period, their times are remotely adjusted several times a day.) But in order to synchronize with the clocks on the ground for rapid determination of distances, we must take into account relativistic corrections. This calculation should make it clear that the proper functioning of the GPS requires our knowledge of relativity, especially general relativity. To investigate such relativistic effects we must first calculate the basic parameters of r_s, the satellite's radial distance (from the center of the earth), and v_s, its speed.

(a) Given the satellite orbit period being 12 h, calculate the speed v_s and distance r_s. For this part, Newtonian formulas will be adequate.

(b) Given the fact that the satellite is moving with high speed, there is a SR time dilation effect $t = \gamma_s \tau$. Calculate the fractional change $[(t/\tau) - 1]$ due to this SR time dilation effect.

(c) Calculate the fractional change due to the gravitational time dilation effect as the satellites are at a different gravitational potential compared to the surface of the earth. Is this GR effect more significant than the SR dilation?

(d) Calculate the error that could be accumulated in one minute if these relativistic corrections were not taken into account. Do these two effects change the satellite time in the same direction, or do they tend to cancel each other? Express your result for the accumulated relativistic effect, during a 1-minute duration, in terms of the distance a light signal would have traversed.

4.4 **Gravitational redshift directly from the Doppler effect** Instead of considering a spaceship in free fall, one can use the equivalence of the spaceship at rest in a gravitational field $-\vec{g}$ to a spaceship moving upward with an acceleration $\vec{a} = \vec{g}$. Use the Lorentz frequency transformation of special relativity as given in (3.47) to derive the gravitational frequency shift (4.24) via (4.22) by noting that the receiver, by the time the signal arrives, will be an observer in motion.

Metric description of a curved space

5

- Einstein's new theory of gravitation is formulated in a geometric framework of curved spacetime. In this chapter, we make a mathematical excursion into the subject of non-Euclidean geometry by way of Gauss's theory of curved surfaces.
- **Generalized (Gaussian) coordinates:** A systematic way to label points in space without reference to any objects outside this space.
- **Metric function:** For a given coordinate choice, the metric determines the intrinsic geometric properties of a curved space.
- **The geodesic equation** describes the shortest and the straightest possible curve in a warped space and is expressed in terms of the metric function.
- **The curvature** is the nonlinear second derivative of the metric. As the deviation from Euclidean relations is proportional to the curvature, it measures how much the space is warped.

By a deep study of the physics results implied by the equivalence principle (Chapter 3), Einstein proposed, as we shall discuss in the next chapter, that the gravitational field is curved spacetime. Curved spacetime being the gravitational field, the proper mathematical framework for general relativity (GR) is Riemannian geometry and tensor calculus. We shall introduce these mathematical topics gradually. The key concepts are Gaussian coordinates, metric functions, and the curvature. Points in space are systematically labeled by Gaussian coordinates; the geometry of space can be specified by length measurements among these points with results encoded in the metric function; the curvature tells us how much space is warped.

Historically Riemann's work on the foundations of geometry was based on an extension of Gauss's theory of curved surfaces. Since it is much easier to visualize two-dimensional (2D) surfaces, we shall introduce Riemannian geometry by first considering various topics in 2D curved space. In particular, we study the description of warped surfaces by a 2D metric g_{ab} with $a = 1, 2$, then suggest how such results can be generalized to higher n dimensions with $a = 1, 2, \ldots, n$.

Mathematically speaking, the algebraic extension to higher dimensional spaces, in particular four-dimensional (4D) spacetime, is relatively straightforward—in most cases it involves an extension of the range of indices.

Nevertheless, the generalization of the concept of "curvature" to higher dimensions is rather nontrivial. A proper study of the curvature in higher dimensional spaces, called the Riemann curvature tensor, will be postponed until Chapter 13 when we present the tensor analysis of GR. For the material in Parts II and III, we only need the metric description of the warped spacetime. Knowing the metric function, which is the relativistic gravitational potential, we can determine the equation of motion of a particle in the curved spacetime and can discuss many GR applications. In Part IV we present the derivation of the Riemann curvature tensor, and this finally allows us to write down the Einstein equation, and to show that the metric functions used in Parts II and III are solutions to this GR field equation.

5.1   Gaussian coordinates

We first need an efficient way to label points in space. This section presents the study of general coordinates in a curved space. For curved surfaces a position vector x^a has two independent components ($a = 1, 2$), and for higher dimensional cases, we concentrate particularly on the 4D spacetime $x^\mu (\mu = 0, 1, 2, 3)$.

Most of us start thinking of curved surfaces in terms of their embedding in three-dimensional (3D) Euclidean space, in which the points can be labeled by a Cartesian coordinate system (X, Y, Z). For illustration, we shall often use the example of a spherical surface, which geometers call a **2-sphere**.

A surface in 3D space is specified by a **constraint condition**:

$$f(X, Y, Z) = 0, \tag{5.1}$$

or equivalently in the form of a relation among the coordinates,

$$Z = g(X, Y). \tag{5.2}$$

For the case of a 2-sphere with radius R, such constraint conditions are

$$X^2 + Y^2 + Z^2 = R^2 \quad \text{or} \quad Z = \pm\sqrt{R^2 - X^2 - Y^2}. \tag{5.3}$$

This discussion of a curved surface being embedded in some larger space is an **extrinsic geometric description**—the physical space (here, the curved surface) is described using entities outside this space. What we are most interested in is an **intrinsic geometric description**—a characterization of the physical space without invoking any embedding. Namely, we are interested in the possibility of a description based solely on the measurement made by an inhabitant who never leaves the 2D surface. Gauss introduced a generalized parametrization, having coordinates (x^1, x^2) that are free to range over their respective domains without constraint:

$$X = X(x^1, x^2), \quad Y = Y(x^1, x^2), \quad Z = Z(x^1, x^2). \tag{5.4}$$

These generalized coordinates (x^1, x^2) are called **Gaussian coordinates**. We note that the employment of Gaussian coordinates avoids such constraint expressions as in (5.2). Using Gaussian coordinates (their number being the dimensionality of the space) the geometric description can be purely intrinsic.

For the case of a 2-sphere (see Fig. 5.1), we illustrate Gaussian coordinate choices by two systems:

1. **The polar coordinate system** We can set up a Gaussian coordinate system $(x^1, x^2) = (\theta, \phi)$ to label points in the 2D surface by first picking a point on the surface (the north pole) and marking the "polar axis" as the line pointing from the center of the spherical surface to this pole, and a longitudinal great circle as the "prime meridian." The polar angle coordinate θ is marked on the spherical surface against the polar axis and the azimuthal angle ϕ is measured against the prime meridian. The coordinate ranges are $0 \leq \theta \leq \pi$ and $0 \leq \phi \leq 2\pi$:

$$X = R \sin\theta \cos\phi, \quad Y = R \sin\theta \sin\phi, \quad Z = R \cos\theta. \tag{5.5}$$

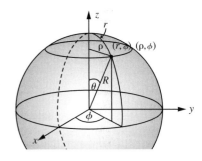

Fig. 5.1 Gaussian coordinates (θ, ϕ), (r, ϕ) and (ρ, ϕ) for the curved surface of a 2-sphere. The dashed line is the prime meridian. NB: the radial coordinate r is taken in reference to the "north pole" on the surface of the sphere, rather than with respect to the center of the sphere.

Closely related to this coordinate choice is to use the polar radial distance $r = R\theta$ instead of the polar angle θ. This (r, ϕ) is essentially the same system as (θ, ϕ), the polar radial coordinate r being just the polar angle θ scale up by the constant factor of R. It resembles the polar coordinate system on a flat plane that we are all familiar with.

2. **The cylindrical coordinate system** We can choose another set of Gaussian coordinates to label points in the 2D surface by having a different radial coordinate: instead of the polar r, we pick the function $\rho = R \sin\theta$ as our radial coordinate with a range of $0 \leq \rho \leq R$. Namely, we now have the system $(x^1, x^2) = (\rho, \phi)$. If the spherical surface is embedded in a 3D Euclidean space, ρ is interpreted as the perpendicular distance to the z-axis as shown[1] in Fig. 5.1:

$$X = \rho \cos\phi, \quad Y = \rho \sin\phi, \quad Z = \pm\sqrt{R^2 - \rho^2}. \tag{5.6}$$

[1] Perhaps the term "cylindrical coordinate" becomes more understandable if, instead of ρ, we use z directly $(x^1, x^2) = (z, \phi)$ with the relation $\rho^2 = R^2 - z^2$.

From now on we will no longer use extrinsic coordinates such as (X, Y, Z). By coordinates, we shall always mean Gaussian coordinates (x^1, x^2) as the way to label points on a 2D space.[2] Since one could have chosen any number of coordinate systems, and at the same time expecting geometric relations to be independent of such choices, a proper formulation of geometry must be such that it is invariant under general coordinate transformations.

[2] We must emphasize the point that the coordinates x^a do not form a vector space (elements of which can be added and multiplies by scalar, etc.). They are labels of points in the curved space and are devoid of any physical significance in their own right.

5.2 Metric tensor

The central idea of differential geometry was that an intrinsic description of space could be accomplished by distance measurements made within physical space. That is, one can imagine labeling various points of space (with a Gaussian coordinate system), then measure the distance among neighboring points. From the resultant "table of distance measurements," one obtains a description of this space. For a given coordinate system, these measurements are encoded in the metric function.

In fact, we have already used this Gaussian prescription in Chapter 3 when we first introduced the notion of a metric in terms of the basis vectors[3] $\{\mathbf{e}_a\}$ defined within the physical space, cf. (3.7):

[3] Here we rely on reader's intuitive conception of a set of "unit vectors" located at every point in the curved surface. We dispense with the proper differential geometry introduction of tangent vectors in a manifold and metrics as the scalar products of tangent vectors, etc. so that this basic introduction can reach GR physics in a timely fashion.

$$g_{ab} = \mathbf{e}_a \cdot \mathbf{e}_b. \tag{5.7}$$

The metric g_{ab} relates the (infinitesimal) length measurement ds to the chosen coordinates $\{dx^a\}$, as shown in (3.13):

$$ds^2 = g_{ab}dx^a dx^b \tag{5.8}$$

$$= g_{11}(dx^1)^2 + g_{22}(dx^2)^2 + 2g_{12}(dx^1 dx^2), \tag{5.9}$$

where in (5.8) Einstein's convention of summing over repeated indices has been employed. We have also used the symmetry property of the metric, $g_{12} = g_{21}$. The above relation can also be written as a matrix equation,

$$ds^2 = (dx^1 \ dx^2) \begin{pmatrix} g_{11} & g_{12} \\ g_{21} & g_{22} \end{pmatrix} \begin{pmatrix} dx^1 \\ dx^2 \end{pmatrix} \tag{5.10}$$

with the metric being represented by a 2×2 matrix.

Metric in polar coordinates To illustrate this for the case of a spherical surface (radius R), one first sets up the latitude/longitude system (i.e. a system of polar coordinates θ and ϕ) to label points on the globe, then measures the distances between neighboring points (Fig. 5.2). One finds that the latitudinal distances ds_ϕ (subtended by $d\phi$ between two points having the same latitude, $d\theta = 0$) become ever smaller as one approaches the poles $ds_\phi = R\sin\theta d\phi$, while the longitudinal distance interval ds_θ between two points at the same longitude ($d\phi = 0$) can be chosen to have the same value over the whole range of θ and ϕ. From such a table of distance measurements, one obtains a description of this spherical surface. Such distance measurements can be compactly expressed in terms of the metric tensor elements. Because we have $g_{\theta\phi} = \mathbf{e}_\theta \cdot \mathbf{e}_\phi = 0$ (i.e. an orthogonal coordinate system), the infinitesimal length between the origin $(0, 0)$ and a nearby point $(d\theta, d\phi)$ can be calculated:[4]

$$ds^2 = (ds_\theta)^2 + (ds_\phi)^2 \tag{5.11}$$

$$= R^2 d\theta^2 + R^2 \sin^2\theta d\phi^2. \tag{5.12}$$

A comparison of (5.12), written as

$$ds^2 = (d\theta \ d\phi) \begin{pmatrix} g_{\theta\theta} & g_{\theta\phi} \\ g_{\phi\theta} & g_{\phi\phi} \end{pmatrix} \begin{pmatrix} d\theta \\ d\phi \end{pmatrix}$$

[4]An infinitesimally small area on a curved surface can be thought of as a (infinitesimally small) flat plane. For such a flat surface ds can be calculated by the Pythagorean theorem as in (5.11).

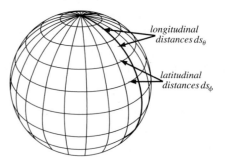

Fig. 5.2 Using distance measurements along longitudes and latitudes to specify the shape of the spherical surface.

longitudinal distances ds_θ

latitudinal distances ds_ϕ

and (5.10) leads to an expression for the metric matrix for this (θ, ϕ) coordinate system to be

$$\left[g^{(\theta,\phi)} \right] = R^2 \begin{pmatrix} 1 & 0 \\ 0 & \sin^2\theta \end{pmatrix}. \tag{5.13}$$

Metric in cylindrical coordinates   We to calculate the metric for a spherical surface with cylindrical Gaussian coordinates (ρ, ϕ) as shown in (5.6). From Fig. 5.1 we see that the cylindrical radial coordinate ρ is related to the polar angle by $\rho = R\sin\theta$, hence $d\rho = R\sqrt{1 - (\rho^2/R^2)}d\theta$. From this and from (5.12) we obtain

$$ds^2 = \frac{R^2 d\rho^2}{R^2 - \rho^2} + \rho^2 d\phi^2, \tag{5.14}$$

corresponding to the metric matrix

$$\left[g^{(\rho,\phi)} \right] = \begin{pmatrix} R^2/(R^2 - \rho^2) & 0 \\ 0 & \rho^2 \end{pmatrix}. \tag{5.15}$$

We are interested in the cylindrical coordinate system also because this offers, as we shall show in Section 5.3.2, a rather compact description of all curved surfaces with constant curvature.

We emphasize it again: the metric g_{ab} is an **intrinsic geometric quantity** because it can be determined without reference to any embedding—a 2D inhabitant on the curved surface can, once the Gaussian coordinates $\{x^a\}$ have been chosen, obtain g_{ab} by various length ds-measurements spanned by dx^a and dx^b. For the 2D case, we have

$$g_{11} = \frac{(ds_1)^2}{(dx^1)^2}, \quad g_{22} = \frac{(ds_2)^2}{(dx^2)^2}, \tag{5.16}$$

$$g_{12} = \frac{(ds_{12})^2 - (ds_1)^2 - (ds_2)^2}{2dx^1 dx^2}, \tag{5.17}$$

where ds_1 and ds_2 are the lengths measured along the 1- and 2-axes, i.e. the two length segments between the origin and the respective points with coordinates $(dx^1, 0)$ and $(0, dx^2)$, while ds_{12} is the length of a segment between the origin and the point with coordinates (dx^1, dx^2). Namely, given the choice of dx^1 and dx^2 (i.e. they are **defined** quantities), the metric elements can be deduced from various length measurements $\{ds_i\}$. From the law of cosines,[5] we see that (5.17) just says that g_{12} is the cosine of the angle subtended by the axes, cf. (5.7) and (5.10). Thus if we had an orthogonal coordinate system the metric matrix would be diagonal ($g_{12} = 0$). It should be emphasized that coordinates $\{x^a\}$ themselves do **not** measure distance. Only through the metric as in (5.8) are they connected to distance measurements.

[5]The flat space geometric relation is applicable because we are working in an infinitesimal region, which can always be described approximately as a flat space, cf. Section 5.2.2.

General coordinate transformation

For a description of the spherical surface one can use, for example, either polar Gaussian coordinates, $(x^1, x^2) = (\theta, \phi)$, or cylindrical Gaussian coordinates, $(x'^1, x'^2) = (\rho, \phi)$. The (intrinsic) geometric properties of the spherical surface are of course independent of the coordinate choice. For example, the length should be unchanged: $ds^2 = g_{ab}dx^a dx^b = g'_{ab}dx'^a dx'^b$. Here, these two coordinate systems are related by $\rho = R \sin\theta$, or $d\rho = R \cos\theta d\theta$. This change of coordinates $(\theta, \phi) \to (\rho, \phi)$ can be expressed as a transformation matrix acting on the coordinate differentials that leaves the infinitesimal length intervals ds^2 unchanged:

$$\begin{pmatrix} d\rho \\ d\phi \end{pmatrix} = \begin{pmatrix} R\cos\theta & 0 \\ 0 & 1 \end{pmatrix} \begin{pmatrix} d\theta \\ d\phi \end{pmatrix}. \tag{5.18}$$

Equation (5.18) can be compared to the similar coordinate transformations of rotation and boost that we discussed in previous chapters, for example, e.g. Eqs. (2.4) and (3.16). We note an important difference: the elements of the transformation matrices in (2.4) and (3.16) are not position-dependent—the same rotation angle θ and the same boost velocity v for every point in space. The matrix in (5.18) has position-dependent elements (here θ is a coordinate). That is, we make a different transformation (from the polar to the cylindrical system) at each position having a different polar angle coordinate θ. This is a key difference between the coordinate transformation in a flat space and those in curved space. To have coordinate changes being dependent on coordinates themselves means that the transformation is nonlinear. This will be discussed extensively in Part IV (see in particular Section 13.1.1) when we present the tensor calculus in GR (curved spacetime) vs. that in SR (flat spacetime).

5.2.1   Geodesic as the shortest curve

So far in our introductory discussion of the metric we have used known curved surfaces such as the sphere to show that its shape can be specified by the metric function. This justifies our subsequent application that a curved space can be represented, for a given coordinate system, by a metric tensor. Once we have the metric, other geometric quantities can then be computed. For example, angles can be determined:

$$\cos\theta = \frac{\mathbf{A} \cdot \mathbf{B}}{AB} = \frac{g_{ab} A^a B^b}{\sqrt{g_{cd} A^c A^d} \sqrt{g_{ef} B^e B^f}}. \tag{5.19}$$

Our main task here is to show that the curve having the extremum length, called the **geodesic line**, can be specified in terms of the metric function. Any curve can be represented by a set of coordinates $x^a(\lambda)$ depending on a single parameter λ, which has some definite range.[6] In a curved space the metric only determines the infinitesimal length as in (5.8):

$$ds = \sqrt{g_{ab}dx^a dx^b}. \tag{5.20}$$

[6]For a concrete example, one can think of the trajectory of a particle as a curve $\vec{x}(t)$ with the curve parameter being the time variable, $\lambda = t$.

For a finite length, we must perform the line-integration,

$$s = \int ds = \int \frac{ds}{d\lambda} d\lambda = \int \sqrt{\left(\frac{ds}{d\lambda}\right)^2} d\lambda = \int L \, d\lambda, \qquad (5.21)$$

where L is the "Lagrangian," which is a function of x and $\dot{x} \equiv dx/d\lambda$,

$$L = \sqrt{g_{ab} \frac{dx^a}{d\lambda} \frac{dx^b}{d\lambda}} = L(x, \dot{x}) \qquad (5.22)$$

where we have made the replacement $ds^2 = g_{ab} dx^a dx^b$ according to (5.8). To determine the shortest (i.e. the extremum) line in the curved space, we impose the extremization condition for variation of the path with end points fixed:

$$\delta s = \delta \int L(x, \dot{x}) d\lambda = 0, \qquad (5.23)$$

which can be translated, by calculus of variations, into a partial differential equation—the Euler–Lagrange equation:

$$\frac{d}{d\lambda} \frac{\partial L}{\partial \dot{x}^a} - \frac{\partial L}{\partial x^a} = 0. \qquad (5.24)$$

To aid the reader in recalling this connection, which is usually learnt in an intermediate mechanics course, we provide a brief derivation for the simple one-dimensional (1D) case. We set out to minimize the 1D integral s with respect to the variation, not of one variable or several variables as in the usual minimization problem, but of a whole function $x(\lambda)$ with initial and final values fixed:

$$\delta x(\lambda) \quad \text{with} \quad \delta x(\lambda_i) = \delta x(\lambda_f) = 0. \qquad (5.25)$$

The variation of the integrand being

$$\delta L(x, \dot{x}) = \frac{\partial L}{\partial x} \delta x + \frac{\partial L}{\partial \dot{x}} \delta \dot{x}, \qquad (5.26)$$

we have

$$0 = \delta s = \delta \int_{\lambda_i}^{\lambda_f} L(x, \dot{x}) d\lambda = \int_{\lambda_i}^{\lambda_f} \left(\frac{\partial L}{\partial x} \delta x + \frac{\partial L}{\partial \dot{x}} \frac{d}{d\lambda} \delta x \right) d\lambda$$

$$= \int_{\lambda_i}^{\lambda_f} \left(\frac{\partial L}{\partial x} - \frac{d}{d\lambda} \frac{\partial L}{\partial \dot{x}} \right) \delta x \, d\lambda. \qquad (5.27)$$

To reach the last expression we have performed an integration-by-parts on the second term, and used the condition in (5.25) to discard the end-point term $(\partial L/\partial \dot{x}) \delta x \big|_{\lambda_i}^{\lambda_f}$. Since δs must vanish for arbitrary variations $\delta x(\lambda)$, the expression in parentheses must vanish. This is the one-dimensional version of the Euler–Lagrange equation (5.24). In mechanics, the curve parameter is time $\lambda = t$ and Lagrangian L is simply the difference between the kinetic and potential energy. For the simplest case of $L = \frac{1}{2} m \dot{x}^2 - V(x)$, the Euler–Lagrange equation is just the familiar equation $F = ma$.

As a mathematical exercise, one can show that the **same** Euler–Lagrange equation (5.24) follows from, instead of (5.22), a Lagrangian of the form:

$$L(x, \dot{x}) = g_{ab}\dot{x}^a\dot{x}^b, \tag{5.28}$$

which without the square-root is much easier to work with than (5.22). With L in this form, the derivatives become

$$\frac{\partial L}{\partial \dot{x}^a} = 2g_{ab}\dot{x}^b, \qquad \frac{\partial L}{\partial x^c} = \frac{\partial g_{ab}}{\partial x^c}\dot{x}^a\dot{x}^b, \tag{5.29}$$

where we have used the fact that the metric function g_{ab} depends on x, but not $\dot{x}$. Substituting these relations back into Eq. (5.24), we obtain the **geodesic equation**,

$$\frac{d}{d\sigma}g_{ab}\dot{x}^b - \frac{1}{2}\frac{\partial g_{cd}}{\partial x^a}\dot{x}^c\dot{x}^d = 0, \tag{5.30}$$

which determines the trajectory of the "shortest curve." One can easily use this equation to check the geodesic lines in the simple surfaces of a flat plane and a spherical surface (Problem 5.4).

5.2.2   Local Euclidean coordinates

We are familiar with the idea that at any point on a curved surface there exists a plane, tangent to the curved surface at that point. The plane in its Cartesian coordinates has a metric δ_{ab}. But this is true only at this point (call it the origin). That is, $\bar{g}_{ab}(0) = \delta_{ab}$. If we are interested in the metric function, we have to be more careful. A more complete statement is given by the flatness theorem.

The flatness theorem   In a curved space with a general coordinate system x^a and a metric value g_{ab} at a given point P, we can always find a coordinate transformation $x^a \to \bar{x}^a$ and $g_{ab} \to \bar{g}_{ab}$ so that the metric is flat at this point: $\bar{g}_{ab} = \delta_{ab}$ and $\partial\bar{g}_{ab}/\partial\bar{x}^c = 0$,

$$\bar{g}_{ab}(\bar{x}) = \delta_{ab} + \gamma_{abcd}(0)\bar{x}_c\bar{x}_d + \cdots. \tag{5.31}$$

That is, the metric in the neighborhood of the origin will differ from δ_{ab} by the **second order derivatives**. This is simply a Taylor series expansion of the metric at the origin—there is the constant $\bar{g}_{ab}(0)$ plus a higher order correction $\gamma_{abcd}(0)\bar{x}_c\bar{x}_d$ being simply the second derivatives. The nontrivial content of (5.31) is the absence of the first derivative. Such a coordinate system $\{\bar{x}^a\}$ is called the **local Euclidean frame** (LEF). That $\bar{g}_{ab}(0) = \delta_{ab}$ should be less surprising: it is not difficult to see that for a metric value at one point one can always find an orthogonal system so that $\bar{g}_{ab}(0) = 0$ for $a \neq b$ and the diagonal elements can be scaled to unity so that the new coordinate bases all have unit length with the metric being an identity matrix. If the original metric has negative determinant, this reduces to the pseudo-Euclidean metric $\eta_{ab} = \text{diag}(1, -1)$ (cf. Problem 5.5).

The theorem can be generalized to n-dimensional space, in particular 4D spacetime. (The proof will be provided in Box 13.1 in Section 13.1.3.) It informs us that the general spacetime metric $g_{ab}(x)$ is characterized at a point

(P) not so much by the value $g_{ab}|_P$ since that can always be chosen to be flat, $\bar{g}_{ab}|_P = \delta_{ab}$, nor by its first derivative which can always be chosen to vanish, $\partial \bar{g}_{ab}/\partial x^c|_P = 0$, but by the second derivatives of the metric, $\partial^2 g_{ab}/\partial x^c \partial x^d$, which are related to the curvature to be discussed in Section 5.3.

Box 5.1 More illustrative calculations of metric tensors for simple surfaces

Here are further examples of metric tensors for two-dimensional surfaces, calculated by using the fact the any surface in the small can be approximated by a plane having Cartesian coordinates. We also take this occasion to discuss the possibility of using the metric tensor to determine whether a surface is curved or not. For a flat surface, we can find a set of coordinates so that the metric tensor is position-independent; this is not the case in a curved surface as $g_{ab} = \mathbf{e}_a \cdot \mathbf{e}_b$ in a curved space must change from point to point.

1. **A plane surface with Cartesian coordinates** For the coordinates $(x^1, x^2) = (x, y)$, we have the infinitesimal length $ds^2 = dx^2 + dy^2$, Fig. 5.3(a). Comparing this to the general expression in (5.10), we see that the metric must be

$$g_{ab} = \begin{pmatrix} 1 & 0 \\ 0 & 1 \end{pmatrix}, \qquad (5.32)$$

which is of course position-independent. This is possible only if the space is not curved.

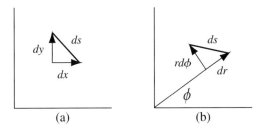

Fig. 5.3 Two coordinate systems in a flat plane: (a) Cartesian coordinates, and (b) polar coordinates.

(a)   (b)

2. **A plane surface with polar coordinates** For the coordinates $(x^1, x^2) = (r, \phi)$, we have the infinitesimal length $ds^2 = (dr)^2 + (rd\phi)^2$, Fig. 5.3(b), thus according to (5.10), a metric

$$g_{ab} = \begin{pmatrix} 1 & 0 \\ 0 & r^2 \end{pmatrix}, \qquad (5.33)$$

which is position dependent! But we can find a coordinate transformation $(x^1, x^2) \rightarrow (x'^1, x'^2)$ so that the metric in the new coordinates is position independent, $g'_{ab} = \delta_{ab}$. Of course, the new coordinates are

(*cont.*)

Box 5.1 (*Continued*)

just the Cartesian coordinates $\left(x'^{1}, x'^{2}\right) = (x, y)$:

$$x = r \cos \phi, \quad y = r \sin \phi. \tag{5.34}$$

3. **A cylindrical surface with cylinder coordinates** Let R be the radius of the cylinder see Fig. 5.4(a). The infinitesimal length for cylinder coordinates $\left(x^{1}, x^{2}\right) = (z, R\phi)$ is then $ds^2 = dz^2 + R^2 d\phi^2 = \left(dx^1\right)^2 + \left(dx^2\right)^2$. This shows that we have a constant metric $g_{ab} = \delta_{ab}$. Thus locally this is a flat surface, even though globally and topologically it is different from a plane surface. For example, a straight line can close onto itself in such a cylinder surface, see Fig. 5.4(b).

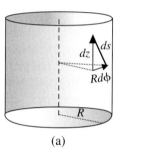 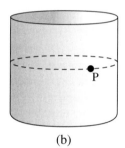

(a)   (b)

4. **A spherical surface with spherical coordinates** For a spherical surface with radius R, we have already calculated the metric: for the polar coordinates (θ, ϕ) in (5.13) and for cylindrical coordinates (ρ, ϕ) in (5.15). They are all position dependent. Furthermore, such position dependence cannot be transformed away by going to any other coordinate system.

5.3 Curvature

From the above discussion, we see that the metric value cannot represent the essence of a curved space because it is coordinate dependent and can always, at a given point, be transformed to a flat space metric. However, this replacement can only be done locally—in an infinitesimally small region. A general metric equates to the flat space metric up to corrections given by the second derivative of the metric function, as shown in (5.31). This suggests that it is the second derivatives that really tell us how curved a curved space is.

Consider the three surfaces in Fig. 5.5. We can usually tell whether a surface is curved by an examination of its relation to the embedding space. Thus the sphere is curved in a fundamental way. By contrast, the curvature of the

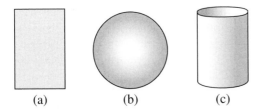

Fig. 5.5 Three kinds of surfaces: (a) flat plane, (b) sphere, and (c) cylindrical surface.

(a)   (b)   (c)

cylinder is less fundamental as we can cut and unroll it into a plane without internal deformation—we say it has zero intrinsic curvature (although such a cylinder has global curvature), cf. Box 5.1, item 3. We are interested in finding a simple **intrinsic** method to determine whether a space is warped or not.

5.3.1 Gaussian curvature

For a flat space we can find a coordinate system such that the metric is position independent, while such coordinates do not exist in the case of a curved space. This way to determine whether a surface is curved or not (for illustrative examples see Box 5.1) is rather unsatisfactory. How can we be sure to have exhausted all possible coordinate systems, none of which have a constant metric? Is there a better way?

Theorema Egregium (a very beautiful theorem)

This is the title of the paper in which Gauss presented his answer to the above questions: he showed that it was possible to define a unique invariant second derivative of the metric tensor ($\partial^2 g$) called the **curvature** K, such that, independent of the coordinate choice, $K = 0$ for a flat and $K \neq 0$ for curved surfaces.

With no loss of generality we shall quote Gauss's result for a diagonalized metric $g_{ab} = \mathrm{diag}(g_{11}, g_{22})$:

$$
K = \frac{1}{2g_{11}g_{22}} \left\{ -\frac{\partial^2 g_{11}}{(\partial x^2)^2} - \frac{\partial^2 g_{22}}{(\partial x^1)^2} \right.
$$
$$
\left. + \frac{1}{2g_{11}} \left[\frac{\partial g_{11}}{\partial x^1} \frac{\partial g_{22}}{\partial x^1} + \left(\frac{\partial g_{11}}{\partial x^2} \right)^2 \right] + \frac{1}{2g_{22}} \left[\frac{\partial g_{11}}{\partial x^2} \frac{\partial g_{22}}{\partial x^2} + \left(\frac{\partial g_{22}}{\partial x^1} \right)^2 \right] \right\}.
$$

$$(5.35)$$

Since this curvature is expressed entirely in terms of the metric and its derivatives, it is also an intrinsic geometric object. To describe such curvature of a 2D surface, it only takes one number—in contrast to the embedded viewpoint, which may lead one to expect that it would take two numbers to characterize the curvature of a 2D space. In fact, there is no curvature for a 1D space; an inhabitant on a line cannot detect any intrinsic curvature. We will not present the derivation of (5.35) since it is contained in the more general result to be discussed in Section 13.3.1 (also Problem 13.11). But, let us check that it indeed has the property as a simple indicator as whether a surface is warped or not.

- For a position-independent metric we automatically have $K = 0$ because the derivatives of the metric vanish. Thus for a plane surface with Cartesian coordinates $g_{ab} = \delta_{ab}$ and a cylindrical surface with cylindrical coordinates (Box 5.1, item 3) we can immediately conclude that they are intrinsically flat surfaces.
- For a plane surface with polar coordinates $(x^1, x^2) = (r, \phi)$, we have a position-dependent metric as shown in (5.33): $g_{11} = 1$ and $g_{22} = r^2 = (x^1)^2$ with $(\partial g_{22}/\partial x^1) = 2x^1$ and $(\partial^2 g_{22}/\partial (x^1)^2) = 2$. However, the curvature (5.35) vanishes:

$$K = \frac{1}{2(x^1)^2} \left\{ -2 + \frac{1}{2(x^1)^2} \left[4(x^1)^2 \right] \right\} = 0 \qquad (5.36)$$

indicating that it is a flat space, even though the corresponding metric (5.33) is position dependent.

- For a spherical surface with polar coordinates $(x^1, x^2) = (\theta, \phi)$, we have Eq. (5.13) having $g_{11} = R^2$ and $g_{22} = R^2 \sin^2 x^1$ with $\partial g_{22}/\partial x^1 = R^2 \sin(2x^1)$ and $\partial^2 g_{22}/(\partial x^1)^2 = 2R^2 \cos(2x^1)$. This leads to

$$K = \frac{1}{2R^2 \sin^2 x^1} \left\{ 2\sin^2 x^1 - 2\cos^2 x^1 + \frac{4\sin^2 x^1 \cos^2 x^1}{2\sin^2 x^1} \right\} = \frac{1}{R^2}. \qquad (5.37)$$

One can easily check that this result holds as well for the cylindrical coordinates with a metric given in (5.15), indicating that $K = R^{-2}$ is the curvature for a spherical surface, independent of coordinate choices.

5.3.2 Curvature measures the deviation from Euclidean relations

On a flat surface, the familiar Euclidean geometrical relations hold. For example, the circumference of a circle with radius r is $S = 2\pi r$, and the angular excess for any polygon equals zero: $\epsilon = 0$. The angular excess ϵ is defined to be the sum of the interior angles in excess of their flat space Euclidean value. For example, in the case of a triangle with angles α, β, and γ, the **angular excess** is defined as

$$\epsilon \equiv \alpha + \beta + \gamma - \pi. \qquad (5.38)$$

The curvature measures how curved a surface is because it is directly proportional to the violation of Euclidean relations. In Fig. 5.6 we show two pictures of circles with radius r drawn on surfaces with nonvanishing curvature.[7] It can be shown (Problem 5.9) that the circular circumference S differs from the flat surface value of $2\pi r$ by an amount controlled by the Gaussian curvature, K:

$$\lim_{r \to 0} \frac{2\pi r - S}{r^3} = \frac{\pi}{3} K. \qquad (5.39)$$

For a positively curved surface the circumference is smaller than, for a negatively curved surface larger than, that on a flat space.

[7] Surfaces with negative curvature will be discussed in the next subsection.

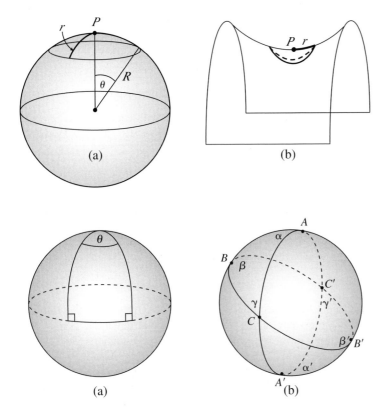

Fig. 5.6 A circle with radius r (but cylindrical radial coordinate ρ) centered on point P, (a) on a spherical surface with curvature $K = 1/R^2$, (b) on the middle portion of a saddle shaped surface, which has negative curvature $K = -1/R^2$.

Fig. 5.7 (a) A triangle with two 90° interior angles on a spherical surface. (b) Three great circles $ACA'C'$, $BCB'C'$ and $ABA'B'$ intersect pair-wise at points $(A$ and $A')$, $(B$ and $B')$, and $(C$ and $C')$. The two identical triangles are ABC with angles α, β, γ on the front hemisphere and $A'B'C'$ with angles α', β', γ' on the back hemisphere.

Angular excess and curvature

We shall also show that the angular excess ϵ is directly proportional to the curvature K with the area of the polygon σ being the proportional constant:

$$\epsilon = K\sigma. \tag{5.40}$$

This relation will be used in Chapter 13 to define the general curvature, the Riemann curvature tensor, for a space of arbitrary dimensions. A contracted form of this Riemann tensor (called the Einstein tensor) enters directly in the GR field equation (the Einstein equation).

Here we shall explicitly prove (5.40) for the case of a spherical surface ($K = 1/R^2$). Let us first illustrate the validity of this relation for a particularly simple example of a triangle with two 90° interior angles and the third one being θ, as shown in Fig. 5.7(a). Clearly, according to the definition of angular excess given in (5.38) we have $\epsilon = \theta$. The triangular area σ is exactly one-half of a **lune** with θ as its vertex angle. A lune, the area in between two great semicircles with an angle θ subtended between them, has an area value being the fraction $\theta/(2\pi)$ of the spherical surface:

$$\sigma_\theta = 2\theta R^2 \qquad \text{(area of a lune with angle } \theta\text{)}. \tag{5.41}$$

Thus the area of this triangle is $\sigma = \frac{1}{2}\sigma_\theta = \theta R^2$ so that $\epsilon = \theta$ is just the relation (5.40) with $K = 1/R^2$.

The proof of (5.40) for a general triangle goes as follows. Draw three great circles (ABA'B'), (ACA'C'), and (BCB'C') as in Fig. 5.7(b). Now consider the three lunes marked out by these geodesic lines, and record their respective areas according to (5.41):

$$\sigma_\alpha = 2\alpha R^2 \qquad \text{(lune AA' with angle } \alpha),$$

$$\sigma_\beta = 2\beta R^2 \qquad \text{(lune BB' with angle } \beta),$$

$$\sigma_\gamma = 2\gamma R^2 \qquad \text{(lune CC' with angle } \gamma).$$

Their sum is

$$\sigma_\alpha + \sigma_\beta + \sigma_\gamma = 2(\alpha + \beta + \gamma)R^2. \tag{5.42}$$

However, an inspection of the diagram in Fig. 5.7(b) shows that the sum of these three lunes covers the entire front hemisphere and, in addition, the triangular areas of (ABC) and (A'B'C'). Thus, another expression for the area sum is

$$\sigma_\alpha + \sigma_\beta + \sigma_\gamma = 2\pi R^2 + \sigma_{\text{ABC}} + \sigma'_{\text{ABC}}. \tag{5.43}$$

For triangles on the same spherical surface, congruity of angles implies congruity of triangles themselves. Hence the angular equalities $\alpha = \alpha'$, $\beta = \beta'$, and $\gamma = \gamma'$ imply the area equality $\sigma_{\text{ABC}} = \sigma'_{\text{ABC}}$. Equations (5.42) and (5.43) then lead to

$$\alpha + \beta + \gamma = \pi + \sigma_{\text{ABC}}/R^2. \tag{5.44}$$

Namely,

$$\alpha + \beta + \gamma - \pi \equiv \epsilon = K\sigma_{\text{ABC}}, \tag{5.45}$$

which is the claimed result (5.40) with $K = 1/R^2$.

Having demonstrated the validity of (5.40) for an arbitrary spherical triangle, it is not difficult to prove its validity for any spherical polygon (Problem 5.10). Furthermore, because a sufficiently small region on any curved 2D surface can be approximated by a spherical surface, Eq. (5.40) must hold for any infinitesimal polygon on any warped 2D space. In Section 13.3, this non-Euclidean relation will be used to generalize the notion of curvature (K) of a 2D space to that of an n-dimensional curved space.

5.3.3 Spaces with constant curvature

In Chapter 9 we shall start our discussion of cosmology. Here spaces (and spacetime) of high symmetry play a very important role. Not surprisingly, this corresponds to a 3D space of constant curvature. Since it is difficult to visualize a warped 3D space, we shall first discuss the 2D surface with constant curvature. The generalization of this result to 3D space will then be presented afterward.

2D surfaces with constant curvature

While the Gaussian curvature in (5.35) is generally a position-dependent function, we have seen in (5.37) that the sphere has a constant $K = 1/R^2$.

Obviously, a flat plane is a surface of constant curvature, $K = 0$. In fact, there are three surfaces having constant curvatures:

$$K = \frac{k}{R^2},\tag{5.46}$$

with the **curvature signature** $k = +1, 0$, and -1. That is, besides the two familiar surfaces of the 2-sphere and flat plane, there is another surface, called a **2-pseudosphere**, with a negative curvature $K = -1/R^2$.

What should be the metric for a pseudosphere so that (5.35) can yield a negative curvature? An inspection of the calculation in (5.37) shows that, in order to obtain a result of $-1/R^2$, we would want the first term in the curly parentheses to change sign (since the next two terms cancel each other). This first term originates from $\cos(2\theta) = -\sin^2\theta + \cos^2\theta$ in the second derivative of g_{22}. This suggests that, to go from the positive curvature for a sphere to a negative curvature for a pseudosphere, the metric term $g_{22} = R^2\sin^2\theta$ should be changed to $g_{22} = R^2\sinh^2\psi$ so that the second derivative of the new g_{22} would have a factor of $\cosh(2\psi) = +\sinh^2\psi + \cosh^2\psi$. Making such a change in Eq. (5.13), we have the metric for the pseudosphere:

$$g_{ab}^{(\psi,\phi)} = R^2 \begin{pmatrix} 1 & 0 \\ 0 & \sinh^2\psi \end{pmatrix}.\tag{5.47}$$

which, compared to (5.37), leads to the curvature

$$K = \frac{1}{2R^2\sinh^2\psi}\left\{-2\sinh^2\psi - 2\cosh^2\psi + \frac{4\sinh^2\psi\cosh^2\psi}{2\sinh^2\psi}\right\} = \frac{-1}{R^2}.$$

Such a negative curvature space is also referred to as a **hyperbolic space**. While for a spherical $k = +1$ surface, the θ coordinate has the interpretation as the polar angle, there is no such simple interpretation of ψ for the hyperbolic surface with negative curvature. However in either case such coordinates are seen to be the dimensionless radial coordinate r/R; we shall collectively name it as the χ coordinate. In this way the infinitesimal separation for the three surfaces with constant curvature in the polar coordinates as shown in (5.13), (5.33), and (5.47), can be expressed as

$$ds_{2D,\chi}^{2(k)} = \begin{cases} R^2(d\chi^2 + \sin^2\chi\, d\phi^2) & \text{for } k = +1, \\ R^2(d\chi^2 + \chi^2 d\phi^2) & \text{for } k = 0, \\ R^2(d\chi^2 + \sinh^2\chi\, d\phi^2) & \text{for } k = -1. \end{cases}\tag{5.48}$$

Unlike the plane and sphere cases, there is no simple way to visualize this whole pseudosphere because the natural embedding is not into a flat 3D space with Euclidean metric of $g_{ij} = \text{diag}(1, 1, 1)$ but into a flat 3D space with a pseudo-Euclidean metric of $g_{ij} = \text{diag}(-1, 1, 1)$. Compared to the embedding of a sphere in a 3D Euclidean space as (5.3), it can be worked out (see Problem 5.7) that the embedding of the 2D $k = -1$ surface in such a 3D pseudo-Euclidean space with coordinate (W, X, Y) corresponds to the condition

$$-W^2 + X^2 + Y^2 = -R^2.\tag{5.49}$$

While we cannot draw the whole pseudosphere in an ordinary 3D Euclidean space, the central portion of a saddle surface does represent a negative curvature surface, see Fig. 5.6(b). We note that the expressions in (5.48) can be written in compact notation as

$$ds^{2(k)}_{2D,\chi} = R^2 \left[d\chi^2 + k^{-1} \left(\sin^2 \sqrt{k}\chi \right) d\phi^2 \right]. \tag{5.50}$$

In the cylindrical coordinates $(x^1, x^2) = (\xi, \phi)$, with the dimensionless radial coordinate $\xi = \rho/R$, the metric for these three surfaces that generalizes (5.14) can be written in a compact form as

$$ds^{2(k)}_{2D,\xi} = R^2 \left(\frac{d\xi^2}{1 - k\xi^2} + \xi^2 d\phi^2 \right). \tag{5.51}$$

We can easily check that for $k = 0$ the metric (5.51) yields $ds^2 = d\rho^2 + \rho^2 d\phi^2$, which is the infinitesimal separation for a flat surface with the familiar polar coordinates, cf. (5.33). For the positive curvature $k = +1$, the metric (5.51) is just the metric (5.15) of a spherical surface in the cylindrical coordinate system. For the pseudosphere, this metric can be easily obtained from (5.48) when we identify the cylindrical radial coordinate by $\rho = R \sinh \chi$.

3D spaces with constant curvature

The 2D spaces with constant curvature have metrics of (5.51) in cylindrical coordinates and metrics of (5.48) in polar coordinates. We now make a heuristic argument for their generalization to 3D spaces with constant curvature. Compared to the 2D coordinates[8] (r, ϕ) or (ρ, ϕ), the 3D spherical coordinate system (r, θ, ϕ) or (ρ, θ, ϕ) involves an additional (polar) angle coordinate. Specifically for the $k = 0$ cases, we have the polar coordinate for a flat 2D surface, and the spherical coordinates for a Euclidean 3D space. Their respective metric relations are well known:

$$ds^{2(k=0)}_{2D} = dr^2 + r^2 d\phi^2$$

and

$$ds^{2(k=0)}_{3D} = dr^2 + r^2 d\Omega^2. \tag{5.52}$$

Namely, to go from a 2D flat space to a 3D flat space, we just replace the angular factor $d\phi^2$ in the 2D interval by the solid angle factor $d\Omega^2 = (d\theta^2 + \sin^2 \theta d\phi^2)$. Here we suggest that, even for the $k \neq 0$ spaces, we can obtain the 3D expressions in the same manner. This way we can "deduce" from (5.48) the metric for the $(k = 0, \pm 1)$ 3D spaces in the spherical polar coordinates (χ, θ, ϕ)

$$ds^{2(k)}_{3D,\chi} = \begin{cases} R^2(d\chi^2 + \sin^2 \chi d\Omega^2) & \text{for } k = +1, \\ R^2(d\chi^2 + \chi^2 d\Omega^2) & \text{for } k = 0, \\ R^2(d\chi^2 + \sinh^2 \chi d\Omega^2) & \text{for } k = -1, \end{cases} \tag{5.53}$$

or,

$$ds^{2(k)}_{3D,\chi} = R^2 \left[d\chi^2 + k^{-1} \left(\sin^2 \sqrt{k}\chi \right) d\Omega^2 \right]. \tag{5.54}$$

[8]For this discussion, we prefer to use the 2D (r, ϕ) polar coordinates instead of (θ, ϕ) polar coordinates as the 2D coordinate $\theta = r/R$ has a different interpretation as the 3D spherical coordinate θ.

Similarly, in terms of the cylindrical coordinates (ξ, θ, ϕ), we have, from (5.51), the metric for 3D with constant curvature,

$$ds^{2\,(k)}_{\text{3D},\xi} = R^2 \left(\frac{d\xi^2}{1 - k\xi^2} + \xi^2 d\Omega^2 \right). \tag{5.55}$$

Equations (5.53) and (5.55) reduce to the respective 2D metric expressions (5.48) and (5.51) when we take a 2D slice of the 3D space with either $d\theta = 0$ or $d\phi = 0$. This means that all the 2D subspaces are appropriately curved. A rigorous derivation of these results would involve the mathematics of symmetric spaces, Killing vectors, and isometry. However, our heuristic deduction will be buttressed in Section 14.4.1 by a careful study of the properties of the curvature tensor for a 3D space with the help of the Einstein equation.

The metrics in (5.53) and (5.55) with $k = +1$ describe a 3-sphere, $k = -1$ a 3-pseudosphere, and the overall distance scale R's are identified with the respective radii of these spheres. See Problem 5.7 for embedding of such 3D spaces in a 4D (pseudo-) Euclidean space—as generalizations of (5.3) and (5.49):

$$\pm W^2 + X^2 + Y^2 + Z^2 = \pm R^2 \tag{5.56}$$

with the plus sign for the space of a 3-sphere, and the negative sign for the 3-pseudosphere. In Part III, we shall study cosmology based on the assumption of a 3D cosmic space having constant curvatures. The flat universe has a cosmic space with $k = 0$ geometry; the closed universe has a positively curved $k = +1$ space, and the open universe has a negatively curved $k = -1$ space.

From Gauss to Riemann

From Gauss's theory of curved surfaces, his student Bernhard Riemann showed that this algebraic approach to geometry can be extended to higher dimensional curved spaces when the spatial index ranges over $1, 2, \ldots, n$ for an n-dimensional space. The virtue of this algebraic method is to make the study of higher dimensional non-Euclidean geometry, which by and large is impossible to visualize, more accessible. A few years before Riemann's presentation of his result in 1854, Bólyai and Lobachevsky already introduced non-Euclidean geometry—geometry without the parallelism axiom. However, their work remained unappreciated until Riemann showed that his larger framework encompassed the Bólyai and Lobachevsky results. This just shows the power of the Riemannian approach.

Our interest in higher dimensional space will mostly be 4D spacetime with the index $\mu = 0, 1, 2, 3$. For such a $(1+3)$-dimensional manifold, the flat metric corresponds to $g_{\mu\nu} = \eta_{\mu\nu} = \text{diag}(-1, 1, 1, 1)$. We should also mention that the extension to higher dimensions is nontrivial (to be presented in Chapter 13), because, beyond two dimensions, the curvature of the space can no longer be described by a single function.

Review questions

1. What does it mean to have an "intrinsic geometric description" (vs. "extrinsic description")?

2. Provide a description of the intrinsic geometric operations to fix the metric elements.

3. How does the geodesic equation represent the curve $x^\alpha (\lambda)$ having an extremum length? (Just say in words the relation of the geodesic equation to the condition of extremum length.)

4. In what sense does the metric function describe all the intrinsic geometric properties of a space? Namely, is the metric an intrinsic quantity? What is the relation between other intrinsic geometric quantities and the metric?

5. Curved surface necessarily has a position-dependent metric. But it is not a sufficient condition. Illustrate this point with an example.

6. What is the fundamental difference between the coordinate transformations in a curved space and those in flat space (e.g. Lorentz transformations in flat Minkowski space)?

7. What is the "flatness theorem?"

8. In what sense is the Gaussian curvature K a good criterion for finding out whether a surface is curved or not?

9. What are the three surfaces of constant curvature?

10. In a 4D space write out the respective embedding equation of a 3-sphere and a 3-pseudosphere. What embedding space did you use for the 3-pseudosphere?

11. The curvature measures how curved a space is because it controls the amount of deviation from Euclidean relations. Give two examples of such non-Euclidean relations, showing the deviation from flatness being proportional to the curvature.

12. What is angular excess? How is it related to the Gaussian curvature? Give a simple example of a polygon on a spherical surface that clearly illustrates this relation.

Problems

5.1 **Metric for the spherical surface in cylindrical coordinates** Show that the metric for the spherical surface with "cylindrical coordinates" of (5.15) follows from (5.6) and the Pythagorean relation $ds^2 = dX^2 + dY^2 + dZ^2$ in the embedding space.

5.2 **Basis vectors on a spherical surface** Equations (5.7) and (5.8) are, respectively, two equivalent and closely related definitions of the metric. In the text we use (5.8) to deduce the metric matrix for a spherical surface. What are the corresponding basis vectors for this coordinate system? Check that they yield the same matrix (5.13) through the definition of (5.7).

5.3 **Coordinate transformation of the metric** Use the metric transformation condition given in sidenote 11 of Chapter 3, $[R^{-1}][g][R] = [g']$, to show explicitly that the transformation (5.18) relates the metric tensors of the two coordinate systems as given in (5.13) and (5.15).

5.4 **Geodesics on simple surfaces** Use the geodesic equation. (5.30) to confirm the familiar results that the geodesic is (a) a straight line on a flat plane and (b) a great circle on a spherical surface.

5.5 **Locally flat metric** Explicitly display a transformation that turns a general 2D metric at a point (5.8) to the Euclidean metric $\bar{g}_{ab} = \delta_{ab}$, the Kronecker delta; that is, $ds^2 = \left(d\bar{x}^1 \right)^2 + \left(d\bar{x}^2 \right)^2$. Similarly, provide the same calculation for the pseudo-Euclidean metric $\bar{g}_{ab} = \eta_{ab}$, where $\eta_{ab} = \text{diag} (1, -1)$. (There are an infinite number of such transformations, just display one.)

5.6 **Checking the Gaussian curvature formula** Check the connection of the metric and the curvature for the three surfaces with constant curvature $K = k/R^2$ by plugging (5.51) and (5.50) into the expression for the curvature in (5.35).

5.7 **3-sphere and 3-pseudosphere**

(a) **3D flat space** Express Cartesian coordinates (x, y, z) in terms of polar coordinates (r, θ, ϕ) as in (5.5). Show that the solid angle factor in the polar coordinate expression for the infinitesimal separation satisfies the relation $r^2 d\Omega^2 = dx^2 + dy^2 + dz^2 - dr^2$.

(b) **3-sphere** Consider the possibility of embedding a 3-sphere in a 4D Euclidian space with Cartesian coordinates (W, X, Y, Z). From (5.53) for the 3-sphere $(k = +1)$ in the coordinate system (r, θ, ϕ) with $\chi = r/R$ and an expression relating Cartesian coordinates to solid angle differential $d\Omega^2$ as suggested by the equation shown in part (a), find the differential for the new coordinate dW. This result should suggest that the 4D embedding space is indeed Euclidian with $W = R \cos(r/R)$. From this, display the entire expression for (W, X, Y, Z) in terms of polar coordinates (r, θ, ϕ). Furthermore, verify that the 3-sphere is a 3D subspace satisfying the constraint

$$W^2 + X^2 + Y^2 + Z^2 = R^2.$$

(c) **3-pseudosphere** Now the fourth coordinate should be $W = R \cosh(r/R)$. Show that the 4D embedding space is pseudo-Euclidian with a Minkowski metric $\eta_{\mu\nu} = \text{diag}(-1, 1, 1, 1)$ and the 3-pseudosphere is a 3D subspace satisfying the condition

$$-W^2 + X^2 + Y^2 + Z^2 = -R^2.$$

5.8 **Volume of higher dimensional space** The general expression for the differential volume is the product of coordinate differentials *and* the square root of the metric determinant:

$$dV = \sqrt{\det g} \prod_i dx^i. \tag{5.57}$$

(a) Verify that for 3D flat space this reduces to the familiar expression for the volume element

$dV = dxdydz$ in Cartesian coordinates, and $dV = r^2 \sin\theta \, dr d\theta d\phi$ in spherical coordinates.

(b) Work out the volume element for a 3-sphere, integrate it to obtain the result $V_3^{(+1)} = 2\pi^2 R^3$. Thus a 3-sphere, much like the familiar spherical surface, is a 3D space having no boundary yet with a finite volume. When applied to cosmology this is a "closed universe" with R being referred to as "the radius of the universe".

5.9 **Non-Euclidean relation between the radius and circumference of a circle** On a curved surface the circumference S of a circle is no longer related to the its radius r by $S = 2\pi r$. The deviation from this flat space relation is proportional to the curvature, as shown in (5.39). Derive this relation for the simple cases of (a) a sphere and (b) a pseudosphere.

5.10 **Angular excess and polygon area** Generalize the proof of (5.40) to the case of an arbitrary polygon. Namely, one still has $\epsilon = \sigma K$ with ϵ being the angular excess over the Euclidean sum of the polygon and σ being the area of the polygon.

5.11 **Local Euclidean coordinates** In this chapter we have obtained the metric g_{ab} for a coordinate system $\{x^a\}$ using the observation that any curved surface is locally flat where, in an infinitesimal region, the Pythagorean theorem holds. In this problem you are asked to formalize this introduction of the local Euclidean coordinates $\{\xi^a\}$ which explicitly obey the Pythagorean relation $ds^2 = \left(d\xi^1\right)^2 + \left(d\xi^2\right)^2$. You are asked to express the metric elements g_{ab} in terms of the derivatives $\partial\xi/\partial x$. From your result, formally derive the metric matrix (5.13) for the spherical polar coordinates (θ, ϕ).

6

GR as a geometric theory of gravity – I

- We first present a **geometric** description of the equivalence-principle physics of gravitational time dilation. In this geometric theory, the metric $g_{\mu\nu}(x)$ plays the role of the relativistic gravitational potential.
- Einstein proposed curved spacetime as the gravitational field. The geodesic equation in spacetime is the GR equation of motion, which is checked to have the correct Newtonian limit.
- At every spacetime point, one can construct a free-fall frame in which gravity is transformed away. However, in a finite-sized region, one can detect the residual tidal force which are second derivatives of the gravitational potential. It is the curvature of spacetime.
- The GR field equation directly relates the mass/energy distribution to spacetime's curvature. Its solution is the metric function $g_{\mu\nu}(x)$, determining the geometry of spacetime.

In Chapter 4 we have deduced several pieces of physics from the empirical principle of equivalence of gravity and inertia. In Chapter 5, elements of the mathematical description of a curved space have been presented. In this chapter, we show how some of equivalence-principle physics can be interpreted as the geometric effects of curved spacetime. Such a study motivated Einstein to propose his general theory of relativity, which is a geometric theory of gravitation, with the equation of motion being the geodesic equation, and the field equation in the form of the curvature being directly given by the mass/energy source fields.

6.1 Geometry as gravity

By a geometric theory, or a geometric description, of any physical phenomenon we mean that the physical measurement results can be attributed directly to the underlying geometry of space and time. This is illustrated by the example we discussed in Section 5.2 in connection with a spherical surface as shown in Fig. 5.2. The length measurements on the surface of a globe are different in different directions: the east-west distances between any pairs of points separated by the same azimuthal angle $\Delta\phi$ become smaller as the pair move away from the equator, while the lengths in the north-south directions for a fixed ϕ remain the same. We could, in principle, interpret such results in two equivalent ways:

1. Without considering that the 2D space is curved, we can say that physics (i.e. dynamics) is such that the measuring ruler changed scale when pointing in different directions—much in the same manner as the FitzGerald–Lorentz length contraction of SR was originally interpreted.
2. The alternative description (the "geometric theory") is that we use a standard ruler with a fixed scale (defining the coordinate distance) and the varying length measurements are attributed to the underlying geometry of a curved spherical surface. This is expressed mathematically in the form of a position-dependent metric tensor $g_{ab}(x) \neq \delta_{ab}$.

Einstein's general theory of relativity is a geometric theory of gravity—gravitational phenomena are attributed as reflecting the underlying curved spacetime. An interval ds, invariant with respect to coordinate transformations, is related to the coordinates dx^μ of the spacetime manifold through the metric $g_{\mu\nu}$:

$$ds^2 = g_{\mu\nu}dx^\mu dx^\nu. \tag{6.1}$$

The Greek indices range over $(0, 1, 2, 3)$ with $x^0 = ct$ and the metric $g_{\mu\nu}$ is a 4×4 matrix. Observers measure with rulers and clocks; the spacetime manifold not only expresses the spatial relations among events but also their causal structure. For special relativity (SR) we have the geometry of a flat spacetime with a position-independent metric $g_{\mu\nu} = \eta_{\mu\nu} = \mathrm{diag}(-1, 1, 1, 1)$. The study of equivalence-principle physics led Einstein to propose that gravity represent the structure of a curved spacetime. GR as a geometric theory of gravity posits that matter and energy cause spacetime to warp $g_{\mu\nu} \neq \eta_{\mu\nu}$, and gravitational phenomena are just the effects of a curved spacetime on a test object.

How did the study of the physics as implied by the equivalence principle (EP) motivate Einstein to propose that the relativistic gravitational field was the curved spacetime? We have already discussed the EP physics of gravitational time dilation—clocks run at different rates at positions having different gravitational potential values $\Phi(\vec{x})$, as summarized in (4.32). This variation of time rate follows a definite pattern. Instead of working with a complicated scheme of clocks running at different rates, this physical phenomenon can be given a geometric interpretation as showing a nontrivial metric, $g_{\mu\nu} \neq \eta_{\mu\nu}$. Namely, a simpler way of describing the same physical situation is by using a stationary clock at $\Phi = 0$ as the standard clock. Its fixed rate is taken to be the time coordinate t. One can then compare the time intervals $d\tau(\vec{x})$ measured by clocks located at other locations (the proper time interval at $\vec{x}$) to this coordinate interval dt. According to EP as stated in (4.38), we should find

$$d\tau(\vec{x}) = \left(1 + \frac{\Phi(\vec{x})}{c^2}\right) dt. \tag{6.2}$$

The geometric approach says that the measurement results can be interpreted as showing a spacetime with a warped geometry having a metric element of

$$g_{00} = -\left(1 + \frac{\Phi(x)}{c^2}\right)^2 \simeq -\left(1 + \frac{2\Phi(x)}{c^2}\right). \tag{6.3}$$

This comes about because (6.1) reduces down to $ds^2 = g_{00}dx^0dx^0$ for $d\vec{x} = 0$, as appropriate for a proper time interval (the time interval measured in the rest frame, hence no displacement) and the knowledge that the line element is just the proper time interval $ds^2 = -c^2d\tau^2$, leading to the expression

$$(d\tau)^2 = -g_{00}(dt)^2, \tag{6.4}$$

and the result in (6.3). It states that the metric element g_{00} in the presence of gravity deviates from the flat spacetime value of $\eta_{00} = -1$ because of the presence of gravity. Thus the geometric interpretation of the EP physics of gravitational time dilation is to say that gravity changes the spacetime metric element g_{00} from -1 to an x-dependent function. Gravity warps spacetime—in this case it warps it in the time direction. Also, since g_{00} is directly related to the Newtonian gravitational potential $\Phi(x)$ as in (6.3), we can say that the ten independent components of the spacetime metric $g_{\mu\nu}(x)$ **are** the "relativistic gravitational potentials."

6.1.1   EP physics and a warped spacetime

Adopting a geometric interpretation of EP physics, we find that the resultant geometry has all the characteristic features of a **warped** manifold of space and time: a position-dependent metric, deviations from Euclidean geometric relations, and at every location we can always transform gravity away to obtain a flat spacetime, just as one can always find a locally flat region in a curved space.

Position-dependent metrics   As we have discussed in Section 5.2, the metric tensor in a curved space is necessarily position dependent. Clearly, (6.3) has this property. In Einstein's geometric theory of gravitation, the metric function is all that we need to describe the gravitational field completely. The metric $g_{\mu\nu}(x)$ plays the role of relativistic gravitational potentials, just as $\Phi(x)$ is the Newtonian gravitational potential.

Non-Euclidean relations   In a curved space Euclidean relations no longer hold (see Section 5.3.2): for example, the sum of the interior angles of a triangle on a spherical surface deviates from $180°$, the ratio of the circular circumference to the radius is different from the value of 2π. As it turns out, EP does imply a non-Euclidean relation among geometric measurements. We illustrate this with a simple example. Consider a cylindrical room in high speed rotation around its axis. This acceleration case, according to EP, is equivalent to a centrifugal gravitational field. (This is one way to produce "artificial gravity.") For such a rotating frame, one finds that, because of SR (longitudinal) length contraction, the radius, which is not changed because velocity is perpendicular to the radial direction, will no longer equal the circular circumference of the cylinder divided by 2π (see Fig. 6.1 and Problem 6.3). Thus Euclidean geometry is no longer valid in the presence of gravity. We reiterate this connection: the rotating frame, according to EP, is a frame with gravity; the rotating frame, according to SR length contraction, has a relation between its radius and circumference that is not Euclidean. Hence, we say the presence of gravity brings about non-Euclidean geometry.[1]

Local flat metric and local inertial frame   In a curved space a small local region can always be described approximately as a flat space. A more precise statement is given by the flatness theorem of Section 5.2.2. Now, if we identify our spacetime as the gravitational field, there should be the physics result corresponding to this flatness theorem. This is Einstein's key insight that EP will always allow us to transform gravity away in a local region. In this region, because of the absence of gravity, SR is valid and the metric is the flat Minkowski metric. General relativity has the same local lightcone structure as SR: $ds^2 < 0$ being time-like, $ds^2 > 0$ space-like, and $ds^2 = 0$ light-like. The relation between local flat and local inertial frames will be further explored in Section 6.3, where we show that the spacetime curvature is the familiar tidal force. Given that SR is a theory of flat Minkowski spacetime, and motivated to seek a new theory of gravity with a built-in EP, Einstein put forward the elegant solution of having gravity identified as the structure of curved spacetime.

[1] Distance measurement in a curved spacetime is discussed in Problem 6.2.

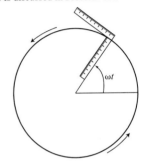

Fig. 6.1 Rotating cylinder with length contraction in the tangential direction but not in the radial direction, resulting in a non-Euclidean relation between circumference and radius.

6.1.2   Curved spacetime as a gravitational field

Recall that a field theoretical description of the interaction between a source and a test particle is a two-step description:

$$\boxed{\text{Source particle}} \xrightarrow[\substack{\text{Field} \\ \text{equation}}]{} \boxed{\text{Field}} \xrightarrow[\substack{\text{Equation of} \\ \text{motion}}]{} \boxed{\text{Test particle}}$$

Instead of the source particle acting directly on the test particle through some instantaneous action-at-a-distance force, the source creates a field everywhere, and the field then acts on the test particle locally. The first step is the field equation which, given the source distribution, determines the field everywhere. In the case of electromagnetism it is Maxwell's equation. The second step is provided by the equation of motion, which allows us to find the motion of the test particle, once the field function is known. The electromagnetic equation of motion follows directly from the Lorentz force law.

Newtonian gravitational field

The field equation in Newton's theory of gravity, when written in terms of the gravitational potential $\Phi(x)$, is given by (4.6)

$$\nabla^2 \Phi = 4\pi G_N \rho, \tag{6.5}$$

where G_N is Newton's constant, and ρ is the mass density function. The Newtonian theory is not a dynamic field theory as it does not provide a description of time evolution. It is the static limit of some field theory, thus has no field propagation. The Newtonian equation of motion is Eq. (4.8):

$$\frac{d^2\vec{r}}{dt^2} = -\vec{\nabla}\Phi. \tag{6.6}$$

The task Einstein undertook was to find the relativistic generalizations of these two sets of equations (6.5) and (6.6). Since in relativity, space and time are treated on an equal footing, a successful relativistic program will automatically yield a dynamical theory as well.

Relativistic gravitational field

The above discussion suggests that the EP physics can be described in geometric language. The resultant mathematics coincides with that describing a warped spacetime. Thus it is simpler, and more correct, to say that the relativistic gravitational field **is** the curved spacetime. The effect of the gravitational interaction between two particles can be described as the source mass giving rise to a curved spacetime which in turn influences the motion of the test mass. Or, put more strongly, EP requires a metric structure of spacetime and particles follow geodesics in such a curved spacetime.

The possibility of using a curved space to represent a gravitational field can be illustrated with the following example involving a 2D curved surface. Two masses on a spherical surface start out at the equator and move along two geodesic lines as represented by the longitudinal great circles. As they move along, the distance between them decreases (Fig. 6.2). We can attribute this to some attractive force between them, or simply to the curved space causing their trajectory to converge. That is to say, this phenomenon of two convergent particle trajectories can be thought of either as resulting from an attractive tidal force, or from the curvature of the space.[2] Eventually we shall write down the relativistic gravitational equations. In Einstein's approach these differential equations can be thought of as reflecting an underlying warped spacetime.

Based on the study of EP phenomenology, Einstein made the conceptual leap (a logical deduction, but a startling leap nevertheless) to the idea that curved spacetime **is** the gravitational field:

$$\boxed{\text{Source}} \quad \xrightarrow[\substack{\text{Einstein Field}\\ \text{equation}}]{} \quad \boxed{\text{Curved specetime}} \quad \xrightarrow[\substack{\text{Geodesic}\\ \text{equation}}]{} \quad \boxed{\text{Test particle}}$$

The mass/energy source gives rise to a warped spacetime, which in turn dictates the motion of the test particle. Plausibly the test particle moves along the shortest and straightest possible curve in the curved manifold. Such a line is the geodesic curve. Hence the GR equation of motion is the geodesic equation (Section 6.2). The GR field equation is the Einstein equation, which relates the mass/energy distribution to the curvature of spacetime (Section 6.3).

While spacetime in SR, like all pre-relativity physics, is fixed, it is dynamic in GR as determined by the matter/energy distribution. GR fulfills Einstein's conviction that "space is not a thing:" the ever changing relation of matter and energy is reflected by an ever changing geometry. Spacetime does not have an independent existence; it is nothing but an expression of the relations among physical processes in the world.

[2]Further reference to gravitational tidal forces vs. the curvature description of the relative separation between two particle trajectories can be found in Section 6.3.1 when we discuss the Newtonian deviation equation for tidal forces. It has its generalization as the "GR equation of geodesic deviation," given in Chapter 14, see Problems 14.4 and 14.5.

Fig. 6.2 Two particle trajectories with decreasing separation can be interpreted either as resulting from an attractive force or as reflecting the underlying geometry of a spherical surface.

6.2 Geodesic equation as GR equation of motion

The metric function $g_{\mu\nu}(x)$ in (6.1) describes the geometry of curved space-time. In GR, the mass/energy source determines the metric function through the field equation. The metric $g_{\mu\nu}(x)$ is the solution of the GR field equation. In this approach, gravity is the structure of spacetime and is not regarded as a force (bringing about acceleration). Thus a test body will move freely in such a curved spacetime, with the equation of motion identified with the geodesic equation.

6.2.1 The geodesic equation recalled

In a geometric theory, the motion of a test body is determined completely by geometry; the GR equation of motion should coincide with the geodesic equation discussed in Chapter 5 on elements of Riemannian geometry. In Section 5.2.1, we have derived the geodesic equation from the property of the geodesic line as the curve with extremum length. We also recall that a point in spacetime is an event and that the trajectory is a worldline (see Box 6.1). The geodesic equation determines the worldline that a test particle will follow under the influence of gravity. The geodesic equation in spacetime is Eq. (5.30) with its Latin indices $a = 1, 2$ (appropriate for the curved 2D space being discussed in Chapter 5) changed into Greek indices $\mu = 0, 1, 2, 3$ with $x^0 = ct$ for a 4D spacetime

$$\frac{d}{d\lambda}\left(g_{\mu\nu}\dot{x}^\nu\right) - \frac{1}{2}\frac{\partial g_{\sigma\rho}}{\partial x^\mu}\dot{x}^\sigma\dot{x}^\rho = 0, \tag{6.7}$$

where $x^\mu = x^\mu(\lambda)$ with λ being the curve parameter, and $\dot{x}^\mu \equiv dx^\mu/d\lambda$.

We can cast (6.7) into a more symmetric form which will also facilitate our later interpretation (in Section 13.2.2) of the geodesic as the straightest possible curve. Carrying out the differentiation of the first term and noting that the metric's dependence on λ is entirely through $x^\mu(\lambda)$:

$$g_{\mu\nu}\frac{d^2 x^\nu}{d\lambda^2} + \frac{\partial g_{\mu\nu}}{\partial x^\sigma}\frac{dx^\sigma}{d\lambda}\frac{dx^\nu}{d\lambda} - \frac{1}{2}\frac{\partial g_{\sigma\rho}}{\partial x^\mu}\frac{dx^\sigma}{d\lambda}\frac{dx^\rho}{d\lambda} = 0 \tag{6.8}$$

Box 6.1 The geodesic is the worldline of a test particle

It may appear somewhat surprising to hear that a test particle will follow a "straight line" in the presence of a gravitational field. After all, our experience is just the opposite: when we throw an object, it follows a parabolic trajectory. Was Einstein saying that the parabolic trajectory is actually straight? All such paradoxes result from confusing the 4D spacetime with the ordinary 3D space. The GR equation of motion tells us that a test particle will follow a geodesic line in spacetime—which is not a geodesic line

(*cont.*)

[3] In fact this discussion can be used as another motivation for a curved spacetime description of gravity: We know that in the absence of gravity an inertial trajectory corresponds to a straight line in a flat spacetime. In the presence of gravity, a free-falling object traces out a curved trajectory in a flat spacetime diagram. Yet, according to EP this can be viewed as an inertial trajectory with a "straight worldline." This is possible only if the spacetime is curved and locally the space can be approximated by a flat space, as EP implies a local equivalence of acceleration and gravity.

Box 6.1 (*Continued*)

in three-dimensional space. Namely, the worldline of a particle should be a geodesic, which generally does not imply a straight trajectory in the spatial subspace.[3] A simple illustration using the spacetime diagram should make this clear.

Let us consider the case of throwing an object to a height of 10 meters over a distance of 10 meters. Its spatial trajectory is displayed in Fig. 6.3(a). When we represent the corresponding worldline in the spacetime diagram we must plot the time-axis ct also, see Fig. 6.3(b). For the case under consideration, this object takes 1.4 seconds to reach the highest point and another 1.4 seconds to come down. But a 2.8 second time interval will be represented by almost one million kilometers of ct in the spacetime diagram (more than the round trip distance to the moon). When the time axis is stretched out in this way, one then realizes this worldline is very straight indeed, see Fig. 6.3(c). The straightness of this worldline reflects the fact that terrestrial gravity is a very weak field (recall $\Phi_\oplus/c^2 \simeq 10^{-10}$)—it curves the spacetime only a tiny amount. In this case the spacetime is practically flat, and thus the geodesic worldline is very close to a straight line.

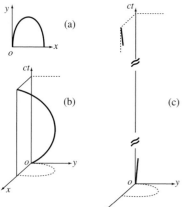

Fig. 6.3 (a) Particle trajectory in the (x,y) plane. (b) Particle worldline with projection onto the (x,y) plane as shown in (a). (c) Spacetime diagram with the time axis stretched a great distance.

Since the product $(dx^\sigma/d\lambda)(dx^\nu/d\lambda)$ in the second term is symmetric with respect to the interchange of indices σ and ν, only the symmetric part of its coefficient:

$$\frac{1}{2}\left(\frac{\partial g_{\mu\nu}}{\partial x^\sigma} + \frac{\partial g_{\mu\sigma}}{\partial x^\nu}\right)$$

can contribute. In this way the geodesic equation (6.7), after factoring out the common $g_{\mu\nu}$ coefficient, can be cast (after relabeling some repeated indices) into the form,

$$\frac{d^2x^\nu}{d\lambda^2} + \Gamma^\nu_{\sigma\rho}\frac{dx^\sigma}{d\lambda}\frac{dx^\rho}{d\lambda} = 0, \tag{6.9}$$

where

$$g_{\mu\nu}\Gamma^\nu_{\sigma\rho} = \frac{1}{2}\left[\frac{\partial g_{\sigma\mu}}{\partial x^\rho} + \frac{\partial g_{\rho\mu}}{\partial x^\sigma} - \frac{\partial g_{\sigma\rho}}{\partial x^\mu}\right]. \tag{6.10}$$

$\Gamma^\nu_{\sigma\rho}$, defined as this particular combination of the first derivatives of the metric tensor, is called the **Christoffel symbol** (also known as the **affine connection**). The geometric significance of this quantity will be studied in Chapter 13. From now on (6.9) is the form of the geodesic equation that we shall use. To reiterate, the geodesic equation is the equation of motion in GR because it is the shortest curve in a warped spacetime. By this we mean that once the gravitational field is given, that is, spacetime functions $g_{\mu\nu}(x)$ and $\Gamma^\mu_{\nu\sigma}(x)$ are known, (6.9) tells us how a test particle will move in such a field: it will always follow the shortest and the straightest possible trajectory in this spacetime. A fuller justification of using the geodesic equation as the GR equation of motion will be given in Section 14.1.2.

In Box 6.2 we demonstrate how the phenomenon of gravitational redshift follows directly from a curve space time description.

6.2.2   The Newtonian limit

Supporting our claim that the geodesic equation is the GR equation of motion, we shall now show that the geodesic equation (6.9) does reduce to the Newtonian equation of motion (6.6) in the **Newtonian limit** of a test particle moving with nonrelativistic velocity $v \ll c$ in a static and weak gravitational field.

- **Nonrelativistic speed** $(dx^i/dt) \ll c$: This inequality $dx^i \ll cdt$ implies that

$$\frac{dx^i}{d\lambda} \ll c\frac{dt}{d\lambda} \left(= \frac{dx^0}{d\lambda} \right). \tag{6.11}$$

Keeping only the dominant term $(dx^0/d\lambda)(dx^0/d\lambda)$ in the double sum over indices λ and ρ of the geodesic equation (6.9), we have

$$\frac{d^2x^\mu}{d\lambda^2} + \Gamma^\mu_{00}\frac{dx^0}{d\lambda}\frac{dx^0}{d\lambda} = 0. \tag{6.12}$$

- **Static field** $(\partial g_{\mu\nu}/\partial x^0) = 0$: Because all time derivatives vanish, the Christoffel symbol of (6.10) takes the simpler form

$$g_{\nu\mu}\Gamma^\mu_{00} = -\frac{1}{2}\frac{\partial g_{00}}{\partial x^\nu}. \tag{6.13}$$

- **Weak field** $h_{\mu\nu} \ll 1$: We assume that the metric is not too different from the flat spacetime metric $\eta_{\mu\nu} = \text{diag}(-1, 1, 1, 1)$

$$g_{\mu\nu} = \eta_{\mu\nu} + h_{\mu\nu} \tag{6.14}$$

where $h_{\mu\nu}(x)$ is a small correction field. Keeping in mind that flat space has a constant metric $\eta_{\mu\nu}$, we have $\partial g_{\mu\nu}/\partial x^\sigma = \partial h_{\mu\nu}/\partial x^\sigma$ and the Christoffel symbols are of order $h_{\mu\nu}$. To leading order, (6.13) is

$$\eta_{\nu\mu}\Gamma^\mu_{00} = -\frac{1}{2}\frac{\partial h_{00}}{\partial x^\nu} \tag{6.15}$$

which, because $\eta_{\nu\mu}$ is diagonal, has for a static h_{00} the following components

$$-\Gamma^0_{00} = -\frac{1}{2}\frac{\partial h_{00}}{\partial x^0} = 0, \quad \text{and} \quad \Gamma^i_{00} = -\frac{1}{2}\frac{\partial h_{00}}{\partial x^i}. \tag{6.16}$$

We can now evaluate (6.12) by using (6.16): the $\mu = 0$ equation leads to

$$\frac{dx^0}{d\lambda} = \text{constant}, \tag{6.17}$$

and the three $\mu = i$ equations are

$$\frac{d^2x^i}{d\lambda^2} + \Gamma^i_{00}\frac{dx^0}{d\lambda}\frac{dx^0}{d\lambda} = \left(\frac{d^2x^i}{c^2dt^2} + \Gamma^i_{00}\right)\left(\frac{dx^0}{d\lambda}\right)^2 = 0, \qquad (6.18)$$

where we have used (6.11) so that $(dx^i/d\lambda) = (dx^i/dx^0)(dx^0/d\lambda)$ and the condition of (6.17) to conclude $(d^2x^i/d\lambda^2) = (d^2x^i/dx^{0\,2})(dx^0/d\lambda)^2$. The above equation, together with (6.16), implies

$$\frac{d^2x^i}{c^2dt^2} - \frac{1}{2}\frac{\partial h_{00}}{\partial x^i} = 0, \qquad (6.19)$$

which is to be compared with the Newtonian equation of motion (6.6). Thus $h_{00} = -2\Phi/c^2$ and using the definition of (6.14) we recover (6.3), first obtained heuristically in Section 6.1:

$$g_{00} = -\left(1 + \frac{2\Phi(x)}{c^2}\right). \qquad (6.20)$$

We can indeed regard the metric tensor as the relativistic generalization of the gravitational potential. This expression also provides us with a criterion to characterize a field being weak as in (6.14):

$$[\ |h_{00}| \ll |\eta_{00}| \ ] \quad \Rightarrow \quad \left[\ \left|\Phi/c^2\right| \ll 1\ \right]. \qquad (6.21)$$

Consider the gravitational potential at the earth's surface. It is equal to the gravitational acceleration times the earth's radius, $\Phi_\oplus = g \times R_\oplus = O(10^7 m^2/s^2)$, or $\Phi_\oplus/c^2 = O(10^{-10})$. Thus a weak field is any gravitational field being less than ten billion $g's$.

Box 6.2 Gravitational redshift reestablished

Previously in Chapter 4 we have shown that the strong equivalence principle implied a gravitational redshift (in a static gravitational field) of light frequency ω

$$\frac{\Delta\omega}{\omega} = -\frac{\Delta\Phi}{c^2}. \qquad (6.22)$$

From this result we heuristically deduced that, in the presence of a nonzero gravitational potential, the metric must deviate from the flat space value. That is, from the gravitation redshift we deduced a curved spacetime. Now we shall reestablish the redshift result, this time going the other way—starting with a curved spacetime, we shall deduce the redshift result.

In this chapter we have seen that Einstein's theory based on a curved spacetime has the result (6.20) in the Newtonian limit. This, as shown in (6.2), can be stated as a relation between the proper time τ and the coordinate time t as follows:

$$dt = \sqrt{-g_{00}}dt \quad \text{with} \quad g_{00} = -\left(1 + 2\frac{\Phi}{c^2}\right). \tag{6.23}$$

Here we wish to see how the gravitational frequency shift result of (6.22) emerges in this curved spacetime description.

In Fig. 6.4, the two curvy lines are the light-like worldlines of two wavefronts emitted at an interval dt_{em} apart. They are curvy because in the presence of gravity the spacetime is curved. (In flat spacetime, they would be two straight 45° lines.) Because we are working with a static gravitational field (hence a time-independent spacetime curvature), this dt_{em} time interval between the two wavefronts is maintained throughout the trip until they are received. That is, these two wavefronts trace out two congruent worldlines. In particular the coordinate time separations at emission and reception are identical,

$$dt_{em} = dt_{rec}. \tag{6.24}$$

On the other hand, the frequency being inversely proportional to the proper time interval $\omega = 1/d\tau$, we can then use (6.23) and (6.24) to derive:

$$\frac{\omega_{rec}}{\omega_{em}} = \frac{d\tau_{em}}{d\tau_{rec}} = \frac{\sqrt{-(g_{00})_{em}}\, dt_{em}}{\sqrt{-(g_{00})_{rec}}\, dt_{rec}} = \left(\frac{1 + 2(\Phi_{em}/c^2)}{1 + 2(\Phi_{rec}/c^2)}\right)^{1/2}$$

$$= 1 + \frac{\Phi_{em} - \Phi_{rec}}{c^2} + O\left(\Phi^2/c^4\right), \tag{6.25}$$

which is the claimed result of (6.22):

$$\frac{\omega_{rec} - \omega_{em}}{\omega_{em}} = \frac{\Phi_{em} - \Phi_{rec}}{c^2}. \tag{6.26}$$

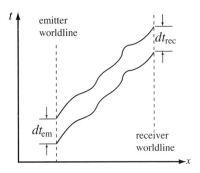

Fig. 6.4 Worldlines for two light wavefronts propagating from emitter to receiver in a static curved spacetime.

6.3 The curvature of spacetime

We have already discussed in Chapter 5 (see especially Section 5.2.2) that in a curved space each small region can be approximated by a flat space, that is, locally one can always perform a coordinate transformation so the new metric is approximately a flat-space metric. This coordinate dependence of the metric shows that the metric value cannot represent the core feature of a curved space. However, as shown in Section 5.3 (and further discussion in Section 13.3), there exits a mathematical quantity involving the second derivative of the metric, called the curvature, which does represent the essence of a curved space: the space is curved if and only if the curvature is nonzero; and, also, the deviations from Euclidean relations are always proportional to the curvature.

If the warped spacetime is the gravitational field, what then is its curvature? What is the physical manifestation of this curvature? How does it enter in the GR gravitational field equation?

6.3.1   Tidal force as the curvature of spacetime

The equivalence principle states that in a freely falling reference frame the physics is the same as that in an inertial frame with no gravity. SR applies and the metric is given by the Minkowski metric $\eta_{\mu\nu}$. As shown in the flatness theorem (Section 5.2.2), this approximation of $g_{\mu\nu}$ by $\eta_{\mu\nu}$ can be done **only** locally, that is, in an appropriately small region. Gravitational effects can always be detected in a **finite-sized** free-fall frame as the gravitational field is never strictly uniform in reality; the second derivatives of the metric come into play.[4]

Consider the lunar gravitational attraction exerted on the earth. While the earth is in free fall toward the moon (and vice versa), there is still a detectable lunar gravitational effect on the earth. This is so because different points on the earth will feel slightly different gravitational pulls by the moon, as depicted in Fig. 6.5(a). The center-of-mass (CM) force causes the earth to "fall towards the moon" so that this CM gravitational effect is "cancelled out" in this freely falling terrestrial frame. After subtracting out this CM force, the remanent forces on the earth, as shown in Fig. 6.5(b), are stretching in the longitudinal direction and compression in the transverse direction. They are just the familiar tidal forces.[5] That is, in the freely falling frame, the CM gravitational effect is transformed away, but, there are still the remnant tidal forces. They reflect the **differences** of the gravitational effects on neighboring points, and are thus proportional to the derivative of the gravitational field.

We can illustrate this point by the following observation. With r_s and r_m being the distances from the earth to the sun and moon, respectively,

[4]The flatness theorem states that in the local inertial frame ($\bar{x}^\mu$) the new metric is, according to (5.31), approximately flat: $\bar{g}_{\mu\nu}(\bar{x}) = \eta_{\mu\nu} + \gamma_{\mu\nu\lambda\rho}(0)\bar{x}^\lambda\bar{x}^\rho + \cdots$ with $\partial\bar{g}_{\mu\nu}/\partial\bar{x}^\lambda = 0$.

[5]The ocean is pulled away in opposite directions giving rise to two tidal bulges. This explains why, as the earth rotates, there are two high tides in a day. This is a simplified account as there are also other effects (e.g. the solar tidal forces).

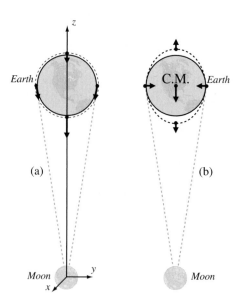

(a)   (b)

Fig. 6.5 Variations of the gravitational field as tidal forces. (a) Lunar gravitational forces on four representative points on the earth. (b) After taking out the center of mass (CM) motion, the relative forces on the earth are the tidal forces giving rise to longitudinal stretching and transverse compression.

we have

$$\left[\mathfrak{g}_s = \frac{G_N M_\odot}{r_s^2} \right] > \left[\mathfrak{g}_m = \frac{G_N M_m}{r_m^2} \right], \tag{6.27}$$

showing that the gravitational attraction of the earth by the sun is much larger than that by the moon. On the other hand, because the tidal force is given by the derivative of the field strength

$$\left[\frac{\partial}{\partial r} \frac{G_N M}{r^2} \right] \propto \left[\frac{G_N M}{r^3} \right], \tag{6.28}$$

and because $r_s \gg r_m$, the lunar tidal forces nevertheless end up being stronger than the solar ones:

$$\left[\mathfrak{T}_s = \frac{G_N M_\odot}{r_s^3} \right] < \left[\mathfrak{T}_m = \frac{G_N M_m}{r_m^3} \right]. \tag{6.29}$$

Since tidal forces cannot be coordinate-transformed away, they should be regarded as the essence of gravitation. They are the variations of the gravitational field, hence the second derivatives of the gravitational potential. From the discussion in this chapter showing that the relativistic gravitational potential being the metric, and that second derivative of the metric being the curvature, we see that Einstein gives gravity a direct geometric interpretation by identifying these tidal forces with the curvature of spacetime. A discussion of tidal forces in terms of the Newtonian deviation equation is given in Box 6.3.

Box 6.3 The equation of Newtonian deviation and its GR generalization

Here we provide a more quantitative description of the gravitational tidal force in the Newtonian framework, which will suggest an analogous GR approach to be followed in Chapter 14.

As the above discussion indicates, the tidal effect concerns the relative motion of particles in a nonuniform gravitational field. Let us consider two particles: one has the trajectory $\vec{x}(t)$ and another has $\vec{x}(t) + \vec{s}(t)$. That is, the locations of these two particles measured at the same time has a coordinate difference of $\vec{s}(t)$. The respective equations of motion ($i = 1, 2, 3$) obeyed by these two particles are:

$$\frac{d^2 x^i}{dt^2} = -\frac{\partial \Phi(x)}{\partial x^i} \quad \text{and} \quad \frac{d^2 x^i}{dt^2} + \frac{d^2 s^i}{dt^2} = -\frac{\partial \Phi(x+s)}{\partial x^i}. \tag{6.30}$$

(*cont.*)

Box 6.3 (*Continued*)

Consider the case where the separation distance $s^i(t)$ is small and we can approximate the gravitational potential $\Phi(x+s)$ by a Taylor expansion

$$\Phi(x+s) = \Phi(x) + \frac{\partial \Phi}{\partial x^j} s^j + \cdots \qquad (6.31)$$

From the difference of the two equations in (6.30), we obtain the **Newtonian deviation equation** that describes the separation between two particle trajectories in a gravitational field

$$\frac{d^2 s^i}{dt^2} = -\frac{\partial^2 \Phi}{\partial x^i \partial x^j} s^j. \qquad (6.32)$$

Thus the relative acceleration per unit separation $(d^2 s^i / dt^2)/s^j$ is given by a tensor having the second derivatives of the gravitational potential (i.e. the tidal force components) as its elements.

We now apply (6.32) to the case of a spherical gravitational source (e.g. the gravity due to the moon on earth as shown in Fig. 6.5), $\Phi(x) = -G_N M/r$, where the radial distance is related to the rectangular coordinates by $r = (x^2 + y^2 + z^2)^{1/2}$. Since $\partial r / \partial x^i = x^i / r$ we have

$$\frac{\partial^2 \Phi}{\partial x^i \partial x^j} = \frac{G_N M}{r^3} \left(\delta_{ij} - \frac{3 x^i x^j}{r^2} \right). \qquad (6.33)$$

Consider the case of the "first particle" being located along the z axis $x^i = (0, 0, r)$. The Newtonian deviation equation (6.32) for the displacement of the "second particle", with the second derivative tensor given by (6.33), now takes on the form

$$\frac{d^2}{dt^2} \begin{pmatrix} s_x \\ s_y \\ s_z \end{pmatrix} = \frac{-G_N M}{r^3} \begin{pmatrix} 1 & 0 & 0 \\ 0 & 1 & 0 \\ 0 & 0 & -2 \end{pmatrix} \begin{pmatrix} s_x \\ s_y \\ s_z \end{pmatrix}. \qquad (6.34)$$

We see that there is an attractive tidal force between the two particles in the transverse direction $\mathfrak{T}_{x,y} = -G_N M r^{-3} s_{x,y}$ which leads to compression; and a tidal repulsion $\mathfrak{T}_z = +2 G_N M r^{-3} s_z$, leading to stretching, in the longitudinal (i.e. radial) direction.

In GR, we shall follow a similar approach (see Problems 14.4 and 14.5): the two equations of motion (6.30) will be replaced by the corresponding geodesic equations; their difference, after a Taylor expansion, leads to the **equation of geodesic deviation**, which is entirely similar[6] to (6.32). Since the metric function is the relativistic potential, the second derivative tensor turns into the curvature tensor of the spacetime (the Riemann curvature tensor). In this geometric language we see that the cause of the deviation from a flat spacetime worldline is attributed to the curvature. Recall the example discussed previously in Section 6.1.2; see especially Fig. 6.2.

[6]We are not quite ready to derive this GR equation as one still needs to learn how to perform differentiations in a curved space (see Chapter 13).

6.3.2   The GR field equation described

We now discuss how the curvature, identified as the tidal forces, enters directly in the field equations of relativistic gravitational theory.

The field equation relates the source distribution to the resultant field; given the source distribution, we can use the field equation to find the field everywhere. For the Newtonian equation (6.5),

$$\nabla^2 \Phi = 4\pi G_N \rho, \tag{6.36}$$

we have the second derivative of the gravitational potential $\nabla^2 \Phi$ being directly proportional to the mass density ρ. What is the relativistic generalization of this equation?

1. For the right-hand side (RHS) of the field equation, from the viewpoint of relativity, mass being just a form of energy (the rest energy) and, furthermore, energy and momentum being equivalent, they can be transformed into each other when viewed by different observers (see (3.37)), the mass density ρ of (6.5) is generalized in relativity to an object called the **"energy–momentum tensor"** $T_{\mu\nu}$. The 16 elements include T_{00}, being the energy density ρc^2, three elements T_{0i} begin the momentum densities, and the remaining 12 elements representing the fluxes associated with the energy and momentum densities—they describe the flow of energy and momentum components. There are actually only 10 independent elements, because it is a symmetric tensor $T_{\mu\nu} = T_{\nu\mu}$. A more detailed discussion of the energy–momentum tensor will be presented in Section 12.3.

2. For the second derivative of the potential on the left-hand side (LHS) of the field equation, we have already seen that the relativistic gravitational potential is the metric $g_{\mu\nu}$ and the curvature in (5.35) is the second derivative of the metric. And, as we shall find in Chapter 13, for higher dimensional spaces, the Gaussian curvature K of Chapter 5 is generalized to the **Riemann curvature tensor**. A particular (contracted) version of the curvature tensor is the **Einstein tensor** $G_{\mu\nu}$, having mathematical properties that match those of the energy–momentum tensor $T_{\mu\nu}$.

This suggests the possible relativistic generalization of the gravitational field equation as having the basic structure of (6.36): the RHS being the energy–momentum tensor, and the LHS being the Einstein tensor involving the second derivative of the metric:

Newton equation	$\nabla^2 \Phi \propto \rho$
Einstein equation	$G_{\mu\nu} \propto T_{\mu\nu}$

In this way, we obtain in Section 14.2 the GR field equation, the **Einstein equation**, in the form of

$$G_{\mu\nu} = \kappa T_{\mu\nu}, \tag{6.37}$$

where κ is a proportionality constant. As we shall show in Section 14.2.2 the nonrelativistic limit of this equation is just Newton's equation (6.36) when we make the identification of

$$\kappa = -\frac{8\pi G_N}{c^4}. \tag{6.38}$$

Since the curvature has a different measurement unit from that for the energy–momentum density,[7] the proportional constant κ, hence Newton's constant G_N, should be interpreted as a **conversion factor**. Just as the speed of light c is the conversion factor between space and time that is fundamental to the special relativistic symmetry of space and time (see Section 3.4), one way of viewing the significance of Newton's constant is that it is the conversion factor[8] fundamental for a geometric description of gravity by GR; it connects the spacetime curvature to the gravitational source of energy and momentum, as in Einstein's equation:

$$\begin{pmatrix} \text{curvature} \\ \text{of spacetime} \end{pmatrix} = (\text{Newton's constant}) \times \begin{pmatrix} \text{energy–momentum} \\ \text{density.} \end{pmatrix}.$$

When worked out in Chapter 14, we shall see that (6.37) represents 10 coupled partial differential equations. Their solution is the metric function $g_{\mu\nu}(x)$, fixing the geometry of spacetime. We emphasize once more that in GR, spacetime is no longer a passive background against which physical events take place. Rather, it is a dynamical entity as it responds to the ever-changing matter/energy distribution in the world. But in more conventional field theory language, the metric is the gravitational field, and energy–momentum density is the "gravity charge."

For the rest of Parts II and III (Chapters 7–11), this Einstein field equation will not be discussed further. Rather, we shall concentrate on investigating its solutions, showing how a curved spacetime description of the gravitational field, that is, knowing the metric $g_{\mu\nu}(x)$, brings about many interesting physical consequences: from bending of light rays, and black holes, to cosmology.

GR and the structure of spacetime

Einstein was motivated to build a new theory of gravity that is compatible with the principle of relativity and has the principle of equivalence fundamentally built into the theory. Remarkably all this could be implemented simply by allowing for a curved spacetime instead of the flat one for SR. In GR gravity is not a force, but a change of structure of spacetime that allows inertial observers to accelerate with respect to each other. The warped spacetime can account for all gravitational phenomena. The Einstein equation shows how the energy–momentum density can change the curvature of spacetime. In this way, GR brings about a radical change in the way we describe physical events: in SR physical events are depicted in a fixed spacetime, while in GR the spacetime itself is a dynamical entity, ever-evolving in response to the presence of energy and momentum.

Box 6.4 Einstein's three motivations: An update

In Chapter 1 we discussed Einstein's motivations for creating general relativity. Now we can see how these issues are resolved in the curved spacetime formulation of a relativistic theory of gravitation.

1. **SR is not compatible with gravity** In the GR formulation, we see that SR is valid only in the local inertial frames in which gravity is transformed away.

2. **A deeper understanding of $m_I = m_G$** The weak EP is generalized to strong EP. The various consequences of the equivalence principle led Einstein to the idea of a curved spacetime as the relativistic gravitational field. At the fundamental level there is no difference between gravity and the "fictitious forces" associated with accelerated frames. Noninertial frames of reference in Newtonian physics are identified in Einstein's theory with the presence of gravity. The GR theory, symmetric with respect to general coordinate transformations (including accelerated coordinates), and the relativistic field theory of gravitation must be one and the same.[9] The equivalence principle is built right into the curved spacetime description of gravitation because any curved space is locally flat.

3. **"Space is not a thing"** The GR equations are covariant under the most general (position-dependent) coordinate transformations.[10] GR physics is valid in any coordinate frame. Furthermore, spacetime (metric) is the solution to the Einstein equation. It has no independent existence except expressing the relation among physical processes in the world.

[9]GR theory respects the symmetry with respect to general coordinate transformation, including the accelerated coordinates. EP teaches us that these accelerated frames are equivalent to the inertial frames with gravity. Hence general relativistic theory must necessarily be a theory of gravity.

[10]See further discussion as the "principle of general covariance" in Section 14.1.

Review questions

1. What does one mean by a "geometric theory of physics?" Use the distance measurements on the surface of a globe to illustrate your answer.

2. How can the phenomenon of gravitational time dilation be phrased in geometric terms? Use this discussion to support the suggestion that the spacetime metric can be regarded as the relativistic gravitational potential.

3. Give the simple example of a rotating cylinder to illustrate how the EP physics implies a non-Euclidean geometric relation.

4. What significant conclusion did Einstein draw from the analogy between the fact that a curved space is locally flat and that gravity can be transformed away locally?

5. How does GR imply a conception of space and time as reflecting merely the relationship between physical events rather than a stage onto which physical events take place.

6. Give a heuristic argument for the GR equation of motion to be the geodesic equation.

7. What is the Newtonian limit? In this limit, what relation can one infer between the Newtonian gravitational potential and a metric tensor component of the spacetime. Use this relation to derive the gravitational redshift.

8. What are tidal forces? How are they related to the gravitational potential? Explain why the solar tidal forces are smaller than lunar tidal forces, even though the gravitational attraction of the earth by the sun is stronger than that by the moon. Explain how in general relativity the tidal forces are identified with the curvature of spacetime.

9. Give a qualitative description of the GR field equation. Explain in what sense we can regard Newton's constant

as a basic "conversion factor" in general relativity. Can you name two other conversion factors in physics that are respectively basic to special relativity and to quantum theory?

10. How are Einstein's three motivations for creating GR resolved in the final formulation of the geometric theory of gravity?

Problems

6.1 **The metric element g_{00}**   From the definitions of metric and proper time, derive the relation between proper time and coordinate time;

$$d\tau = \sqrt{-g_{00}}dt.$$

6.2 **Spatial distance and spacetime metric**   Einstein suggested the following definition of the spatial distance dl between two neighboring points (A, B) with a coordinate difference of dx^i (where $i = 1, 2, 3$). A light pulse is sent to B and reflected back to A. If the elapsed proper time (according to A) is $d\tau_A$, then $dl \equiv cd\tau_A/2$. The square of the spatial distance should also be quadratic in dx^i

$$dl^2 = \gamma_{ij}dx^i dx^j.$$

How is this spatial metric γ_{ij} related to the spacetime metric $g_{\mu\nu}$ as defined by $ds^2 = g_{\mu\nu}dx^\mu dx^\nu$ (where $\mu = 0, 1, 2, 3$)? Is it just $\gamma_{ij} = g_{ij}$? *Suggestion*: Solve the $ds^2 = 0$ equation in terms of the various elements of $g_{\mu\nu}$, for the coordinate time required for the light signal to make the round trip. See Landau and Lifshitz (1975, §84).

6.3 **Non-Euclidean geometry of a rotating cylinder**   In Section 6.1.1 we used the example of a rotating cylinder to motivate the need for non-Euclidean geometry. Use the formalism derived in Problem 6.2 to work out the spatial distance, showing this violation of the Euclidean relation between radius and circumference.

6.4 **Geodesic equation in a rotating coordinate**   Knowing the metric for a rotating coordinate from Problem 6.3, work out the corresponding Christoffel symbols and geodesic

equation. This can be taken as the relativistic version of the centrifugal force.

6.5 **The geodesic equation and light deflection**   Use the geodesic equation, rather than Huygens' principle, to derive the expression of gravitational angular deflection given by (4.44) and (4.45), if the only warped metric element is $g_{00} = -1 - 2\Phi(x)/c^2$. One approach is to note (see Fig. 4.6) that the infinitesimal angular deflection of a photon with momentum $\vec{p} = p\hat{x}$ is related to momentum change by $d\phi = dp_y/p$. In turn we can always choose a curve parameter τ for the light geodesic so that the photon momentum equals the derivative of displacement with respect to such a parameter $p^\mu = dx^\mu/d\tau$ with $\mu = 0, 1, 2, 3$. For the photon momentum 4-vector in the $\hat{x}$ direction, we have $p^\mu = (p, p, 0, 0)$ (see Section 3.2.2). The deflection can be calculated from the geodesic equation by its determination of $dx^\mu/d\tau$, hence p^μ.

6.6 **Symmetry property of the Christoffel symbols**   From the definition of (6.10), check explicitly that

$$\Gamma^\lambda_{\mu\nu} = \Gamma^\lambda_{\nu\mu}.$$

6.7 **The matrix for tidal forces is traceless**   One notes that the matrix in (6.34) is traceless (vanishing sum of the diagonal elements). Why should this be so?

6.8 **G_N as a conversion factor**   From Newton's theory we know that Newton's constant has the dimension of energylengthmass^{-2}. With such a G_N in the proportional constant (6.38) of the Einstein equation (6.37), check that it yields the correct dimension for the curvature on the LHS of the Einstein equation.

Spherically symmetric spacetime – GR tests

7

- A spherically symmetric metric has two unknown scalar metric functions: g_{00} and g_{rr}.
- The Schwarzschild solution to the GR field equation yields $g_{00} = -g_{rr}^{-1} = -(1 - r^*/r)$ with $r^* = 2G_N M/c^2$. An embedding diagram can be used to visualize such a warp space.
- GR predicts a solar deflection of a light ray that is twice as large as that implied by the equivalence principle.
- We present a brief discussion of gravitational lensing, with the lens equation derived.
- The precession of Mercury's perihelion and the Shapiro time delay of a light ray are worked out as successful tests of general relativity.

In the previous chapter we presented some preliminaries for a geometric description of gravity. The gravitational field as curved spacetime can be expressed (once a coordinate system has been chosen) in terms of the metric solution of the Einstein field equation, $g_{\mu\nu}(x)$, which may be regarded as the relativistic gravitational potential. The region outside a spherically symmetric source (e.g. the sun) has the (exterior) **Schwarzschild geometry**. Its nonrelativistic analog is the gravitational potential

$$\Phi(r) = -\frac{G_N M}{r}, \qquad (7.1)$$

which is the solution to the Newtonian field equation $\nabla^2 \Phi(x) = 4\pi G_N \rho(x)$ with a spherically symmetric mass density $\rho(x)$ and a total mass M inside a sphere with radius less than the radial distance r.

The GR field equation with a spherical source will be solved in Chapter 14. Given the source mass distribution, we can find the metric $g_{\mu\nu}(x)$, and therefore the spacetime geometry, outside a spherical star. In this chapter we shall only quote the solution, called the Schwarzschild metric, and concentrate on the description of a test particle's trajectory in this geometry. We study three interesting applications of GR: the precession of the planet Mercury's perihelion, and the deflection (gravitational lensing) and the time delay of a light ray as it passes near a massive body. In this chapter, the applications are by and large restricted to the solar system, for which gravity is relatively weak (and for which, therefore, general relativity is not absolutely necessary). In Chapter 8, we study the strong gravity of black holes, for which the GR theory of gravitation is indispensable.

7.1   Description of Schwarzschild spacetime

We shall first show that, in a spherical coordinate system (t, r, θ, ϕ), the metric tensor for a spherically symmetric spacetime has only two unknown scalar functions. These metric functions $g_{00}(t, r)$ and $g_{rr}(t, r)$ can be obtained by solving the GR field equation, which we will not do explicitly until Section 14.3. Here we just present the result: the metric is diagonal with $g_{\theta\theta} = r^2$, $g_{\phi\phi} = r^2 \sin^2 \theta$, and

$$g_{00}(t, r) = -\frac{1}{g_{rr}(t, r)} = -1 + \frac{2G_N M}{rc^2}. \tag{7.2}$$

The implication of such a geometry for various physical situations will be discussed in subsequent sections.

7.1.1   Properties of a spherically symmetric metric tensor

Because the source is spherically symmetric, the spacetime it generates (as the solution to the Einstein equation for such a source) must have this symmetry. The corresponding metric function $g_{\mu\nu}(x)$ must be isotropic in the spatial coordinates. It is natural to pick a spherical coordinate system (t, r, θ, ϕ) having the center coincident with that of the spherical source. As shown in Box 7.1, the form of such an isotropic metric has only two unknown scalar functions, g_{00} and g_{rr}:

$$g_{\mu\nu} = \text{diag}\left(g_{00}, g_{rr}, r^2, r^2 \sin^2 \theta\right). \tag{7.3}$$

For points that are far away from the gravitational source, $g_{\mu\nu}$ approaches the flat spacetime limit:

$$\lim_{r \to \infty} g_{00}(t, r) \to -1 \qquad \text{and} \qquad \lim_{r \to \infty} g_{rr}(t, r) \to 1. \tag{7.4}$$

Box 7.1 The standard form of an isotropic metric

Here we shall explicitly show that a spherically symmetric metric tensor has only two unknown scalar functions.

General considerations of isotropy   The infinitesimal invariant interval $ds^2 = g_{\mu\nu}dx^\mu dx^\nu$ must be quadratic in $d\vec{r}$ and dt without singling out any particular spatial direction.[1] That is, ds^2 must be composed of terms having two powers of $d\vec{r}$ and/or dt; furthermore, the vectors $\vec{r}$ and $d\vec{r}$ must appear in the form of dot products so as not to spoil the spherical symmetry. (A dot product is a scalar, not singling out any particular

[1] Here we adopt the notation $x^\mu = (ct, \vec{r})$.

direction.) The vector $\vec{r}$ can appear because the metric is a function of position and time, $g_{\mu\nu} = g_{\mu\nu}(\vec{r}, t)$

$$ds^2 = A d\vec{r} \cdot d\vec{r} + B\,(\vec{r} \cdot d\vec{r})^2 + C dt\,(\vec{r} \cdot d\vec{r}) + D dt^2 \qquad (7.5)$$

where A, B, C, and D are scalar functions of t and $\vec{r} \cdot \vec{r}$. In a spherical coordinate system (r, θ, ϕ),

$$\vec{r} = r\hat{r} \qquad \text{and} \qquad d\vec{r} = dr\hat{r} + r d\theta\hat{\theta} + r \sin\theta d\phi\hat{\phi}, \qquad (7.6)$$

where $\hat{r}$, $\hat{\theta}$, and $\hat{\phi}$ are orthonormal basis vectors. Thus

$$\vec{r} \cdot \vec{r} = r^2, \qquad \vec{r} \cdot d\vec{r} = rdr,$$

$$\text{and} \quad d\vec{r} \cdot d\vec{r} = dr^2 + r^2\left(d\theta^2 + \sin^2\theta d\phi^2\right).$$

In terms of spherical coordinates, the invariant separation is now

$$ds^2 = A\left[dr^2 + r^2\left(d\theta^2 + \sin^2\theta d\phi^2\right)\right] + Br^2 dr^2 + Cr dr dt + D dt^2, \qquad (7.7)$$

or, equivalently, with some relabeling of scalar functions (for example, $A + Br^2 = B'$, which can be labeled again as B),

$$ds^2 = A\left[r^2\left(d\theta^2 + \sin^2\theta d\phi^2\right)\right] + B dr^2 + Cr dr dt + D dt^2. \qquad (7.8)$$

Simplification by coordinate choices From our discussion in Chapter 5, we learned that Gaussian coordinates, as labels of points in space, can be freely chosen. In the same way, the names given to coordinates in Riemannian geometry have no intrinsic significance until their connection to the physical length ds^2 is specified by the metric function. Thus we are free to make a new choice of coordinates (with a corresponding modification of the metric) until the metric takes on the simplest form. Of course, in our particular case, the process of changing to new coordinates must not violate spherical symmetry.

1. **Pick a new time coordinate so that there is no cross term** $dt dr$: we introduce a new coordinate t' such that

$$t \to t' = t + f\,(r). \qquad (7.9)$$

Differentiating, we have $dt' = dt + (df/dr)dr$, and squaring both sides and solving for dt^2,

$$dt^2 = dt'^2 - \left(\frac{df}{dr}\right)^2 dr^2 - 2\frac{df}{dr}dr dt. \qquad (7.10)$$

Now the cross-term $dt dr$ has a coefficient $Cr - 2D(df/dr)$, which can be eliminated[2] by choosing a $f(r)$ that satisfies the differential equation

$$\frac{df}{dr} = +\frac{Cr}{2D}. \qquad (7.11)$$

(cont.)

[2] The absence of any linear factor dt means that the metric is also time-reversal invariant (i.e. changing t to $-t$ does not affect ds^2). In fact one can understand why the metric should be a diagonal matrix over all because the spherical symmetry of the problem implies that the solution should be symmetric under the reflection $(r, \theta, \phi) \to (-r, -\theta, -\phi)$ for each coordinate.

Box 7.1 (*Continued*)

Similarly, other terms such as $(df/dr)^2 dr^2$ in (7.10) can be absorbed into a new scalar function $B' = B - (df/dr)^2 D$, etc.

2. **New radial coordinate so that the angular coefficient is simple:** We can set the function A in (7.8) to unity by choosing a new radial coordinate

$$r^2 \to r'^2 = A(r, t) r^2$$

so that the first term on the RHS of (7.8) is just $r'^2 (d\theta^2 + \sin^2 \theta d\phi^2)$. The effect of this new radial coordinate on other terms in ds^2 can again be absorbed by relabeling of scalar functions. We leave this as an exercise for the reader to complete.

With these new coordinates, we are left with only two unknown scalar functions in the metric. In this way, the line element takes on the form of

$$ds^2 = g_{00}(r, t) c^2 dt^2 + g_{rr}(r, t) dr^2 + r^2 \left(d\theta^2 + \sin^2 \theta d\phi^2 \right),$$

$$(7.12)$$

which agrees with the metric $g_{\mu\nu}$ shown in (7.3).

Interpreting the coordinates

Our spherically symmetric metric (7.3) is diagonal. In particular, $g_{i0} = g_{0i} = 0$ (with $i = 1, 2, 3$ being a spatial coordinate index), which means that for a given t, we can discuss the spatial subspace separately. That is, the nine elements g_{ij} of the metric completely describe what happens to dx^1, dx^2, and dx^3. For a fixed t, we can visualize this spherically symmetric coordinate system as a series of 2-spheres having different radial coordinate values of r, with their center at the origin of the spherically symmetric source. Each 2-sphere, having a surface area of $4\pi r^2$ and a volume of $4\pi r^3/3$, can be thought of as being made up of rigid rods arranged in a grid corresponding to various (θ, ϕ) values, with synchronized clocks attached at each grid point (i.e. each point in this subspace has the same coordinate time).

- Before the gravitational source is "turned on," the coordinate r is the proper radial distance ρ defined as

$$dr^2 = ds_r^2 \equiv d\rho^2,$$

$$(7.13)$$

where ds_r^2 is the invariant interval ds^2 with $dt = d\theta = d\phi = 0$, and the coordinate t is the proper time τ for an observer at a fixed location,

$$-c^2 dt^2 = ds_t^2 \equiv -c^2 d\tau^2,$$

$$(7.14)$$

where ds_t^2 is the invariant interval ds^2 with $dr = d\theta = d\phi = 0$. Thus, the coordinates (r, t) have the physical interpretation of being the radial distance and time as measured by an observer far away from the (spherical) gravitational source. That is, the far-away spacetime should be flat.

- After the gravitational source is "turned on," the spacetime is warped. In particular, $g_{rr} \neq 1$. There is a curvature of the spacetime in the spatial radial direction so that the proper radial distance, $\rho \neq r$, is

$$d\rho = \sqrt{g_{rr}}dr. \tag{7.15}$$

Consequently, the spherical surface area $4\pi r^2$ and the volume $4\pi r^3/3$ no longer bear the Euclidean relation with their radius ρ. Similarly, the proper time

$$d\tau = \sqrt{-g_{00}}dt, \tag{7.16}$$

differs from the coordinate time because $g_{00} \neq -1$. This signifies the warping of the spacetime in the time direction.

7.1.2 The Schwarzschild geometry and the embedding diagram

In general relativity, spacetime is not a passive background against which physical processes take place. Rather the geometry of spacetime is determined by the distribution of mass and energy. Given the source distribution, we can solve the GR field equation to find the metric $g_{\mu\nu}(x)$. The solution of Einstein's equation for the spacetime **exterior** to a spherical source[3] will be carried out in Chapter 14. Here we quote this result:[4]

$$g_{00}(t,r) = -\frac{1}{g_{rr}(t,r)} = -1 + \frac{r^*}{r}, \tag{7.17}$$

where r^* is some constant distance. We see that the deviation from the flat spacetime of (7.4) is determined by the size of the ratio r^*/r. The resultant metric is called the **Schwarzschild metric**,

$$g_{\mu\nu} = \text{diag}\left[\left(-1+\frac{r^*}{r}\right),\left(1-\frac{r^*}{r}\right)^{-1},r^2,r^2\sin^2\theta\right]. \tag{7.18}$$

The Schwarzschild radius

We can relate this constant distance r^* to familiar quantities by considering the Newtonian limit. Equations (6.20) and (7.1) lead to

$$g_{00} = -\left(1+\frac{2\Phi}{c^2}\right) = -1 + \frac{2G_N M}{c^2 r}. \tag{7.19}$$

Comparing this to (7.17), we obtain an expression for the **Schwarzschild radius**

$$r^* = \frac{2G_N M}{c^2}. \tag{7.20}$$

This is generally a very small distance. For example, the solar and terrestrial Schwarzschild radii are respectively:

$$r^*_\odot \simeq 3\,\text{km} \quad\text{and}\quad r^*_\oplus \simeq 9\,\text{mm}. \tag{7.21}$$

Hence, in general, the ratio r^*/r, which signifies the modification of the flat Minkowski metric, is a very small quantity. For the Schwarzschild exterior solution to be applicable, r must have the same value at or outside of the radius

[3] As we shall discuss in Chapter 14, the Einstein equation for the geometry exterior to a source with mass M corresponds to a vanishing Ricci tensor $R_{\mu\nu} = 0$.

[4] The reader may wish to take an early look at Problem 15.2 (and its solution) for the derivation of the Schwarzschild solution from the **linearized** Einstein equation. While this metric is an approximation, up to an $O(r^{*2})$ correction, of the exact solution, the calculation is simple and it shows clearly how the spacetime is warped by the presence of a spherically symmetric $1/r$ gravitational potential.

of mass, and so the smallest value that r can take is the radius R of the spherical source: the above r^* values translate into r^*/r of

$$\frac{r^*_\odot}{R_\odot} = O\left(10^{-6}\right) \qquad \text{and} \qquad \frac{r^*_\oplus}{R_\oplus} = O\left(10^{-9}\right), \tag{7.22}$$

for the case of the sun and the earth, respectively.

The metric (7.18) is singular at the Schwarzschild radius. That is, when $r = r^*$, g_{rr} is infinite, which means that ds^2 is infinite. This singular feature was not extensively studied until the 1950s because many early relativists thought $r = r^*$ was not physically realizable. Since r must be outside of the radius of the mass (otherwise, the Schwarzschild exterior solution doesn't apply), $r = r^*$ implies that the radius of the mass, r_{source}, is less than the Schwarzschild radius r^*. For the earth, for instance, this would mean that the entire mass of the earth would have to be contained within a small ball with a radius of ~ 1 cm. Such a gravitational mass would be extraordinarily dense. Only gradually was the possibility of such a dense, massive object taken seriously. In Chapter 8, we will discuss black holes, for which $r_{source} < r^*$. The above discussion suggests that a black hole is expected to have an extremely high mass density. In fact a stellar-mass black hole, for which $M = O(M_\odot)$, indeed has a high density $O(10^{19} \text{ kg/m}^3)$, comparable to nuclear density. But we must keep in mind that the black hole density[5] $M/(r^*)^3$ actually has an inverse dependence on it mass (density $\propto M^{-2}$) because the Schwarzschild radius $r^* = 2G_N M/c^2$ increases with mass. Thus for a supermassive galactic back hole, for which $M = O(10^9 M_\odot)$, the density is less than that of water!

[5] Based on the fact that no measurements can be made inside of r^*, we chose to define the density of a black hole (regardless of the size and structure of the source) as the ratio of its mass to its spherical volume with radius r^*.

Embedding diagram

A helpful way to visualize warped space is to use an **embedding diagram**. Since it is difficult to work with the full three-dimensional curved space, we shall concentrate on the two-dimensional subspace corresponding to a fixed polar angle $\theta = \pi/2$ (and at some given instant of time). That is, we will focus on the 2D space slicing across the middle of the source. In the absence of gravity, this is just a flat plane as depicted in Fig. 7.1(a).

In the presence of gravity, this is a curved 2D space. We would like to have a way to visualize the warped nature of this 2D subspace (outside the source). A helpful way to do this is to imagine that this curved 2D surface is embedded in a fictitious 3D Euclidean space, Fig. 7.1(b). For the particular case of the Schwarzschild geometry, the line element with $dt = d\theta = 0$ is

$$ds^2 = \left(1 - \frac{r^*}{r}\right)^{-1} dr^2 + r^2 d\phi^2. \tag{7.23}$$

We now imagine this 2D surface as a subspace of a 3D Euclidean space with a cylindrical polar coordinate system: $x^i = (r, \phi, z)$. With the cylindrical axis being the z coordinate, we have chosen the polar radial (perpendicular) distance from the z axis of this 3D space to coincide with the Schwarzschild r coordinate, and the polar angle with the Schwarzschild ϕ angle. Just as a line-curve is represented in 3D space by $x^i(\lambda)$, with a single curve parameter λ, a 2D surface is represented by $x^i(\lambda, \sigma)$, with two surface parameters. In our case, we naturally choose the two parameters to be the Schwarzschild

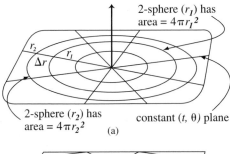

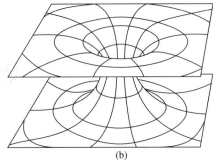

Fig. 7.1 (a) The $\theta = \pi/2$ plane (r, ϕ) cutting across the spherical source. (b) In a fictitious 3D embedding space, the physical 2D subspace of (a) is shown as a curved surface. In this example, we have used the Schwarzschild solution of $g_{rr} = (1 - r^*/r)^{-1}$.

coordinates (r, ϕ). Because of cylindrical symmetry (coming from the spherical symmetry with a fixed θ value), there is no ϕ-dependence in the metric. Thus, the only nontrivial relation is $z(r)$, which can be worked out as follows. The 3D space being Euclidean, we have the line element

$$ds^2 = dz^2 + dr^2 + r^2 d\phi^2$$

$$= \left[\left(\frac{dz}{dr} \right)^2 + 1 \right] dr^2 + r^2 d\phi^2. \qquad (7.24)$$

A comparison of (7.24) with (7.23) leads to $dz = \pm r^*/(r - r^*)^{1/2} dr$, which can be integrated to yield $z = \pm 2r^*(r - r^*)^{1/2}$, or

$$z^2 = 4r^* \left(r - r^* \right), \qquad (7.25)$$

which is a sideways parabola in the (r, z) plane with a fixed ϕ. After folding in the trivial ϕ dependence (so that the equal-r lines are simple circles in the r-ϕ plane with a fixed z), we have the 3D representation of the curved surface as shown in Fig. 7.1(b). This is a picture of the Einstein–Rosen bridge, which we shall encounter when discussing the wormhole of Schwarzschild geometry in Section 8.1.2.

Isotropic metric is time independent

We note from (7.17) that the scalar metric functions are time independent even though we have not assumed a constant source. This turns out to be a general result (**Birkhoff's theorem**): whenever the source is isotropic the resultant spacetime must necessarily be time independent. The theorem will be proven in Box 14.3. In the meantime it is worthwhile to point out that the same result holds for the Newtonian theory as well. Recall that the gravitational field outside a spherical source is identical to the gravitational field of a point source having all the mass at the center of the spherical source. This proof depends only on the symmetry property of the problem, and is not affected by any possible time dependencies. Thus, regardless of whether the spherical mass is pulsating or exploding, etc. the resultant field is the same, as long as the spherical symmetry is maintained. The analogous situation in electromagnetism is the statement that there is only dipole, quadrupole, etc., but no monopole, radiation.[6]

[6]By radiation we mean the production of wave fields: electromagnetic or gravitational waves.

7.2   Gravitational lensing

From the consideration of the equivalence principle, Einstein deduced (see Section 4.3.3) that there will be a bending of the starlight grazing past the sun. This effect is closely related to the idea of gravitational time dilation, expressed as a deviation from the flat space metric, see (6.3) and (7.19):

$$g_{00} = -\left(1 + \frac{2\Phi}{c^2}\right) = -1 + \frac{2G_N M}{c^2 r} = -1 + \frac{r^*}{r}, \tag{7.26}$$

which we see is part of the exact Schwarzschild solution (7.18). In the full GR theory, the warping of spacetime takes place not only in the time direction, $g_{00} \neq -1$, but also in the radial spatial direction, $g_{rr} \neq 1$. Here we calculate the effect of this extra warping on the bending of the light ray, finding a doubling of the deflection angle. The bending of a light ray by a massive object is analogous to the bending of a light ray by a lens. In Section 7.2.2 we shall present the lens equation, and discuss gravitational lensing as an important tool for modern astronomy.

7.2.1   Light ray deflection: GR vs. EP

[7]We can impose $d\theta = d\phi = 0$ here because the light-deflection calculation, as shown in Section 4.3.2, involves two parts. In the first, we calculate the speed of light $c(r)$ in terms of the gravitational potential; in the second part, the deflection $d\phi$ is related to this result for $c(r)$. To calculate the speed we can work with any propagation direction.

Let us consider the ligh-like worldline ($ds^2 = 0$) in a fixed direction,[7] $d\theta = d\phi = 0$:

$$ds^2 = g_{00}c^2 dt^2 + g_{rr} dr^2 = 0. \tag{7.27}$$

To an observer far from the source, using the coordinate time and radial distance (t, r), the effective speed of light according to (7.27) is

$$c(r) \equiv \frac{dr}{dt} = c\sqrt{-\frac{g_{00}(r)}{g_{rr}(r)}}. \tag{7.28}$$

A slightly different way of arriving at the same result is by noting that the speed of light is absolute in terms of local physical quantities, $c = d\rho/d\tau$, which is just (7.28) with the proper distance and time (ρ, τ) being related to the coordinate distance and time (r, t) by (7.15) and (7.16).

In the previous EP discussion, we ignored the possibility that the spatial direction might be curved; thus we effectively set $g_{rr} = 1$. Now the Schwarzschild solution (7.17) informs us that $g_{rr} = -g_{00}^{-1}$. The influence of $g_{00} \neq -1$ and $g_{rr} \neq 1$ in (7.28) is of the same size and in the same direction. Thus the deviation of the vacuum index of refraction $n(r)$ from unity is **twice** as large when spatial curvature is taken into account as it was when only the EP effect was taken into account as in (4.40):

$$n\left(r\right) = \frac{c}{c\left(r\right)} = \sqrt{-\frac{g_{rr}\left(r\right)}{g_{00}\left(r\right)}} = \frac{1}{-g_{00}\left(r\right)} = \left(1 + 2\frac{\Phi\left(r\right)}{c^2}\right)^{-1}. \qquad (7.29)$$

Namely, the retardation of a light signal (from c) is twice as large as that given in (4.39):

$$c\left(r\right) = \left(1 + 2\frac{\Phi\left(r\right)}{c^2}\right)c. \qquad (7.30)$$

According to Eqs. (4.39)–(4.45), the deflection angle $\delta\phi$, being directly proportional to this deviation from c, is twice as large[8] as that given by (4.47):

$$\delta\phi_{\text{GR}} = 2\delta\phi_{\text{EP}} = \frac{4G_N M_\odot}{c^2 r_{\text{min}}}, \qquad (7.31)$$

where r_{min} is the closest distance that the light ray comes to the massive object. We should apply $r_{\text{min}} = R_\odot$, the solar radius, for the case of a light ray grazing the edge of the sun.

The deflection of 1.74 arcseconds (about $1/4000$ of the angular width of the sun as seen from earth) that was predicted by Einstein's equations in 1915 was not easy to detect and therefore to test. One needed a solar eclipse against the background of several bright stars (so that some could be used as reference points). The angular position of a star with light grazing past the (eclipsed) sun would appear to have moved to a different position when compared to the location in the absence of the sun (see Fig. 4.6). On May 29, 1919 there was such a solar eclipse. Two British expeditions were mounted: one to Sobral in northern Brazil, and another to the island of Principe, off the coast of West Africa. The report by F.W. Dyson, A.S. Eddington, and C. Davidson that Einstein's prediction was successful in these tests created a worldwide sensation, partly for scientific reasons, and partly because the world was amazed that so soon after World War I the British should finance and conduct an expedition to test a theory proposed by a German citizen.

[8]This result will be re-derived (Problem 7.3) by using the geodesic equation for a light ray in Schwarzschild spacetime. For another calculation of the light geodesic, see Section 7.3.2 on the Shapiro time delay.

7.2.2   The lens equation

The Gravitational deflection of a light ray discussed above, Fig. 7.2(a), has some similarity to the bending of light by a glass lens, Fig. 7.2(b). The difference is that while a convex lens can focus images at a single point, gravitational lenses "focus" on a line, Fig. 7.3(b). This is so because the gravitational deflection angle is a decreasing function of the impact parameter $b = r_{\text{min}}$, as seen in (7.31), whereas for a light ray passing through a glass lens, the deflection angle is an increasing function of b. When the source and observer are sufficiently far from the lensing mass, bending from both sides of the lensing mass can produce double images, $S'_\pm$ in Fig. 7.4(a), and even a ring image, as in Fig. 7.3(a). Thus in general one finds stretched arc images of the light source, see, e.g. the photo in Fig. 7.5.

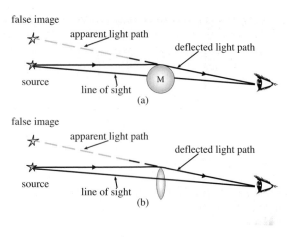

Fig. 7.2 Bending of a light ray as a lensing effect. (a) Light from a distant star (source) is deflected by a lensing mass M lying close to the line of sight from the observer to the source. As a result, the source star appears to be located at a different location. (b) Bending of light in (a) is analogous to that caused by a glass lens.

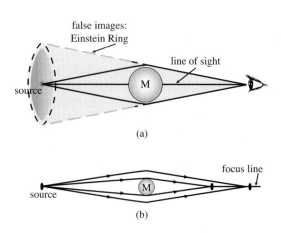

Fig. 7.3 Gravitational lensing of distant stars. (a) When the source and lensing mass are sufficiently far, double images can result. If the line of sight passes directly through the center of the symmetrical lensing mass distribution, the false image appears as a ring, the Einstein ring. (b) Gravitational lens "focuses" images on a line.

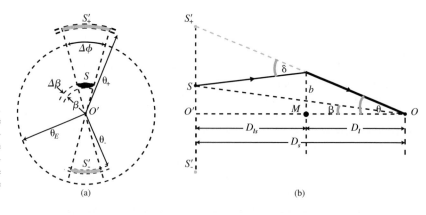

Fig. 7.4 Geometry of gravitational lensing. (a) Azimuthal and polar angle labels. (b) The observer and source are at O and S, respectively. The light ray is deflected by an angle δ by the lensing mass M. The true and apparent positions S and S' of the distant star are located by angles β and θ, respectively. The impact parameter is $b = r_{\min}$.

Fig. 7.5 Gravitational lensing effects due to the galaxy cluster Abell 2218. Nearly all of the bright objects in this picture taken by the Hubble Space Telescope are galaxies in this cluster, which is so massive and so compact that it lenses the light from galaxies that lie behind it into multiple images that appear as long faint arcs.

Solar deflection has been shown in our previous discussion to yield a bending angle of $\delta = 1.74''$. Since the deflection angle is so small, the earth is too close to the sun[9] to see lensing features such as multiple images of background stars near the sun. The minimum distance needed in order to see multiple images is $R_\odot/(1.74''$ in radians$) \simeq 0.8 \times 10^{14}$ m $\gtrsim 500$ AU.

To the extent that we can approximate the light trajectory by its asymptotes,[10] we can derive the lens equation by the simple geometrical consideration of Fig. 7.4(a). The distance SS' can be arrived at in two ways: (i) It subtends the angle δ to yield $SS' = D_{ls}\delta$, where D_{ls} is the distance between the lensing mass and the source light. (ii) It is the length difference $SS' = S'O' - SO' = D_s(\theta - \beta)$, where D_s is the distance from the observer to the source light and $(\theta - \beta)$ is the angular separation between the image and source points. Equating these two expressions and plugging in the previous result of (7.31) with $b = r_{min}$:

$$\delta = \frac{D_s}{D_{ls}}(\theta - \beta) = \frac{4G_N M}{bc^2}. \qquad (7.32)$$

Approximating the impact parameter $b \approx D_l\theta$, we then have the **lens equation**

$$\beta = \theta - \frac{D_{ls}}{D_s D_l}\frac{4G_N M}{\theta c^2}. \qquad (7.33)$$

After multiplication by θ, (7.33) is a quadratic equation in the variable θ. Such an equation usually has two solutions corresponding to the two images, resulting from bending on two sides of the lensing mass, Fig. 7.3(a).

In the special case of $\beta = 0$, i.e. perfect alignment of the source, lens and observer, the azimuthal axial symmetry of the problem yields a ring image, the **Einstein ring**, with angular radius

$$\theta_E = \sqrt{\frac{D_{ls}}{D_s D_l}\frac{4G_N M}{c^2}}. \qquad (7.34)$$

For the general case in which the source is not exactly behind the lens, the lens equation (7.33) for a single point lens (as when the lensing mass can be taken

[9]One astronomical unit AU $= 1.5 \times 10^{13}$ cm is the average distance from the earth to the sun.

[10]This is the *thin lens approximation*: all the action of the deflection is assumed to take place at one position. It is valid only if the relative velocities of the lens, source and observer are small compared to the velocity of light, $v \ll c$.

as a point)

$$\beta = \theta - \theta_E^2/\theta \tag{7.35}$$

has two solutions:

$$\theta_\pm = \frac{1}{2}\left(\beta \pm \sqrt{\beta^2 + 4\theta_E^2}\right), \tag{7.36}$$

corresponding to the presence of two images. One image is inside the would-be Einstein ring, $\theta_- < \theta_E$, and the other is outside. With the azimuthal angular width $\Delta\phi$ of the source unchanged, these two images are distorted into arcs, Fig. 7.4(a). Furthermore, if the distance to the source D_s and to the lens D_l can be estimated, the mass of the lens can be deduced via (7.34) by measurements of $\theta_\pm$ (hence θ_E).

Since 1979, astronomers have discovered several dozen "double quasars," or what appears to be two quasars[11] having the same properties but separated by a few arcseconds or less. Actually, a "double quasar" is two images of the same quasar. The lensing masses that are responsible for such sizable image separations are expected to be galaxies with masses of a billion or more solar masses. Even more dramatically, if the lens is not a single galaxy but an entire cluster of galaxies, the images can be clusters of distorted arcs. In Fig. 7.5, we display the distorted images of distant galaxies as lensed by the cluster Abell 2218.

[11] Quasars (quasi-stellar objects) are very luminous sources of small angular size at great cosmic distances. See the further discussion in Section 8.3.2.

Box 7.2 Microlensing and the search for MACHOs

Gravitational lensing by stellar objects is typically too small to produce multiple images (i.e. the separate images cannot be resolved). Such **microlensing** events show up, because of the overlap of images, as an increase of luminosity flux of lensed sources. For the point lens discussed here, we can calculate the magnification factor by noting that the light intensity (flux per unit solid angle of the source/image) is the same for each of the images $I = I_+ = I_-$ because they have the same source energy. Thus their flux is proportional to their respective subtended solid angles Ω, and the magnification is the ratio of the combined image to that of the original flux in the absence of lensing mass:

$$\mu = \frac{f_+ + f_-}{f} = \frac{I_+ d\Omega_+ + I_- d\Omega_-}{I d\Omega} = \frac{\theta_+ d\theta_+ + \theta_- d\theta_-}{\beta d\beta}, \tag{7.37}$$

where we have used $d\Omega = \sin\theta d\theta d\phi \simeq \theta d\theta d\phi$ and the fact that the azimuthal angular width is unchanged, $d\phi = d\phi_+ = d\phi_-$. We can calculate the individual magnification (either $+$ or $-$ image) by a simple differentiation of (7.35):

$$\frac{\theta d\theta}{\beta d\beta} = \left[1 - \left(\frac{\theta_E}{\theta}\right)^4\right]^{-1}. \tag{7.38}$$

Plugging in the $\theta = \theta_\pm$ solutions of (7.36), we then obtain (Problem 7.4), using $\hat{\beta} = \beta/\theta_E$, the total magnification:

$$\mu = \frac{\hat{\beta}^2 + 2}{\hat{\beta}\sqrt{\hat{\beta}^2 + 4}} > 1. \tag{7.39}$$

Especially when $\hat{\beta} \to 0$, the magnification $\mu \propto 1/\hat{\beta}$ due to the whole Einstein ring-image can be quite significant. As an example of gravitational lensing being a powerful tool of modern astronomy, evidence for the existence of "massive compact halo objects" or MACHOs has been obtained this way. Such nonluminous objects may be dead stars, black holes, or other massive objects that do not shine, and are collectively called MACHOs. If a MACHO drifts in front of a background star, it will act as a lensing mass, and thus it will enhance the brightness of that star temporarily. Several astronomical teams undertook the search for MACHOs by simultaneously monitoring millions of stars in the Large Magellanic Cloud (a small satellite galaxy of the Milky Way). In 1997, the MACHO collaboration project (Alcock *et al.*, 1997) announced the discovery of several "lensing events," lasting a few weeks to several months, indicating the presence of such massive nonluminous objects. In Section 9.2.2 we shall present another application of lensing in the study of nonluminous astrophysical objects. In that case, the Bullet Cluster, one can clearly see that a major portion of the nonluminous matter has no electromagnetic interaction (called dark matter).

7.3 Geodesics in Schwarzschild spacetime

A particle's motion in a curved spacetime (e.g. the Schwarzschild spacetime) is determined by the GR equation of motion, the geodesic equation. As we recall our discussion in Section 5.2.1, this equation is the Euler–Lagrange equation

$$\frac{\partial L}{\partial x^\mu} = \frac{d}{d\tau} \frac{\partial L}{\partial \dot{x}^\mu}, \tag{7.40}$$

with the Lagrangian $L(x, \dot{x})$ being the particle's 4-velocity squared,

$$L = \left(\frac{ds}{d\tau}\right)^2 = g_{\mu\nu}\dot{x}^\mu \dot{x}^\nu. \tag{7.41}$$

As is appropriate for a massive test particle, we have picked its proper time τ to be the curve parameter and used the notation $\dot{x}^\mu = dx^\mu/d\tau$. The dependence of L on the coordinates is through the metric tensor $g_{\mu\nu}$.

To find the particle orbit, we first use Eq. (7.40) to identify the constants of motion (i.e. the conserved quantities along the particle path): If the metric $g_{\mu\nu}$ is independent of a particular coordinate, say x^0, then $\partial L/\partial x^0 = 0$, which can be translated, through (7.40), into a conservation statement of $d(\partial L/\partial \dot{x}^0)/d\tau = 0$. However, the constant of motion written this

[12]The subscript "(t)" on $K_{(t)}^{\mu}$ is not a tensor index. It is a label indicating which coordinate the metric is independent of. While $K_{(t)}^{\mu} = (1, 0, 0, 0)$ is clearly a coordinate-dependent expression, the variation of its "length" $g_{\mu\nu}K_{(t)}^{\mu}K_{(t)}^{\nu}$ reflects the evolving underlying geometry (as encoded in the metric).

[13]Only for a massive particle do we have $L = -c^2$; for massless particles (see the discussion of 4-momentum in Section 3.2.2) we have $L = 0$. Exactly the same approach can be used for the orbits of a light ray, but with the equation $L = 0$. Such examples as the Shapiro time delay and gravitational light deflection are worked out in Section 7.3.2 and in Problem 7.3.

way, $\partial L/\partial \dot{x}^0 = 2g_{\mu 0}\dot{x}^{\mu}$, is coordinate dependent. To overcome this shortcoming, we introduce the **Killing vector**, $K_{(t)}^{\mu} = (1, 0, 0, 0)$, for a t-independent metric and the above conserved quantity can then be expressed as a coordinate scalar $g_{\mu\nu}\dot{x}^{\mu}K_{(t)}^{\nu}$, as all the Lorentz indices are contracted.[12] We shall adopt the following systematic procedure: (1) identify the coordinate symmetry (i.e which coordinates $x_{(a)}^{\mu}$ the metric is independent of), and (2) list the corresponding Killing vector(s) $K_{(a)}^{\mu}$. The constants of motion can then be identified as the coordinate scalars:

$$\gamma_{(a)} = g_{\mu\nu}\dot{x}^{\mu}K_{(a)}^{\nu}. \tag{7.42}$$

With some of the $\dot{x}$'s being fixed, we can often express the remaining $\dot{x}$ in terms of these constants of motion using the simple equation[13] expressing the normalization of the particle's 4-velocity:

$$L = \left(\frac{ds}{d\tau}\right)^2 = -c^2 \tag{7.43}$$

because $ds^2 = -c^2 d\tau^2$. In this way we can obtain an energy balance equation, from which we can eventually arrive at the orbit equation.

We shall apply this procedure in the following applications of finding the geodesics in the Schwarzschild geometry, first the geodesics of a massive particle in Section 7.3.1, and then that of a light ray in Section 7.3.2. In particular we shall calculate the perihelion precession of the planet Mercury; this is the first GR success in explaining an astronomical puzzle.

7.3.1   Precession of Mercury's perihelion

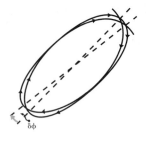

Fig. 7.6 A perturbed $1/r^2$ attraction leads to an open elliptical orbit which may be described as an elliptical orbit with a precessing axis. For planetary motion, this is usually stated as the precession of the minimal-distance point from the sun, the perihelion.

In this section we shall discuss the motion of a test mass in Schwarzschild spacetime. A particle, under the Newtonian $1/r^2$ gravitational attraction, has an elliptical orbit. Here we calculate the GR correction to such a trajectory.

Celestial mechanics based on the Newtonian theory of gravitation has been remarkably successful. However, it had been realized around 1850 that there was a discrepancy between the theory and the observed **precession of the perihelion of the planet Mercury**. The pure $1/r^2$ force law of Newton predicts a closed elliptical orbit for a planet, i.e. an orbit with an axis **fixed** in space. However, the perturbations due to the presence of other planets and astronomical objects lead to a trajectory that is no longer closed. Since the perturbation is small, such a deviation from the closed orbit can be described as an ellipse with a **precessing** axis, or equivalently, a precessing perihelion (the point of the orbit closest to the sun), Fig. 7.6.

[14]Most of the raw observational value of $5600''$ is due to the effect of rotation of our earth-based coordinate system and it leaves the planetary perturbations to account for the remaining $574''$, to which Venus contributes $277''$, Jupiter $153''$, Earth $90''$, Mars and the rest of the planets $10''$.

For the case of Mercury, such planetary perturbation could account for most[14] of the planetary perihelion advances of 574 arcseconds per century. However there was still the discrepancy of 43 arcseconds left unaccounted for. Following a similar situation involving Uranus that eventually led to the prediction and discovery of the outer planet Neptune in 1846, a new planet, named Vulcan, was predicted to lie inside Mercury's orbit. But it was never found. This is the "perihelion precession problem" that Einstein solved by applying his new theory of gravitation. As we shall presently see, GR implies a small correction to the $1/r^2$ force law, which just accounts for the missing $43''$ advance of Mercury's orbit.

As in Newtonian mechanics, because angular momentum is conserved under the action of a central force, the trajectory will always remain in the plane spanned by the particle's initial velocity and the vector $\vec{r}$ connecting the force center to the test particle. By setting $\theta = \pi/2$, the Lagrangian (7.43) with the Schwarzschild metric for a particle with mass takes the form of

$$L(x, \dot{x}) = -\left(1 - \frac{r^*}{r}\right)c^2\dot{t}^2 + \left(1 - \frac{r^*}{r}\right)^{-1}\dot{r}^2 + r^2\dot{\phi}^2 = -c^2. \tag{7.44}$$

Two constants of motion L does not explicitly depend on t and ϕ. The two corresponding Killing vectors in the Schwarzschild coordinates of $x^\mu = (ct, r, \theta, \phi)$ are:

$$K_{(t)}^\mu = (1, 0, 0, 0) \quad \text{and} \quad K_{(\phi)}^\mu = (0, 0, 0, 1), \tag{7.45}$$

leading to two constants of motion, as shown in (7.42),

$$\kappa \equiv -g_{\mu\nu}\dot{x}^\mu K_{(t)}^\nu = g_{00}c\dot{t} = \left(1 - \frac{r^*}{r}\right)c\dot{t}, \tag{7.46}$$

and, for $\theta = \pi/2$,

$$l \equiv mg_{\mu\nu}\dot{x}^\mu K_{(\phi)}^\nu = mg_{\phi\phi}\dot{\phi} = mr^2\dot{\phi}. \tag{7.47}$$

Energy balance equation Plugging in the constants l and κ, the $L = -c^2$ equation (7.44) becomes

$$-\frac{\kappa^2}{1 - r^*/r} + \frac{\dot{r}^2}{1 - r^*/r} + \frac{l^2}{m^2r^2} = -c^2. \tag{7.48}$$

After multiplying by $\frac{1}{2}m(1 - r^*/r)$, and using $r^* = 2G_N M/c^2$ on the RHS, this may be written as

$$-\frac{m\kappa^2}{2} + \frac{1}{2}m\dot{r}^2 + \left(1 - \frac{r^*}{r}\right)\frac{l^2}{2mr^2} = -\frac{1}{2}mc^2\left(1 - \frac{2G_N M}{c^2 r}\right) \tag{7.49}$$

or

$$\frac{1}{2}m\dot{r}^2 + \left(1 - \frac{r^*}{r}\right)\frac{l^2}{2mr^2} - \frac{G_N mM}{r} = \frac{m\kappa^2}{2} - \frac{1}{2}mc^2. \tag{7.50}$$

Renaming the constant κ as

$$\frac{\mathcal{E}}{m} = \frac{\kappa^2 - c^2}{2}, \tag{7.51}$$

this equation takes on the form

$$\frac{1}{2}m\dot{r}^2 + \left(1 - \frac{r^*}{r}\right)\frac{l^2}{2mr^2} - \frac{G_N mM}{r} = \mathcal{E}. \tag{7.52}$$

Except for the $(1 - r^*/r)$ factor, this is just the energy balance equation for the nonrelativistic central force problem—kinetic energy plus potential energy equals the total (Newtonian) energy $\mathcal{E}$. The extra factor of

$$-\frac{r^*}{r}\frac{l^2}{2mr^2} = -\frac{G_N M l^2}{mc^2 r^3} \tag{7.53}$$

may be regarded, for the problem of the particle's orbit, as a small GR correction to the Newtonian potential energy $-G_N m M r^{-1}$ by a r^{-4} type of force.

The precessing orbit   This relativistic energy equation (7.52) can be cast in the form of an orbit equation. We can solve for $r(\phi)$ by standard perturbation theory (see Box 7.3). With e being the eccentricity of the orbit, $\alpha = l^2/G_N M m^2 = (1+e)r_{\min}$ and $\epsilon = 3r^*/2\alpha$, the solution is

$$r = \frac{\alpha}{1 + e \cos\left[(1-\epsilon)\phi\right]}. \tag{7.54}$$

Thus the planet returns to its perihelion $r_{\min}$ not at $\phi = 2\pi$ but at $\phi = 2\pi/(1-\epsilon) \simeq 2\pi + 3\pi r^*/\alpha$. The perihelion advances (i.e. the whole orbit rotates in the same sense as the planet itself) per revolution by (Fig. 7.6)

$$\delta\phi = \frac{3\pi r^*}{\alpha} = \frac{3\pi r^*}{(1+e)\,r_{\min}}. \tag{7.55}$$

With the solar Schwarzschild radius $r_\odot^* = 2.95\,\mathrm{km}$, Mercury's eccentricity $e = 0.206$, and its perihelion $r_{\min} = 4.6 \times 10^7\,\mathrm{km}$, we have the numerical value of the advance as

$$\delta\phi = 5 \times 10^{-7}\ \text{radian per revolution} \tag{7.56}$$

or $5 \times 10^{-7} \times \frac{180}{\pi} \times 60 \times 60 = 0.103''$ (arcsecond) per revolution. In terms of the advance per century,

$$0.103'' \times \frac{100\,\text{years}}{\text{Mercury period of 0.241 years}} = 43''\ \text{per century.} \tag{7.57}$$

This agrees with the observational evidence.

This calculation explaining the perihelion advance of the planet Mercury from first principles, and the correct prediction for the bending of starlight around the sun, were all obtained by Albert Einstein in an intense two week period in November, 1915. Afterwards, he wrote to Arnold Sommerfeld in a, by now, famous letter.

This last month I have lived through the most exciting and the most exacting period of my life; and it would be true to say this, it has been the most fruitful. Writing letters has been out of the question. I realized that up until now my field equations of gravitation have been entirely devoid of foundation. When all my confidence in the old theory vanished, I saw clearly that a satisfactory solution could only be reached by linking it with the Riemann variations. The wonderful thing that happened then was that not only did Newton's theory result from it, as a first approximation, but also the perihelion motion of Mercury, as a second approximation. For the deviation of light by the Sun I obtained twice the former amount.

We have already discussed the doubling of the light deflection angle. The topics of the Riemannian curvature tensor and the GR field equation (having the correct Newtonian limit) will be taken up in Part IV—in Section 13.3 and 14.2, respectively.

Box 7.3 The orbit equation and its perturbation solution

We solve this relativistic energy equation (7.52) as a standard central force problem. The relevant kinematic variables are shown in Fig. 7.7.

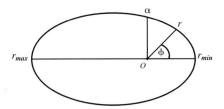

Fig. 7.7 Points on an elliptical orbit are located by the coordinates (r, ϕ), with some notable positions at $(r_{min}, 0)$, (r_{max}, π), and $(\alpha, \pi/2)$.

The orbit equation To obtain the orbit equation $r(\phi)$ we first change all the time derivatives into differentiation with respect to the angle ϕ by using the angular momentum equations (7.47):

$$d\tau = \frac{mr^2}{l} d\phi \qquad (7.58)$$

and then making the change of variable $u \equiv 1/r$ (and thus $u' \equiv du/d\phi = -u^2(dr/d\phi)$). In this way (7.52) turns into

$$u'^2 + u^2 - \frac{2}{\alpha}u - r^*u^3 = C \qquad (7.59)$$

where $\alpha = \lambda^2/G_N M$, and C is some definite constant. This is the equation we need to solve in order to obtain the planet orbit $r(\phi)$.

Zeroth-order solution Split the solution $u(\phi)$ into an unperturbed part u_0 and a small correction: $u = u_0 + u_1$ with

$$u_0'^2 + u_0^2 - \frac{2}{\alpha}u_0 = C. \qquad (7.60)$$

This unperturbed orbit equation can be solved by differentiating with respect to ϕ and dividing the resultant equation by $2u'$:

$$u_0'' + u_0 = \alpha^{-1} \qquad (7.61)$$

which is a simple harmonic oscillator equation in the variable $(u_0 - \alpha^{-1})$, with ϕ the "time" variable and $\omega = 1$ the "angular frequency" . It has the solution $(u_0 - \alpha^{-1}) = D \cos \phi$. We choose to write the constant $D \equiv e/\alpha$ so that the solution takes on the well-known form of a conic section,

$$r = \frac{\alpha}{1 + e \cos \phi}. \qquad (7.62)$$

(*cont.*)

Box 7.3 (*Continued*)

It is clear (see Fig. 7.7) that we have $r = \alpha/(1+e) = r_{\min}$ (perihelion) at $\phi = 0$ and $r = \alpha/(1-e) = r_{\max}$ (aphelion) at $\phi = \pi$. Geometrically, e is called the eccentricity of the orbit. The radial distance at $r(\phi = \pi/2) = \alpha$ can be expressed in terms of the perihelion and eccentricity as

$$\alpha = (1+e)\, r_{\min}. \tag{7.63}$$

Relativistic correction   We now plug $u = u_0 + u_1$ into (7.59)

$$\left(u_0' + u_1'\right)^2 + (u_0 + u_1)^2 - \frac{2}{\alpha}(u_0 + u_1) - r^* (u_0 + u_1)^3 = C$$

and separate out the leading and the next leading terms, with $u_1 = O(r^*)$:

$$\left(u_0'^2 + u_0^2 - \frac{2}{\alpha}u_0 - C\right) + \left(2u_0'u_1' + 2u_0u_1 - \frac{2}{\alpha}u_1 - r^*u_0^3\right) = 0, \tag{7.64}$$

where we have neglected writing out higher order terms proportional to u_1^2, r^{*2} or $u_1 r^*$. After using (7.60), we can then pick out the first-order equation:

$$2u_0'u_1' + 2u_0u_1 - \frac{2}{\alpha}u_1 = r^*u_0^3, \tag{7.65}$$

where

$$u_0 = \frac{1 + e\cos\phi}{\alpha}, \qquad u_0' = -\frac{e}{\alpha}\sin\phi. \tag{7.66}$$

The equation for u_1 is then given by

$$-e\sin\phi\frac{du_1}{d\phi} + e\cos\phi\, u_1 = \frac{r^*(1 + e\cos\phi)^3}{2\alpha^2}. \tag{7.67}$$

One can verify that it has the solution

$$u_1 = \frac{r^*}{2\alpha^2}\left[\left(3 + 2e^2\right) + \frac{1 + 3e^2}{e}\cos\phi - e^2\cos^2\phi + 3e\phi\sin\phi\right]. \tag{7.68}$$

The first two terms have the form of the zeroth-order solution, $(A + B\cos\phi)$; thus they represent unobservably small corrections. The third term, being periodic in ϕ in the same way as the zeroth-order term, is also unimportant. We only need to concentrate on the fourth term[15] which is ever-increasing with ϕ (modulo 2π). Plugging this into $u = u_0 + u_1$, we obtain

$$r = \frac{\alpha}{1 + e\cos\phi + \epsilon e\phi\sin\phi}. \tag{7.69}$$

The denominator factor $\epsilon = 3r^*/2\alpha$ being a small quantity, the angular terms in the denominator can be cast in the same form as that for the zeroth-order solution (7.62): after approximating $\cos\epsilon\phi \simeq 1$ and $\sin\epsilon\phi \simeq \epsilon\phi$ so

[15]While the detection of the higher-order effects represented by the first three terms would require measurements at impossibly high accuracy, the fourth term, while equally small, is a new effect (it is a "correction to a zero") that has a much better chance for its measurement.

that

$$e \cos (\phi - \epsilon \phi) \simeq e \, (\cos \phi + \epsilon \phi \sin \phi) \,, \tag{7.70}$$

we have the solution (7.69) in the more transparent form of

$$r = \frac{\alpha}{1 + e \cos [(1 - \epsilon) \, \phi]} \tag{7.71}$$

as shown in (7.54).

7.3.2   The Shapiro time delay of a light signal

We now consider the geodesic of a light ray in the Schwarzschild spacetime. The effect of light deflection has already been calculated in Section 4.3.3 (and 7.2.1) using Huygen's principle. It can also be derived by a procedure entirely similar to that used in our discussion of perihelion precession:[16] here we study the geodesics for a (massless) light ray, while in Section 7.3.1 we studied the geodesic for a particle with mass. In this section we shall work out the related phenomenon of the time delay of a light ray in the solar gravitational field. Irwin I. Shapiro first proposed, and then led the effort to measure, this effect; hence it is called "Shapiro time delay" (Shapiro, 1964, 1968).

Time delay defined   As we have already discussed in Chapter 4, the speed of light propagation is affected by a gravitational field, which can be thought of as giving rise to an effective index of refraction in the vacuum. In the language of gravity as curved spacetime, the light trajectory in a curved spacetime will be different from that in a flat spacetime, Fig. 7.8. Consequently the duration for a light signal making the round trip between two points A and B in the Schwarzschild spacetime (due to a mass M) will be longer than that in the flat spacetime. Let r_0 be the radial distance of the "point of closest approach to M" (see Fig. 7.8). We label the time that it takes a light ray to propagate from the emission point A to r_0 as $t(r_A, r_0)$; the total elapsed time for a round-trip between A and the reflection point B is then given by

$$\Delta t_{tot} = 2 \, [t \, (r_A, r_0) + t \, (r_B, r_0)] \,. \tag{7.72}$$

The "time-delay," $\Delta t_\delta = \Delta t_{tot} - \Delta t_0$, is defined as the difference of this total time and the time over the flat-space duration,

$$\Delta t_0 = \frac{2}{c} \left(\sqrt{r_A^2 - r_0^2} + \sqrt{r_B^2 - r_0^2} \right), \tag{7.73}$$

calculated from a simple inspection of the geometry as shown in Fig. 7.8(b).

Killing vectors and constants of motion   We follow closely the procedure used in Section 7.3.1. The light-like worldline $x^\mu(\lambda)$ is chosen to reside on the equatorial plane with $\theta = \pi/2$. Even though the curve parameter σ cannot be identified with the proper time, the derivative $\dot{x}^\mu = dx^\mu/d\lambda$ can still be

[16]The reader is invited to employ this approach to rederive the light deflection result (Problem 7.3).

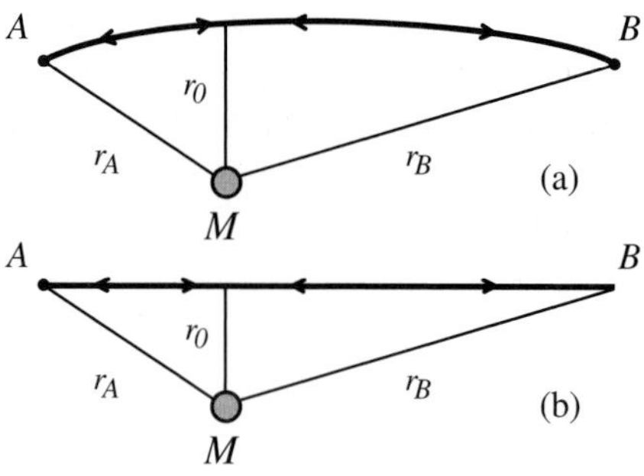

Fig. 7.8 (a) Light ray sent from A to B and reflected back in the gravitational field of a spherical mass M follows a curved path. The elapsed time is longer than that of (b) with no gravity, as the light follows a straight path in a flat spacetime.

thought of as the photon's "4-velocity," and the same Killing vectors (7.45) lead, just as in (7.46) and (7.47), to the conserved quantities

$$\kappa = -g_{\mu\nu}\dot{x}^\mu K^\nu_{(t)} = -g_{00}c\dot{t} = \left(1 - \frac{r^*}{r}\right)c\frac{dt}{d\lambda} \tag{7.74}$$

and

$$j = g_{\mu\nu}\dot{x}^\mu K^\nu_{(\phi)} = g_{\phi\phi}\dot{\phi} = r^2\frac{d\phi}{d\lambda}. \tag{7.75}$$

Time duration as a radial integral   We can express the time duration for a light ray traveling from r_1 to r_2 as an integral:

$$t(r_1, r_2) = \int dt = \int_{r_1}^{r_2} \frac{dt}{dr}dr = \int_{r_1}^{r_2} \frac{dt/d\lambda}{dr/d\lambda}dr. \tag{7.76}$$

The numerator of the integrand $dt/d\lambda$ can be directly related to the energy constant κ of (7.74); the denominator $dr/d\lambda$ can be expressed as a function of the radial distance r through the light-like condition $L = g_{\mu\nu}\dot{x}^\mu\dot{x}^\nu = 0$. Plugging in the explicit form of the Schwarzschild metric and the two constants of motion (κ, j), after some rearrangement this $L = 0$ equation becomes

$$\frac{1}{b^2} = \frac{1}{j^2}\left(\frac{dr}{d\lambda}\right)^2 + \frac{1}{r^2}\left(1 - \frac{r^*}{r}\right), \tag{7.77}$$

where $b \equiv j/\kappa$ is the impact parameter.[17] Combining this equation in the form of

$$\frac{dr}{d\lambda} = j\left[\frac{1}{b^2} - \frac{1}{r^2}\left(1 - \frac{r^*}{r}\right)\right]^{1/2} \tag{7.78}$$

with (7.74) solved for $dt/d\lambda$, the integrand in (7.76) becomes

$$\frac{dt}{dr} = \frac{dt/d\lambda}{dr/d\lambda} = \frac{1}{c}\left(1 - \frac{r^*}{r}\right)^{-1}\left[1 - \frac{b^2}{r^2}\left(1 - \frac{r^*}{r}\right)\right]^{-1/2}. \tag{7.79}$$

[17]When a light ray is far away from the source M, it travels along a straight line at $dr/dt = c$. The impact parameter b is defined to be the perpendicular distance between this straight trajectory and the parallel line passing directly through M. In this asymptotic region we have $\phi = b/r$, or $|d\phi/dr| = b/r^2$, and $|d\phi/dt| = |(d\phi/dr)(dr/dt)| = bc/r^2$. Through this result and the respective definitions of κ by (7.74) in the limit of $r \gg r^*$ and of j by (7.75), we see that their ratio $|j/\kappa|$ is the impact parameter: $|j/\kappa| = (r^2/c)|\dot{\phi}/\dot{t}| = (r^2/c)|d\phi/dt| = b$.

The $O(r^*)$ time delay We are interested in calculating the leading GR correction, which is $O(r^*)$, to the time duration. The zeroth-order terms will represent the flat-space time duration, and the first-order terms will represent the difference between the curved-space time duration and the flat-space time duration, or the time delay. For this purpose we can show that the impact parameter b is just r_0 with a small correction: the point of closest approach r_0 being the turning point $(dr/d\lambda)_{r_0} = 0$, we have from (7.77):

$$b^2 = r_0^2 \left(1 - \frac{r^*}{r_0}\right)^{-1} \simeq r_0^2 + r_0 r^*. \tag{7.80}$$

After a systematic expansion (Problem 7.6), we find an expression for the derivative (7.79) that clearly separates the zeroth- and the first-order terms in r^*:

$$c\frac{dt}{dr} = \left(1 - \frac{r_0^2}{r^2}\right)^{-1/2}\left(1 + \frac{r^*}{r}\frac{r + \frac{3}{2}r_0}{r + r_0}\right). \tag{7.81}$$

The zeroth-order term leads to the flat-space duration:

$$ct_0(r_1, r_2) = \int_{r_1}^{r_2} \frac{dr}{\sqrt{1 - \left(\frac{r_0}{r}\right)^2}} = \sqrt{r_2^2 - r_0^2} - \sqrt{r_1^2 - r_0^2}, \tag{7.82}$$

which is just the expected result of $ct_0(r, r_0) = \sqrt{r^2 - r_0^2}$ displayed in (7.73). The time delay is obtained by integrating over the r^* term in (7.81):

$$ct_\delta(r, r_0) = \int_{r_0}^{r} \frac{dr}{\sqrt{1 - \left(\frac{r_0}{r}\right)^2}}\left(\frac{r^*}{r}\frac{r + \frac{3}{2}r_0}{r + r_0}\right)$$

$$= \frac{r^*}{2}\sqrt{\frac{r - r_0}{r + r_0}} + r^* \ln\left(\frac{r + \sqrt{r^2 - r_0^2}}{r_0}\right), \tag{7.83}$$

which for $r \gg r_0$ becomes

$$ct_\delta(r, r_0) = \frac{r^*}{2} + r^* \ln\left(\frac{2r}{r_0}\right) \simeq r^* \ln\left(\frac{2r}{r_0}\right). \tag{7.84}$$

Thus, when the distance from the gravitational source M to the closest point r_0 is much smaller than either of the distances to A or to B, the total time delay for a light pulse traveling round trip between A and B in the gravitation field due to the spherical mass M, obtained by using (7.84) four times, once for each of the four intervals involved in the round-trip path of the light ray, has the simple logarithmic form:

$$\Delta t_\delta = \frac{4G_N M}{c^3} \ln\left(\frac{4r_A r_B}{r_0^2}\right). \tag{7.85}$$

[18]This is possible because earth's orbiting plane is slightly different from those of the planets used to reflect the electromagnetic wave.

Experimental test   Intense effort was mounted to verify this time-delay feature by detecting the echo of the radar signals bouncing back to earth from another planet, e.g. Venus, when the electromagnetic pulses pass close to the sun. Even for smallest $r_0 = R_\odot$, the solar radius, when the earth, the sun and the planet are all lined up (the superior conjunction[18]), the predicted delay for the inner planets is still small: $\Delta t_\delta = O(200\,\mu s)$ with $\Delta t_\delta / \Delta t_{\text{tot}} \simeq 10^{-9}$. Furthermore, there are other background effects (most notably the solar corona) that can obscure this result. However, one very effective way to pick out this tiny effect is by using its characteristic logarithmic dependence on r_A, r_B, and r_0, which are all varying (Fig. 7.9) over a period of sending radar signals from the circulating earth and reflecting these signals by a radar transponder on another circulating planet. In Fig. 7.10, we show the observational result obtained by Shapiro *et al.* in a radar echo experiment (off the planet Venus) performed at the Haystack and Arecibo radio observatories. This verified the GR result to an accuracy of 2%. In the intervening years, using satellites and other radio wave receivers the time delay experiments have been improved to a tenth of a percent accuracy.

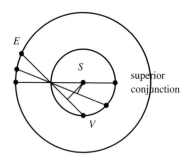

Fig. 7.9 Schematic representation of light pulse exchange between earth and Venus while both are circulating around the sun. The distance of closest approach r_0 varies, attaining the smallest value at superior conjunction.

Fig. 7.10 Earth–Venus time-delay experiment. The GR result (solid line) is compared to the observation result obtained at the Haystack and Arecibo observatories. (From Shapiro *et al.* 1971.)

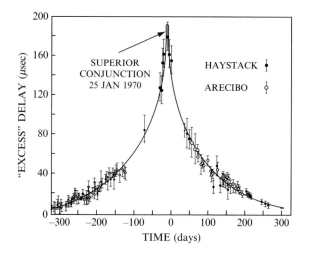

Review questions

1. What is the **form** of the spacetime metric (when written in terms of the spherical coordinates) for a spherically symmetric space? Explain very briefly how such a spacetime is curved in space as well as in time.

2. Present a simple proof of **Birkhoff's theorem** for Newtonian gravity. Explain how one then concludes that there is no monopole radiation.

3. Write down the metric elements for the Schwarzschild spacetime (i.e. the Schwarzschild solution). If in addition one is told that the metric element g_{00} is related to the gravitational potential as $g_{00} = -(1 + 2\Phi/c^2)$ in the Newtonian limit, demonstrate that the Newtonian result $\Phi = -G_N M/r$ is contained in this Schwarzschild solution.

4. A light ray is bent in the presence of a gravitational field, with the angle of deflection $(d\phi)$ being proportional to the deviation of the index of refraction (dn) from unity, as shown in (4.43). How does the feature $g_{rr} = -g_{00}^{-1}$ in the Schwarzschild metric lead to a bending of the light ray in GR which is twice as much as that predicted by the equivalence principle alone, $\delta\phi_{GR} = 2\delta\phi_{EP}$?

5. In simple qualitative terms, explain how gravitational lensing can, in some circumstances, give rise to "Einstein rings," and, in some cases, an enhancement of the brightness of a distant star.

6. In simple qualitative terms, explain how the GR effect causes the perihelion of an elliptical orbit to precess.

7. If the metric does not depend on a certain coordinate, say the ϕ angle of the (t, r, θ, ϕ) coordinates, what is the associated Killing vector $K^\mu_{(\phi)}$? Use the geodesic equation to show that this leads to a constant of motion: $\gamma_{(\phi)} = g_{\mu\nu}\dot{x}^\mu K^\nu_{(\phi)}$.

8. Once the constants of motion $\kappa \sim dt/d\lambda$ and $j \sim d\phi/d\lambda$ have been identified, how would you go about obtaining the total angular change,

$$\Delta\phi = \int_{r_1}^{r_2} \frac{d\phi}{dr}dr,$$

of a light ray traveling from r_1 to r_2? Outline your approach for finding the integrand as a function of the radial distance r.

Problems

7.1 **Energy relation for a particle moving in the Schwarzschild spacetime** Show that (7.44), expressing the invariant line element for a material particle $ds^2 = -c^2 d\tau^2$, can be interpreted as the Schwarzschild spacetime generalization of the familiar SR relation (3.37) between energy and momentum. That is, show that the flat spacetime version of (7.44) with $r^* = 0$ can be written as $E^2 = p^2 c^2 + m^2 c^4$.

7.2 **Gravitational redshift via energy conservation** Use the concept of the Killing vector $K^\mu_{(t)}$ and the associated constant of motion $g_{\mu\nu} p^\mu K^\nu_{(t)}$ for a light geodesic to rederive the gravitational redshift $(\Delta\omega)/\omega = -(\Delta\Phi)/c^2$ by way of $\omega_{em}/\omega_{rec} = \sqrt{(g_{00})_{rec}/(g_{00})_{em}}$ (as first shown in Section 6.2.2). Here one should choose the light geodesic curve parameter σ so that $\dot{x}^\mu = dx^\mu/d\lambda = p^\mu$, the photon 4-momentum. We note that the photon frequency (being proportional to its energy) as measured by an observer (at rest with respect in a static gravitational field) is related to the 4-velocity U^μ of the observer as $\omega_{em} \sim g_{\mu\nu} p^\mu U^\nu_{em} = (g_{00} p^0 U^0)_{em}$ and $\omega_{rec} \sim (g_{00} p^0 U^0)_{rec}$ with $(g_{\mu\nu} U^\mu U^\nu)_{em} = (g_{\mu\nu} U^\mu U^\nu)_{rec} = -c^2$.

7.3 **Light deflection via the geodesic equation** We have used Huygen's principle to derive in Section 4.3.2 the equivalence principle expression of gravitational angular deflection $\delta\phi_{EP}$. This is then modified in Section

7.2.1 when we also take into account the bending of space to obtain $\delta\phi_{GR}$ of (7.31). Here you are asked to obtain this result following the more standard procedure, as in Sec 7.3.2, of using directly the geodesic equation of a light ray. Plugging in the constants of motion as given in (7.74) and (7.75), the photon Lagrangian $L = 0$ equation becomes (7.77). Following the same steps as given in Box 7.3, you can change the differentiation with respect to the curve parameter λ to that of the orbit angle, $d\lambda = (r^2/j)d\phi$, and use the variable $u = r^{-1}$ to obtain the equation, equivalent to (7.59), for the light ray geodesic:

$$u'' + u - \epsilon u^2 = 0$$

where $u'' = d^2 u/d\phi^2$ and $\epsilon = 3r^*/2$. A perturbation solution $u = u_0 + \epsilon u_1$ should lead to the result accurate up to first order in r^*:

$$\frac{1}{r} = \frac{\sin\phi}{r_{min}} + \frac{3 + \cos 2\phi}{4}\frac{r^*}{r_{min}^2}.$$

From this expression for the light trajectory $r(\phi)$, one can compare the directions of the initial and final asymptotes to deduce the angular deflection to be $\delta\phi_{GR} = 2r^*/r_{min}$ of (7.31).

7.4 **Lens equation**   Carry out the calculations for (7.38) and (7.39).

7.5 **Total energy in curved spacetime**   Equation (7.52) suggests that $\mathcal{E} \equiv m(\kappa^2 - c^2)/2$ should be interpreted as the total nonrelativistic (NR) energy. That is in the NR limit, energy can be approximated by $E = mc^2 + \mathcal{E}$. Show that, with this interpretation of $\mathcal{E}$, the conserved quantity κ has the interpretation of being the total energy per unit rest energy in the Schwarzschild spacetime $\kappa/c = E/mc^2$.

7.6 **Details for time-delay calculation**   Work out the systematic expansion of (7.79) in order to obtain the result of (7.81).

7.7 **Four-velocity of a particle in a circular orbit**   For the simple configuration of a test body in circular orbit (radius R) in the equatorial plane in the Schwarzschild spacetime

(a) Find the explicit expression, in the Schwarzschild coordinates (ct, r, θ, ϕ), the 4-velocity of a particle along a circular orbit $U^\mu = (U^t, 0, 0, U^\phi)$ where the components $U^\phi = d\phi/d\tau$ and $U^t = dt/d\tau$ are two constants of motion.

(b) From these two quantities, show that Kepler's third law $(d\phi/dt)^2 = \Omega^2 = G_N M/R^3$ also holds in GR.

(c) As a consistency check, show that the 4-velocity under discussion does satisfy the invariant relation $g_{\alpha\beta} U^\alpha U^\beta = -c^2$.

Black holes

- A **black hole** is an object so compact that it is inside its **event horizon**, a one-way surface through which particles and light can only traverse inward, and an exterior observer cannot receive any signal sent from inside.

- To understand the full physical content allowed by GR of the geometry exterior to a spherical source, we need to view this geometry using different coordinate systems. We present the Schwarzschild geometry in Eddington–Finkelstein coordinates as well as in Kruskal coordinates.

- Besides a black hole, the GR field equation also allows the time-reversed solution of a **white hole**: Particles and light cannot be stationary inside the event horizon and must pass through it outward.

- The GR field equation also allows the dynamical structure of a **wormhole,** which connects two asymptotically flat spacetime regions that are otherwise disconnected.

- The gravitational energy released when a particle falls into a tightly bound orbit around a black hole can be enormous. The percentage of the unleashed particle rest-energy can be more than 10 times greater than that released in a nuclear fusion reaction. This powers some of the most energetic phenomena observed in the universe.

- The physical reality of, and observational evidence for, black holes are briefly discussed.

- Quantum fluctuation around the event horizon brings about thermal emission of particles and light from a black hole. This **Hawking radiation,** as well as the **Penrose process** in a rotating black hole, comes about because of the possibility of negative energy particles falling into a black hole.

In Chapter 7 we presented the Schwarzschild geometry exterior to a spherical source. From this solution to the GR field equation we worked out the geodesics of particle (perihelion precession) and light (bending trajectory and time-delay). However, the GR examples we worked out by and large represent small corrections to the Newtonian result. In this chapter, we study the spacetime structure exterior to any object with its mass so compressed that its radius is smaller than $r^* = 2G_N M/c^2$ of the Schwarzschild geometry. Such objects have been given the evocative name **black holes**, because it is impossible to transmit outwardly any signal, any light, from the region inside the $r = r^*$ surface. This necessarily involves such strong gravity and curved spacetime that the GR framework is indispensable.

In this chapter we shall mostly concentrate on the simplest case of a Schwarzschild black hole. However, in two separate appendices, which the reader may choose to skip upon a first reading, we also provide a brief introduction to some topics beyond: the Kerr geometry exterior to a rotating source, Hawking radiation and black hole thermodynamics. In Appendix A, we show that a rotating black hole has a structure that is more complex than a Schwarzschild black hole. In particular there exists the ergosphere region outside the horizon that allows one to extract the rotational energy of the black hole through the Penrose process. In Appendix B, we present a glimpse of the more advanced topics related to the intersection of black holes and quantum physics.

8.1   Nonrotating black holes

We shall first study black holes of the Schwarzschild geometry exterior to a nonrotating spherical source, concentrating on its causal (lightcone) structure. We shall show that this solution to the GR field equation also implies the curious possibilities of a "white hole" as well as a "wormhole."

Singularities of the Schwarzschild metric   The Schwarzschild geometry in Schwarzschild coordinates (t, r, θ, ϕ) has the metric,

$$ds^2 = -\left(1 - \frac{r^*}{r}\right)c^2dt^2 + \left(1 - \frac{r^*}{r}\right)^{-1} dr^2 + r^2d\theta^2 + r^2\sin^2\theta d\phi^2,$$

$$(8.1)$$

which has singularities at $r = 0$ and $r = r^* \equiv 2G_{\mathrm{N}}M/c^2$, and[1] $\theta = 0$ and π. We are familiar with the notion that $\theta = 0$ and π are **coordinate singularities** associated with our choice of the spherical coordinate system. They are not physical, do not show up in physical measurements at $\theta = 0$ and π, and can be removed by a coordinate transformation. However, the $r = 0$ singularity is real. This is not surprising as the Newtonian gravitational potential for a point mass already has this feature: $-GM/r$. What about the $r = r^*$ surface? As we shall demonstrate, it is actually a coordinate singularity,[2] i.e. it is not physical and can be transformed away by coordinate transformation (e.g. Kruskal coordinates, discussed in Section 8.1.2).

The event horizon   While physical measurements are not singular at $r = r^*$, it does not mean that this surface is not special. It is an **event horizon**, separating events that can be viewed from afar, from those which cannot (no matter how long one waits). That is, the $r = r^*$ surface is a boundary of a region, from within it is impossible to send out any signal. It is a boundary of communication, much like earth's horizon is a boundary of our vision. The key

[1] By "metric" we include the consideration also of the inverse metric which has a $\sin^2\theta^{-1}$ term.

[2] In Chapter 13 we shall study the Riemann curvature tensor $R_{\mu\nu\lambda\rho}$, which is a nonlinear second derivative of the metric and is non-trivial only in a curved space. In the case of Schwarzschild geometry, the coordinate-independent product of the curvature tensor $R_{\mu\nu\lambda\rho}R^{\mu\nu\lambda\rho} = 12r^{*2}/r^6$ is only singular at $r = 0$. This means that the singularity at $r = r^*$ that we encounter in Schwarzschild coordinates must be associated with our choice of coordinate system.

property of an event horizon is that any time-like worldline can pass through it only toward the $r = 0$ direction. Particles and light rays cannot move outward from the $r < r^*$ region. Traveling inside the horizon, a particle inexorably moves towards the physical singularity at $r = 0$.

We shall begin the discussion of such properties of an event horizon with a simple examination of the elapsed time for a particle to travel inward across the $r = r^*$ boundary, showing that, according to a far away observer, this crossing would take an infinite amount of time. Thus no signal can be reached to such an observer if sent from the horizon surface or its interior. Only in later sections do we discuss the geometry around the horizon showing that lightcones tip over in such a way that all time-like worldlines cannot cross the $r = r^*$ surface in the outward direction.

8.1.1 Time measurements around a black hole

In Chapter 4 we used different coordinate frames (e.g. a freely falling spaceship or an observer in the gravitational field watching the spaceship in acceleration, etc.) in order to get different viewpoints of spacetime. Similarly, useful insights of the Schwarzschild geometry can be had by using different coordinate systems. Here we give the respective descriptions: first according to an observer in a spaceship falling towards the center of the gravitational source, then an observer viewing such event from afar.

The local proper time

We already mentioned that $r = r^*$ is not a physical singularity. Here we will display a specific case of time measurement by an observer traveling across the Schwarzschild surface. The result shows that such physical measurement is **not** singular at $r = r^*$.

Let τ be the time measured on the surface of a collapsing star (alternatively, the proper time onboard a spaceship travelling radially towards the origin $r = 0$). Recall from Section 7.3.1 that, for a particle (with mass) in the Schwarzschild spacetime, we can write a generalized energy balance equation (7.52). This equation can be simplified further when we specialize to the radial motion along some fixed azimuthal angle ϕ (i.e. $d\phi/d\tau = 0$, thus the angular momentum $l = 0$) with zero energy ($\mathcal{E} = 0$) at $r = \infty$—i.e. the collapsing star or the infalling spaceship start from rest at $r = \infty$, so that

$$\frac{1}{2}\dot{r}^2 - \frac{G_N M}{r} = 0 \tag{8.2}$$

or

$$\frac{1}{c^2}\left(\frac{dr}{d\tau}\right)^2 = \frac{2G_N M}{c^2 r} = \frac{r^*}{r}, \quad \text{or} \quad c d\tau = \pm\sqrt{\frac{r}{r^*}}dr. \tag{8.3}$$

The $+$ sign corresponds to an exploding star (or, an outward-bound spaceship), while the—sign to a collapsing star (or, an inward-bound probe). We pick the

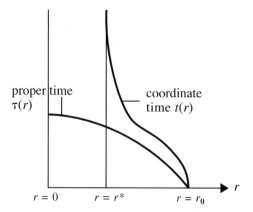

Fig. 8.1 The contrasting behavior of proper time $\tau(r)$ vs. coordinate time $t(r)$ at the Schwarzschild surface. For r and $r_0 \gg r^*$, these two functions approach each other.

minus sign. A straightforward integration yields

$$\tau(r) = \tau_0 - \frac{2r^*}{3c}\left[\left(\frac{r}{r^*}\right)^{\frac{3}{2}} - \left(\frac{r_0}{r^*}\right)^{\frac{3}{2}}\right] \tag{8.4}$$

where τ_0 is the time when the probe is at some reference point r_0

Thus the proper time $\tau(r)$ is perfectly smooth at the Schwarzschild surface, (see Fig. 8.1). The time for the star to collapse from $r = r^*$ to the singular point at $r = 0$ is $\Delta\tau = 2r^*/3c$ which is of the order of 10^{-4} s for a star with a mass ten times the solar mass (i.e. $r^* \simeq 30$ km). Thus an observer on the surface of the collapsing star would not feel anything peculiar when the star passed through the Schwarzschild radius. And it will take both the star and the observer about a tenth of a millisecond to reach the origin, which is a physical singularity.

The Schwarzschild coordinate time

While the time measurement by an observer traveling across the Schwarzschild surface is perfectly finite, this is not the case according to the observer far away from the source. Recall that the coordinate t of the Schwarzschild coordinates is the time measured by an observer far away from the source, where the spacetime approaches the flat Minkowski space. Here we will show that the event's Schwarzschild coordinate time becomes singular (it blows up) when it approaches the $r = r^*$ surface. To find the coordinate time as a function of the radial coordinate in the $r > r^*$ region we start with $dt/dr = (dt/d\tau)/(dr/d\tau)$. We already have an expression for $dr/d\tau$ from (8.3), while $dt/d\tau$ is, according to (7.46), directly related to the constant of motion κ, which is fixed to be c because we are considering a geodesic with zero kinetic energy at infinity, $\mathcal{E} = m(\kappa^2 - c^2)/2 = 0$. In this way, we find $dt/dr = \dot{t}/(dr/d\tau) = -(1 - r^*/r)^{-1}/(r^*/r)^{1/2}$ so that

$$cdt = -\sqrt{\frac{r}{r^*}}\frac{dr}{1 - r^*/r}, \tag{8.5}$$

which shows clearly the singularity at r^*. For $r \simeq r^*$, we can integrate $cdt = -r^* dr/(r - r^*)$ to display the logarithmic singularity:

$$t - t_0 = -\frac{r^*}{c} \ln \frac{r - r^*}{r_0 - r^*}. \tag{8.6}$$

Equivalently, we have $(r - r^*) \sim \exp{-c(t - t_0)}/r^*$; it takes an infinite amount of coordinate time to reach $r = r^*$. The full function of $t(r)$ in the region outside the Schwarzschild surface can be calculated (Problem 8.2) and is displayed in Fig. 8.1.

Infinite gravitational redshift The above-discussed phenomenon of a distant observer seeing the collapsing star slow down to a standstill can also be interpreted as representing an infinite gravitational time dilation. The relation (7.16) between coordinate and proper time interval for a light signal is given by

$$dt = \frac{d\tau}{\sqrt{-g_{00}}} = \frac{d\tau}{\sqrt{1 - r^*/r}}. \tag{8.7}$$

Clearly the coordinate time interval dt will blow up as as r approaches r^*. In terms of wave peaks, this means that it takes an infinite time interval for the next peak to reach the far away receiver. This can be equivalently phrased as an "infinite gravitational redshift." Our discussion in Box 6.2 has the ratio of the received frequency to the emitted frequency as

$$\frac{\omega_{\text{rec}}}{\omega_{\text{em}}} = \sqrt{\frac{(g_{00})_{\text{em}}}{(g_{00})_{\text{rec}}}} = \sqrt{\frac{1 - r^*/r_{\text{em}}}{1 - r^*/r_{\text{rec}}}}. \tag{8.8}$$

When $r_{\text{em}} \to r^*$, the received frequency approaches zero, as it would take an infinite interval to receive the next photon (i.e. the peak-to-peak time is proportional to ω^{-1}). Thus no signal transmission from the black hole is possible.[3]

8.1.2 Eddington–Finkelstein coordinates: Black holes and white holes

To have a deeper understanding of the Schwarzschild surface as an event horizon, we need to study the causal structure of the geometry exterior to a spherical source. This means the study of lightcones, hence the light geodesics, in a Schwarzschild geometry. It is helpful to recall that in a flat spacetime, the radial (i.e. with fixed θ and ϕ) light-like worldlines, corresponding to the solutions of $ds^2 = -c^2 dt^2 + dr^2 = 0$, or $cdt = \pm dr$, are the 45° straight lines in the (r, t) spacetime diagram for a flat geometry:

$$ct = \pm r + \text{constant}. \tag{8.9}$$

The plus sign is for the outgoing (r increasing with t) and minus sign for ingoing light, as shown in Fig. 3.4. As we shall demonstrate presently, even in the Schwarzschild geometry, it is possible to choose a coordinate system

[3] In the eighteenth century John Michell (and independently, Pierre Laplace) speculated on the possibility of a "black star," when its mass/radius ratio was so large, its gravitational attraction so strong, that even a particle travelling at the speed c could not escape. The well-known Newtonian expression (see Eq. (10.15)) for the escape velocity being $v_{\text{esc}} = \sqrt{2 G_N M/r} = c\sqrt{r^*/r}$, thus $v_{\text{esc}} \geq c$ if $r \leq r^*$. Of course, this speculation was based (from our modern perspective) on the erroneous assumption that light carried a gravitational mass.

so that the worldlines for light are such $45°$ lines in the entire spacetime. Furthermore, when displayed in such new coordinates, the Schwarzschild solution contains, besides black holes, other geometric structures such as white holes, wormholes, etc.

Schwarzschild coordinates and their limitation

A radial ($d\theta = d\phi = 0$) worldline for a photon in Schwarzschild spacetime in Schwarzschild coordinates has a null line-element of

$$ds^2 = -\left(1 - \frac{r^*}{r}\right)c^2 dt^2 + \left(1 - \frac{r^*}{r}\right)^{-1} dr^2 = 0. \qquad (8.10)$$

We then have[4]

$$cdt = \pm \frac{dr}{1 - r^*/r}. \qquad (8.11)$$

[4]This relation differs from that in (8.5) because we are now considering a light-like worldline.

This can be integrated to obtain, for some reference spacetime point of (r_0, t_0),

$$c\,(t - t_0) = \pm\left(r - r_0 + r^* \ln\left|\frac{r - r^*}{r_0 - r^*}\right|\right), \qquad (8.12)$$

or simply,

$$ct = \pm\left(r + r^* \ln\left|r - r^*\right| + \text{constant}\right). \qquad (8.13)$$

The +sign stands for the outgoing, and the −sign infalling, light-like geodesics, as shown in Fig. 8.2. To aid our viewing of this spacetime diagram we have drawn in several lightcones in various spacetime regions. We note that for the region far from the source where the spacetime becomes flat, the lightcone approaches the expected form with $\pm 45°$ sides.

The most prominent feature we notice is that the lightcones "tip over" when crossing the Schwarzschild surface. Instead of opening towards the $t = \infty$ direction, they tip towards the $r = 0$ line. This can be understood by noting

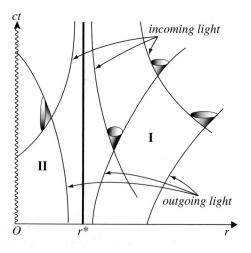

Fig. 8.2 Lightcones in Schwarzschild spacetime. Regions I and II are separated by the Schwarzschild surface. Different light rays correspond to (8.13) with different constant. Note that the outgoing light rays in region II also end at the $r = 0$ line.

that the roles of space and time are interchanged when crossing over the $r = r^*$ surface in Schwarzschild geometry.

- In the spacetime region I outside the Schwarzschild surface $r > r^*$, the time and space coordinates have the usual property being time-like $ds_t^2 < 0$ and space-like $ds_r^2 > 0$ cf. (7.13) and (7.14). In particular, lines of fixed r (parallel to the time axis) are time-like, and lines of fixed t are space-like. Since the trajectory for any particle must necessarily be time-like (i.e. every subsequent segment must be contained inside the lightcone at a given worldpoint), the lightcones open upwards in the ever-larger t direction. In this region I of Fig. 8.2, an observer is fated to have an ever-increasing t. (See discussion related to Fig. 3.4.)

- In region II, inside the Schwarzschild surface, the roles of t and r are reversed. A worldline of fixed time is now time-like. This comes about because the $(1 - r^*/r)$ factor changes sign in g_{00} and g_{rr}. For a world-line to remain time-like $(ds^2 < 0)$, the particle can no longer stay put at one position $(dr \neq 0)$, but is forced to move inward towards the $r = 0$ singularity. For the worldline to be contained within a lightcone, the lightcones themselves must tip over when crossing the $r = r^*$ surface. The tipping over of the lightcones also suggests that, once inside region II, there is no way one can send a signal to the outside region. Hence, the Schwarzschild surface is an event horizon.

- The key feature that defines the $r = r^*$ surface as an event horizon is its "no return" property. Particles and light signal can proceed only in the inward direction, not outward. Once inside the event horizon, a particle is driven to $r = 0$ inexorably. The geometric property that makes the Schwarzschild surface an one-way membrane is that it is a "null surface," which will be discussed at the end of this section, see, particularly, Fig. 8.5 below.

The fact that the metric becomes singular at the $r = r^*$ surface means that the Schwarzschild coordinates (t, r, θ, ϕ) are not convenient for the discussion of events near the Schwarzschild surface. In our description of the "tipping-over" of the lightcones in Fig. 8.2 the use of Schwarzschild coordinates is suspect as the function $t(r)$ is singular across the $r = r^*$ surface. In the remainder of this section we shall demonstrate that it is possible, through a series of coordinate changes, to extend the spherically symmetric GR solutions to a domain which is devoid of any coordinate singularities (such as the ones at $r = r^*$), so that all geodesics can be extended to all regions unless they end up at the physical singularity of $r = 0$. Such a spacetime manifold is said to be **geodesic complete**.

The Eddington–Finkelstein coordinates

We now search for coordinates such that the Schwarzschild geometry can be displayed without the presence of the r^* singularity, and in such coordinates the lightcones tip over smoothly. We start with the Eddington–Finkelstein coordinates. This choice can be motivated as follows. Recall the proper time of an infalling particle into the black hole is smooth for all values of r, see (8.4). Instead of setting up the coordinate system using a static observer far from the gravitational source (as is the case of the Schwarzschild coordinates) one can describe the Schwarzschild geometry from the viewpoint of an infalling observer. Mathematically, an even simpler choice is to use an infalling photon as the "observer" to set up the new time coordinate $\bar{t}$. The infalling null geodesic $(ds^2 = 0)$ in the new spacetime diagram with $(\bar{t}, r)$ axes should be a $-45°$ straight line—just as (8.9) is for the infalling photon worldline in the flat spacetime:

$$c\bar{t} = -r + \text{constant}. \tag{8.14}$$

This should be compared to the equation for infalling light in the Schwarzschild coordinates given by (8.13),

$$ct + r^* \ln|r - r^*| = -r + \text{constant}. \tag{8.15}$$

A comparison of the LHSs of (8.14) and (8.15) suggests that we make the coordinate transformation of

$$ct \longrightarrow c\bar{t} \equiv ct + r^* \ln|r - r^*|, \tag{8.16}$$

or, equivalently,

$$cd\bar{t} = cdt + \frac{r^*}{r - r^*}dr. \tag{8.17}$$

Substituting into the Schwarzschild line element (8.1), we find (Ω being the solid angle with $d\Omega^2 = d\theta^2 + \sin^2\theta d\phi^2$)

$$ds^2 = -\left(1 - \frac{r^*}{r}\right)c^2 d\bar{t}^2 + \frac{r^*}{r} 2cd\bar{t}dr + \left(1 + \frac{r^*}{r}\right)dr^2 + r^2 d\Omega^2,$$

$$= -c^2 d\bar{t}^2 + dr^2 + \frac{r^*}{r}\left(c^2 d\bar{t}^2 + 2cd\bar{t}dr + dr^2\right) + r^2 d\Omega^2 \tag{8.18}$$

which no longer has the r^* singularity and is regular in both regions I and II.[5]

Let us investigate the lightcone structure in the Eddington–Finkelstein coordinates. Since we expect the ingoing light to be a straight worldline as in (8.14), the line element ds^2 (for fixed angles $d\theta = d\phi = 0$) must be proportional to a factor of $(cd\bar{t} + dr)$ so that $ds^2 = 0$ has (8.14) as one of its solutions. This suggests that we can simplify our calculation by introducing the variable

$$p \equiv c\bar{t} + r, \quad \text{that is,} \quad dp = cd\bar{t} + dr. \tag{8.19}$$

The line element in (8.18) can then be written as

$$ds^2 = -\left(1 - \frac{r^*}{r}\right)dp^2 + 2dp dr + r^2 d\Omega^2. \tag{8.20}$$

For the worldline of radially $(d\Omega = 0)$ infalling light $(ds^2 = 0)$, we must have

$$\left[-\left(1 - \frac{r^*}{r}\right)dp + 2dr\right]dp = 0. \tag{8.21}$$

[5]One can object that this extension is achieved by a transformation (8.16) that itself becomes singular at $r = r^*$. However, the relevant point is that we have found a set of coordinates, as defined by the line element (8.18) which also describes the geometry outside a spherical source. How one found such a coordinate set is immaterial.

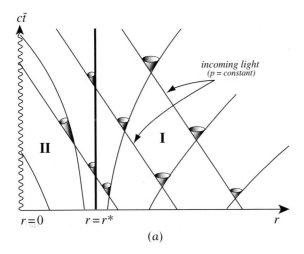

(a)

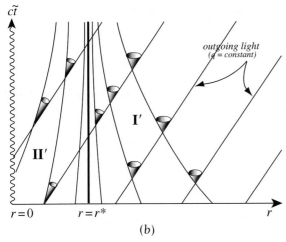

(b)

Fig. 8.3 Lightcones in Eddington–Finkelstein spacetime. (a) In advanced EF coordinates $(\bar{t}, r)$ with a black hole, showing two regions I and II with $r = 0$ as a line of future singularities. (b) In retarded EF coordinates $(\tilde{t}, r)$ with a white hole, showing two regions I' and II' with $r = 0$ as past singularities.

This equation has two solutions: one being $dp = 0$ which is the straight infalling $-45°$ line of (8.14), forming the infalling side of lightcones; the other solution

$$dp = \frac{2dr}{1 - r^*/r} = cd\bar{t} + dr \tag{8.22}$$

can be integrated to yield

$$c\bar{t} = \int \frac{r + r^*}{r - r^*} dr = r + 2r^* \ln \left| r - r^* \right| + \text{constant}, \tag{8.23}$$

which resembles the outgoing null lines (8.13) in the Schwarzschild coordinates, and it is the outgoing side of a lightcone. Plotting them in Fig. 8.3(a) we see now that lightcones tip over smoothly across the Schwarzschild surface. Inside the horizon, both sides of lightcones bend towards the $r = 0$ line so that it is the future singularity. In Fig. 8.4, which is similar to Fig. 8.3(a), we display the spacetime diagram of a imploding star with an observer on its surface sending out light signals at a regular interval.

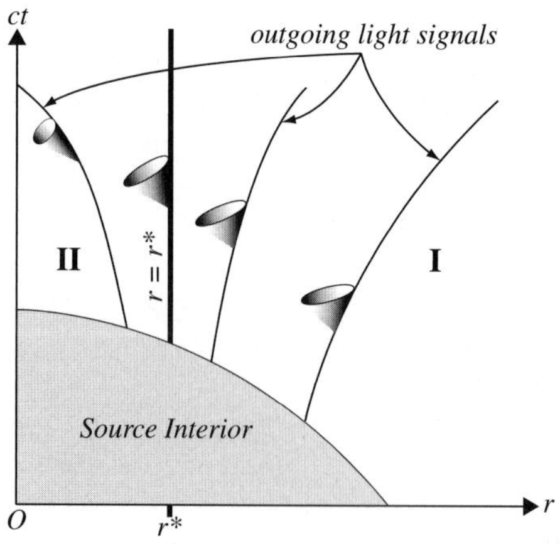

Fig. 8.4 A star undergoing gravitational collapse (two spatial dimensions suppressed). The points on the surface of the collapsing star correspond to radially moving particles. One such worldline, the same as that shown in Fig. 8.1, is displayed above. The region outside the collapsing star has the Schwarzschild geometry.

[6]We have suppressed the angular term $r^2 d\Omega$ in the ds^2 entries.

Advanced vs. retarded EF coordinates

The GR field equation is unchanged under the time reversal transformation of $t \leftrightarrow -t$. This means that we could just as well pick a set of coordinates, like the Eddington–Finkelstein coordinates discussed above, except with a different sign for the time coordinate. We shall call the above-discussed coordinates "advanced EF coordinates" and introduce now "retarded EF coordinates." This can be put in more physical terms: the advanced EF time coordinate $\bar{t}$ is the proper time for ingoing light geodesics; we now introduce the retarded EF time $\tilde{t}$ which is the proper time for outgoing light. That is, instead of straightening the ingoing null geodesics, we now straighten instead the outgoing null geodesics. Since the two sets of coordinates closely resemble each other, we present them together, in Table 8.1.[6]

One can follow exactly the same steps to find the outgoing and ingoing light geodesics in this retarded EF coordinates. They are depicted in Fig. 8.3(b). Again we see that lightcones tip over smoothly across the Schwarzschild surface. But, instead of tipping inward, as in Fig. 8.3(a), they tip outward, away from the $r = 0$ singularity. That is, while the Schwarzschild geometry depicted in the advance EF coordinates has a future singularity at $r = 0$, the geometry depicted in retarded EF coordinates shows a past singularity at $r = 0$. Once again the $r = r^*$ surface is a one-way membrane allowing transmission of particles and light only in one direction—in the case of a black hole (containing

Table 8.1 Advanced vs. retarded Eddington–Finkelstein coordinates.

Advanced EF coordinate	Retarded EF coordinate
$c\bar{t} = ct + r^* \ln \lvert r - r^* \rvert$	$c\tilde{t} = ct - r^* \ln \lvert r - r^* \rvert$
$cd\bar{t} = cdt + \frac{r^*}{r - r^*} dr$	$cd\tilde{t} = cdt - \frac{r^*}{r - r^*} dr$
$p \equiv c\bar{t} + r$	$q \equiv c\tilde{t} - r$
$ds^2 = -\left(1 - \frac{r^*}{r}\right) dp^2 + 2dp\,dr$	$ds^2 = -\left(1 - \frac{r^*}{r}\right) dq^2 - 2dq\,dr$

the future singularity), it only allows particles to cross inward, and now we have a **white hole** (containing the past singularity): particles can only move outward from the interior $r < r^*$ region II$'$ to the $r > r^*$ region I$'$, and not the other way around.

Properties of the Schwarzschild surface

The geometric property that makes the $r = r^*$ Schwarzschild surface an event horizon is that all points on such a surface have their lightcones entirely on one side of the surface as shown in Fig. 8.5. In the case of a black hole, they are on the side of $r < r^*$ as shown in Fig. 8.3(a); in the case of a white hole, they are on the side of $r > r^*$ as shown in Fig. 8.3(b). Any timelike worldline, being contained inside the lightcone, can cross the surface only in one direction. The 3-surface with such a property is called a **null surface** because it is a "side" of the lightcones. That is, one can find a light-like segment on the surface; it is a 4-vector $\boldsymbol{n}$ with null length $\boldsymbol{n}\cdot\boldsymbol{n} = 0$. Being a 3D surface, at any point there are three independent, hence mutually orthogonal, tangent vectors ($\boldsymbol{t}_{1,2,3}$) and one can choose $\boldsymbol{n}$ as one of the independent tangents $\boldsymbol{t}_1 = \boldsymbol{n}$. Thus, the $\boldsymbol{n}$ vector, with the property of $\boldsymbol{n}\cdot\boldsymbol{n} = 0$, must be orthogonal to all the tangent vectors $\boldsymbol{n}\cdot\boldsymbol{t}_{1,2,3} = 0$—it is a normal of the 3-surface. A null surface is defined as a surface with its normal being null. It has the interesting property that its normal lies in the surface itself. Such a geometry gives rise to its "one-way transmission" property. It can be shown that the necessary condition for a null surface is for the metric element $g_{rr} = \infty$.

Fig. 8.5 A null surface is an event horizon. The lightcones of all points on the null surface are on one side of the surface. All timelike worldlines (samples shown as heavy arrowed lines) being contained inside lightcones can cross the null surface only in one direction. Thus, a null surface is a "one-way membrane."

The Schwarzschild surface is a null surface   For the Schwarzschild geometry, it is easy to see that the Schwarzschild surface is a null surface when we use the Eddington–Finkelstein coordinates (whether advanced or retarded).[7] From Table 8.1, it is easy to find (Problem 8.3) a light-like line element $ds^2 = 0$ lying on the Schwarzschild surface $r = r^*$ with $dr = d\Omega = 0$.

We have already said above in Section 8.1.1 that the frequency shift becomes infinite at $r = r^*$ because $g_{00}(r^*) = 0$. The Schwarzschild surface is a **surface of infinite redshift**. We also suggested that any particle cannot be stationary inside the $r = r^*$ surface: they must move, in the case of a black hole, towards (or in the case of a white hole, away from) the $r = 0$ singularity. We call such a surface a **stationary limit surface**. Thus, the Schwarzschild surface is a null 3-surface $g_{rr}(r^*) = \infty$, a stationary limit surface and a surface of infinite redshift with $g_{00}(r^*) = 0$. As we shall show in Section 8.4, for a rotating black hole surfaces with these different properties do not in general coincide.

[7]We also see the difficulty one would have encountered using Schwarzschild coordinates when discussing the Schwarzschild surface as a null surface because of the ambiguity with $ds^2 = 0/0$ in (8.1) when $r = r^*$ with $dr = d\Omega = 0$.

8.1.3   Kruskal coordinates and the wormhole

The advanced EF coordinate straightens the ingoing light geodesic, and the retarded EF coordinate, the outgoing lights. This naturally suggests we try to find a coordinate system in which both ingoing and outgoing null geodesics are, just like the case of flat spacetime, 45° straight lines. Our goal is to find a coordinate system in which there is no singularity at r^* and the Schwarzschild solution is analytic in all regions of spacetime other than the physical singularity at $r = 0$. Clearly, with both infalling and outgoing light geodesics being straight lines, we anticipate a depiction similar to that of a flat spacetime. Such a spacetime is said to be conformally flat—it is a flat geometry except for a changed scale.

At the first stage we make use of the coordinate parameters p and q shown in the above table. Since $dp = 0$ and $dq = 0$ represent respectively the straight ingoing and outgoing light, we expect that, in a system having both p and q coordinates, ds^2 (when $d\Omega = 0$) is proportional to $dpdq$. A simple calculation can demonstrate that this is indeed the case. We first express dp in terms of dt and dr,

$$dp = cd\bar{t} + dr = cdt + \frac{r^*}{r - r^*}dr + dr = cdt + \left(1 - r^*/r\right)^{-1} dr.$$

Similarly, we have $dq = cdt - (1 - r^*/r)^{-1}dr$. It then follows that

$$dpdq = c^2 dt^2 - \left(1 - r^*/r\right)^{-2} dr^2$$

$$= -\left(1 - \frac{r^*}{r}\right)^{-1} \left[-\left(1 - \frac{r^*}{r}\right) c^2 dt^2 + \left(1 - \frac{r^*}{r}\right)^{-1} dr^2\right].$$

The expression inside the square bracket being just the 2D line element ds^2 in the Schwarzschild coordinates (8.1), we obtain

$$ds^2 = -\left(1 - \frac{r^*}{r}\right) dpdq, \tag{8.24}$$

where we have suppressed again the angular factors. In order to remove the $(1 - r^*/r)$ factor in ds^2, we make further coordinate changes $(p, q) \to (p', q')$ with

$$p' = \exp\left(p/2r^*\right) \quad \text{and} \quad q' = -\exp\left(-q/2r^*\right) \tag{8.25}$$

so that

$$dpdq = 4r^{*2} \exp\left[-(p - q)/2r^*\right] dp'dq'.$$

Replacing[8] the $(p - q)$ factor by the Schwarzschild radial coordinate r, we can now write

$$ds^2 = -\frac{4r^{*3}}{r} e^{-r/r^*} dp'dq'; \tag{8.26}$$

the only singularity is now at $r = 0$. To have a coordinate system more like the familiar space and time coordinates, we replace (p', q') by (V, U):

$$V = \frac{1}{2}\left(p' + q'\right) \quad \text{and} \quad U = \frac{1}{2}\left(p' - q'\right) \tag{8.27}$$

so that $-dV^2 + dU^2 = -dp'dq'$, and, putting back the angular factor, we have the line element in the **Kruskal coordinates**[9] (V, U):

$$ds^2 = \frac{4r^{*3}}{r} e^{-r/r^*} \left(-dV^2 + dU^2\right) + r^2 d\Omega. \tag{8.28}$$

[8] The radial distance r is implicitly determined by its relation to p and q: from $p = c\bar{t} + r = ct + r^* \ln|r - r^*| + r$ and $q = c\bar{t} - r = ct - r^* \ln|r - r^*| - r$, we have $(p - q) = 2r + 2r^* \ln|r - r^*|$.

[9] Such coordinates are also known as Kruskal–Szekeres coordinates.

Because $(-dV^2 + dU^2)$ has the same form as $(-c^2dt^2 + dr^2)$, this metric represents a conformally flat geometry. The radial variable r can be fixed by $V^2 - U^2 = p'q' = -\exp(p-q)/2r^*$ so that

$$V^2 - U^2 = \left(1 - \frac{r}{r^*}\right) e^{r/r^*}. \tag{8.29}$$

Tracing the coordinate transformations all the way back to (t, r), we can, through a straightforward calculation (Problem 8.4), find the relation between the Kruskal and the Schwarzschild coordinates as

$$V = \left(\frac{r}{r^*} - 1\right)^{\frac{1}{2}} e^{r/2r^*} \sinh\left(\frac{ct}{2r^*}\right) \tag{8.30}$$

$$U = \left(\frac{r}{r^*} - 1\right)^{\frac{1}{2}} e^{r/2r^*} \cosh\left(\frac{ct}{2r^*}\right) \tag{8.31}$$

in the $r > r^*$ region, and

$$V = \left(1 - \frac{r}{r^*}\right)^{\frac{1}{2}} e^{r/2r^*} \cosh\left(\frac{ct}{2r^*}\right) \tag{8.32}$$

$$U = \left(1 - \frac{r}{r^*}\right)^{\frac{1}{2}} e^{r/2r^*} \sinh\left(\frac{ct}{2r^*}\right) \tag{8.33}$$

in the $r < r^*$ region.

The Kruskal diagram

We now discuss the spacetime diagram in Kruskal coordinates. With the angular degrees of freedom (θ, ϕ) suppressed, the axes are (U, V) as shown in Fig. 8.6. Let us first note that the lines of constant r are hyperbolas in the (U, V) plane because Eq. (8.29) shows that they must have constant $V^2 - U^2$. Particularly the $r = 0$ singularity is mapped into two hyperbolas $V = \pm\sqrt{U^2 + 1}$. Because the ratio V/U, according to Eqs. (8.30)–(8.33), depends only on the Schwarzschild time

$$\frac{V}{U} = \tanh\left(\frac{ct}{2r^*}\right) \quad \text{for } r > r^*$$

$$\frac{V}{U} = \coth\left(\frac{ct}{2r^*}\right) \quad \text{for } r < r^*,$$

the lines of constant t are straight lines passing through the origin of $V = U = 0$. In particular $t = 0$ is mapped into the U and V axes, and $t = +\infty$ into the $45°$ straight line $V = +U$ while $t = -\infty$ into $V = -U$ line. Since the metric is conformally flat the lightcones are just like flat Minkowski space lightcones and the V coordinate is time-like. Any time-like worldline would point towards ever increasing V.

How are the regions I, II, I′ and II′ of Fig. 8.3 mapped into this Kruskal diagram? The boundary between region I and region II, as well as the boundary between region I′ and region II′, is the Schwarzschild surface $r = r^*$ which according to (8.29) is mapped into two $45°$ straight lines of $V = \pm U$. (They

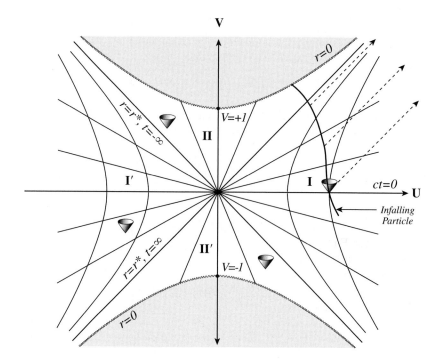

Fig. 8.6 Kruskal diagram.

coincide with the $t = \pm\infty$ lines.) It is clear that the regions I and I' corresponding to $r > r^*$ are mapped into the left and right quadrants. The region II containing the future $r = 0$ singularity (black hole) corresponds to the top quadrant, and region II' containing the past $r = 0$ singularity (white hole) corresponds to the bottom quadrant.

The wormhole

The left and right quadrants (regions I and I') in the Kruskal diagram in Fig. 8.6 appear to be identical; both of them approach a flat spacetime for large r and they are outside each other's forward lightcones, although time-like worldlines originating from two asymptotically flat regions can enter the same black hole (region II) and terminate at the same future singularity. What exactly is their relation? What is the physical significance of these two regions?

To study the relation between regions I and I' we consider the $V = 0$ subspace. With this condition Eq. (8.29) fixes the relation between the U coordinate in terms of the Schwarzschild radial distance r:

$$U = \left(\frac{r}{r^*} - 1\right)^{1/2} e^{r/2r^*}$$

or

$$dU = \frac{r^{1/2}}{2\,(r^*)^{3/2}} \frac{dr}{\left(1 - \frac{r^*}{r}\right)^{1/2}} e^{r/2r^*}.$$

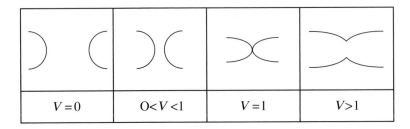

| $V=0$ | $0<V<1$ | $V=1$ | $V>1$ |

Fig. 8.7 Wormhole time evolution. Here we display only the $V \geq 0$ region as the the $V \leq 0$ region is identical.

The line element (8.28), for a fixed polar angle θ, reduces in this $V = 0$ subspace to

$$ds^2 = \left(1 - \frac{r^*}{r}\right)^{-1} dr^2 + r^2 d\phi^2$$

which is just Eq. (7.23) considered in Section 7.1.2 showing it to be an Einstein–Rosen bridge of Fig. 7.1(b): two asymptotically flat surfaces connected by a narrow throat with a minium radial distance of $r_{\min} = r^*$. We say that regions I and I′ are connected by a **wormhole**.

Questions such as "what exactly is the relation between these two flat spacetime region?", "are they just two separate flat regions of the same spacetime?", or "are they two separate universes" are questions beyond the GR field equation, which can only determine the local geometry of spacetime, not its global or topological structure.

Let us examine this bridge structure a bit more closely. When we move the line of $V = $ constant up from $V = 0$ towards $V = +1$, the throat of the wormhole narrows from $r_{\min} = r^*$ to $r_{\min} = 0$; for $V > 1$, the two universes separate. (See Fig. 8.7.) The Kruskal solution being symmetric with respect to exchange of $V \leftrightarrow -V$, the same sequence of configurations holds when one varies V from $V = 0$ down to $V < -1$. The key point to keep in mind is that this wormhole structure is dynamic. Recall that in the Schwarzschild coordinates, the metric functions are independent of t but depend on r. Since the role of (t, r) is interchanged going from regions I and I′ to regions II and II′, and since the nontrivial configurations discussed here are in the range of $-1 \leq V \leq +1$ are in regions II and II′, the metric functions have nontrivial dependence on the time-like coordinate r. Thus the configurations shown in Fig. 8.7 represent a sequence of time evolution of the wormhole. Since the relevant time-scale must be extremely small, many argue that such fluctuations of the spacetime can properly understood only in the context of a quantum theory of gravitation. Thus, at present the theoretical status of wormholes is quite uncertain.

8.2 Orbits and accretion disks around a black hole

In this section we shall present a brief discussion of particle orbits in the gravitational field just outside the horizon of a black hole. Our main purpose is to show that gravitational binding can be extraordinarily tight. Thus, the energy that can be released when a particle falls into such orbits (before they spiral into the black hole) can be enormous—much more than the thermonuclear energy released in a nuclear fusion reaction. Energy radiation associated with the accretion disks around black holes brings about some of the most energetic phenomena observed in the universe. Here we study the case of a Schwarzschild black hole. Results for a rotating black hole will also be outlined in Section 8.4.4.

8.2.1   Effective potential of the Schwarzschild spacetime

The formalism of Section 7.3.1 in our study of the relativistic orbit of a planet can also be applied to the study of the motion of a particle around a black hole. Given[10] that the constant of motion κ is directly related to the total particle energy when far away from the source, $\kappa = E(\infty)/mc$, the energy balance equation (7.52) may be written, through Eq. (7.51), as

$$\frac{1}{2}\left(\frac{[E(\infty)]^2}{mc^2} - mc^2\right) \equiv \mathcal{E} = \frac{1}{2}m\dot{r}^2 + m\Phi_{\text{eff}} \tag{8.34}$$

with the effective potential being

$$\Phi_{\text{eff}} = -\frac{r^*c^2}{2r} + \frac{l^2}{2m^2r^2} - \frac{r^*l^2}{2m^2r^3}, \tag{8.35}$$

where l is the second constant of motion, the orbital angular momentum. While the first term in Φ_{eff} is the familiar gravitational potential of $-GM/r$ and the second term the centrifugal barrier, the last term is a new GR contribution. It is a small correction for situations such as planet motion discussed in Chapter 7, but can be very important when the radial distance r is comparable to the Schwarzschild radius r^* as in the case of an orbit just outside the horizon. We can find the extrema of this potential by setting $\partial\Phi_{\text{eff}}/\partial r = 0$:

$$\frac{r^*c^2}{2r^2} - \frac{l^2}{m^2r^3} + \frac{3r^*l^2}{2m^2r^4} = 0 \tag{8.36}$$

or

$$r^2 - 2\left(\frac{l}{mc}\right)^2\frac{r}{r^*} + 3\left(\frac{l}{mc}\right)^2 = 0. \tag{8.37}$$

The solutions R_+ and R_- specify the locations where Φ_{eff} has a maximum and minimum, respectively, see Fig. 8.8,

$$R_\pm = \frac{1}{r^*}\left(\frac{l}{mc}\right)^2\left[1 \mp \sqrt{1 - 3\left(\frac{r^*mc}{l}\right)^2}\right]. \tag{8.38}$$

We note the important difference with the Newtonian Φ_{eff} where the centrifugal barrier always dominates (with $\Phi_{\text{eff}} \to \infty$) in the small r region; we have effectively in that case $R_+ = 0$. This means that a particle cannot fall into the $r = 0$ center as long as $l \neq 0$. In the relativistic Schwarzschild geometry, in the small r regime, the last (GR correction) term in (8.35) becomes the most important one when $r \to 0$ and $\Phi_{\text{eff}} \to -\infty$. When $\mathcal{E} \geq m\Phi_{\text{eff}}(R_+)$, a

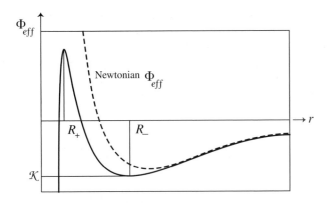

Fig. 8.8 Schwarzschild vs. Newtonian effective potential

particle can plunge into the gravity center even if $l \neq 0$. For $\mathcal{E} = m\Phi_{\text{eff}}(R_-)$, just like the Newtonian case, we have a stable circular orbit with a radius equal to R_-.

The innermost stable circular orbit

The circular radius cannot be arbitrarily small. The condition for the circular orbit having the smallest radius is the vanishing of the square root in (8.38), which fixes the particle angular momentum to be l_0,

$$3\left(\frac{r^* mc}{l_0}\right)^2 = 1 \quad \text{or} \quad \left(\frac{l_0}{mc}\right)^2 = 3\left(r^*\right)^2, \tag{8.39}$$

so that the innermost stable circular orbit (ISCO) has radius

$$R_0 = \frac{1}{r^*}\left(\frac{l_0}{mc}\right)^2 = 3r^*. \tag{8.40}$$

The plasma in an accretion disk around a black hole settles into these stable orbits and will lose its orbital angular momentum through turbulent viscosity (due to magnetohydrodynamical instability). These particles will eventually spiral into the black hole.[11]

8.2.2   The binding energy of a particle around a black hole

We are familiar with the fact that thermonuclear fusion, when compared to chemical reactions, is a very efficient process for releasing the rest energy of a particle in a reaction. Here we show that binding of a particle to a compact mass, like a black hole, can be an even more efficient mechanism. The thermonuclear reactions taking place in the sun can be summarized as fusing four protons (hydrogen nuclei each with a rest energy of 938 MeV) into a helium nucleus (having a rest energy smaller that the sum of the four proton rest energies) with a released energy of 27 MeV, which represents $27/(4 \times 938) = 0.7\%$ of the input energy. Here we discuss the energy that can be released when a particle first falls into stable orbits around a black hole before it eventually spirals through the event horizon.

For gravitational binding by a Schwarzschild black hole, consider a free particle that falls towards a black hole, and ends up bound in an ISCO outside

[11] It may be useful to remind ourselves of the origin of this centripetal force. Recall that force is related to the gradient of the potential energy $F_i = -\partial_i (m\Phi_{\text{eff}})$. In particular the radial component is given, according to (8.35), as

$$F_r = -m\frac{\partial \Phi_{\text{eff}}}{\partial r} = -\frac{A}{r^2} + B\frac{l^2}{r^3} - C\frac{l^2}{r^4},$$

where A, B and C are positive constants. In the Newtonian limit the C term is absent and a particle is prevented from reaching $r = 0$ by the centrifugal barrier repulsive force term of B. The GR "correction" C term is the dominant attractive force that pulls the particle into the black hole.

the Schwarzschild radius. Thus, according to (8.40) and (8.39) obtained above, the particle is circulating at a radial distance of $r = R_0 = 3r^*$ with angular momentum $l_0 = \sqrt{3}r^*mc$. According to the energy balance equation (8.34) with $\dot{r} = 0$, we have $\mathcal{E} = m\Phi_{\text{eff}} = -mc^2/18$. This solution gives the total energy for the gravitationally bound particle to be

$$\frac{E(\infty)}{mc^2} = \sqrt{\frac{2\mathcal{E}}{mc^2} + 1} = \sqrt{\frac{8}{9}} = 0.94. \tag{8.41}$$

That is, 6% of the rest energy is released—almost 10 times larger than the 0.7% from thermonuclear fusion.[12]

[12]In Section 8.4.4 we shall demonstrate that the gravitational binding energy of a particle around an extreme-spinning black hole is even greater—it can be as much a 42% of its rest energy!

8.3   Physical reality of black holes

Because of the extraordinary feature of the strongly warped spacetime near the Schwarzschild surface, it took a long time for the physics community to accept the reality of the black hole prediction by the Schwarzschild solution. Here is a short summary of the developments during the fifty years 1915–65, leading to the recognition of the true physical nature of black holes and the modern astronomical observation of such objects.

8.3.1   The long road to the acceptance of the black hole's reality

There have been two parallel, and intertwined, lines of study:

1. GR study of the Schwarzschild solution and warped spacetime, much along the lines discussed in our presentation here.
2. Study of gravitational collapse of massive stars—In a normal star, gravitational contraction is balanced by the thermal pressure of the gas, which is large enough if it is hot enough, due to the thermonuclear reactions at the core. The question naturally presents itself: After the exhaustion of nuclear fuel, what will be the fate of a massive star?

We present some of the highlights of this development:

- 1920s and 1930s: No one was willing to accept the extreme predictions that Schwarzschild gave for highly compact stars. Einstein and Eddington, the opinion setters, openly expressed the view that such gravitational features could not be physical. Calculations were done and results were interpreted as indicating the impossibility of black holes, instead of interpreting them correctly as indicating that no force could resist the gravitational contraction in such a situation.
- 1930: S. Chandrasekhar used the new quantum mechanics to show that, for a stellar mass $M > 1.4M_\odot$, the electron's **degenerate pressure** will not be strong enough to stop the gravitational contraction. (Electrons obey Pauli's exclusion principle. This effect gives rise to a repulsive force, the degenerate pressure, that resists the gravitational attraction.) Stars having masses under this limit so that the gravitational collapse can

be resisted by the electron's fermionic repulsion become **white dwarfs**. In 1932 Chadwick discovered the neutron, which is also a fermion. Zwicky suggested that the remnent of a supernova explosion, associated with the final stage of gravitational collapse, was a **neutron star**. Oppenheimer and Volkov, and independently Landau, studied the upper mass limit for neutron stars and found it to be a few solar masses. If greater than this limit, the neutron repulsion would not be large enough to resist the gravitational collapse all the way towards the $r = 0$ singularity, forming a black hole.

- In the meantime (1939) Oppenheimer and Snyder performed a GR study and made most of the points as presented in our discussion in Section 8.1.1. But the physics community remained skeptical as to the reality of black holes. The reservations were many. For example, one questioned whether the spherical symmetrical situation was too much an idealization. How do we take account of the realistic complications such as stellar rotation (the spin causing the star to bulge), deformation to form lumps, shock waves leading to mass ejection, and effects of electromagnetic, gravitational, and neutrino radiation, etc.?

- 1940s and 1950s: The development of atomic and hydrogen bombs during World War II and the cold war period involved a similar type of physics and mathematical calculations as the study of realistic stellar collapse. From such experience, groups led by Wheeler (USA), Zel'dovich (USSR) and others carried out realistic simulation calculations of stellar collapse. By the end of the 1950s, the conclusion had been reached that, despite the complications of spin, deformation, radiation, etc. the implosion proceeded much the way as envisioned in the idealized Oppenheimer and Snyder calculation. Even with some uncertainty in the nuclear physics involved, this maximum value is determined to be $\approx 2\ M_\odot$. Any star with a mass $M \gtrsim 2M_\odot$ would contract all the way to become a black hole.

- One development that had a significant impact on the thinking of theorists was the rediscovery in 1958 by Finkelstein of the coordinate system first invented by Eddington (1924) in which the Schwarzschild singularity does not appear, showing clearly that it is a coordinate singularity (see Section 8.1.2).

8.3.2 Observational evidence of black holes

Black holes being small black disks in the sky far away, it would seem rather hopeless to ever observe them. But by taking account of the gravitational effects of a black hole on its surroundings (such as gravitational lensing, the orbits of nearby stars, and heating gas orbiting the black hole to very high temperatures), we now have fairly convincing evidence for a large number of black holes. The basic approach is to determine that the mass of the object is greater than the maximum allowed mass of a neutron star ($\approx 2\ M_\odot$); since no known mechanism can stop such a massive system from gravitational collapse into a black hole, it must be a black hole. The "observed" black holes can be classified into several categories:

Black holes in X-ray binaries The majority of all stars are members of binary systems orbiting each other. If the black hole is in a binary system with another visible star, by observing the Kepler motion of the visible companion, one can obtain some limit on the mass of the invisible star as well. If it exceeds $2M_\odot$, it is a black hole candidate. Even better, if the companion star produces significant gas (as is the case of solar flares), the infall of such gas (called **accretion**) into the black hole will produce intense X-rays. A notable example is Cygnus X-1, which is now generally accepted as a black hole binary system with the visible companion being a supergiant star[13] having a mass $M_{vis} \approx 30 M_\odot$ and the invisible compact object, presumably a black hole, having a mass $M \approx 8.7 M_\odot$. Altogether, close to 10 such binary black holes have been identified in our galaxy.

[13]The supergiant star having a radius of about 20 $R_\odot$ cannot be the source of the observed X–rays.

Quasars It has also been discovered (again by detecting the gravitational influence on visible nearby matter) that at the centers of most galaxies are supermassive black holes, with masses ranging from 10^6 to $10^{12} M_\odot$. Even though the initial finding had been a great surprise, once the discovery was made, it is not too difficult to understand why we should expect such super-massive centers. The gravitational interaction between stars is such that they "swing and fling" past each other: one flies off outward while the other falls inward. Thus we can expect many stars and dust to be driven inward towards the galactic core, producing a supermassive gravitational aggregate. It has been observed that some of these galactic nuclei emit huge amounts of X-rays and visible light, becoming a thousand times brighter than the stellar light of a galaxy. Such galactic centers are called **AGN** (active galactic nuclei). The well-known astrophysical objects, **quasars** (quasi-stellar objects), are interpreted as AGNs in the early stage of cosmic evolution. Observations suggest that an AGN is composed of a massive center surrounded by a molecular accretion disk. They are thought to be powered by rotating supermassive black holes at the cores of such disks. To power such a huge emission one needs extremely efficient mechanisms for releasing the energy associated with the matter surrounding the black hole. Recall our discussion in Section 8.2.1 of the huge gravitational energy of particles orbiting close to the black hole horizon. Thus, besides the electromagnetic extraction of rotational energy as alluded to above, an important vehicle is the gravitational binding energy of accreting matter—the gravitational energy is converted into radiation when free particles fall into lower-energy centrally bound states in the formation of the accretion disk around the black hole. From a whole host of such observation and deduction, we conclude that the galactic centers have objects having tens of millions of solar masses. They must be black holes because no known object other than a black hole could be so massive and so small.[14]

[14]For a report of a recent direct observation of the structure on the scale four times the Schwarzschild radius of the supermassive black hole at our galactic center, see Doeleman *et al.* (2008).

Gamma-ray bursts Gamma-ray bursts (GRBs) are the most luminous electromagnetic events observed in the universe. They are a short, intense flash of gamma rays that can last from a few milliseconds to about a hundred seconds.

This is followed by an afterglow of longer wavelength radiation that can last for weeks or even years. These gamma rays seem to emanate from random places deep in space at random times. The leading theoretical explanation[15] is the collapsar model first proposed by Woosley (MacFadyen and Woosley, 1999). The model posits that the core of an extremely massive rapidly rotating star collapses into a black hole, and the infalling material from the star onto the black hole powers an energetic jet that blasts outward through the stellar envelope producing the gamma-ray burst. This theory has gained support by detailed observations of the afterglow light curves of GRBs.

[15]This explanation applies to long-duration GRBs. The short GRBs are generally less energetic and are not associated with massive stars. They may originate from processes such as merging of neutron stars.

8.4 Appendix A: Rotating source of gravity

In this appendix, we shall offer a brief introduction to some of the results that have been discovered about black holes, beyond the simplest nonrotating spherical case, mainly the curved spacetime exterior to a rotating source.

8.4.1 Properties of an axially symmetric metric tensor

Most realistic stars are rotating objects having only axial symmetry. The simplest of such sources is characterized not just by its mass M but also by its angular momentum J. The spacetime exterior to such a rotating source is given by the **Kerr geometry** which reduces to the Schwarzschild geometry in the limit of $J = 0$. The Kerr geometry is considerably more complicated and we shall only present a brief introduction to some of its salient features.

We first study a generic axially symmetric metric tensor and from its form we can already deduce some characteristic features of such a geometry. Again, we label the coordinates as $x^\mu = (ct, r, \theta, \phi)$. Here we are considering a stationary axially symmetric geometry. The metric is independent of the time coordinate t and azimuthal angle ϕ so that the metric is $g_{\mu\nu} = g_{\mu\nu}(r, \theta)$. Since one cannot distinguish a rotating source from another source rotating in the opposite direction, the stationary exterior gravitational field, hence the line element, ds^2 associated with this geometry, should be unchanged under the simultaneous reversals of $t \rightarrow -t$ and $\phi \rightarrow -\phi$. The metric elements $g_{tr} = g_{t\theta} = g_{r\phi} = g_{\theta\phi}$ should be absent because $dtdr$, $dtd\theta$, $drd\phi$, and $d\theta d\phi$ would change sign under such a reflection transformation. Furthermore, we can choose a new radial coordinate (r) so that[16] $g_{r\theta} = 0$. Thus, an axially symmetric metric has the form

$$ds^2 = g_{tt}dt^2 + 2g_{t\phi}dtd\phi + g_{\phi\phi}d\phi^2 + g_{rr}dr^2 + g_{\theta\theta}d\theta^2. \qquad (8.42)$$

The most interesting feature is the presence of the nondiagonal metric element $g_{t\phi}$. Here we will discuss some striking physical consequences following from such a term.

[16]This can be understood by the observation that with $g_{tr} = g_{t\theta} = g_{r\phi} = g_{\theta\phi} = 0$ the line element contains three terms $g_{rr}(dr)^2 + g_{r\theta}drd\theta + g_{\theta\theta}(d\theta)^2$ that can be interpreted as the line element of the 2D (x^1, x^2) subspace. As we have shown in our discussion of the Kruskal coordinates, any 2D space is a conformally flat space and there must exist a new set of coordinates (x^1, x^2) so that the line element is proportional $(dx^1)^2 + (dx^2)^2$, i.e. $g_{12} = 0$.

The dragging of inertial frames

For a particle moving in this axially symmetric geometry with metric elements $g_{\mu\nu}$ having no ϕ dependence, its angular momentum $l = g_{\phi\mu}\dot{x}^\mu = g_{t\phi}\dot{t} + g_{\phi\phi}\dot{\phi}$ (with $\dot{x}^\mu = dx^\mu/d\tau$, etc.), according to (7.47), is a constant of

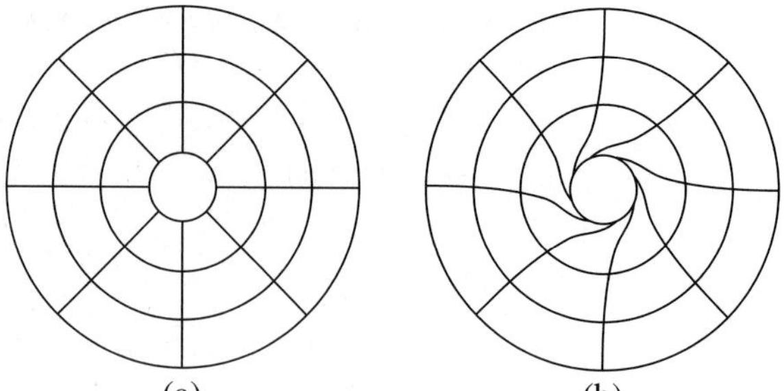

Fig. 8.9 Dragging of the inertial frame by a rotating source shown as turning (a) some initial spacetime (before source rotation) to (b) some "twisted" geometry (after source rotation).

(a)   (b)

the motion along its geodesic. If it starts with a vanishing angular momentum $l = 0$, such a particle (or photon) would fall straight towards the center of the gravitational attraction. Even so, such a zero-angular-momentum particle still develops an angular velocity because the condition $l = 0$ implies $\dot{\phi}/\dot{t} = -g_{t\phi}/g_{\phi\phi}$ so that

$$\omega \equiv \frac{d\phi}{dt} = \frac{\dot{\phi}}{\dot{t}} = -\frac{g_{t\phi}}{g_{\phi\phi}} \tag{8.43}$$

is nonvanishing as $g_{t\phi}$ is nonvanishing. We interpret this phenomenon as the "dragging of inertial frames." That is, as the particle falls freely in a gravitational field, it should follow a geodesic path with respect to the inertial frame of reference. That it acquires an angular velocity means that the inertial frame must itself be rotating with respect to a remote observer located at the beginning of the particle path. We say that the spacetime is being dragged along by the rotating source (Fig. 8.9). A gyroscope will be precessing in such a rotating spacetime. In fact, an experiment, the **Gravity Probe B** experiment, has been launched into space with the aim of measuring this GR prediction of a tiny gyroscopic precession brought about by earth's rotation.

Stationary limit surface and surface of infinite redshift

No material particle can travel faster than light. Now consider a light ray being emitted at a particular point (r, θ, ϕ) in either the $+\phi$ or the $-\phi$ direction. Hence, initially, only $dt \neq 0$ and $d\phi \neq 0$. This light-like line element ($ds^2 = 0$) in the axially symmetric geometry of (8.42) has the form of

$$ds^2 = g_{tt}dt^2 + 2g_{t\phi}dtd\phi + g_{\phi\phi}d\phi^2 = 0. \tag{8.44}$$

This is a quadratic equation in $d\phi/dt$ having the solutions

$$\frac{d\phi}{dt} = -\left(\frac{g_{t\phi}}{g_{\phi\phi}}\right) \pm \sqrt{\left(\frac{g_{t\phi}}{g_{\phi\phi}}\right)^2 - \frac{g_{tt}}{g_{\phi\phi}}}$$

$$= \omega \pm \sqrt{\omega^2 - (g_{tt}/g_{\phi\phi})}. \tag{8.45}$$

where we have plugged in the angular velocity ω of (8.43), which is the angular velocity a light ray attains when falling towards a black hole even though it has zero angular momentum initially, i.e. the angular momentum of the inertial frame dragged along by the rotating source.

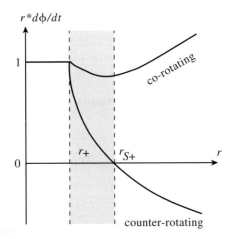

Fig. 8.10 Tangential velocities of circulating light around a rotating black hole. Inside the stationary limit surface $(r < r_{S+})$ even the "counter-rotating light" has to rotate in the same direction, $d\phi/dt > 0$, as the source. Both corotating and counter-rotating light have the same tangential velocity at the event horizon $r = r_+$. Our depiction is for the case of an extreme spinning black hole with angular parameter $a = r^*/2$ so that $r_{S+} = 2r_+ = r^*$. (See Problem 8.7.)

- For the "normal" situation of the coordinate time be time-like $g_{tt}(r, \theta) < 0$, the square root quantity has a greater magnitude than ω, and the light angular velocity can be positive (rotating along with the source) or negative (rotating in the opposite direction to the source). See Fig. 8.10.
- However, on the $g_{tt}(r, \theta) = 0$ surface, while one solution is positive $(d\phi/dt) = 2\omega$, the other corresponds to a vanishing angular velocity. That is, the dragging of the frame is so strong that even a light ray emitted in a direction counter to the source rotation direction, cannot travel in the emitted direction. Clearly, in the region inside this surface where $g_{tt}(r, \theta) > 0$ and the time coordinate is spacelike, no light, hence no particle, can be stationary; they all rotate in the same direction as the source. Hence, the surface where $g_{tt}(r, \theta) = 0$ is called the **stationary limit surface**.
- As we shall discuss below, the event horizon of a rotating black hole is enclosed inside the stationary limit surface. It can be shown (Problem 8.8) that at the event horizon we have $(g_{t\phi}/g_{\phi\phi})^2 = g_{tt}/g_{\phi\phi}$ so that both the corotating and counter-rotating light circulate at the same velocity (as depicted in Fig. 8.10).

Let us recall our discussion of redshift in Section 8.1.1—see in particular Eq. (8.8). The received light frequency ω_{rec} being proportional to $\sqrt{g_{tt}}$, this stationary limit surface having $g_{tt}(r, \theta) = 0$ is thus also the surface of infinite redshift. This is consistent with the result obtained for the Schwarzschild black hole; in that case the $g_{tt} = 0$ condition corresponds to $r = r^*$. This r^*-surface is indeed the **surface of infinite redshift** as well as the stationary limit surface as no particle can stand still in the $r < r^*$ region, but must move towards the $r = 0$ singularity. On the other hand, for the Schwarzschild geometry this surface is also an event horizon. As we shall see below, for a rotating black hole, these two surfaces (the $g_{tt} = 0$ surface and the event horizon with $g_{rr} = \infty$) do not in general coincide. A particle moving in the region between these two surfaces (being outside the horizon, can still escape to infinity) can sometimes appear to have negative total energy. This in-between region is the **ergosphere**.[17]

[17] *Ergo* is the Greek word for energy and work.

[18]The Einstein equation for the geometry exterior to a source with mass M and angular momentum J is a set of coupled partial differential equations that correspond to the vanishing Ricci tensor $R_{\mu\nu} = 0$. From this set of equations, one solves for g_{tt}, $g_{t\phi}$, $g_{\phi\phi}$, g_{rr}, and $g_{\theta\theta}$.

8.4.2   Kerr geometry and the Penrose process

Solving the Einstein field equation[18] for a rotating axially symmetric source with mass M and angular momentum J, Kerr found the scalar metric functions of (8.42) leading to an expression for the line element of:

$$ds^2 = -\left(1 - \frac{rr^*}{\rho^2}\right)c^2dt^2 - \frac{2arr^*\sin^2\theta}{\rho^2}cdtd\phi + \frac{\rho^2}{\triangle}dr^2 + \rho^2 d\theta^2$$

$$+ \left(r^2 + a^2 + \frac{a^2 rr^* \sin^2\theta}{\rho^2}\right)\sin^2\theta d\phi^2. \tag{8.46}$$

Besides the Schwarzschild radius $r^* = 2G_N/c^2$, we have the new parameters:

$$a = J(cM)^{-1}, \tag{8.47}$$

$$\rho^2 = r^2 + a^2 \cos^2\theta, \tag{8.48}$$

$$\triangle = r^2 - rr^* + a^2. \tag{8.49}$$

We now check some simple limits of this Kerr metric (8.46).

Kerr geometry is asymptotically flat   In the limit of $r \gg r^*$ and $r \gg a$, we have, up to $O(r^*/r)$ and $O(a/r)$, the approximation

$$\rho \simeq r, \quad \triangle \simeq r^2\left(1 - \frac{r^*}{r}\right), \quad \text{hence,} \quad \frac{\rho^2}{\triangle} \simeq \left(1 + \frac{r^*}{r}\right),$$

and thus

$$ds^2 = -\left(1 - \frac{r^*}{r}\right)c^2dt^2 - \frac{2ar^*\sin^2\theta}{r}cdtd\phi$$

$$+ \left(1 + \frac{r^*}{r}\right)dr^2 + r^2\left(d\theta^2 + \sin^2\theta d\phi^2\right), \tag{8.50}$$

making it clear that, when $r^*/r = 0$, it is the Minkowski flat metric in spherical coordinates.

The nonrotating limit is the Schwarzschild geometry   When the source is not rotating, thus angular momentum $J = 0$, dropping the term proportional to the angular momentum parameter a and setting $\rho = r$ and $\triangle = r(r - r^*)$ in the Kerr line element in (8.46), we recover the Schwarzschild metric (7.18).

The Boyer–Lindquist coordinates in the $M = 0$ flat spacetime limit   From the above discussion it is clear that in the $a = 0$ limit the coordinates (r, θ, ϕ) in (8.46) coincide with the Schwarzschild coordinates, which are the familiar spherical polar axes. However, one must keep in mind that this is not true for the general case when $a \neq 0$. In fact the coordinates used in our presentation of the Kerr metric represent a distinct set, called the **Boyer–Lindquist coordinates**. We can see this more clearly by turning off the gravity by setting $M = 0$, hence $r^* = 0$ and $\Delta = r^2 + a^2$, in (8.46), so that

$$ds^2 = -c^2 dt^2 + \frac{\rho^2}{r^2 + a^2} dr^2 + \rho^2 d\theta^2 + \left(r^2 + a^2\right) \sin^2 \theta d\phi^2. \qquad (8.51)$$

It can be shown that, after making the substitution of

$$x = \sqrt{r^2 + a^2} \sin \theta \cos \phi$$
$$y = \sqrt{r^2 + a^2} \sin \theta \sin \phi \qquad (8.52)$$
$$z = r \cos \theta,$$

the line element in (8.51) is revealed as that for a flat Minkowski space:

$$ds^2 = -c^2 dt^2 + dx^2 + dy^2 + dz^2, \qquad (8.53)$$

which is consistent with our expectation that, in the absence of a source, the spacetime should be flat. From the relations in (8.52) we see that the (r, θ) coordinates differ from the usual spherical coordinates. The lines of fixed θ and ϕ are not straight lines passing through the origin, but are hyperbolas; the surfaces of $r = $ constant are not spheres centered around the origin but ellipsoids of revolution (i.e. ellipses in the *x-z* or *y-z* planes, but circles in the *x-y* cross-sectional surfaces). In particular $r = 0$ does not correspond to a point, but a disk of radius a lying in the *x-y* plane (as it is made up of rings with radii ranging from zero to a as θ ranges from 0 to $\pi/2$)

$$x = (a \sin \theta) \cos \phi, \quad y = (a \sin \theta) \sin \phi. \qquad (8.54)$$

Particularly, the loci of $r = 0$ with $\theta = \pi/2$ form a ring of radius a.

Rotating black holes   The Schwarzschild black hole has a relatively simple configuration: a spherical horizon surface at $r = r^*$ (which is also the surface of infinite redshift and stationary limit surface) encloses the physical singularity at the point $r = 0$. The black hole due to a rotating mass, the Kerr black hole, has a more complicated structure (Fig. 8.11).

Singularities of the Kerr metric and event horizon

An inspection of (8.46) shows that the Kerr metric is singular at $\rho = 0$ and $\Delta = 0$. In fact, $\rho = 0$ is a physical singularity while $\Delta = 0$, is a coordinate singularity.[19]

The physical singularity is a ring with radius a Since $\rho^2 = r^2 + a^2 \cos^2 \theta$, the condition $\rho = 0$ can only be satisfied by

$$r = 0 \quad \text{and} \quad \theta = \pi/2. \qquad (8.55)$$

We have shown in the above that $r = 0$ is a disk with radius a. For $\theta = \pi/2$ we see that the physical $\rho = 0$ singularity is, interestingly, a ring with a radius (a) proportional to the angular momentum of the rotating source.

[19]In sidenote 2 of this chapter, we mentioned that in the case of Schwarzschild geometry, the coordinate-independent product of the curvature tensor $R_{\mu\nu\lambda\rho} R^{\mu\nu\lambda\rho}$ is only singular at $r = 0$. This means that the singularity at $r = r^*$ that we encounter in the Schwarzschild coordinates is a coordinate singularity. Similarly, using the Kerr metric one can calculate the invariant $R_{\mu\nu\lambda\rho} R^{\mu\nu\lambda\rho}$ showing that it is singular at $\rho = 0$ but not at $\Delta = 0$.

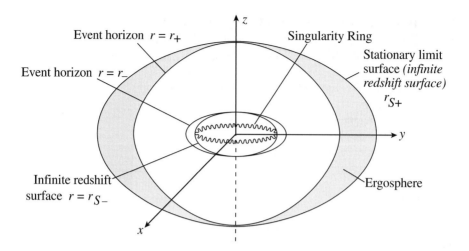

Fig. 8.11 The structure of a rotating black hole.

The coordinate singularity at $\triangle = 0$ We have discussed at the end of Section 8.1.2 that an event horizon corresponds to a "null 3-surface"—all points on such a surface have their lightcones being entirely on one side of the surface. That is, while a light ray can travel on this surface, all other particles can only cross this surface and cross it only in one direction. For the case of a black hole they can only cross towards the $r = 0$ singularity, not away from it. While we will not present the geometric derivation, our intuition that the condition for such a surface corresponds to the singularity of $g_{rr} = \infty$ (as is the case with the nonrotating black hole) turns out to be correct. Thus the event horizon of a Kerr black hole is the surface correspondings to the solution of the quadratic equation $\triangle = r^2 - rr^* + a^2 = 0$. That is

$$ r = r_\pm = \frac{r^*}{2} \pm \sqrt{\left(\frac{r^*}{2}\right)^2 - a^2}. \tag{8.56} $$

Like all surfaces with $r = $ constant in the Boyer–Lindquist coordinates, the $r = r_\pm$ surface are not spheres, but are ellipsoids of revolution. In the absence of source rotation, the r_+ surface reduces to the Schwarzschild spherical surface at $r = r^*$, while the r_- surface reduces to the $r = 0$ point singularity. In the following, we shall concentrate on the r_+ surface as the event horizon. It is also interesting to note from (8.56) that the angular momentum parameter of a rotating black hole is restricted to $a \leq (r^*/2)$. The limiting case of $a_{max} = (r^*/2)$, called an **extreme-spinning black hole**, is of considerable interest because such a limit is often realized in astrophysical situations. Most massive stars and galactic nuclei can collapse to form black hole; as more material is accreted onto a black hole, it spins faster and faster to reach a final state with an angular momentum parameter very close to a_{max}.

The ergosphere: Extracting energy from a rotating black hole

For a rotating black hole, the horizon (a null 3-surface determined by the condition $g_{rr} = \infty$) no longer coincides, as is the case of a Schwarzschild black hole, with the $g_{tt} = 0$ surface of infinite redshift (also, the stationary limit surface), across which the metric element g_{tt} changes sign. The event horizon (at $r = r_+$) is inside the stationary limit surface ($r_+ \leq r_{S+}$). In the Kerr geometry the stationary-limit condition $g_{tt} = 0$, according to (8.46) and (8.48), corresponds to

$$1 - \frac{rr^*}{r^2 + a^2 \cos^2 \theta} = 0, \tag{8.57}$$

which is a quadratic equation in r:

$$r^2 - r^* r + a^2 \cos^2 \theta = 0 \tag{8.58}$$

with the solutions

$$r_{S\pm} = \frac{r^*}{2} \pm \sqrt{\left(\frac{r^*}{2}\right)^2 - a^2 \cos^2 \theta}. \tag{8.59}$$

We again concentrate on the r_{S+} solution. This is an axially symmetric ellipsoidal surface flattened at $\theta = 0, \pi$, see Fig. 8.11. As $\cos^2 \theta \leq 1$, a comparison of (8.56) and (8.59) shows that $r_{S+} \geq r_+$ (the equality holds only at the two poles of $\theta = 0, \pi$); the horizon is enclosed inside the stationary limit surface. The space between these two surfaces is called the **ergosphere**. Particles moving in this region have the interesting possibility of having a negative total energy. While no particle can stand still in the ergosphere, since this region is outside the event horizon, it can still escape to $r = \infty$. As discussed in Box 8.1, this gives rise to the possibility of extracting energy from a rotating black hole through the **Penrose process**.

Box 8.1 Negative-energy particles and the Penrose process

Particle energy measured by an observer In our discussion of special relativity in Section 3.2.2, we introduced the 4-momentum of a particle $p^\mu = (E/c, p^i)$ where the time component (energy E) must be nonnegative, while the spatial components can be either positive or negative. In general relativity with a curved spacetime, we need to be more explicit as to the observer who measures the energy and momentum of the particle. The energy measured by an observer with 4-velocity $U^\mu_{\text{obs}} = \gamma \left(c, v^i\right)$ is given by

$$E_{\text{obs}} = -g_{\mu\nu} p^\mu U^\nu_{\text{obs}}, \tag{8.60}$$

which checks with the (special) relativistic energy quoted above for an observer at rest $U^\mu_{\text{obs}} = (c, 0, 0, 0)$ in a flat spacetime $g_{\mu\nu} = \eta_{\mu\nu} = \text{diag}(-1, 1, 1, 1)$. In a general geometry, while the energy measured by a local observer is always positive (since geometry is locally flat we can always go to such a local frame to recover the SR result), the energy

(*cont.*)

[20] This just means that it takes more than mc^2 amount of energy to move a particle to the site of the observer. For simplicity imagine a particle at rest bound to a gravitational system so that $E_{tot} = mc^2 + E_{bin}$. If the (negative) binding energy E_{bin} has a magnitude greater than mc^2, the total energy is negative.

[21] This is possible because the ergosphere is outside the event horizon.

[22] We recall that the invariant length of the Killing vector is $g_{\mu\nu} K^\mu_{(t)} K^\nu_{(t)} = g_{tt}$.

Box 8.1 (*Continued*)

as measured by a distant observer in a certain situation can actually be negative.[20]

Conserved energy along a geodesic   As discussed in Section 7.3.1, a particle moving in a spacetime with stationary geometry has, from (7.46), the constant of motion $\kappa = -g_{\mu\nu} \dot{x}^\mu K^\nu_{(t)}$ with the Killing vector $K^\mu_{(t)} = (1, 0, 0, 0)$. Given that the 4-momentum is $p^\mu = m\dot{x}^\mu$, we have $\kappa = E(\infty)/mc$; that is, the constant of motion κ is the energy (in units of mc) as measured by an observer at infinity (where the geometry is flat). See also Problem 7.5.

The Penrose process   Imagine that a particle, traveling into the ergosphere, decays into two particles: one gets out again[21] to $r = \infty$ and the other with 4-momentum $\tilde{p}^\mu$ falls into the black hole. Energy and momentum must be conserved in this decay process:

$$p^\mu_{\text{in}} = p^\mu_{\text{out}} + \tilde{p}^\mu. \tag{8.61}$$

We can contract every term in this equation with $-g_{\mu\nu} K^\nu_{(t)}$ to obtain, with obvious notation, $(m\kappa)_{\text{in}} = (m\kappa)_{\text{out}} + (\tilde{m}\tilde{\kappa})$, or

$$E_{\text{in}}(\infty) = E_{\text{out}}(\infty) + \tilde{E}(\infty). \tag{8.62}$$

The key point is that it is possible for $\tilde{\kappa}$, hence $\tilde{E}(\infty)$, to be negative. We first note that because the in-particle starts out from, and the out-particle ends up at, infinity, $E_{\text{in}}(\infty)$ and $E_{\text{out}}(\infty)$, being respectively the energies observed by a local observer at infinity, must be positive. This means that we have positive $(m\kappa)_{\text{in}}$ and $(m\kappa)_{\text{out}}$ along the particle worldlines connecting the decay site and $r = \infty$. Such a restriction does not apply to $\tilde{\kappa}$ because this $\tilde{p}$ particle falls into the black hole and never travels outside the ergosphere. Since the decay takes place in the ergosphere (a region with $g_{tt} > 0$), where the t-coordinate and Killing vector $K^\mu_{(t)}$ are space-like,[22] $\tilde{\kappa}$ is proportional to a component of the spatial momentum so that $\tilde{\kappa}$, hence $\tilde{E}(\infty)$, can be either positive or negative. Through (8.62), a negative $\tilde{E}(\infty)$ implies $E_{\text{out}}(\infty) > E_{\text{in}}(\infty)$: the output energy is greater than the input. We can extract energy from a rotating black hole through the Penrose process!

Negative energy particle must have negative angular momentum   We now show that if a particle has negative energy, it must also have negative angular momentum—that is, its angular momentum has opposite sign to that of the rotating gravitational source. Thus, when the negative-energy $\tilde{p}$ particle falls into the rotating black hole, it not only decreases the black hole's energy, it also diminishes its angular momentum. Recall from our discussion in Section 7.3.1 that for a particle moving in an axially symmetric geometry we have, from (7.47), the conserved particle angular momentum $l = m\lambda = -g_{\mu\nu} p^\mu K^\nu_{(\phi)}$ with $K^\mu_{(\phi)} = (0, 0, 0, 1)$. The particle's angular momentum must be negative if $\tilde{\kappa}$ is negative. This restriction can be deduced from the fact the particle energy measured by a local observer must be positive. Here we must first specify a local observer. Since no particle can be stationary inside the ergosphere, a simple trajectory (not necessarily a geodesic) for an observer can be a circulating orbit (i.e. $\dot{r} = \dot{\theta} = 0$) with an angular speed of $\omega_{\mathrm{obs}} = d\phi/dt = \dot{\phi}/\dot{t}$ so that the 4-velocity of this observer is

$$U^{\mu}_{\mathrm{obs}} = \left(c\dot{t}, 0, 0, \dot{\phi}\right) = \dot{t}\left(c, 0, 0, \omega_{\mathrm{obs}}\right). \tag{8.63}$$

This can be written as a linear combination of the two Killing vectors,

$$U^{\mu}_{\mathrm{obs}} = c\dot{t}\left(K^{\mu}_{(t)} + \frac{\omega_{\mathrm{obs}}}{c}K^{\mu}_{(\phi)}\right). \tag{8.64}$$

The energy of the $\tilde{p}$-particle measured by this local observer (8.60) must be positive:

$$E_{\mathrm{obs}} = -g_{\mu\nu}\tilde{p}^{\mu}U^{\nu}_{\mathrm{obs}} = c\dot{t}\left(-g_{\mu\nu}\tilde{p}^{\mu}K^{\nu}_{(t)} - \frac{\omega_{\mathrm{obs}}}{c}g_{\mu\nu}\tilde{p}^{\mu}K^{\nu}_{(\phi)}\right)$$

$$= \dot{t}\left(\tilde{E}\left(\infty\right) - \omega_{\mathrm{obs}}\tilde{l}\right) > 0. \tag{8.65}$$

The above inequality, with $\dot{t}$ being positive, implies that

$$\tilde{E}\left(\infty\right) > \omega_{\mathrm{obs}}\tilde{l}. \tag{8.66}$$

Since the angular speed ω_{obs} of any observer in the ergosphere must be positive—see discussion following equation (8.45)—a negative $\tilde{E}(\infty)$ implies a negative particle angular momentum $\tilde{l}$. Thus, when the $\tilde{p}$-particle falls into the rotating black hole, it not only reduces the energy of the black hole but also its angular momentum. The Penrose process extracts energy from the rotational energy of a black hole.

The particle binding energy by a rotating black hole

In this section we shall demonstrate that the gravitational binding energy of a particle around a Kerr black hole is even greater than that around a Schwarzschild black hole—it can be as much a 42% of its rest energy!

Effective potential of a particle in the equatorial plane   We follow the same procedure used in the study of the particle orbit in the Schwarzschild geometry as shown in Section 8.2.1. We have two constants of motion. The geometry being stationary, the Killing vector $K^{\mu}_{(t)} = (1, 0, 0, 0)$ leads to the conserved quantity

$$-\kappa = g_{\mu\nu}\dot{x}^{\mu}K^{\nu}_{(t)} = g_{tt}\dot{t} + g_{t\phi}\dot{\phi}, \tag{8.67}$$

while the Killing vector $K^{\mu}_{(\phi)} = (0, 0, 0, 1)$ of the axial symmetry leads to

$$l = mg_{\mu\nu}\dot{x}^{\mu}K^{\nu}_{(\phi)} = mg_{t\phi}\dot{t} + mg_{\phi\phi}\dot{\phi}. \tag{8.68}$$

The particle trajectories in the Kerr geometry are generally more complicated. Here we choose to work with the simple case when they are confined to the equatorial plane $\theta = \pi/2$. With this restriction, the Kerr metric of (8.46) is simplified, the metric elements needed in (8.67) and (8.68) are

$$g_{tt} = -\left(1 - \frac{r^*}{r}\right), \quad g_{t\phi} = -\frac{ar^*}{r}, \quad g_{\phi\phi} = \left(r^2 + a^2 + \frac{a^2 r^*}{r}\right). \quad (8.69)$$

We recall that these two constants (κ, l) have the physical interpretation as the total energy $\kappa = E/(mc)$ and angular momentum of the particle. From the two equations (8.67) and (8.68), we can solve for $\dot{t}$ and $\dot{\phi}$ and substitute them into the Lagrangian $L = g_{\mu\nu}\dot{x}^\mu \dot{x}^\nu = -c^2$ to obtain the energy balance equation of a particle in the Kerr geometry

$$\mathcal{E} = m\frac{\kappa^2 - c^2}{2} = \frac{1}{2}m\dot{r}^2 + m\Phi_{\text{eff}}^{(K)} \quad (8.70)$$

with the effective potential

$$\Phi_{\text{eff}}^{(K)} = -\frac{c^2 r^*}{2r} + \frac{l^2 - a^2\left(\kappa^2 - c^2\right)}{2m^2 r^2} - \frac{r^*\left(l - a\kappa\right)^2}{2m^2 r^3}, \quad (8.71)$$

which clearly checks with (8.35) in the nonrotating Schwarzschild limit of $a = 0$.

The binding energy for an extreme-spinning black hole As discussed in the paragraph following Eq. (8.59), the limit when the source rotates at maximal angular momentum, $a = a_{\max} = r^*/2$, the extreme-spinning black hole, is of considerable astrophysical interest. Just as in Section 8.2, we can again find the innermost stable circular orbit of a particle around such a black hole. From $\partial \Phi_{\text{eff}}^{(K)} = 0$, just like in (8.36), and the condition like (8.40) for the innermost orbit radius, as well as the energy balance equation (8.70) with $\dot{r} = 0$, we can solve (Problem 8.9), for the extreme-spinning black hole case, the three unknowns of the ISCO radius r_0, angular momentum l_0, and particle energy κ_0:

$$r_0 = \frac{r^*}{2}, \quad l_0 = \frac{mcr^*}{\sqrt{3}}, \quad \kappa_0 = \frac{c}{\sqrt{3}}. \quad (8.72)$$

This yields the energy of the circulating particle:

$$E_0 = mc\kappa_0 = \frac{mc^2}{\sqrt{3}} = 0.58mc^2, \quad (8.73)$$

corresponding to an extraordinarily large binding energy of $0.42mc^2$. The possibility to release a comparable amount of gravitational binding energy of circulating matter gives rise to some of the most energetic phenomenon observed in the cosmos.

8.4.3 Beyond the Schwarzschild and Kerr black holes

Here we shall offer brief remarks on some of the GR results that have been discovered about black holes, beyond the simplest nonrotating spherical and rotating axial symmetrical cases discussed in previous sections.

- The Schwarzschild black hole is characterized by a single parameter, the stellar mass M, while the Kerr black hole is characterized by an additional parameter, the angular momentum J of the source. The GR solutions for the warped spacetime outside an electrically charged mass sources is also known: it is the **Reissner–Nordström geometry**. This set of solutions is characterized by at most three parameters: the total mass M, the angular momentum J, and the electric charge Q. The lack of any detailed feature on a black hole has been described as "**black holes have no hair.**"
- The **singularity theorem** of GR states that any gravitational collapse that has proceeded far enough along its destiny will end in a physical singularity. Thus the $r = 0$ singularity encountered in the Schwarzschild exterior spacetime is not a peculiar feature of the spherical coordinate system.
- It is conjectured, but has not been proven in generality, that for the general nonspherically symmetric collapse GR also predicts the formation of an event horizon, shielding the physical singularity from all outside observers. This is called the **cosmic censorship conjecture.**

8.5   Appendix B: Black holes and quantum physics

In this appendix, we shall offer some brief comments on the intersection between black holes and quantum physics: Hawking radiation, black hole thermodynamics, and quantum gravity, etc. Any detailed discussion of these advanced topics is beyond the scope of this introductory exposition. Our purpose here is merely to alert readers to the existence of a vast body of knowledge on such topics, which are at forefront of current research.

8.5.1   The Planck scale

General relativity is a classical macroscopic theory. For a microscopic description, we would need to combine GR with quantum mechanics, called **quantum gravity**. The scale at which physics must be described by quantum gravity is the **Planck scale**.

Soon after the 1900 discovery of **Planck's constant** $\hbar$ in fitting the blackbody spectrum, Planck noted that a self-contained system of units of **mass–length–time** can be obtained from various combinations of Newton's constant G_N (gravity), $\hbar$ (quantum theory), and the velocity of light c (relativity). When we recall that G_N (mass)^2, has the unit of **energy · length**, and in relativistic quantum theory the natural scale of **energy·length** is $\hbar c$, we have the natural mass scale for quantum gravity, the **Planck mass**,

$$M_{\text{Pl}} = \left(\frac{\hbar c}{G_N} \right)^{1/2}. \tag{8.74}$$

From this we can immediately deduce the other Planck scales:

$$\text{Planck energy } E_{\text{Pl}} = M_{\text{Pl}}c^2 = \left(\frac{\hbar c^5}{G_N}\right)^{1/2} = 1.22 \times 10^{19} \text{ GeV}$$

$$\text{Planck length } l_{\text{Pl}} = \frac{\hbar c}{E_{\text{Pl}}} = \left(\frac{\hbar G_N}{c^3}\right)^{1/2} = 1.62 \times 10^{-33} \text{ cm}$$

$$\text{Planck time } t_{\text{Pl}} = \frac{l_{\text{Pl}}}{c} = \left(\frac{\hbar G_N}{c^5}\right)^{1/2} = 5.39 \times 10^{-44} \text{ s.} \tag{8.75}$$

Such extreme scales are vastly beyond the reach of any laboratory setups. (Recall that the rest energy of a nucleon is about 1 GeV, and the highest energy the current generation of accelerators can reach is less than 10^4 GeV.) The natural phenomena that can reach such an extreme density of $M_{\text{Pl}}/(l_{\text{Pl}})^3 = c^5/(\hbar G_N) = 5.16 \times 10^{96}$ g/cm^3 are the physical singularities in GR: endpoints of gravitational collapse hidden inside a black hole horizon and the origin of the cosmological big bang. It is expected that quantum gravity will modify such singularity features of GR.

Quantum gravity is a union of general relativity with quantum mechanics having the Planck scale as its natural scale. The union of special relativity with quantum mechanics is quantum field theory (QFT). Namely, SR embodies classical fields such as Maxwell's fields and the quantum description of field systems (such as electromagnetic fields) is quantum field theory (such as quantum electrodynamics). The quanta of a field are generally viewed as particles. For example, the quanta of an electromagnetic field are photons. QFT is a natural language to describe interactions that include the possibility of particle creation and annihilation allowed by the relativistic energy and mass relation of $E = mc^2$. Since one is not dealing with a quantum description of gravity, QFT does not require the extreme Planck scale for its description to be relevant.

8.5.2   Hawking radiation

The surprising theoretical discovery by Stephen Hawking that a black hole (contrary to the general expectation that nothing can come out) can radiate was made in the context of a quantum description of particle fields in the background Schwarzschild geometry. That is, the relevant theoretical framework involves only a partial unification of gravity with quantum theory: while the fields of photons, electrons, etc., are treated as quantized systems, gravity is still described by the classical (nonquantum) theory of GR. Thus, the relevant context is **quantum field theory in a curved spacetime**.

The quantum uncertainty principle of energy and time, $\triangle E \, \triangle t \gtrsim \hbar/2$, implies that processes violating energy conservation can occur, provided they take place in a sufficiently short time interval $\triangle t$. Such quantum fluctuations cause the empty space to become a medium with particle and antiparticle pairs appearing and disappearing. In normal circumstances such energy-nonconserving processes cannot survive on the macroscopic time-scale.

(Hence the temporarily created and destroyed particles are called **virtual particles**.) However, as Hawking showed, if such random quantum fluctuations take place near an event horizon of a black hole, the virtual particles can become real because, in such a situation, energy conservation can be maintained on the observable scale.

Consider the simplest quantum fluctuation of creating a particle and antiparticle pair from the vacuum. Energy–momentum conservation requires $0 = p^\mu + \tilde{p}^\mu$. We can contract every term in this equation with the Killing vector $-g_{\mu\nu}K^\nu_{(t)}$ to obtain the relation between the constants of motion along their respective geodesics after their creation,[23] $0 = \kappa + \tilde{\kappa}$, or

$$0 = E\left(\infty\right) + \tilde{E}\left(\infty\right). \tag{8.76}$$

If both particles could reach $r = \infty$, then $E(\infty)$ and $\tilde{E}(\infty)$ would be the energies measured by local observers at infinity. Such energies must be positive and the equality (8.76) cannot be satisfied. However, if this quantum fluctuation takes place sufficiently close to the event horizon of a black hole, one particle can be outside (and eventually travels to $r = \infty$) and the other particle inside[24] the event horizon (and falls into the singularity). One of the particle, say the p particle, reaches $r = \infty$ thus $E(\infty)$ must be positive; if $\tilde{E}(\infty)$ of the other $\tilde{p}$ particle is negative, then the conservation relation (8.76) can be satisfied on the macroscopic time-scale. Just as we have detailed in Box 8.1 on the Penrose process, $\tilde{\kappa}$, hence $\tilde{E}(\infty)$, can be negative because in the region inside the stationary limit surface (which coincides with the event horizon for the nonrotating case) g_{tt} is positive so that the t-coordinate is space-like. The result is that, to a distant observer, the black hole emits a particle with positive energy. This is known as the **Hawking effect** or **Hawking radiation**. Clearly, the above discussion is applicable for either particle to reach infinity. Thus we expect the Hawking radiation to be composed of an equal number of particles and antiparticles. This radiation can take place because of an addition to the black hole of particles of negative energy, i.e. a decrease of the black hole's positive energy and mass.

8.5.3   Black hole thermodynamics

QFT (in curved spacetime) calculations by Hawking have shown that this radiation has a blackbody temperature and a thermal energy inversely proportional to its mass M:

$$k_B T = \frac{1}{4\pi}\frac{\hbar c}{r^*} = \frac{1}{8\pi}\frac{E_{Pl}^2}{Mc^2}, \tag{8.77}$$

where k_B is Boltzmann's constant, r^* the Schwarzschild radius, and $E_{Pl}^2 = \hbar c^5/G_N$ is the Planck energy squared, see (8.75). It is easy to understand this temperature magnitude by dimensional analysis. It is a quantum effect involving gravity: the natural scale of (energy·length) in relativistic quantum theory is $\hbar c$ and it must be divided by a length quantity to produce the thermal energy of $k_B T$; the natural length in GR is $r^* = 2G_N M/c^2$. The second expression can be understood in the quantum gravity framework. There is the natural Planck

[23] Here we use the same notation defined in previous sections. See, especially, the discussion in Box 8.1 on negative energy particles and the Penrose process.

[24] During a fluctuation, one cannot locate any particle to such precision. If one insists on a classical picture with exact particle location, one can say that after the particle creation one of them, during the fluctuation time Δt, travels across the event horizon.

energy scale of E_{Pl}; since we expect the energy to be proportional to Planck's constant $\hbar$, we must have E_{Pl}^2 divided by another energy factor. The only other natural energy scale in this problem is Mc^2. Of course one needs the detailed calculation to fix the unknown coefficients in front.

Evaporation of mini black holes   The temperature of (8.77) is inversely proportional to the mass $T \sim M^{-1}$. This means that for astrophysical black holes (large M) the Hawking radiation is rather weak and this emission is rather insignificant compared to the addition of energy by the in-falling material. On the other hand, for small-mass "mini black holes," because any loss of energy/mass will result in a hotter black hole (hence an even faster radiation), Hawking radiation will lead to their eventual total evaporation.

Entropy and black hole area increasing theorem   Knowing the blackbody temperature, one can associate a thermodynamic entropy S with a black hole (the **Bekenstein–Hawking entropy**) by a straightforward application of thermodynamic formulas: as $T dS = dU$ with $U = Mc^2$ for energy, we have from (8.77)

$$dS = \frac{8\pi G_N k_B M}{\hbar c} dM. \tag{8.78}$$

Integrate this relation to obtain the entropy associated with a black hole in units of Boltzmann's constant (recall: S and k_B have the same units):

$$\frac{S}{k_B} = \frac{4\pi G_N M^2}{\hbar c}. \tag{8.79}$$

This can be expressed very simply in terms of the **horizon area** A^* in units of Planck's length squared $l_{Pl}^2 = \hbar G_N / c^3$, see (8.75). The horizon area for a nonrotating black hole is $A^* = 4\pi r^{*2}$, and we have

$$\frac{S}{k_B} = \frac{1}{4} 4\pi \left(\frac{2 G_N M}{c^2} \right)^2 \frac{c^3}{\hbar G_N} = \frac{1}{4} \frac{A^*}{l_{Pl}^2}. \tag{8.80}$$

Since entropy is an ever-increasing function, the relation (8.80) between the black hole entropy and area also implies that the black hole's horizon area is ever-increasing. This Bekenstein–Hawking "**area increasing theorem**" was in fact discovered before the advent of Hawking radiation. Even then it had been speculated that black hole formulas had a thermodynamic interpretation.

8.5.4   Black holes and quantum gravity

GR, as a classical field theory, is the $\hbar \to 0$ limit of the quantum theory of gravity. Quantum gravity, as the quantum description of space, time, and the universe, represents the union of GR with quantum mechanics. All indications are that it is **the** fundamental theory of physics because its candidate theories (such as the superstring M theory) also encompass the description of strong, weak, and electromagnetic interactions. Thus, it is a theory of quantum gravity and unification. Although impressive advances have been made, this program is still very much a work-in-progress. It represents a major forefront in current theoretical physics research.

The black hole physics of GR is a classical macroscopic description. For a microscopic theory we would need quantum gravity. We are familiar with the macroscopic physics as given by thermodynamics, which is obtained by averaging over atomic motions. While we do not have a fully developed theory of quantum gravity, one is curious to see whether the various candidate theories of quantum gravity have the correct features that can be checked by this micro–macro connection. It turns out that the blackbody radiation emitted by black holes, the Hawking radiation, provides such a handle.

Black holes and advances in current study of quantum gravity Black holes are a unique arena to study quantum gravity. The singularity hidden behind the horizon demands both GR and quantum descriptions. Thus, a black hole is an ideal laboratory for thought experiments relating to quantum gravity. For instance, the Schwarzschild event horizon is seen to give rise to the phenomenon of infinite gravitational redshift, infinitely stretching the wavelength. Thus, Planck length phenomena near the horizon can be greatly amplified. It should be fruitful to investigate quantum gravitational processes associated with black holes. In this connection, we mention two important developments:

1. **The number of quantum states in a black hole**. Recall the well-known connection in statistical mechanics between entropy and information: $S = k_B \ln W$, where W is the number of microstates in the system under study. For a black hole, one can assume that W is given by the number of ways a quantum black hole can be formed. In superstring M theory, for example, W has been calculated and found to be in perfect agreement with the Bekenstein–Hawking entropy of (8.80). That such theories can give the correct count of quantum states of a black hole certainly encourages us to believe that they contain essentially correct ingredients for a true quantum gravity theory.
2. **The holographic principle**. The fact that the entropy of a black hole is proportional to its area, as in (8.80), has led to the conjecture that states in a spacetime region can equally well be represented by bits of information contained in its surface boundary. This "holographic principle" by Gerard 't Hooft and Leonard Susskind has become one of the principles in studies of theories of quantum gravity and unification.

Review questions

1. What does it mean to say that the Schwarzschild surface at $r = r^*$ is only a coordinate singularity?

2. What is an event horizon associated with a black hole?

3. At the $r = r^*$ surface of a Schwarzschild black hole, the proper time is finite while the coordinate time is infinite. What time measurements do we refer to in these two descriptions? What is an alternative (but equivalent) description of this phenomenon of infinite coordinate time?

4. An event horizon is geometrically a null 3-surface. What is a null surface? Why does such a surface allow particles and light to traverse only in one direction?

5. What is the basic property of the time coordinate in the advanced Eddington–Finkelstein coordinate system that allows the lightcone to tip over smoothly inward across the $r = r^*$ surface? Also answer the same question for the retarded Eddington–Finkelstein system. This property of the coordinates allows their respective spacetime diagram to display the black hole and white hole solutions. What is a white hole?

6. In the Schwarzschild geometry, the metric conditions in the Schwarzschild coordinates $g_{tt} = 0$ and $g_{rr} = \infty$ are satisfied simultaneously by the $r = r^*$ surface. In the Kerr geometry exterior to a rotating source, they correspond to distinct surfaces. What are the distinct properties of the spacetime region as defined by these two metric conditions?

7. In the Kruskal coordinates both the ingoing and outgoing light geodesics are $45°$ straight lines so that both the black

hole and white hole can be displayed on the same space-time diagram. Describe very briefly the four quadrants of a Kruskal diagram. In this connection, what is a wormhole? Why must a wormhole be a dynamical structure of the geometry?

8. The effective potential for a particle in the Schwarzschild spacetime has the form of

$$V_{\text{eff}} = -\frac{A}{r} + \frac{Bl^2}{r^2} - \frac{Cl^2}{r^3}$$

with the positive coefficients A, B, and C. Use this expression to explain why, unlike the Newtonian central force problem, a particle can spiral into the center even with nonzero angular momentum $l \neq 0$.

9. Black holes are linked with many of the most energetic phenomena observed in the cosmos. What is the energy source associated with a black hole that can power such phenomena?

10. List three or more astrophysical phenomena that are thought to be associated with black holes.

11. Black holes, white holes and wormholes are all solutions of GR field equations. We have plenty of evidence for black holes; what do you think are the reasons that we do not hear as much (except in science fiction) about white holes and wormholes?

Problems

8.1 **Effective speed of light coming out of a black hole vanishes** Follow the discussion of the gravitational index of refraction in Section 4.3.2 to show that, according to an observer far away, the light coming out of a black hole has zero speed.

8.2 **The coordinate time across an event horizon** Carry out the integration on both sides of (8.5) to obtain the Schwarzschild coordinate time $t(r)$ that takes a particle to reach a radial distance r. Demonstrate that your result has the limit value given in (8.6) as r approaches the Schwarzschild surface $r = r^*$.

8.3 **Null 3-surface** In the text around Fig. 8.5, to show that the Schwarzschild surface ($dr = 0$ with $r = r^*$) is a null surface, we have displayed a light-like tangent vector $\mathbf{t}_1 = \mathbf{n}$ with $\mathbf{n} \cdot \mathbf{n} = 0$. Since it should be a 3-surface, display two other independent tangents of the 3D space.

8.4 **Kruskal coordinates** Starting from the definition of (8.27), derive the expressions (8.30)–(8.33) that relate the Kruskal coordinates (V, U) to the Schwarzschild coordinates (t, r).

8.5 **Circular orbits** For the simplest case of circular orbits in the Schwarzschild geometry, show that the two conserved constants κ and l of (7.46) and (7.47) are fixed to be

$$l^2 = GMm^2 r \left(1 - \frac{3}{2}\frac{r^*}{r}\right)^{-1}$$

$$\kappa^2 = \left(1 - \frac{r^*}{r}\right)^2 \left(1 - \frac{3}{2}\frac{r^*}{r}\right)^{-1}. \qquad (8.81)$$

Use this result to show that a particle in the innermost stable circular orbit (ISCO) with $r = 3r^*$ has a total energy of $E(\infty) = \sqrt{8/9}\, mc^2$. *Suggestion*: Use (7.46) for the bound state total energy $\mathcal{E}/m = (\kappa^2 - c^2)/2$ and write the effective potential (8.35) as

$$\Phi_{\text{eff}} = \frac{c^2}{2}\left[\left(1 - \frac{r^*}{r}\right)\left(1 + \frac{l^2}{m^2 r^2 c^2}\right) - 1\right].$$

8.6 **No stable circular orbit for light** Show that there is no stable circular trajectory for a photon going around a Schwarzschild black hole. One can follow the same procedure as when we discussed the orbit for a particle with mass, where the energy balance equation (8.34) follows from the Lagrangian $L = -c^2$. Similarly we can prove this result of no circular orbit for light by using an effective potential Φ_{eff} defined through the $L = 0$ equation for the case of a light ray orbit, cf. Eq. (7.77).

8.7 **No counter-rotating light is possible in the ergosphere** In the case of an "extreme spinning black hole" with angular momentum parameter $a = r^*/2$, the horizon radius is at $r_h = r^*/2$, while the stationary limit surface is at $r_S = r^*$ on the equatorial plane $\theta = \pi/2$. You are asked to demonstrate that a light ray cannot rotate counter to the direction of the rotating source by calculating the angular velocity $d\phi/dt$ for the corotating and counter-rotating light at $r = r_S$ and $r = r_h$.

8.8 **Circulating light at the horizon** Demonstrate that for a general rotating Kerr black hole (not just the extreme spinning one) the corotating and counter-rotating light circulate, on the equatorial plane $\theta = \pi/2$, at the same angular velocity at the event horizon.

8.9 **Binding energy of a particle in the ISCO of a rotating black hole** Starting from the effective potential $\Phi_{\text{eff}}^{(K)}$ as in (8.71), find the radius, angular momentum and total energy of a particle in the ISCO around an extreme spinning black hole ($a = r^*/2$). That is, you are asked to write down the three equations and check that the results of (8.72) satisfy these three equations.

COSMOLOGY

Part III

The homogeneous and isotropic universe

9

- The framework required to study the whole universe as a physical system is general relativity.
- The universe, when observed on distance scales $\gtrsim 100$ Mpc, is homogeneous and isotropic.
- Hubble's discovery that the universe is expanding suggests strongly that it had a beginning when all objects were concentrated in a tiny region of extremely high density. The estimate of the age of the universe by astrophysics from observational data is $12 \text{ Gyr} \lesssim t_0 \lesssim 15 \text{ Gyr}$.
- The mass density of (nonrelativistic) matter in the universe has around a quarter of the "critical density" $\Omega_M \simeq 0.25$. There is strong evidence showing that most of the mass in the universe does not shine: while the luminous mass ratio Ω_{lum} is only half a percent, the nonluminous matter consists of ordinary (baryonic) matter $\Omega_B \simeq 0.04$ (mostly as the intergalactic medium) and nonrelativistic exotic dark matter $\Omega_{DM} \simeq 0.21$.
- The spacetime satisfying the cosmological principle (the universe is homogeneous and isotropic at each epoch) is described by the Robertson–Walker metric in comoving coordinates (the cosmic rest frame).
- In an expanding universe with a space that may be curved, any treatment of distance and time must be carried out with care. We study the relations between cosmic redshift and proper, as well as luminosity, distances.

Cosmology is the study of the whole universe as a physical system: What is its matter–energy content? How is this content organized? What is its history? How will it evolve in the future? We are interested in a "smeared" description with the galaxies being the constituent elements of the system. On the cosmic scale the only relevant interaction among galaxies is gravitation; all galaxies are accelerating under their mutual gravity. Thus the study of cosmology depends crucially on our understanding of the gravitational interaction. Consequently, the proper framework for cosmology is general relativity. The solution of Einstein's equation describes the whole universe because it describes the whole of spacetime.

From Chapter 7 we learnt that, for a given gravitational system (M and R being the respectively characteristic mass and length dimensions), one could use the dimensionless parameter

$$\frac{G_{\mathrm{N}} M}{c^2 R} \equiv \varepsilon \tag{9.1}$$

to decide whether Einstein's theory was required, or a Newtonian description would be adequate. In the context of the spatially isotropic solution, it is just the relative size of the Schwarzschild radius to the distance scale R. Recall $\varepsilon_\odot = O(10^{-6})$ for the sun, cf. Eq. (7.22). Typically the GR effects are small at the level of an ordinary stellar system. On the other hand, we have also considered the case of stellar objects that were so compact that they became black holes when the distance scale is comparable to the Schwarzschild radius, $\varepsilon_{\mathrm{bh}} = O(1)$. For the case of cosmology, the mass density is very low. Nevertheless, the distance involved is so large that the total mass M, which increases faster than R, is even larger. This also results in a sizable ε (Problem 9.1). Thus, to describe events on cosmic scales, we must use GR concepts.[1]

Soon after the completion of his papers on the foundation of GR, Einstein proceeded to apply his new theory to cosmology. In 1917 he published his paper, "Cosmological considerations on the general theory of relativity". Since then almost all cosmological studies have been carried out in the framework of GR.

[1] Given that the theory had been tested only within the solar system, applying GR to cosmology would involve an extraordinary (something like 15 orders of magnitude) extrapolation. This is a bold assumption indeed.

9.1   The cosmos observed

We begin with the observational features of the universe: the organization of its matter content, the large-scale motion of its components, its age and mass density.

9.1.1   Matter distribution on the cosmic distance scale

The distance unit traditionally used in astronomy is the parsec (pc). This is defined, see Fig. 9.1(a), as the distance to a star having a parallax of one arcsecond[2] for a base-line equal to the (mean) distance between the earth and the sun (called an AU, the **astronomical unit**). Thus 1pc = $(1''$ in radian$)^{-1} \times$ AU = 3.1×10^{16} m = 3.26 light-years. Here we first introduce the organization of stars on the cosmic scales of kpc, Mpc, and even hundreds of Mpc.

[2] One arcsec equals 4.85×10^{-6} rad.

The distance from the solar system to the nearest star is 1.2 pc. Our own galaxy, the Milky Way, is a typical spiral galaxy. It comprises $O\,(10^{11})$ stars in a disk with a diameter of 30 kpc and a disk thickness of about 2 kpc, see Fig. 9.1(b). Galaxies in turn organize themselves into bodies of increasingly large sizes—into a series of hierarchical clusters. Our galaxy is part of a small cluster, called the Local Group, comprising about 30 galaxies in a volume measuring 1 Mpc across; the distance, for example, to the Andromeda Galaxy (M31) is 0.7 Mpc. This cluster is part of the Local, or Virgo, Supercluster over a volume measuring 50 Mpc across, with the Virgo Cluster comprising 2000 galaxies over a distance scale of 5 Mpc as its physical center. (The Virgo

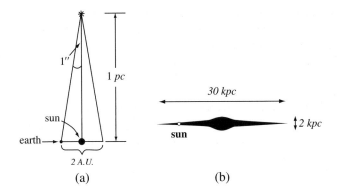

Fig. 9.1 (a) The astronomical distance unit **parsec** (parallax second) defined, see text. (b) Side view of Milky Way as a typical spiral galaxy.

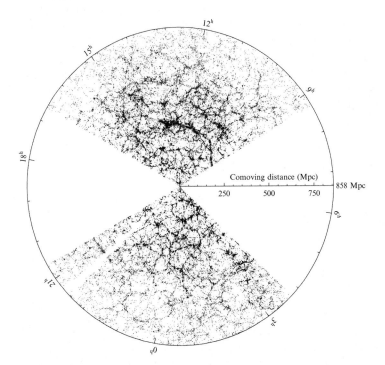

Fig. 9.2 Galaxy distribution out to 858 Mpc, compiled by Gott *et al.* (2005) based on data collected by the SDSS and 2dF surveys.

Cluster is about 15 Mpc from us.) This and other clusters of galaxies, such as the Hydra–Centaurus Supercluster, appear to reside on the edge of great voids. In short, the distribution of galaxies about us is not random, but rather clustered together in coherent patterns that can stretch out up to 100 Mpc. The distribution is characterized by large voids and a network of filamentary structures (see Fig. 9.2). However, beyond this distance scale the universe does appear to be fairly uniform. In fact the largest observable item in the universe is the cosmic microwave background (CMB) radiation which appears to be remarkably uniform.

9.1.2   Cosmological redshift: Hubble's law

Olbers' paradox: Darkness of the night sky   Up until about 100 years ago, the commonly held view was that we lived in a static universe (comprising essentially our Milky Way galaxy) that was infinite in age and infinite in size. However, such a cosmic picture is contradicted by the observation that the night sky is dark. If the average luminosity (emitted energy per unit time) of a star is $\mathcal{L}$, then the brightness (i.e. flux) seen at a distance r would be $f(r) = \mathcal{L}/4\pi r^2$. The resultant flux from integrating over all the stars in the infinite universe would be unbounded:

$$B = \int n\, f(r) dV = n\mathcal{L} \int_{r_{\min}}^{\infty} dr = \infty, \qquad (9.2)$$

where n, the number density of stars, has been assumed to be a constant. This result of infinite brightness is an over-estimate because stars have finite angular sizes, and the above calculation assumes no obstruction by foreground stars. The correct conclusion is that the night sky in such a universe would have the brightness as if the whole sky were covered by shining suns. Because every line-of-sight has to end at a shining star, although the flux received from a distant star is reduced by a factor of r^{-2}, for a fixed solid angle, the number of unobstructed stars increases with r^2. Thus, there would be an equal amount of flux from every direction. It is difficult to find any physical mechanism that will allow us to evade this result of a night sky ablaze. For example, one might suggest that interstellar dust would diminish the intensity for light having traveled a long distance. But this does not help, because over time, the dust particles would be heated and radiate as much as they absorb. Maybe our universe is not an infinite and static system?[3]

[3]As we shall see, according to modern cosmology, our universe has a finite age and all distant stars in an expanding universe are receding away from us with their emitted light progressively shifted to lower frequencies. Cf. Problem 9.5.

Hubble's discovery

Astronomers have devised a whole series of techniques that can be used to estimate the distances ever farther into space. Each new one, although less reliable, can be used to reach out further into the universe. During the period 1910–1930, the "cosmic distance ladder" reached out beyond 100 kpc. The great discovery was made that our universe was composed of a vast collection of galaxies, each resembling our own Milky Way. One naturally tried to study the motions of these newly discovered "island universes" by using the Doppler effect. When a galaxy is observed at visible wavelengths, its spectrum typically has absorption lines because of the relatively cool upper stellar atmosphere. For a particular absorption line measured in the laboratory as having a wavelength λ_{em}, the received wavelength by the observer may, however, be different. Such a wavelength shift

$$z \equiv \frac{\lambda_{\text{rec}} - \lambda_{\text{em}}}{\lambda_{\text{em}}} \qquad (9.3)$$

is related to the emitter motion by the Doppler effect (cf. Box 3.3), which, for nonrelativistic motion, can be stated as

$$z = \frac{\Delta\lambda}{\lambda} \simeq \frac{v}{c}, \qquad (9.4)$$

where v is the recession velocity of the emitter (away from the receiver).

A priori, for different galaxies one expects a random distribution of wavelength shifts: some positive (redshift) and some negative (blueshift). This is more or less true for the Local Group. But beyond the few nearby galaxies, the measurements by Vesto Slipher of some 40 galaxies, over a 10 year period at Arizona's Lowell Observatory, showed that all, except a few in the Local Group, were redshifted. Edwin Hubble (Mt. Wilson Observatory, California) then attempted to correlate these redshift results to the more difficult measurements of the distances to these galaxies. He found that the redshift was proportional to the distance d to the light-emitting galaxy. In 1929, Hubble announced his result:

$$z = \frac{H_0}{c}d \qquad (9.5)$$

or, substituting in the Doppler interpretation[4] of (9.4),

$$v = H_0 d, \qquad (9.6)$$

with a positive H_0. Namely, we live in an expanding universe. On distance scales greater than 10 Mpc, all galaxies obey Hubble's law: they are receding from us with speed linearly proportional to the distance. The proportional constant H_0, the **Hubble constant**, gives the recession speed per unit separation (between the receiving and emitting galaxies). It is the expansion rate. To obtain an accurate account of H_0 has been a great challenge as it requires one to ascertain great cosmic distances. Only recently has it become possible to yield consistent results among several independent methods. We have the convergent value[5]

$$H_0 = (72 \pm 5 \text{ km/s}) \text{ Mpc}^{-1}, \qquad (9.7)$$

where the subscript 0 stands for the present epoch $H_0 \equiv H(t_0)$. An inspection of Hubble's law (9.6) shows that H_0 has the dimension of inverse time, the **Hubble time** $t_H \equiv H_0^{-1}$. Similarly, we can also define a **Hubble length** $l_H = ct_H$.

Hubble's law and the Copernican principle

That all galaxies are receding away from us may lead one to suggest erroneously that our location is the center of the universe. The correct interpretation is in fact just the opposite. The Hubble relation in fact follows naturally from a straightforward extension of the **Copernican principle**: our galaxy is not at a privileged position in the universe. The key observation is that this is a **linear relation** between distance and velocity at each cosmic epoch. As a result, it is compatible with the same law holding for all observers at every galaxy. Namely, observers on every galaxy would see all the other galaxies receding away from them according to Hubble's law.

Let us write Hubble's law in vector form:

$$\vec{v} = H_0 \vec{r}. \qquad (9.8)$$

That is, a galaxy G, located at position $\vec{r}$, will be seen by us (at the origin O) to recede at velocity $\vec{v}$ proportional to $\vec{r}$. Now consider an observer on another galaxy O′ located at $\vec{r}'$ from us as in Fig. 9.3. Then, according to Hubble's law,

[4]A Doppler redshift comes about because of the increase in the distance between the emitter and the receiver of a light signal. In the familiar situation, this is due to the relative motion of the emitter and the receiver. This language is being used here in our initial discussion of Hubble's law. However, as we shall show in Section 9.3, especially Eq. (9.43), the proper description of this enlargement of the cosmic distance is reflecting the expansion of the space itself, rather than the motion of the emitter in a static space.

[5]Throughout Chapters 9–11, we shall quote the cosmological parameters as presented by Tegmark *et al.* (2006), cf. Table 11.1 in Section 11.5.

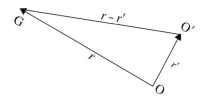

Fig. 9.3 Relative positions of a galaxy G with respect to two observers located at two other galaxies: O and O'.

it must be receding from us according to

$$\vec{v}' = H_0 \vec{r}' \tag{9.9}$$

with the **same** Hubble constant H_0 which is independent of distance and velocity. The difference of these two equations yields

$$(\vec{v} - \vec{v}') = H_0 (\vec{r} - \vec{r}'). \tag{9.10}$$

But $(\vec{r} - \vec{r}')$ and $(\vec{v} - \vec{v}')$ are the respective location and velocity of G as viewed from O'. Since $\vec{v}$ and $\vec{v}'$ are in the same direction as $\vec{r}$ and $\vec{r}'$, the vectors $(\vec{v} - \vec{v}')$ and $(\vec{r} - \vec{r}')$ must also be parallel. That is, the relation (9.10) is just Hubble's law valid for the observer on galaxy O'. Clearly such a deduction would fail if the velocity and distance relation, at a given cosmic time, were nonlinear (i.e. if H_0 depends either on position and/or on velocity).

Distance measurement by redshift

We can turn the Hubble relation around and use it as a means to find the distance to a galaxy by its observed redshift. In fact, the development of new techniques of multi-fiber and multi-slip spectrographs allowed astronomers to measure redshifts for hundreds of galaxies simultaneously. This made large surveys of galaxies possible. In the 1980s there was the Harvard–Smithsonian Center for Astrophysics (CfA) galaxy survey, containing more than 15 000 galaxies. Later, the Las Campanas mapping eventually covered a significantly larger volume and found the "greatness limit" (i.e. cosmic structures have a maximum size and on any larger scale the universe would appear to be homogeneous). But this was still not definitive. The modern surveys culminated in two recent parallel surveys: the Anglo-Australian Two-Degree Field Galaxy Redshift Survey (2dF) and the Sloan Digital Sky Survey (SDSS) collaborations have measured some quarter of a million galaxies over a significant portion of the sky, confirming the basic cosmological assumption that the universe of a large distance $\gtrsim 100$ Mpc is homogeneous and isotropic. (For further discussion see Sections 9.3 and 9.4.) In fact, an important tool for modern cosmology is just such large-structure study. Detailed analysis of survey data can help us to answer questions such as whether the cosmic structure observed today came about in a top–down (i.e. the largest structure formed first, then the smaller ones by fragmentation) or in a bottom–up process. (The second route is favored by observational data.) In fact many of the cosmological parameters, such as Hubble's constant and the energy density of the universe, etc. can also be extracted from such analysis.

9.1.3　Age of the universe

If all galaxies are rushing away from each other now presumably they must have been closer in the past. Unless there was some new physics involved, extrapolating back in time there would be a moment, "the big bang", when all objects were concentrated at one point of infinite density.[6] This is taken to be the origin of the universe. How much time has evolved since this fiery beginning? What is then the age of our universe?

[6]See Problem 9.10 for a brief description of the alternative cosmology called the **steady-state theory** which avoids the big bang beginning by having a constant mass density, maintained through continuous spontaneous matter creation as the universe expands.

It is useful to note that the inverse of the Hubble's constant at the present epoch, the **Hubble time**, has the value of

$$t_H \equiv H_0^{-1} = 13.6 \text{ Gyr.} \tag{9.11}$$

By Hubble "constant," we mean that, at a given cosmic time, H is independent of the separation distance and the recessional velocity—the Hubble relation is a linear relation. The proportional coefficient between distance and recessional speed is not expected to be a constant with respect to time: there is matter and energy in the universe, and their mutual gravitational attraction will slow down the expansion, leading to a monotonically decreasing expansion rate $H(t)$—a **decelerating universe**. Only in an "empty universe" do we expect the expansion rate to be a constant throughout its history, $H(t) = H_0$. In that case, the age t_0 of the empty universe is given by the Hubble time

$$t_{0\text{empty}} = \frac{d}{v} = \frac{1}{H_0} = t_H. \tag{9.12}$$

For a decelerating universe full of matter and energy, the expansion rate must be larger in the past: $H(t) > H_0$ for $t < t_0$. Because the universe was expanding faster than the present rate, this would imply that the age of the decelerating universe must be shorter than the empty universe age: $t_0 < t_H$. Nevertheless, we shall often use the Hubble time as a rough benchmark value for the age of the universe, which has a current horizon[7] of $ct_H = l_H \simeq 4300$ Mpc.

Phenomenologically, we can estimate the age of the universe from observational data. For example, from astrophysical calculation, we know the relative abundance of nuclear elements when they are produced in a star. Since they have different decay rates, their present relative abundance will be different from the initial value. The difference is a function of time. Thus, from the decay rates, the initial and observed relative abundance, we can estimate the time that has elapsed since their formation. Typically, such a calculation gives the ages of stars to be around 13.5 Gyr. This only gives an estimate of time when stars were first formed, thus only a lower bound for the age of the universe. However, our current understanding informs us that the formation of stars started a hundred million years or so after the big bang, thus such a lower limit still serves as a useful estimate of t_0.

An important approach to the study of the universe's age has been the research work on systems of 10^5 or so old stars known as **globular clusters**. These stars are located in the halo, rather than the disk, of our Galaxy. It is known that a halo lacks the interstellar gas for star formation. These stars must be created in the early epochs after the big bang (as confirmed by their lack of elements heavier than lithium, cf. Section 10.4). Stars spend most of their lifetime undergoing nuclear burning. From the observed brightness (flux) and the distance to the stars, one can deduce their intrinsic luminosity (energy output per unit time). From such properties, astrophysical calculations based on established models of stellar evolution, allowed one to deduce their ages (Krauss and Chaboyer, 2003):

$$12 \text{ Gyr} \lesssim t_{0\text{gc}} \lesssim 15 \text{ Gyr.} \tag{9.13}$$

[7] Two objects, separated by a distance of ct_H, would recede from each other, according to the Hubble relation of (9.6), at the speed of light c.

For reference, we note that the age of our earth is estimated to be around 4.6 Gyr.

9.2   Mass density of the universe

It is useful to express the mass density in terms of a benchmark value for a universe with expansion rate given by the Hubble constant H. One can check that the ratio, with H^2 divided by Newton's constant G_N, has the units of mass density. With an appropriate choice[8] of the coefficient, we have the expression of the **critical density**

$$\rho_c \equiv \frac{3H^2}{8\pi G_N}. \tag{9.14}$$

The significance of this quantity will be discussed in Chapter 10 when the Einstein equation for cosmology will be presented. In the meantime, we introduce the notation for the **density ratio**

$$\Omega \equiv \frac{\rho}{\rho_c}. \tag{9.15}$$

Since the Hubble constant is a function of cosmic time, the critical density also evolves with time. We denote the values for the present epoch with the subscript 0. For example, $\rho(t_0) \equiv \rho_0$, $\rho_c(t_0) \equiv \rho_{c,0}$, and $\Omega(t_0) \equiv \Omega_0$, etc. For the present Hubble constant H_0 as given in (9.7), the critical density has the value

$$\rho_{c,0} = (0.97 \pm 0.08) \times 10^{-29} \text{ g/cm}^3 \tag{9.16}$$

or, equivalently, a **critical energy density**[9] of

$$\rho_{c,0}c^2 \simeq 0.88 \times 10^{-10} \text{ J/m}^3 \simeq 5500 \text{ eV/cm}^3. \tag{9.17}$$

In the following we shall discuss the measurement of the universe's various mass densities (averaged over volumes on the order of 100 Mpc^3) for both luminous and nonluminous matter. In recent years, these parameters have been deduced rather accurately by somewhat indirect method—including detailed statistical analysis of the temperature fluctuation in the cosmic microwave background (CMB) radiation and from large-structure studies by the 2dF and SDSS galaxy surveys mentioned above. The large-structure study involves advanced theoretical tools that are beyond the scope of this introductory presentation. In the following we choose to present a few methods that involve rather simple physical principles, even though they may be somewhat "dated" in view of recent cosmological advances. Our discussion will in fact be only semiquantitative. Subtle details of derivation as well as qualification of the stated results will be omitted. The purpose is to provide some general idea as to how cosmological parameters can in principle be deduced phenomenologically.

[8]We can remember ρ_c as the density of a universe with its radius $R = ct_H = c/H$ just equal to the Schwarzschild radius: $R = 2G_N M/c^2$ where $M = (4\pi R^3/3)\rho_c$.

[9]In the natural unit system of quantum field theory, this critical density is approximately $\rho_c c^2 \approx (2.5 \times 10^{-3} eV)^4/(\hbar c)^3$, where $\hbar$ is Planck's constant (over 2π) with $\hbar c \approx 1.9 \times 10^{-5}$ eV $\cdot$ cm. Also, $\rho_{c,0}c^2 \simeq 5.5$ GeV/m^3 is equivalent to the rest energy of $\simeq 6$ protons per cubic meter.

9.2.1 Luminous matter and the baryonic density

Luminous matter

The basic idea of measuring the mass density for luminous matter is through its relation to the luminosity $\mathcal{L}$, which is the energy emitted per unit time,

$$\rho_{\text{lum}} = \left(\begin{array}{c} \text{luminosity} \\ \text{density} \end{array} \right) \times \left(\frac{M}{\mathcal{L}} \right). \qquad (9.18)$$

(Here we omit the subscript 0 for the present epoch.) That is, one finds it convenient to decompose the mass density into two separate factors: the luminosity density and the mass-to-luminosity ratio. The luminosity density can be obtained by a count of galaxies per unit volume, multiplied by the average galactic luminosity. Several surveys have resulted in a fairly consistent conclusion of 200 million solar luminosity $\mathcal{L}_\odot$ per Mpc^3 volume,

$$\left(\begin{array}{c} \text{luminosity} \\ \text{density} \end{array} \right) \approx 2 \times 10^8 \frac{\mathcal{L}_\odot}{(\text{Mpc})^3}. \qquad (9.19)$$

The ratio $(M/\mathcal{L})$ is the amount of mass associated, on the average, with a given amount of light. This is the more difficult quantity to ascertain. Depending on the selection criteria one gets a range of values for the mass-to-luminosity ratio. The average of these results came out to be $(M/\mathcal{L}) \approx 4M_\odot/\mathcal{L}_\odot$. Plugging this and (9.19) into (9.18) we obtain an estimate of the density for luminous matter $\rho_{\text{lum}} \approx 8 \times 10^8 M_\odot/\text{Mpc}^3 \approx 5 \times 10^{-32}$ g/cm^3, or in terms of the density ratio defined in (9.15)

$$\Omega_{\text{lum}} \approx 0.005. \qquad (9.20)$$

Total amount of baryonic matter and the intergalactic medium

We designate the type of matter, for which we cannot directly detect its presence through its electromagnetic emissions, as nonluminous matter. This includes such matter as neutrinos which have no electromagnetic interaction, as well as matter such as intergalactic hydrogen molecules, which, although they do not "shine," can be detected through their absorption of electromagnetic radiation.

Ordinary matter made of baryons (protons and neutrons) and electrons is referred to in cosmology as **baryonic matter**.[10] Baryons is the particle physics name for strongly interacting particles, composed of quark triplets, that carry nontrivial **baryon numbers**—as are the cases of protons and neutrons. For our purpose here, the baryon number is just the proton plus neutron numbers. Other types of particles, such as photons, electrons, and neutrinos, carry zero baryon number. Baryon matter (protons, neutrons and electrons) can clump to form atoms and molecules, leading to large astronomical bodies. Luminous matter (shining stars) is baryonic matter; but some of the baryonic matter, such as interstellar or intergalactic gas, may not shine—they are nonluminous baryonic matter.[11] That is, baryonic matter can be luminous stars or optically nonluminous gas[12] of ordinary atoms:

$$\Omega_{\text{B}} = \Omega_{\text{lum}} + \Omega_{\text{gas}}. \qquad (9.21)$$

[10]Such a name neglects electrons (one species of leptons), which constitute less than 0.1% of the baryonic matter masses.

[11]Besides the interstellar gas around galaxies, nonluminous baryonic matter can be planets or stellar remnants such as black holes, white dwarfs, and brown dwarfs (the last category being stars of the size of Jupiter, with not enough mass to trigger the thermonuclear reaction to make it shine).

[12]This includes X-ray emitting hot gas.

As it turns out, we have methods that can deduce the total baryonic abundance Ω_B regardless of whether they are luminous or nonluminous. The light nuclear elements (helium, deuterium, etc.) were produced predominantly in the early universe at the cosmic time $O(10^2 \, \text{s})$, cf. Section 10.4. Their abundance (in particular deuterium) is sensitive to the baryonic abundance. From such considerations one deduces the result (Burles *et al.* 2001; Tegmark *et al.* 2006)

$$\Omega_B = 0.042 \pm 0.002, \tag{9.22}$$

which is confirmed by the latest CMB anisotropy measurements (see Chapter 11), as well as gravitational microlensing (see Box 7.2).

From (9.20), we see that $\Omega_B \gg \Omega_{lum}$. This means that most of the "ordinary matter" is not visible to us. Theoretical studies, backed up by detailed simulation calculations, indicate that a major portion of it is in the form of unseen neutral gas in galaxies as well as in the space between galaxies. Such an intergalactic medium (IGM) is in the form of wispy filaments that connect galaxy clusters. Their presence has been verified by careful examination of quasar spectra. Quasars are among the most powerful light sources in the universe. Their light reaches us after passing through successive layers of IGM at various distances resulting in a quasar spectrum imprinted with neutral hydrogen absorption lines. From the line depths one can infer the amount and distribution of the absorbing gas. Such studies were able to account[13] for the difference between Ω_B and Ω_{lum}; namely, an optically nonluminous $\Omega_{gas} \approx 0.038$.

9.2.2   Dark matter and the total mass density

One of the great discoveries of modern cosmology has been the finding that there is more mass in the universe than just baryonic matter. That is, the bulk of the nonluminous matter is not baryonic. We call such nonluminous and nonbaryonic matter, **dark matter**.

Dark matter vs. baryonic matter

Dark matter is supposedly made up of exotic particles that have neither electromagnetic emission nor absorption. Namely, they have no electromagnetic interaction at all (i.e. they do not have strong or electromagnetic charges). Neutrinos are cases in point. They only feel the weak nuclear force. With their masses being extremely small, neutrinos are expected to be in relativistic motion. They are examples of "**hot dark matter**." Also, there may exist "**cold dark matter**" composed of nonrelativistic heavy particles. Hot and cold dark matter have distinctly different effects on the formation of galaxies and clusters of galaxies from the initial density inhomogeneity in the universe. Research in the past decade favors the possibility of cold dark matter.[14] The prime examples of CDM are the "weakly interacting massive particles" (WIMPs) predicted by the various extensions of the Standard Model of particle interactions.[15] WIMPs are expected to be much more massive than nucleons (in the 50–1000 GeV/c^2 range) but interact weakly—a particle with such a mass and interaction rate can produce just the correct CDM abundance in the big bang cosmology, to be discussed in Chapter 10 (see in particular the related subject of primordial

---

[13] Until very recently, such an IGM has been detected in the early universe; finding such nonluminous atoms in the nearby universe had not been successful. However, theoretical studies (e.g. Cen and Ostriker, 1999) suggest such IGM baryons should have been shock-heated by the large-scale collapsing and squeezing that formed the foamy cosmic structure. The corresponding absorption lines of the heated atoms move up to the far-ultraviolet and X-ray region. Measurements of such absorption lines with the expected intensity has recently been reported (Danforth and Shull, 2008).

[14] If the dark matter had been fast moving (hot) particles, they would be able to stream away from high density regions, thus smooth out small density perturbations. This would have left only the large-scale perturbations, leading to the formation of the largest structure (superclusters) first, with the smaller structures (galaxies) being produced from fragmentation. This top-down scenario is inconsistent with observation.

[15] It has been suggested that the Standard Model of particle physics be extended by the inclusion of supersymmetry (cf. discussion in Section 11.7.3). Every known elementary particle must then have a supersymmetric partner, with a spin differing by half a unit. The lightest of such hypothesized supersymmetry particles are expected to be **neutralino** fermions (partners to the neutral Higgs scalar and weak gauge bosons) and should be stable against spontaneous decay. They can in principle make up the bulk of the required dark matter WIMPs.

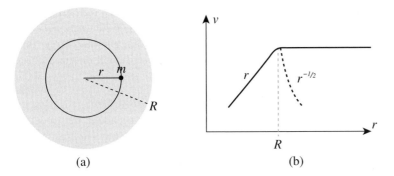

(a)   (b)

Fig. 9.4 (a) Gravitational attraction on a mass m due to a spherical mass distribution (shaded disk). The circle passing through m represents the Gaussian spherical surface. (b) The solid line is the observed rotation velocity curve $v(r)$. It does not fall as $r^{-1/2}$ beyond R, the edge of the visible portion of a galaxy.

neutrinos in Section 10.5.3). For a recent review,[16] see for example Bertone, Hooper and Silk (2005).

The total mass of the universe can thus be divided into two categories: baryonic, which may be luminous or nonluminous, and dark matter,[17] which has only weak interaction:

$$\Omega_M = \Omega_B + \Omega_{DM}. \tag{9.23}$$

Although the dark matter does not emit electromagnetic radiation, it still feels gravitational effects. In the following we first list several methods of detecting the total amount of masses, whether due to luminous or nonluminous matter, through their gravitational interaction.

Galactic rotation curves

The most direct evidence of dark matter's existence comes from measured "rotation curves" in galaxies. Consider the gravitational force that a spherical (or ellipsoidal) mass distribution exerts on a mass m located at a distance r from the center of a galaxy, see Fig. 9.4(a). Since the contribution outside the Gaussian sphere (radius r) cancels out, only the interior mass $M(r)$ enters into the Newtonian formula for gravitational attraction. The object is held by this gravity in circular motion with centripetal acceleration v^2/r. Hence

$$v(r) = \sqrt{\frac{G_N M(r)}{r}}. \tag{9.24}$$

In this way, the tangential velocity inside a galaxy is expected to rise linearly with the distance from the center ($v \sim r$) if the mass density is approximately constant. For a light source located outside the galactic mass distribution (radius R), the velocity is expected to decrease as $v \sim 1/\sqrt{r}$, see Fig. 9.4(b).

The velocity of particles located at different distances (the rotation curves) can be measured through the 21-cm lines of the hydrogen atoms. The surprising discovery was that, beyond the visible portion of the galaxies ($r > R$), instead of this fall-off, they are observed to stay at the constant peak value (as far as the measurement can be made). See, for example, Cram *et al.* (1980). This indicates that the observed object is gravitationally pulled by other than the luminous matter; hence it constitute direct evidence for the existence of dark matter. Many subsequent studies confirm this discovery. The general picture of a galaxy that has emerged is that of a disk of stars and gas embedded in a large halo of dark matter, see Fig. 9.5. According to (9.24), the flatness

[16] Among other examples of speculated CDM particles are **axions** and **Kaluza–Klein particles**. Axions are associated with our effort to explain how the strong interaction theory of QCD avoids having a large violation of the combined symmetry of charge conjugation and parity. The KK particles are associated with the speculated existence of compactified extra spatial dimensions.

[17] Hot dark matter such as neutrinos contribute a negligible amount.

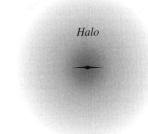

Fig. 9.5 The dark matter halo surrounding the luminous portion of the galaxy. In our simple presentation, we take the halo to be spherical. In reality the dark matter halo may not be spherical and its distribution may not be smooth.

of the rotation curve means that $M \propto r$. We can think of the halo as a sphere with mass density decreasing as r^{-2}. Measurements of the rotational curve for spiral galaxies have shown that halo radii are at least ten times larger than the visible radii of the galaxies. This leads to a lower bound on the dark matter density of $\Omega_{DM} \gtrsim 0.1$.

Use of the virial theorem to infer gravitational mass

Because the rotation curves cannot be measured far enough out to determine the extent of the dark matter halo, we have to use some other approach to fix the mass density of the dark matter in the universe. Here we discuss one method which allows us to measure the total (luminous and dark) mass in a system of galaxies (binaries, small groups, and large clusters of galaxies), that are bound together by their mutual gravitational attraction. This involves measurements of the mean-square of the galactic velocities $\langle v^2 \rangle$ and the average galactic inverse separation $\langle s^{-1} \rangle$ of the luminous components of the system. These two quantities, according to the **virial theorem** of statistical mechanics, $-2\langle T \rangle = \langle V \rangle$, relating the average kinetic and potential energy, are proportional to each other—with the proportional constant given by the total gravitational mass M (luminous and dark) of the system,

$$\langle v^2 \rangle = G_N M \left\langle \frac{1}{s} \right\rangle. \tag{9.25}$$

The proof of this theorem is left as an exercise (Problem 9.6). Here we shall merely illustrate it with a simple example. Consider a two-body system (M, m), with $M \gg m$, separated by distance s. The Newtonian equation of motion $G_N M m / s^2 = m v^2 / s$ immediately yields the result in (9.25). From such considerations,[18] one obtains a total mass density that is something like 50 times larger than the luminous matter. Thus the luminous matter, being what we can see when looking out into space, represents only a tiny fraction of the mass content of the universe.

There are now several independent means to determine the mass density at the present era $\Omega_{M,0}$: one approach is through gravitational lensing by galaxies, and clusters of galaxies[19] (see Section 7.2); another is by comparing the number of galaxy clusters in galaxy superclusters throughout the cosmic age; yet another is from measured CMB temperature fluctuations. A value for the total mass density that is generally consistent with the above discussed results has been obtained (Tegmark *et al.* 2006):

$$\Omega_{M,0} = 0.245 \pm 0.025. \tag{9.26}$$

We shall show in the next chapter that the whole universe is permeated with radiation. However, its energy density is considerably smaller so that $\Omega_{R,0} \ll \Omega_{M,0}$.

A historical note   That there might be a significant amount of dark matter in the universe was first pointed out by Fritz Zwicky in the 1930s. The basis of this proposal is just the method we have outlined here. Zwicky noted that, given the observed radial velocities of the galaxies, the combined mass of the visible stars and gases in the Coma Cluster was simply not enough to hold

[18] While the argument involving the virial theorem may appear to be somewhat abstract, its result can be understood crudely as saying that the constituents of a system held gravitationally cannot move too fast so as to exceed the escape velocity $v_{esc}^2 = 2G_N M/r$. For a review of the simple concept of escape velocity, see Eq. (10.15) in Section 10.1.2.

[19] After subtracting out the peaks corresponding to the stars and galaxies from the mass distribution as deduced by gravitational lensing (e.g. Fig. 7.5), one is still left with a huge smooth bulged piece that can only be accounted for by the existence of dark matter and optically nonluminous gas.

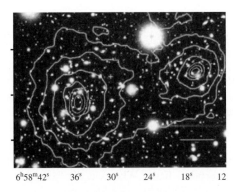 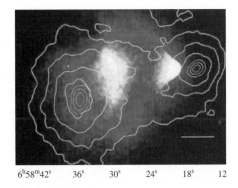

| 6^{h}58^{m}42^s | 36^s | 30^s | 24^s | 18^s | 12 |

| 6^{h}58^{m}42^s | 36^s | 30^s | 24^s | 18^s | 12 |

Fig. 9.6 Images of the "Bullet Cluster" 1E0657−558 from Clowe, *et al.* (2006) showing it as having been produced by the collision of two galactic clusters, resulting in the separation of hot gas and dark matter (with their embedded stars and galaxies). (a) Optical image from the Hubble Space Telescope; (b) X-ray image from the Chandra telescope. Mass density contours from gravitational lensing reconstruction, showing two mass peaks separated from the hot gaseous regions.

them together gravitationally. That is, what is holding together a galaxy or a cluster of galaxies must be some form of dark matter. The modern era began in 1970 when Vera Rubin and W. Kent Ford, using more sensitive techniques, were able to extend the rotation curve measurements far beyond the visible edge of gravitating systems of galaxies and clusters of galaxies.

Bullet Cluster offers direct empirical evidence of dark matter

In all the above discussions, the presence of dark matter was deduced through its gravitational effects (finding total $\Omega_M > \Omega_B$). On might wonder whether it is possible to explain the observation, instead of postulating the existence of a new form of matter, by modifying the law of gravity. Here we present a piece of evidence for dark matter that simply cannot be evaded by this alternative explanation. This is the observational result shown in Fig. 9.6. Three images of the galaxy cluster 1E0657-558, the "Bullet Cluster," are displayed here. The picture on the left shows galaxies that make up a few percent of the cluster mass; the picture on the right is the X-ray image from the Chandra telescope showing where the bulk of the hot gas is located. Superimposed on top of these two pictures is the mass contours as derived from gravitational lensing. These counters have two mass peaks which, while they more or less track the locations of observed galaxies, are situated at very different positions from the atomic gas. Such an observation cannot be explained by any modified law of gravity but is consistent with the interpretation that this Bullet Cluster came about because of a collision of two clusters of galaxies. The dark matter and baryonic gas are separated because the dark matter (having small interaction cross-section) passes through[20] "like a bullet" while the baryonic gases are left behind.

[20]Most of the galxies track the deep dark matter gravitational potentials.

Matter densities in the universe: A summary   Dark matter is mostly nonrelativistic particles having only gravitational and weak interactions. It does not emit or absorb electromagnetic radiation. Its presence has been deduced from the velocity distribution of a gravitationally bound system. The most direct empirical evidence is the mass distribution in the Bullet Cluster. On an even larger scale the abundance of dark matter can be quantified from the study of large cosmic structure and CMB. All this leads to a total mass density equal approximately to a quarter of the critical density:

$$\Omega_M = \Omega_B + \Omega_{DM} \simeq 0.25. \tag{9.27}$$

The total baryonic (atomic) density can be deduced from the observed amount of light nuclear elements and the big bang nucleosynthesis theory or from the observed CMB temperature anisotropy:

$$\Omega_B \simeq 0.04, \tag{9.28}$$

The bulk of which is in the intergalactic medium and has been detected through its electromagnetic absorption lines. What we can see optically, stars and galaxies, is only a small part of this baryonic density:

$$\Omega_B = \Omega_{gas} + \Omega_{lum} \quad \text{with} \quad \Omega_{gas} \gg \Omega_{lum} \simeq 0.005. \tag{9.29}$$

Thus the luminous matter associated with the stars we see in galaxies represents about 2% of the total mass content. Most of the matter is dark ($\Omega_{DM} \simeq 0.21$) composed mostly of exotic nonrelativistic particles such as WIMPs. The exact nature of these exotic nonbaryonic CDM particles remains one of the unsolved problems in physics.

9.3   The cosmological principle

That the universe is homogeneous and isotropic on the largest scale of hundreds of Mpc has been confirmed by direct observation only recently (cf. the discussion at the end of Section 9.1.2). Other evidence for its homogeneity and isotropy came in the form of the extremely uniform CMB radiation. This is the relic thermal radiation left over from an early epoch when the universe was only 10^5 years old. The nonuniformity of CMB is on the order of 10^{-5} (cf. Sections 10.5 and 11.3.1). This shows that the "baby universe" can be described as being highly homogeneous and isotropic.

But long before obtaining such direct observational evidence, Einstein had adopted the **strategy** of starting the study of cosmology with a basic working hypothesis called **the cosmological principle** (CP): at each epoch (i.e. each fixed value of cosmological time t) the universe is homogeneous and isotropic. It presents the **same** aspects (except for local irregularities) from each point: the universe has no center and no edge.

- This statement that there is no privileged location in the universe (hence homogeneous and isotropic) is sometimes referred to as the **Copernican cosmological principle**. It is in essence the ultimate generalization of the Copernican principle.
- This is a priori the most reasonable assumption, as it is difficult to think of any other alternative. Also, in practice, it is also the most "useful," as it involves the least number of parameters. There is some chance for the theory to be predictive. Its correctness can then be checked by observation. Thus CP was invoked in the study of cosmology long before there was any direct evidence for a homogeneous and isotropic universe, but it is now fully supported by observation.

- The observed irregularities, i.e. the structure, in the universe (stars, galaxies, clusters of galaxies, superclusters, voids, etc.) are assumed to arise because of gravitational clumping around some initial density unevenness. Various mechanisms for seeding such density perturbation have been explored. Most of the efforts have been concentrated around the idea that, in the earliest moments, the universe passed through a phase of extraordinarily rapid expansion, the "**cosmic inflationary epoch.**" The small quantum fluctuations were inflated to astrophysical size and they seeded the cosmological density perturbation (cf. Sections 11.2.3 and 11.3.1).

The cosmological principle gives rise to a picture of the universe as a physical system of a "cosmic fluid." The fundamental particles of this fluid are galaxies, and a fluid element has a volume that contains many galaxies, yet is small compared to the whole system of the universe. Thus, the motion of a cosmic fluid element is the smeared-out motion of the constituent galaxies. It is determined by the gravitational interaction of the entire system—the self-gravity of the universe. This means that each element is in free-fall; all elements follow geodesic worldlines. (In reality, the random motions of the galaxies are small, on the order of 10^{-3}.)

Such a picture of the universe allows us to pick a privileged coordinate frame, the **comoving coordinate system**, where

$t \equiv$ the proper time of each fluid element

$x^i \equiv$ the spatial coordinates carried by each fluid element.

A comoving observer flows with a cosmic fluid element. The comoving coordinate time can be synchronized over the whole system. For example, t is inversely proportional to the temperature of the cosmic background radiation (see Section 10.3) which decreases monotonically. Thus, we can in principle determine the cosmic time by a measurement of the background radiation temperature. This property allows us to define space-like slices, each with a fixed value of the coordinate time, and each is homogenous and isotropic.

Because each fluid element carries its own position label the comoving coordinate is also the cosmic rest frame—as each fluid element's position coordinates are unchanged with time. But we must remember that in GR the coordinates do not measure distance, which is a combination of the coordinates and the metric. As we shall detail below, viewed in this comoving coordinate, the expanding universe, with all galaxies rushing away from each other, is described not by changing position coordinates, but by an ever-increasing metric. This emphasizes the physics underlying an expanding universe not as something exploding in the space, but as the expansion of space itself.

9.4   The Robertson–Walker spacetime

9.4.1   The metric in the comoving coordinate system

The cosmological principle says that, at a fixed cosmic time, each space-like slice of the spacetime is homogeneous and isotropic. In Section 7.1

spherical symmetry has been found to restrict the metric to the form of $g_{\mu\nu} = \text{diag}(g_{00}, \ g_{rr}, \ r^2, \ r^2 \sin^2\theta)$ with only two scalar functions (g_{00} and g_{rr}). In this section we discuss the geometry resulting from the cosmological principle: when expressed in comoving coordinates, it has a Robertson–Walker metric.

The time components

Because the coordinate time is the proper time of fluid elements, we must have $g_{00} = -1$. The fact that the spacelike slices for fixed t can be defined means that the spatial axes are orthogonal to the time axes:[21]

$$g_{00} = -1 \quad \text{and} \quad g_{0i} = g_{i0} = 0. \tag{9.30}$$

The self-consistency of this choice of coordinates can be checked as follows. A particle at rest in the comoving frame is a particle in free fall under the mutual gravity of the system; it should follow a geodesic worldline obeying Eq. (6.9):

$$\frac{d^2x^\mu}{d\tau^2} + \Gamma^\mu_{\alpha\beta} \frac{dx^\alpha}{d\tau} \frac{dx^\beta}{d\tau} = 0. \tag{9.31}$$

Being at rest, $dx^i = 0$ with $i = 1, 2, 3$, we only need to calculate the Christoffel symbol Γ^μ_{00}. But the metric properties of (9.30) imply that $\Gamma^\mu_{00} = 0$ for all μ. Thus these fluid elements at rest with respect to the comoving frame ($dx^i/d\tau = d^2x^i/d\tau^2 = 0$) do satisfy (trivially) the geodesic equation.

The metric for a 3D space with constant curvature

Let g_{ij} be the spatial part of the metric; the relations in (9.30) imply that the 4D metric that satisfies the cosmological principle is block-diagonal:

$$g_{\mu\nu} = \begin{pmatrix} -1 & 0 \\ 0 & g_{ij} \end{pmatrix}. \tag{9.32}$$

The invariant interval expressed in terms of the comoving coordinates is

$$ds^2 = -c^2 dt^2 + g_{ij} dx^i dx^j$$
$$\equiv -c^2 dt^2 + dl^2. \tag{9.33}$$

Because of the CP requirement (i.e. no preferred direction and position), the time dependence in g_{ij} must be an **overall** length factor $R(t)$, sometimes referred to as the radius of the universe, with no dependence on any of the indices:

$$dl^2 = R^2(t)d\tilde{l}^2 \tag{9.34}$$

where the reduced length element $d\tilde{l}$ is both t-independent and dimensionless. It is also useful to define a dimensionless **scale factor**

$$a(t) \equiv \frac{R(t)}{R_0}, \tag{9.35}$$

---

[21] Because fixed-time space-like slices of space exist, we can consider an event as separated from two other events in two distinctive ways. The first connects the event to another on a space-like space containing all events with the same cosmic time: $da^\mu = (0, dx^i)$ for a definite spatial index i; the second is an interval connecting the event to another one along the worldline of a comoving observer: $db^\mu = (dt, 0)$. The inner product of these two intervals $g_{\mu\nu}da^\mu db^\nu = g_{i0}dx^i dt$ (the repeated μ and ν indices are summed, but not the i indices) is an invariant, valid in any coordinate system including the local Minkowski frame. This makes it clear that the left-hand side vanishes. The above equality then implies $g_{i0} = 0$.

with denominator on the right-hand side (RHS) $R_0 \equiv R(t_0)$ so that the scale factor is normalized at the present epoch by $a(t_0) = 1$.

One has the picture of the universe as a three-dimensional (3D) map with cosmic fluid elements labeled by the fixed comoving coordinates $\tilde{x}_i$. Time evolution enters entirely through the time dependence of the map scale $R(t) = a(t)R_0$, see Fig. 9.7,

$$x_i(t) = a(t)R_0\tilde{x}_i \qquad (9.36)$$

with $\tilde{x}_i$ being the fixed (t-independent) dimensionless map coordinates, while $a(t)$ is the size of the grids and is independent of the map coordinates. As the universe expands, the **relative distance** relations (i.e. the shape of things) are not changed.

The Robertson–Walker metric

The Robertson–Walker (RW) metric is for a spacetime which, at a give time, has a 3D homogeneous and isotropic space. One naturally expects this space to have a constant curvature. In Section 5.3.2 we have already written down the metric[22] for the 3D spaces with constant curvature in two spherical coordinate systems, with the dimensionless radial coordinates being $\chi = r/R_0$ and $\xi = \rho/R_0$, respectively, and the differential solid angle $d\Omega^2 = d\theta^2 + \sin^2\theta d\phi^2$:

- Equation (5.53) for the comoving "polar" coordinates (χ, θ, ϕ):

$$dl^2 = R_0^2 a^2(t)d\chi^2 + k^{-1}(\sin^2\sqrt{k}\chi)d\Omega^2. \qquad (9.37)$$

- Equation (5.55) for the comoving "cylindrical" coordinates (ξ, θ, ϕ)

$$dl^2 = R_0^2 a^2(t)\left(\frac{d\xi^2}{1-k\xi^2} + \xi^2 d\Omega^2\right). \qquad (9.38)$$

The parameter k in g_{ij} can take on the values $\pm 1, 0$ with $k = +1$ for a 3-sphere, $k = -1$ for a 3-pseudosphere, and $k = 0$ for a 3D Euclidean (flat) space. Some properties of such spaces, such as their embedding and their volume evaluation, were also discussed in Problems 5.7 and 5.8. In the context of cosmology, the universe having a $k = +1$ positively curved space is called a "**closed universe**," a $k = -1$ negatively curved space an "**open universe**," and $k = 0$ a "**flat universe**."

In practice, one can use either one of the two coordinates displayed in (9.37) and (9.38); they are equivalent. In the following, for definiteness, we shall work with the (ξ, θ, ϕ) coordinate system of (9.38).

9.4.2 Distances in the RW geometry

In an expanding universe with a space that may be curved, we must be very careful in any treatment of distance. In the following sections we shall deal with several kinds of distance, starting with conceptually the simplest: the proper distance.

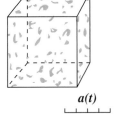

$a(t)$

Fig. 9.7 A three-dimensional map of the cosmic fluid with elements labeled by t-independent $\tilde{x}_i$ comoving coordinates. The time dependence of any distance is entirely determined by the t-dependent scale factor.

[22]While the deduction of the 3D spatial metric given in Section 5.3.2 is only heuristic, in Section 14.4.1 we shall provide an independent derivation of the same result.

The proper distance

The proper distance $d_p(\xi, t)$ to a point at the comoving radial distance ξ and cosmic time t can be calculated from the metric (9.38) with $d\Omega = 0$ and $dt = 0$:

$$d_p(\xi, t) = a(t)R_0 \int_0^\xi \frac{d\xi'}{(1 - k\xi'^2)^{1/2}} \tag{9.39}$$

so that the time dependence (due to expansion of the universe) on the RHS is entirely contained in the scale factor $a(t)$, as already indicated in (9.36):

$$d_p(\xi, t) = a(t)d_p(\xi, t_0) \tag{9.40}$$

where the fixed (comoving) distance at the present epoch is

$$d_p(\xi, t_0) = R_0 \int_0^\xi \frac{d\xi'}{(1 - k\xi'^2)^{1/2}} = \left(\frac{R_0}{\sqrt{k}}\right) \sin^{-1}(\sqrt{k}\xi). \tag{9.41}$$

Namely, for a space with positive curvature $k = +1$, we have $d_p(\xi, t_0) = R_0 \sin^{-1} \xi$; negative curvature, $R_0 \sinh^{-1} \xi$, and a flat space $R_0\xi = \rho$.

Hubble's law follows from the cosmological principle   The relation (9.40) implies a proper velocity of

$$v_p(t) = \frac{d(d_p)}{dt} = \frac{\dot{a}(t)}{a(t)} d_p(t). \tag{9.42}$$

Evidently the velocity is proportional to the separation. This is just Hubble's law with the Hubble constant expressed in terms of the scale factor:

$$H(t) = \frac{\dot{a}(t)}{a(t)} \quad \text{and} \quad H_0 = \dot{a}(t_0). \tag{9.43}$$

Recall that the appearance of an overall scale factor in the spatial part of the Robertson–Walker metric follows from our imposition of the homogeneity and isotropy condition. The result in (9.42) confirms our expectation that in any geometrical description of a dynamical universe which satisfies the cosmological principle,[23] hence the distance scaling relation (9.40), Hubble's law emerges automatically. We emphasize that, in the GR framework, the expansion of the universe is described as the expansion of space, and "big bang" is not any sort of "explosion of matter in space," but rather it is an "expansion of space itself." Space is a dynamic quantity, which is expanding; that is, the metric function of spacetime is the solution to Einstein equation and its scale factor increases with time.

[23] As discussed in Section 9.3, the cosmological principle can be viewed as a generalized Copernican principle. Thus the present result gives further support to the argument, made in p.185, showing that Hubble's law is compatible to the Copernican principle.

Relating distance to the scale factor at emission   To relate distance to the redshift of a light source located at the comoving distance ξ_{em}, we use the fact that the observer and emitter are connected by a light ray along a radial path ($d\Omega = 0$),

$$ds^2 = -c^2dt^2 + R_0^2 a^2(t)\frac{d\xi^2}{1 - k\xi^2} = 0. \tag{9.44}$$

Moving $c^2 dt^2$ to one side and taking the minus sign for the square-root for incoming light, we have

$$-\int_{t_0}^{t_{em}} \frac{c\,dt}{a(t)} = R_0 \int_0^{\xi_{em}} \frac{d\xi}{(1 - k\xi^2)^{1/2}} = d_p(\xi_{em}, t_0) \tag{9.45}$$

where (9.41) has been used to express the second integral in terms of the proper distance at $t = t_0$. The first integral can be put into a more useful form by changing the integration variable to the scale factor,

$$-\int_{t_0}^{t_{em}} \frac{c\,dt}{a(t)} = -\int_1^{a_{em}} \frac{c\,da}{a(t)\dot{a}(t)} = -\int_1^{a_{em}} \frac{c\,da}{a^2(t)H(t)}, \tag{9.46}$$

where we used (9.43) to reach the last expression. In this way (9.45) becomes the relation tbetween the proper distance and scale factor at the emission time

$$d_p(\xi_{em}, t_0) = -\int_1^{a_{em}} \frac{c\,da}{a^2 H(a)}. \tag{9.47}$$

Once again, this is the distance between us and the light emitter located at comoving radial coordinate ξ_{em} with light emitted when the scale factor was a_{em}.

Redshift and the scale factor We see that the scale factor $a(t)$ is the key quantity in our description of the time evolution of the universe. In fact, because $a(t)$ is generally a monotonic function, it can serve as a kind of cosmic clock. How can the scale factor be measured? The observable quantity that has the simplest relation to $a(t)$ is the wavelength shift of a light signal.

The spectral shift, according to (9.3), is

$$z = \frac{\Delta\lambda}{\lambda} = \frac{\lambda_{rec}}{\lambda_{em}} - 1. \tag{9.48}$$

We expect that the wavelength (in fact any length) scales as $a(t)$ (see Problem 9.8 for a more detailed justification):

$$\frac{\lambda_{rec}}{\lambda_{em}} = \frac{a(t_{rec})}{a(t_{em})}. \tag{9.49}$$

Since the "received time" is at t_0 with $a(t_0) = 1$, we have the basic relation

$$1 + z = \frac{1}{a(t_{em})}. \tag{9.50}$$

For example, at the redshift of $z = 1$, the universe had a linear size half as large as at the present one. In fact a common practice in cosmology is to refer to "the redshift of an era" instead of its cosmic time. For example, the "photon decoupling time," when the universe became transparent to light (cf. Section 10.5), is said to occur at $z = 1100$, etc.

Distance in terms of redshift   Changing the integration variable in (9.47) to the redshift, we have the relation between proper distance and redshift in the Robertson–Walker spacetime:

$$d_p(z) = \int_0^z \frac{c\,dz'}{H(z')}.$$

(9.51)

The functional dependence of distance on the redshift is, of course, the Hubble relation. Different cosmological models having a Hubble constant with different z dependence would yield a different distance–redshift relation. Thus the Hubble curve can be used to distinguish between different cosmological scenarios. As we shall discuss in the next chapter, our universe has been discovered to be in an accelerating expansion phase. By fitting the Hubble curve we shall deduce that the universe's dominant energy component is some unknown "dark energy," which provides the repulsion in causing the expansion to proceed at an ever faster rate.

Luminosity distance and standard candle

The principal approach in calculating the distance to any stellar object is to estimate its true luminosity and compare that with the observed flux (which is reduced by the squared distance). Thus it is important to have stars with known intrinsic luminosity that can be used to gauge astronomical distances. Stars with luminosity that can be deduced from other properties are called "standard candles." A well-known class of standard candles is the Cepheid variable stars, which have a definite correlation between their intrinsic luminosity and their pulse rates. In fact, Edwin Hubble used Cepheids to deduce the distances of the galaxies collected for his distance vs. redshift plot. Clearly, the reliability of the method depends on one's ability to obtain the correct estimate of the intrinsic luminosity. A famous piece of history is that Hubble underestimated the luminosity of his Cepheids by almost a factor of 50, leading to an underestimation of the distances, hence an overestimate of the Hubble constant H_0 by a factor of seven. This caused a "cosmic age problem" because the resultant Hubble time (which should be comparable to the age of the universe) became much shorter than the estimated ages of many objects in the universe. This was corrected only after many years of further astronomical observation and astrophysical modeling. Here, we assume that the intrinsic luminosity of a standard candle can be reliably obtained.

   In this section, we study the distance that can be obtained by measuring the light flux from a remote light source with known luminosity. Because we use observations of light emitted in the distant past of an evolving universe, this requires us to be attentive in dealing with the concept of time.

   The measured flux of watts per unit area is related to the intrinsic luminosity $\mathcal{L}$, which is the total power-radiated by the emitting object, as

$$f \equiv \frac{\mathcal{L}}{4\pi d_L^2}.$$

(9.52)

This defines the **luminosity distance** d_L. Let us note that in space with any constant curvature the area of a "sphere" is given by $4\pi d_p^2$ where d_p is the proper radial distance, cf. (9.41). In a static universe, the luminosity distance equals the proper distance to the source: $d_p = d_L$.

$$f_{(\text{static})} = \frac{\mathcal{L}}{4\pi d_p^2}.$$

(9.53)

In an expanding universe the observed flux, being proportional to the energy transfer per unit time, is reduced by a factor of $(1 + z)^2$: one power of $(1 + z)$ comes from energy reduction due to wavelength lengthening of the emitted light, and another power due to the increasing time interval. Let us explain: The energy being proportional to frequency ω, the emitted energy, compared to the observed one, is given by the ratio,

$$\frac{\omega_{\text{em}}}{\omega_0} = \frac{\lambda_0}{\lambda_{\text{em}}} = \frac{1}{a(t_{\text{em}})} = 1 + z,$$

(9.54)

where we have used $a(t_0) = 1$ and (9.49) and (9.50). Just as frequency is reduced by $\omega_0 = \omega_{\text{em}}(1 + z)^{-1}$, the time interval must be correspondingly increased by $\delta t_0 = \delta t_{\text{em}}(1 + z)$, leading to a reduction of energy transfer rate by another power of $(1 + z)$:

$$\frac{\omega_0}{\delta t_0} = \frac{\omega_{\text{em}}}{\delta t_{\text{em}}}(1 + z)^{-2}.$$

(9.55)

Thus the observed flux in an expanding universe, in contrast to the static universe result of (9.53), is given by

$$f = \frac{\mathcal{L}}{4\pi d_p^2 (1 + z)^2}.$$

(9.56)

Namely, the luminosity distance (9.52) differs from the proper distance by

$$d_L = d_p(1 + z).$$

(9.57)

In Chapter 10 the cosmological equations will be solved to obtain the epoch-dependent Hubble's constant in terms of the energy/mass content of the universe. In this way we can find how the proper distance d_p (thus also the luminosity distance) depends on the redshift z via (9.51) for the general relation. (Problem 9.11 works out the case of small z.) In Box 9.1 we explain the astronomy practice of plotting the Hubble diagrams of redshift vs. **distance modulus** (instead of luminosity distance), which is effectively the logarithmic luminosity distance.

Box 9.1 Logarithmic luminosity and distance modulus

Ancient Greek astronomers classified the brightness (observed flux) of stars as having "first magnitude" to "sixth magnitude" for the brightest to the faintest stars visible to the naked eye—the brighter a star is, the smaller its magnitude. Since for this magnitude range of $m_{(6)} - m_{(1)} = 5$ the apparent luminosities span roughly a factor of 100, namely, $f_{(1)}/f_{(6)} \simeq 100$,

(*cont.*)

Box 9.1 (*Continued*)

a definition of **apparent magnitude** m is suggested:

$$m \equiv -2.5 \log_{10} \frac{f}{f_0} \qquad (9.58)$$

so that $m_{(6)} - m_{(1)} = 2.5 \log_{10} f_{(1)}/f_{(6)} = 5$. The reference flux is taken to be $f_0 \equiv 2.52 \times 10^{-8} W\ m^{-2}$ so that the brightest visible stars correspond to $m = 1$ objects. In this scale, for comparison, the sun has an apparent magnitude $m_\odot = -26.8$.

Similar to (9.58), we can define a logarithmic scale, called **absolute magnitude**, for the intrinsic luminosity of a star:

$$M \equiv -2.5 \log_{10} \frac{\mathcal{L}}{\mathcal{L}_0} \qquad (9.59)$$

where the reference luminosity $\mathcal{L}_0$ is defined so that a star with this power output will be seen at a distance 10 pc away to have a flux f_0:

$$f_0 = \frac{\mathcal{L}_0}{4\pi\,(10\text{ pc})^2}. \qquad (9.60)$$

This works out to be $\mathcal{L}_0 = 78.7\mathcal{L}_\odot$. Using the definition of luminosity distance as given in (9.52), Eq. (9.60) can be translated into an expression for the luminosity ratio

$$\frac{f}{\mathcal{L}} = \frac{f_0}{\mathcal{L}_0} \left(\frac{10\text{ pc}}{d_L} \right)^2. \qquad (9.61)$$

Taking the logarithm of this equation leads to the definition of **distance modulus** $(m - M)$, which can be related to luminosity distance by taking the difference of (9.58) and (9.59) and substituting in (9.61):

$$m - M = 5 \log_{10} \frac{d_L}{10\text{ pc}}. \qquad (9.62)$$

In the astronomy literature one finds the common practice of plotting the Hubble diagram with one axis being the redshift z and another axis, instead of luminosity distance, its logarithmic function, the distance modulus (e.g. Fig. 11.8 and Fig. 11.11)

Review questions

1. What does it mean that Hubble's law is a linear relation? What is the significance of this linearity? Support your statement with a proof.

2. What is the Hubble time t_H? Under what condition is it equal to the age of the universe t_0? In a universe full of matter and energy, what would be the expected relative

magnitude of these two quantities ($t_H > t_0$ or $t_H < t_0$)? What is the lower bound for t_0 deduced from the observation data on globular clusters?

3. What are "galaxy rotation curves?" What feature would we expect if the luminous matter were a good representation of the total mass distribution? What observational feature of the rotation curve told us that there were significant amounts of nonluminous matter associated with galaxies and clusters of galaxies?

4. Give a simple example that illustrates the content of the virial theorem for a gravitational system. How can this be used to estimate the total mass of the system?

5. What is baryonic matter? The bulk of baryonic matter resides in the intergalactic medium (IGM) and does not shine. Why don't we count it a part of dark matter?

6. What are the values that we have for the total mass density Ω_M, for the luminous matter Ω_{lum}, and for the baryonic matter Ω_B? From this deduce an estimate of the dark matter density Ω_{DM}. All values are for the present epoch, and list them only to one significant figure.

7. What is the cosmological principle? What are comoving coordinates?

8. Write out the form of the Robertson–Walker metric for two possible coordinate systems. What is the input (i.e. the assumption) used in the derivation of this metric?

9. What are the physical meanings of the scale factor $a(t)$ and the parameter k in the Robertson–Walker metric? How is the epoch-dependent Hubble constant $H(t)$ related to the scale factor $a(t)$?

10. What is the scaling behavior of wavelength? From this derive the relation between the scale factor $a(t)$ and the redshift z.

11. Derive the integral expression for the proper distance $d_p = c \int H^{-1} dz$ to the light source with redshift z.

12. What is luminosity distance? How is it related to the proper distance?

Problems

9.1 The universe as a strong gravitational system One can check that the universe as a whole corresponds to a system of strong gravity that requires a GR description by making a crude estimate of the parameter ε in Eq. (9.1). For this calculation you can assume a static Euclidean universe having a finite spherical volume with radius given by a horizon length $c H_0^{-1}$ and having a mass density comparable to the critical density as given in (9.14).

9.2 Luminosity distance to the nearest star The nearest star appears to us to have a brightness $f_* \simeq 10^{-11} f_\odot$ ($f_\odot$ being the observed solar flux). Assuming that it has the same intrinsic luminosity as the sun, estimate the distance d_* to this star, in the distance unit of **parsec**, as well as in the **astronomical unit** AU $\simeq 5 \times 10^{-6}$pc.

9.3 Gravitational frequency shift contribution to the Hubble redshift Hubble's linear plot of redshift vs. distance relies on spectral measurement of galaxies beyond the Local Group with redshift $z \gtrsim 0.01$. A photon emitted by a galaxy suffers not only a redshift because of cosmic recession, but also a gravitational redshift. Is the latter a significant factor when compared to the recessional effect? *Suggestion*: compare the gravitational redshifts of light from a galaxy with mass $M_G = O(10^{11} M_\odot)$ and linear dimension $R_G = O(10^{12} R_\odot)$ to the redshift for light leaving the surface of the sun, $z_\odot = O(10^{-6})$.

9.4 Energy content due to starlight By assuming the stars have been shining with the same intensity since the beginning of the universe and always had the luminosity density as given in (9.19), estimate the density ratio $\Omega_* = \rho_*/\rho_c$ for starlight. For this rough calculation you can take the age of universe to be the Hubble time t_H.

9.5 Night sky as bright as day Olbers' paradox is solved in our expanding universe because the age of the universe is not infinite $t_0 \simeq t_H$ and, having a horizon length $\simeq ct_H$, it is effectively not infinite in extent. Given the present luminosity density of (9.19), with the same approximation as Problem 9.4, estimate the total flux due to starlight. Compare your result with the solar flux $f_\odot = \mathcal{L}_\odot/(4\pi \, (\mathrm{AU})^2)$. We can increase the star light flux by increasing the age of the universe t_0. How much older does the universe have to be in order that the night sky is as bright as day?

9.6 The virial theorem Given a general bound system of mass points (located at $\mathbf{r}_n$) subject to gravitational forces (central and inverse square) $\mathbf{F}_n = -\nabla V_n$ with $V_n \propto r_n^{-1}$, by considering the time derivative, and average, of the sum of dot-products of momentum and position $G \equiv \Sigma_n \mathbf{p}_n \cdot \mathbf{r}_n$ (called the **virial**), show that the time-averages of the kinetic and potential energy are related by $2 \langle T \rangle = - \langle V \rangle$.

9.7 Proper distance from comoving coordinate χ　In the text we worked out the proper distance from a point with radial coordinate ξ as in (9.40). Now perform the same calculation (and obtain a similar result) for a point labeled by the alternative radial coordinate χ with a metric given by (9.37).

9.8 Wavelength in an expanding universe　By a careful consideration of the time interval between emission and reception of two successive wavecrests, prove that in an expanding universe with a scale factor $a(t)$, the wavelength scales as expected:

$$\frac{\lambda_{\mathrm{rec}}}{\lambda_{\mathrm{em}}} = \frac{a(t_{\mathrm{rec}})}{a(t_{\mathrm{em}})}.$$

Suggestion: Apply Eq. (9.45) to two successive emissions/receptions of waves.

9.9 The deceleration parameter and Taylor expansion of the scale factor　Display the Taylor expansion of the scale factor $a(t)$ and $[a(t)]^{-1}$ around $t = t_0$, up to $(t - t_0)^2$, in terms of the Hubble's constant H_0 and the **deceleration parameter** defined by

$$q_0 \equiv \frac{-\ddot{a}(t_0)\, a(t_0)}{\dot{a}^2(t_0)}. \tag{9.63}$$

9.10 The steady-state universe　The conventional interpretation of an ever increasing scale factor (expanding universe) means that all objects must have been closer in the past, leading to a big bang beginning. We also mentioned in Section 9.4.2 that, because of an initial overestimate of the Hubble constant (by a factor of seven), there was a "cosmic age problem." To avoid this difficulty, an alternative cosmology, called the **steady-state universe** (SSU), was proposed by Hermann Bondi, Thomas Gold, and Fred Hoyle. It was suggested that, consistent with the Robertson–Walker description of an expanding universe, all cosmological quantities besides the scale factor (the expansion rate, deceleration parameter, spatial curvature, matter density, etc.) are time independent. A constant mass density means that the universe did not have a big hot beginning; hence there cannot be a cosmic age problem. To have a constant mass density in an expanding universe requires the continuous, energy-nonconserving, creation of matter. To SSU's advocates, this spontaneous mass creation is no more peculiar than the creation of

all matter at the instant of big bang. In fact, the name "big bang" was invented by Fred Hoyle as a somewhat disparaging description of the competing cosmology.

(a) Supporters of SSU find this model attractive on theoretical grounds—because it is compatible with the "**perfect cosmological principle**." From the above outline of SSU and the cosmological principle in Section 9.3, can you infer what this "perfect CP" must be?

(b) RW geometry, hence (9.43), also holds for SSU, but with a constant expansion rate $H(t) = H_0$. From this, deduce the explicit t-dependence of the scale factor $a(t)$. What is the SSU prediction for the deceleration parameter q_0 defined in (9.63)?

(c) SSU has a 3D space with a curvature $K = k/R^2$ that is not only constant in space but also in time. Does this extra requirement fix its spatial geometry? If so, what is it?

(d) Since the matter density is a constant $\rho_M(t) = \rho_{M,0} \simeq 0.3\rho_{c,0}$ and yet the scale factor increases with time, SSU requires spontaneous matter creation. What must be the rate of this mass creation per unit volume? Express it in terms of the number of hydrogen atoms created per cubic kilometer per year.

9.11 z^2 correction to the Hubble relation　The Hubble relation (9.5) is valid only in the low velocity limit. Namely, it is the leading term in the power series expansion of the proper distance in terms of the redshift. Use the definition of deceleration parameter introduced in (9.63) to show that, including the next order, the Hubble relation reads as

$$d_p(t_0) = \frac{cz}{H_0}\left(1 - \frac{1 + q_0}{2}z\right). \tag{9.64}$$

(a) One first uses (9.45) to calculate the proper distance up to the quadratic term in the "look-back time" $(t_0 - t_{\mathrm{em}})$.

(b) Use the Taylor series of Problem 9.10 to express the redshift in terms of the look-back time up to $(t_0 - t_{\mathrm{em}})^2$.

(c) Deduce the claimed result of (9.64) by using the result obtained in (a) and inverting the relation between the redshift and look-back time obtained in (b).

The expanding universe and thermal relics

- The dynamics of a changing universe are determined by Einstein's equation, which for a Robertson–Walker geometry with the ideal fluid as its mass/energy source takes on the form of the Friedmann equations. Through these equations matter/energy determines the scale factor $a(t)$ and the curvature constant k in the metric description of the cosmic spacetime.
- The Friedmann equations have simple quasi-Newtonian interpretations.
- The universe began hot and dense (the big bang), and thereafter expanded and cooled. The early universe had undergone a series of thermal equilibria—it had been a set of "cosmic soups" composed of different particles.
- The observed abundance of the light nuclear elements match well with their being the product of the big bang nucleosynthesis.
- When the universe was 360 000 years old, photons decoupled, and they remain today as the primordial light having a blackbody spectrum with temperature $T = 2.725\,\text{K}$.
- The cosmic microwave background (CMB) is not perfectly uniform. The dipole anisotropy is primarily determined by our motion in the rest frame of the CMB; higher multipoles contain much information about the geometry and matter/energy content of the universe, as well as the initial density perturbation out of which grew the cosmic structure we see today.

In the previous chapter, we studied the kinematics of the standard model of cosmology. The requirement of a homogeneous and isotropic space fixes the spacetime to have the Robertson–Walker metric in comoving coordinates. This geometry is specified by a curvature signature k and a t-dependent scale factor $a(t)$. Here we study the dynamics of such a universe. The unknown quantities k and $a(t)$ are to be determined by its matter/energy content through the Einstein field equation.

We live in an expanding universe: all the galaxies are now rushing away from each other. This also means that they must have been closer, hence denser and hotter, in the past. Ultimately, at the cosmic beginning $a(0) = 0$, everything must have been right on top of each other. Thus, the standard model of cosmology makes the remarkable prediction that our universe started with a hot big bang.

This prediction that there existed a hotter and denser period in the early universe received strong empirical support, notably by the following discoveries:

1. The 1964 discovery of an all-pervasive microwave background radiation, which is the "after-glow" of the big bang, or the primordial light shining from the early universe, see Section 10.5.
2. The agreement found in the observed abundance of the light nuclear elements, ^{4}He, D, Li, . . . , etc. with the predicted values by the big bang cosmology. Big bang nucleosynthesis will be discussed in Section 10.4.

In Chapter 11, we shall discuss speculations about the nature of the big bang itself, as described by the inflationary cosmology, as well as the recent discovery that the expansion of our universe is accelerating because of the presence of "dark energy," which exerts a repulsive gravitational force.

10.1 Friedmann equations

The Einstein equation relates spacetime's geometry on one side and the mass/energy distribution on the other, $G_{\mu\nu} = \kappa T_{\mu\nu}$, cf. Section 6.3.2. For a description of the universe as a physical system that satisfies the cosmological principle, we have learnt in Sections 9.3 and 9.4 that spacetime must have the Robertson–Walker metric in comoving coordinates. This fixes $G_{\mu\nu}$ on the geometry side of Einstein's equation; the source side should also be compatible with a homogeneous and isotropic space. The simplest plausible choice is to have the energy–momentum tensor $T_{\mu\nu}$ being given by that of an ideal fluid. It should be "ideal" because thermal conductivity and viscosity is unimportant in the cosmic fluid. The proper tensor description of an ideal fluid will be given in Section 12.3 and it is specified by two parameters: the mass density ρ and pressure p. Thus the GR field equation relates the geometric parameters of the curvature signature k and the scale factor $a(t) = R(t)/R_0$ to the cosmic fluid density $\rho(t)$ and pressure $p(t)$.

10.1.1 The GR field equations for cosmology

The Einstein equation with the Robertson–Walker metric and ideal fluid source leads to a basic set of cosmic equations. They are called the **Friedmann equations**, after the Russian mathematician and meteorologist who was the first, in 1922, to appreciate that the Einstein equation admitted cosmological solutions leading to an expanding universe.[1]

One component of the Einstein equation becomes "the first Friedmann equation,"

$$\frac{\dot{a}^2(t)}{a^2(t)} + \frac{kc^2}{R_0^2 a^2(t)} = \frac{8\pi G_N}{3}\rho. \tag{10.1}$$

[1] After Einstein's 1917 cosmology paper, one notable contribution was by W. de Sitter, who studied a dynamical model with nonzero cosmological constant $\Lambda \neq 0$ but devoid of ordinary matter and energy, see Section 11.2.2.

Another component becomes "the second Friedmann equation,"

$$\frac{\ddot{a}(t)}{a(t)} = -\frac{4\pi G_N}{c^2}\left(p + \frac{1}{3}\rho c^2\right).$$

(10.2)

Because the pressure and density factors are positive we have a negative second derivative $\ddot{a}(t)$: the expansion must decelerate because of mutual gravitational attraction among the cosmic fluid elements. It can be shown (Problem 10.1) that a linear combination of these two Friedmann Eqs. (10.1) and (10.2) leads to

$$\frac{d}{dt}(\rho c^2 a^3) = -p\frac{da^3}{dt},$$

(10.3)

which, having the form of the first law of thermodynamics $dE = -pdV$, is the statement of energy conservation. Since it has such a simple physical interpretation, we shall often use Eq. (10.3) instead of (10.2), and by "Friedmann equation" one usually means the first Friedmann equation (10.1).

Because there are only two independent equations, yet there are three unknown functions $a(t)$, $\rho(t)$, and $p(t)$, we need one more relation. This is provided by the "equation of state," relating the pressure to the density of the system. Usually such a relation is rather complicated. However, since cosmology deals only with a dilute gas, the equation of state we need to work with can usually be written simply as

$$p = w\rho c^2,$$

(10.4)

which defines w as the constant parameter that characterizes the material content of the system. For example, for nonrelativistic matter the pressure is negligibly small compared to the rest energy of the material particles, hence $w = 0$, and for radiation we have $w = \frac{1}{3}$, etc. (cf. Problem 12.16).

While the discussion of the precise relation between the Friedmann and the Einstein field equations will be postponed till Section 14.4.2, here in the following subsection we shall present a quasi-Newtonian approach, which gives the Friedmann equations a more transparent physical interpretation.

Critical density of the universe

The first Friedmann equation (10.1) can be rewritten as

$$-k = \left(\frac{\dot{a}R_0}{c}\right)^2\left(1 - \frac{\rho}{\rho_c}\right),$$

(10.5)

where the critical density is defined as in (9.14)

$$\rho_c(t) = \frac{3}{8\pi G_N}\frac{\dot{a}^2}{a^2} = \frac{3H(t)^2}{8\pi G_N}$$

(10.6)

after using the expression for Hubble's constant $H = \dot{a}/a$ of Eq. (9.43). Denoting the density ratio by $\Omega = \rho/\rho_c$ as in (9.15), we can express (10.5) as

$$-\frac{kc^2}{\dot{a}^2R_0^2} = 1 - \Omega.$$

(10.7)

In particular at $t = t_0$ with $H_0 = \dot{a}$, it becomes, for $\Omega_0 = \rho_0/\rho_{c,0}$,

$$\frac{kc^2}{R_0^2} = H_0^2(\Omega_0 - 1). \tag{10.8}$$

This clearly expresses the GR connection between matter/energy distribution (Ω_0) and geometry (k): if our universe has a mass density greater than the critical density, the average curvature must be positive $k = +1$ (a closed universe); if the density is less than the critical density, then $k = -1$, the geometry of an open universe having a negative curvature; and if $\rho = \rho_c$, we have the $k = 0$ flat geometry:

$$\begin{aligned}
\Omega_0 > 1 &\longrightarrow k = +1 &&\text{closed universe,}\\
\Omega_0 = 1 &\longrightarrow k = 0 &&\text{flat universe,}\\
\Omega_0 < 1 &\longrightarrow k = -1 &&\text{open universe.}
\end{aligned} \tag{10.9}$$

That is, the critical density is the value that separates the positively curved, high-density universe from the negatively curved, low-density universe. From the phenomenological values stated in Chapter 9, $\Omega_0 = \Omega_{M,0} \simeq 0.3$, it would seem that we live in a negatively curved open universe. In Chapter 11 we shall discuss this topic further, and conclude that we need to modify Einstein's equation (by the addition of the cosmological constant). This theoretical input, together with new observational evidence, now suggests that we in fact live in a $k = 0$ flat universe, with the energy/mass density in the universe just equal to the critical value.

10.1.2   The quasi-Newtonian interpretation

The derivation of the Friedmann equations involves rather long and tedious calculations (see Section 14.4.2) of the Einstein tensor components $G_{\mu\nu}$ from unknown factors k and $a(t)$ of the metric $g_{\mu\nu}$ and relating them via the Einstein equation $G_{\mu\nu} = \kappa T_{\mu\nu}$ to the density ρ and pressure p from the energy–momentum tensor $T_{\mu\nu}$. Having stated this connection of the Friedmann equations (10.1) and (10.2) to the GR Einstein equation, we now show that they actually have rather simple Newtonian interpretations.

Applicability of Newtonian interpretations

At the beginning of Chapter 9, we presented arguments for the necessity of GR as the proper framework to study cosmology. Indeed, the Friedmann equations are for the scale factor $a(t)$ and the curvature signature k, which are the fundamental concepts of a curved spacetime description of gravity. Nevertheless, as we shall show, these equations have rather simple Newtonian interpretations when supplemented with global geometric concepts at appropriate junctures. There is no contradiction that cosmological equations must fundamentally be relativistic and yet have a Newtonian interpretation. The cosmological principle states that every part of the universe, large or small, behaves in the same way. When we concentrate on a small region, the Newtonian description should be valid, because the gravity involved is not strong and small space can always be approximated by a flat geometry.

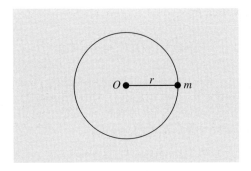

Fig. 10.1 The effect on the separation *r* between two galaxies due to the gravitational attraction by all the mass in the cosmic fluid. The net force on *m* is as if all the mass inside the sphere is concentrated at the center *O*.

Thus, we should be able to understand the cosmological equation with a Newtonian approach when it is carried out in an overall GR framework.

Interpretations of the Friedmann equations

Equation (10.3) is clearly the statement of energy conservation as expressed in the form of the first law of thermodynamics $dE = -pdV$. That is, the change in energy E (with energy per unit volume ρc^2) is equal to the work done on the system with the volume V being proportional to a^3.

The Friedmann equation (10.1) has a straightforward interpretation as the usual energy balance equation (the total energy being the sum of the kinetic and potential energy) for a central force problem. Recall that in our homogeneous and isotropic cosmological models we ignore any local motions of the galaxies. The only dynamics we need to consider is the change in separation due to the change of the scale factor $a(t)$. Namely, the only relevant dynamical question is the time-dependence of the separation between any two fluid elements.

To be specific, let us consider a cosmic fluid element (i.e. an element composed of a collection of galaxies), in a homogeneous and isotropic fluid (Fig. 10.1), with mass m at the radial distance $r(\xi, t) = a(t)r_0(\xi)$, cf. (9.40), with ξ being the dimensionless time-independent comoving radial coordinate with respect to an arbitrarily selected comoving coordinate origin O, and the radial distance $r_0(\xi) = r(\xi, t_0)$. We wish to study the effect of gravitational attraction on this mass point m by the whole fluid, which may be treated as spherically symmetric[2] centered around O. The gravitational attractions due to the mass outside the sphere (radius r) mutually cancel. To understand this you can imagine the outside region as composed of a series of concentric spherical shells and the interior gravitational field inside each shell vanishes. This is the familiar Newtonian result. Here we must use GR because the gravitational attraction from the mass shells at large distances is not Newtonian. But it turns out this familiar nonrelativistic solution is also valid in GR, related to the validity of Birkhoff's theorem (as stated at the end of Section 7.1 and Box 14.3). Consequently, the mass element m feels only the total mass M inside the sphere.

This is a particularly simple central force problem, as we have only the radial motion $\dot{r} = \dot{a}r_0$ to consider. The energy balance equation has no orbital angular momentum term:

$$\frac{1}{2}m\dot{r}^2 - \frac{G_N mM}{r} = E_{\text{tot}}, \quad (10.10)$$

[2]Homogeneity and isotropy imply spherical symmetry with respect to every point.

which may be rewritten as

$$\frac{1}{2}m\dot{a}^2 r_0^2 - \frac{G_N m M}{a r_0} = E_{tot}.$$ (10.11)

The expansion of the universe means an increasing $a(t)$ and hence an increasing potential energy (i.e. it is less negative). This necessarily implies a decreasing $\dot{a}(t)$, namely, a slowdown of the expansion. The total mass inside the (flat space) sphere being $M = \rho 4\pi a^3 r_0^3/3$, we get

$$\dot{a}^2 - \frac{8\pi G_N}{3}\rho a^2 = \frac{2E_{tot}}{mr_0^2}.$$ (10.12)

Remember that this calculation is carried out for an arbitrary center O. Different choices of a center correspond to different values of r_0 and thus different E_{tot}. The assumption of a homogeneous and isotropic space leads to the GR conclusion that the right-hand side (RHS) of Eq. (10.12) is a constant with respect to any choice of O. Furthermore, GR interprets this constant as the curvature parameter (k):

$$\frac{2E_{tot}}{mr_0^2} \equiv \frac{-kc^2}{R_0^2}.$$ (10.13)

In this way we see that (10.12) is just the first Friedmann equation (10.1).

Similarly one can show the Friedmann equation (10.2) as the $F = ma$ equation of this system (Problem 10.2).

Mass density also determines the fate of the universe

With this interpretation of the Friedmann Eq. (10.1) as the energy balance equation and with the identification of the curvature signature k as being proportional to the total energy of (10.13), it is clear that the value of k, hence also that of density ρ by way of (10.5), determines not only the geometry of the 3D space, but also the fate of the cosmic evolution.[3]

For the central force problem, we recall that whether the motion of the test mass m is bound or not is determined by the sign of the total energy E_{tot}. An unbound system allows $r \to \infty$ where the potential energy vanishes and the total energy is given by the kinetic energy, which must be positive: $E_{tot} > 0$ namely, $k < 0$, cf. (10.13). Also, the same equation (10.10) shows that the sign of E_{tot} reflects the relative size of the positive kinetic energy as compared to the negative potential energy. We can phrase this relative size question in two equivalent ways:

1. **Compare the kinetic energy to a given potential energy: the escape velocity**   Given a source mass (i.e. the potential energy), one can decide whether the kinetic energy term (i.e. test particle velocity) is big enough to have an unbound system ($E_{tot} > 0$?). To facilitate this comparison, we write the potential energy term in the form

$$G_N \frac{mM}{r} \equiv \frac{1}{2}mv_{esc}^2$$ (10.14)

[3]Even though this connection between density and the outcome of time evolution is broken when the Einstein equation is modified by the presence of a cosmological constant term (Chapter 11), the following presentation can still give us some insight to the meaning of the critical density.

with the **escape velocity** being

$$v_{\text{esc}} = \sqrt{\frac{2G_{\text{N}}M}{r}}.$$ (10.15)

The energy equation (10.10) then takes the form of

$$G_{\text{N}}\frac{mM}{r}\left(\frac{v^2}{v_{\text{esc}}^2} - 1\right) = E_{\text{tot}}.$$ (10.16)

When $v < v_{\text{esc}}$, thus $E_{\text{tot}} < 0$, the test mass m is bound and can never escape.

2. **Compare the potential energy to a given kinetic energy: the critical mass**
   Given the test particle's velocity (i.e. the kinetic energy), one can decide whether the potential energy term (i.e. the amount of mass M) is big enough to overcome the kinetic energy to bind the test mass ($E_{\text{tot}} < 0$?). Writing the kinetic energy term as

$$\frac{1}{2}m\dot{r}^2 \equiv G_{\text{N}}\frac{mM_c}{r}$$ (10.17)

with the **critical mass** being

$$M_c = \frac{r\dot{r}^2}{2G_{\text{N}}} = \frac{a\dot{a}^2 r_0^3}{2G_{\text{N}}},$$ (10.18)

the energy equation (10.10) then takes the form of

$$\frac{1}{2}m\dot{r}^2\left(1 - \frac{M}{M_c}\right) = E_{\text{tot}}.$$ (10.19)

When $M > M_c$, thus $E_{\text{tot}} < 0$, the test mass m is bound and can never escape.

The analogous question of whether, given an expansion rate $H(t)$, the test galaxy m is bound by the gravitational attraction of the cosmic fluid is determined by whether there is enough mass in the arbitrary sphere (on its surface m lies) to prevent m from escaping completely. Namely, the question of whether the universe will expand forever, or its expansion will eventually slow down and recollapse, will be determined by the value of $k \sim -E_{\text{tot}}$. Since the sphere is arbitrary, what matters is the density of the cosmic fluid. We will divide the critical mass (10.18) by the volume of the sphere, and use the Hubble constant relation $H = \dot{a}/a$ of (9.43) to obtain

$$\frac{a\dot{a}^2 r_0^3/2G_{\text{N}}}{4\pi a^3 r_0^3/3} = \frac{3H^2(t)}{8\pi G_{\text{N}}},$$ (10.20)

which is just the critical density ρ_c defined in (10.6). With M/M_c being replaced by ρ/ρ_c, Eq. (10.19) with E_{tot} given by (10.13) is just the Friedmann equation as written in (10.5) and (10.7).

10.2 Time evolution of model universes

We now use the Friedmann equations (10.1) and (10.3) and the equation of state (10.4) to find the time dependence of the scale factor $a(t)$ for a definite value of the curvature k. Although in a realistic situation we need to consider several different energy/mass components $\rho = \Sigma_w \rho_w$ with their respective pressure terms $p_w = w\rho_w c^2$, we shall at this stage consider mostly single component systems. To simplify notation we shall omit the subscript w in the density and pressure functions.

Scaling of the density function

Before solving $a(t)$, we shall first study the scaling behavior of the density and pressure as dictated by the energy conservation condition (10.3). Carrying out the differentiation in this equation, we have

$$\dot{\rho}c^2 = -3(\rho c^2 + p)\frac{\dot{a}}{a}, \tag{10.21}$$

which, after using the equation of state (10.4), turns into

$$\frac{\dot{\rho}}{\rho} = -3(1 + w)\frac{\dot{a}}{a}. \tag{10.22}$$

This can be solved by straightforwardly integrating $d\rho/\rho = -3(1 + w)da/a$ and then exponentiating the result to yield[4]

$$\rho(t) = \rho_0 a(t)^{-3(1+w)}. \tag{10.23}$$

For a matter dominated universe $w = 0$, and a radiation dominated universe $w = \frac{1}{3}$, the respective densities scale as

$$\rho_M(t) = \rho_{M,0} a(t)^{-3} \quad \text{and} \quad \rho_R(t) = \rho_{R,0} a(t)^{-4}. \tag{10.24}$$

While the first equation displays the expected scaling behavior of an inverse volume, the second relation can be understood because radiation energy is inversely proportional to wavelength, hence scales as a^{-1}, which is then divided by the volume factor a^3 to get the density. For the special case of negative pressure $p = -\rho c^2$ with $w = -1$, Eq. (10.23) leads to a constant energy density $\rho(t) = \rho(t_0)$ even as the universe expands. As we shall discuss in the next chapter the newly discovered "dark energy" seems to have just this property.

Model universe with k = 0

We proceed to solve Eq. (10.1) for the time evolution of the scale factor $a(t)$ for some simple situations. We first consider a class of model universes with $k = 0$. As we shall see, a spatially flat geometry is particularly relevant for the universe we live in. The Friedmann equation (10.1), combining with (10.23), reads

$$\left(\frac{\dot{a}}{a}\right)^2 = \frac{8\pi G_N}{3}\rho_0 a^{-3(1+w)}, \tag{10.25}$$

---

[4]One can also check easily that this result follows directly from the energy conservation equation (10.3), and the definition (10.4).

Assuming a power law growth for the scale factor

$$a(t) = \left(\frac{t}{t_0}\right)^x \tag{10.26}$$

thus

$$\frac{\dot{a}}{a} = \frac{x}{t}. \tag{10.27}$$

We can immediately relate the age of the universe t_0 to the Hubble time $t_H = H_0^{-1}$ as follows:

$$H_0 \equiv \left(\frac{\dot{a}}{a}\right)_{t_0} = \frac{x}{t_0} \quad \text{or} \quad t_0 = x t_H. \tag{10.28}$$

Furthermore, by substituting (10.27) into (10.25), we see that in order to match the powers of t on both sides, there must be the relation

$$x = \frac{2}{3(1+w)}. \tag{10.29}$$

For the matter-dominated and radiation-dominated cases, Eqs. (10.29), (10.26) and (10.28) lead to

$$\text{MDU} \;\; (w = 0) \quad x = \tfrac{2}{3} \quad a = \left(\frac{t}{t_0}\right)^{2/3} \quad t_0 = \tfrac{2}{3} t_H$$

$$\text{RDU} \;\; \left(w = \tfrac{1}{3}\right) \quad x = \tfrac{1}{2} \quad a = \left(\frac{t}{t_0}\right)^{1/2} \quad t_0 = \tfrac{1}{2} t_H. \tag{10.30}$$

Also, (10.29) informs us that x is singular for the $w = -1$ cosmic fluid, which possesses, we have already noted, negative pressure and constant energy density. This means that for this case the scale factor has a nonpower-growth t-dependence, that is, assumption (10.26) is not applicable. We also point out the general situation of $t_0 < t_H$ for $w > -\tfrac{1}{3}$, and $t_0 > t_H$ for $w < -\tfrac{1}{3}$.

Here we have considered the specific case of a flat geometry $k = 0$. But we note that the result also correctly describes the early epoch even in a universe with curvature $k \neq 0$. This is so because in the $t \to 0$ limit, the curvature term in the Friedmann equation (10.1) is negligible when compared to the $\dot{a}$ term which grows as some negative power of the cosmic time.

Time evolution in single-component universes

Here we shall consider the time evolution of toy universes with only one component of energy/matter. They can be thought of as approximations to a more realistic multi-component universe if the energy content is dominated by one component. Thus, a matter-dominated universe means that the energy of the universe resides primarily in the form of (nonrelativistic) matter, even though there may be many more relativistic radiation particles than matter particles.

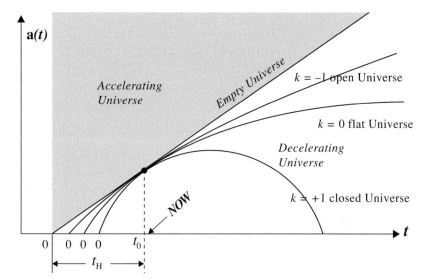

Fig. 10.2 Time dependence of the scale factor $a(t)$ for the open, flat and closed universe. The qualitative features of these curves are the same for radiation- or matter-dominated universes. All models must have the same radius a_0 and slope $\dot{a}_0$ at t_0 in order to match the Hubble constant $H_0 = \dot{a}_0/a_0$ at the present epoch. The origin of the cosmic time $t = 0$ is different for each curve.

Radiation-dominated universe   The radiation energy density scales as $\rho \sim a^{-4}$, and the Friedmann equation (10.1) can be written, with A being some constant, as

$$\dot{a}^2(t) = \frac{A^2}{a^2(t)} - \frac{kc^2}{R_0^2}. \tag{10.31}$$

With a change of variable $y = a^2$ this equation is simplified to

$$\dot{y}^2 + \frac{4kc^2}{R_0^2} y = 4A^2,$$

which has the solutions:

$$a^2(t) = 2At - \frac{kc^2}{R_0^2} t^2. \tag{10.32}$$

We note that this expression for the scale factor does have the correct early universe limit of $a \sim t^{1/2}$ for a radiation-dominated universe, as derived in (10.30). The different time dependence of the scale factor $a(t)$ for $k = 0, \pm 1$ is plotted in Fig. 10.2. Note that the straight ($\dot{a} = $ constant) line corresponds to an empty universe, and all other curves lie **below** this, reflecting the fact that in all cases the expansion undergoes deceleration because of gravitational attraction.

Matter-dominated universe   The matter density scales as $\rho \sim a^{-3}$ and the first Friedmann equation (10.1) becomes, with B being also a constant,

$$\dot{a}^2(t) = \frac{B}{a(t)} - \frac{kc^2}{R_0^2}. \tag{10.33}$$

The solution is more complicated. We merely note that the qualitative behavior of $a(t)$ as depicted in Fig. 10.2 is again obtained. Namely, for density less than ρ_c the expansion of the open universe ($k = -1$) will continue forever; for $\rho > \rho_c$ the expansion of a closed universe ($k = +1$) will slow down to a stop then start to recollapse—all the way to another $a = 0$ "big crunch;" for the flat universe ($k = 0$) the expansion will slow down but not enough to stop.

10.3 Big bang cosmology

During the epochs immediately after the big bang, the universe was much more compact, and the energy associated with the random motions of matter and radiation is much larger. Thus, we say, the universe was much hotter. As a result, elementary particles could be in thermal equilibrium through their interactions. As the universe expanded, it also cooled. With the lowering of the particle energy, particles (and antiparticles) would either disappear through annihilation, or combine into various composites of particles, or "decouple" to become free particles. As a consequence, there would be different kinds of thermal relics left behind by the hot big bang. One approach to study the universe's history is to start with some initial state which may be guessed based on our knowledge (or speculation) of particle physics. Then we can evolve the universe forward in the hope of ending up with something like the observed universe today. That we can speak of the early universe with any sort of confidence rests with the idea that the universe had been in a series of equilibria, cf. (10.45). At such stages, the property of the system was determined, independent of the details of the interactions, by a few parameters such as the temperature, density, pressure, etc. A thermodynamical investigation of the cosmic history was pioneered by Tolman (1934). This approach to extracting observable consequences from big bang cosmology was first vigorously pursued by Gamow (1949) and his collaborators, Alpher and Herman. Here, we shall give an overview of the thermal history of the universe, in particular, the scale-dependence of the radiation temperature.

Once again, it should be pointed out that the calculations carried out in this chapter are rather crude, and they are for illustrative purposes only—to give us a flavor of how in principle cosmological predictions can be made. Typically, realistic calculations would be far more complicated, involving many reaction rates with numerous conditions.

10.3.1 Scale-dependence of radiation's temperature

For the radiation component of the universe, we can neglect particle masses and chemical potentials[5] compared to $k_B T$, where k_B is the Boltzmann constant. The number distributions[6] with respect to the energy E of the radiation system composed of bosons (for the minus sign) and fermions (for the positive sign) are, respectively,

$$dn = \frac{4\pi g}{h^3 c^3} \frac{E^2 dE}{e^{E/k_B T} \pm 1} \tag{10.34}$$

[5]Except that for photons, there is no strong theoretical ground to set the chemical potential μ to zero. Yet, since there is nothing that requires a sizable μ, we shall for simplicity set $|\mu| \ll k_B T$ in our discussion.

[6]The number density is the number per unit volume. We can check the various factors in (10.34) by noting that the number is a count of the phase space elements, which is an integral over $d^3 x d^3 p / h^3$, as dictated by Heisenberg's uncertainty principle that no particle can be located in a region smaller than h^3. While spatial integration leads to the total volume, the momentum integration measure can be expressed for an isotropic distribution in terms of the energy variable as $d^3 p = 4\pi E^2 dE/c^3$.

[7]Particles, with mass and spin s, have $2s + 1$ spin states (e.g. spin $\frac{1}{2}$ electrons have two spin states), but massless spin particles (e.g. spin 1 photons or spin 2 gravitons) have only two spin states.

with h the Planck constant, and g the number of spin states of the particles making up the radiation; for photons and electrons, we have $g(\gamma) = 2$ and $g(e) = 2$, respectively,[7] neutrinos have $g(\nu) = 1$ because only the left-handed spin states participate in interactions. Carrying out the integration, we get the number density $n = \int dn \sim T^3$:

$$n_b = \frac{4}{3} n_f = 2.404 \frac{g}{2\pi^2} \left(\frac{k_B T}{hc} \right)^3 \qquad (10.35)$$

for the respective boson and fermion systems. We can derive the thermodynamic relation (the **Stefan–Boltzmann law**) between radiation energy density (u) and its temperature by performing the integration of $u = \int E \, dn \sim T^4$:

$$\rho_R c^2 \equiv u_R = \frac{g^*}{2} a_{SB} T^4, \qquad (10.36)$$

where a_{SB} is the Stefan–Boltzmann constant

$$a_{SB} = \frac{\pi^2 k_B^4}{15 c^3 \hbar^3} = 7.5659 \times 10^{-16} \, \text{J/m}^3/\text{K}^4. \qquad (10.37)$$

We have summed over the energy contribution by all the constituent radiation particles so that g^* is the "effective number" of spin states of the particles making up the radiation:

$$g^* = \sum_i (g_b)_i + \frac{7}{8} \sum_i (g_f)_i \qquad (10.38)$$

where $(g_b)_i$ and $(g_f)_i$ are the spin factors of the ith species of boson and fermion radiation particles, respectively. The $\frac{7}{8}$ factor reflects the different integral values for the fermion distribution, with a plus sign in (10.34), vs. the minus sign for the boson case.

Knowing the number and energy densities we can also display the average energy of the constituent radiation particles $\bar{E} = \rho_R c^2 / n$. In particular we have the photon average energy

$$\bar{E}_R = 2.7 k_B T. \qquad (10.39)$$

Combining this result (10.36) of $\rho_R \sim T^4$ with our previous derived relation (10.24) for a radiation-dominated system $\rho_R \sim a^{-4}$, we deduce the scaling property for the radiation temperature

$$T \propto a^{-1}. \qquad (10.40)$$

This expresses, in precise scaling terms, our expectation that temperature is high when the universe is compact, or equivalently, when it expands, it also cools. Under this temperature scaling law, the distributions in (10.34) are unchanged (Tolman, 1934), because the radiation energy was inverse to the wavelength, $E \sim \lambda^{-1} \sim a^{-1}(t)$, the combinations $V E^2 dE$ and $E/k_B T$ were invariant under the scale changes.

Remark In the context of the Newtonian interpretation of the cosmological (Friedmann) equations, we can understand energy conservation in an expanding universe as follows: while the total number of radiation particles $N = nV$

does not change during expansion, the total radiation energy $(k_B T)$ scales as a^{-1}. This loss of radiation energy, because of an increase in a, is balanced by the increase of gravitational energy of the universe. The gravitational potential energy is also inversely proportional to distance, hence $\sim a^{-1}$, but it is negative. Thus, it increases with an increase in a because it becomes less negative.

Relation between radiation temperature and time The early universe is dominated by radiation with the scale factor $a \propto t^{1/2}$, cf. (10.30). We can drop the curvature term in the Friedmann equation (10.1), and replace $\dot{a}/a$ by $(2t)^{-1}$ so that the radiation energy density is related to cosmic time as[8]

$$\rho_R c^2 = \frac{3}{32\pi} \frac{c^2}{G_N} t^{-2}. \tag{10.41}$$

We can rewrite this relation in a way making it easier to remember, by using quantum gravity units: the Planck energy density and time as defined in Section 8.5.1:

$$\rho_R c^2 = \frac{3}{32\pi} (\rho c^2)_{Pl} \left(\frac{t_{Pl}}{t}\right)^2. \tag{10.42}$$

Namely, in the natural unit system the radiation density is about one thirtieth per unit cosmic time squared. The radiation density can be related to temperature by the Stefan–Boltzmann law of (10.36), leading to

$$k_B T \simeq 0.46\, E_{Pl} \left(\frac{t}{t_{Pl}}\right)^{-1/2} \tag{10.43}$$

or equivalently an easy to remember numerical relation:

$$t\,(s) \simeq \frac{10^{20}}{[T\,(K)]^2}. \tag{10.44}$$

From this estimate we shall see that the big bang nucleosynthesis, taking place at temperature $T_{bbn} \simeq 10^9$ K, cf. (10.53), corresponded to a cosmic age of $t_{bbn} = O(10^2 \text{ s})$ after the big bang.

[8]The time scaling property of ρ_R checks with our previously derived results of (10.24) and (10.30): $\rho_R \sim a^{-4} \sim (t^{1/2})^{-4} = t^{-2}$.

10.3.2   Different thermal equilibrium stages

Subsequent to the big bang, the cooling of the universe allowed for the existence of different composites of elementary particles. When the falling thermal energy $k_B T$ could no longer produce various types of particle–antiparticle pairs, this lack of fresh supply of antiparticles caused their disappearance from equilibrium states as their annihilation with particles continued. Quarks combined into protons and neutrons (collectively called nucleons). The latter would in turn join into atomic nuclei. At a time some 360 000 years after the big bang, the lower temperature would allow electrons to combine with hydrogen and other light nuclei to form electrically neutral atoms without being immediately blasted apart by high energy electromagnetic radiation.

As a result, the universe became transparent to photons. No longer being pushed apart by radiation, the gas of atoms (mostly hydrogen), was free to collapse under its own gravitational attraction, and thus began the process to form stars and galaxies in a background of free photons as we see them today.

In the early universe, the energy density was high. This implies a rapidly expanding and cooling universe, cf. (10.1). To determine what kinds of particle reactions would be taking place to maintain thermal equilibrium involves dynamical calculations, taking into account the reaction rate in an expanding and cooling medium. The basic requirement is that the time interval between particle scatterings be much shorter than the age of the cosmos. This can be expressed as the "Gamow condition" that the reaction rate Γ must be faster than the expansion rate of the universe as measured by Hubble's constant H:

$$\Gamma \geq H, \tag{10.45}$$

where Γ is given by the product of the number density n of the reactant particles, the relative particle velocity v, and the reaction cross-section σ, which gives the probability for the reaction to take place:[9]

$$\Gamma = n\sigma v. \tag{10.46}$$

The particle velocity entered because it is the flux of the interacting particles and was given by the velocity distribution as determined by the thermal energy of the system. The condition for a new equilibrium stage to take place is given by the condition of $\Gamma = H$. The cross-section, being laboratory measured or theory predicted quantities, is assumed to be given, and this condition can be used to solve for the thermal energy and the redshift value at which a new equilibrium stage starts.

Epochs of neutrino and positron decoupling

A convenient reference point may be taken when the thermal energy was about 1 GeV corresponding to the age of the universe at $t \simeq 10^{-5}$ s. Prior to this, all the stable particles (protons, neutrons, electrons, neutrinos, and their antiparticles, as well as photons) were in thermal equilibrium.

As the universe cooled, different particles would go out of equilibrium. We mention two examples: neutrino decoupling and the disappearance of positrons.

1. The neutrinos started out to be in thermal equilibrium, through the (reversible) weak interaction reactions, which also allowed proton and neutron to transform into each other.

$$\nu + n \rightleftarrows e^- + p, \qquad \bar{\nu} + p \rightleftarrows e^+ + n \tag{10.47}$$

where n, p, e^-, e^+, ν and $\bar{\nu}$ stand for neutron, proton, electron, positron, neutrino and antineutrino, respectively. But as the system cooled, the particle energy was reduced. These weak interaction cross-sections σ of (10.46), which had a strong energy-dependence, fell rapidly, and eventually the reaction rate Γ would fall below the expansion rate H.

[9]Properly, the reaction rate should be thermal-averaged $\Gamma = n\overline{\sigma v}$. Since the product σv is close to a constant, this average for our purpose can be trivially done.

The neutrinos (both neutrinos and antineutrinos) no long interacted with matter. Putting it another way, the above listed weak interaction processes which maintained the neutrinos in thermal equilibrium and the exchange between protons and neutrons effectively switched off when the universe cooled below a certain temperature ($k_B T_\nu \approx 3$ MeV). In this way, the neutrinos decoupled from the rest of matter[10] and the proton–neutron ratio was 'frozen'.[11] The neutrinos, which participate in weak interactions only, evolved subsequently as free particles. These free neutrinos cooled down as the universe expanded. In the present epoch the universe should be filled everywhere with these primordial neutrinos (and antineutrinos) having a thermal spectrum (with $T_{\nu,0} \approx 1.9$ K) corresponding to a density $n_\nu \approx 150$ cm^{-3}. Because the neutrino interaction cross-section is so small, it does not seem possible to detect them with the present technology. Nevertheless, if neutrinos have even a small mass, they can potentially be an important contributor to the dark matter mass density.[12]

2. Similarly, the disappearance of the electron's antiparticles, positrons, proceeded as follows. Initially we had the reversible reaction of

$$e^+ + e^- \rightleftarrows \gamma + \gamma. \tag{10.48}$$

However, as the universe cooled, the photons became less energetic. The rest energy of an electron or positron being just over $\frac{1}{2}$ MeV, when $k_B T$ fell below their sum of 1 MeV, the reaction could no longer proceed from right to left. Because there were more electrons than positrons, the positrons would be annihilated by this reaction (going from left to right). They disappeared from the universe.[13]

When we go back in time, before the age of the nucleon, to such high energies the strong interaction underwent a "QCD deconfinement phase transition," when all the quarks inside the nucleons were released.[14] Initially we had mostly the "up" and "down" quarks. As we go further back in time, energy got higher and other heavy flavors of quarks and leptons ("strange" quarks and muons, etc.) would be present. See the **chronology of the universe** (Table 10.1).

[10]That is, the average interval between neutrino–baryon scattering became longer than the age of the universe. See Box 10.1 below and a similar discussion in Section 10.5 of decoupled photons.

[11]Free neutrons can decay into protons (plus electrons and antineutrinos) with a half-life just over 10 minutes. However, soon after nucleon freeze-out, before any significant fraction of the neutrons disappeared, they fused into deuterons (each composed of a neutron and proton) and other light nuclei, see Section 10.4. Neutrons bound inside stable nuclei are safe against spontaneous decay if the nuclear binding energy is greater than the neutron–proton rest energy difference.

[12]The latest results on neutrino oscillation do indicate that $m_\nu \neq 0$. However, it is so small, $\lesssim 10^{-3}$ eV, that neutrinos cannot possibly be the principal component of dark matter. Furthermore, current understanding of the structure formation in the universe disfavors such "hot dark matter" as the dominant form of dark matter.

[13]The matter–antimatter asymmetry, showing up as an excess of electrons over positrons in the universe, will be discussed briefly in Box 10.2 at the end of this chapter.

[14]The fundamental strong interaction of quantum chromodynamics (QCD), under a normal low energy environment, binds quarks into nucleons and other strongly interacting particles. At extremely high energy and density, quarks are set free—deconfined.

Table 10.1 A chronology of the universe.

	Radiation dominated universe			Matter dominated universe		
Cosmic time	10^{-5} s		10^2 s	10^{13} s	10^{15} s	10^{17} s
Thermal energy	1 GeV	3 MeV / 1 MeV	0.7 MeV	1 eV		
Age of...	Quarks-leptons Nucleons	ν decouple / e^+ decouple	Nuclear synthesis	Photon decouple	Galaxies	Now
Physics	Particle physics		Nuclear physics	Atomic physics	Astronomy	

In the following sections, we shall discuss two particular epochs in the history of the universe which had left observable features on our present-day cosmos.

(1) In Section 10.4 we study the epoch of big bang nucleosynthesis $t_{\mathrm{bbn}} \simeq 10^2$ s, when protons and neutrons combined into charged nuclei (ions) at the end of the age of nucleons. But the lack of stable nuclei with mass number at 5 and 8 prevents the formation of elements heavier than lithium. Thus the abundance of light nuclear elements in the universe can be deduced via cosmological considerations.

(2) In Section 10.5 we study the epoch at $t_\gamma \simeq 360\,000$ years when photons no longer interacted with matter. Having been decoupled, they survived to the present era as the CMB radiation.

10.4   Primordial nucleosynthesis

When we look around our universe, we see mostly hydrogen, and very little of the heavy elements. The abundances of heavy elements can all be satisfactorily accounted for by the known nuclear reactions taking place inside stars and supernovae. On the other hand, everywhere we look, besides hydrogen we also see a significant amount of helium. (The helium abundance has been deduced by measurements of the intensities of spectral lines of ^{4}He in stars, planetary nebulas, and H-II regions of galaxies.) The observational data indicate a helium-4 **mass fraction** being close to 24%:

$$ y \equiv \left(\frac{^4\mathrm{He}}{\mathrm{H} + {}^4\mathrm{He}} \right)_{\mathrm{mass}} \qquad \text{with } y_{\mathrm{obs}} \simeq 0.24. \qquad (10.49) $$

Similarly, we observe a uniform density, at a much smaller abundance, for the other light elements of deuterium (D), helium-3 (^{3}He), and lithium-7 (^{7}Li). Gamow (1946) and Alpher and Herman (1948) were the first to suggest, in the late 1940s, that these light nuclear elements were synthesized in the early universe. The primordial processes were theorized to follow the path described below.

[15] A nucleus is composed of Z number of protons and N number of neutrons, giving it the mass number $A = Z + N$. Since chemical properties are determined by the proton number, we can identify Z from the name of the element, e.g. hydrogen has $Z = 1$ and helium $Z = 2$. Nuclei having the same Z but different number of neutrons are isotopes. From the mass number, usually denoted by a superscript on the left side of the nucleus symbol, we can figure out the number of neutrons. The most abundant helium isotope is helium-4 (^{4}He) having two neutrons, followed by helium-3 (^{3}He) having one neutron. Hydrogen's isotopes have their distinctive names: the deuteron has one proton and one neutron ^{2}H $\equiv$ D, and the tritium neuclenus ^{3}H has two neutrons.

The age of nucleons

During this epoch, the cosmic soup was composed of protons, neutrons, electrons, positrons, neutrinos, antineutrinos, and photons. There was a tendency for the protons and neutrons to bind (through strong interactions) into nuclear bound states.[15] However, as soon as they were formed, they were blasted apart by energetic photons (photodissociation). We can categorize the dominant reactions during the age of nucleons into two types:

1. The transitions between protons and neutrons p $\leftrightarrow$ n via prototypical weak interaction processes involving neutrinos as given in (10.47).
2. The protons and neutrons could fuse into light-nuclei ions via strong interaction processes (by adding, one at a time, a proton or a neutron):

The key reaction is

$$p + n \rightleftarrows D^+ + \gamma, \tag{10.50}$$

where D^+ is the deuteron (i.e. the singly-charged deuterium ion, comprising one proton and one neutron), an isotope of hydrogen, and γ denotes, as before, an energy-carrying photon. As the universe cools there are fewer photons energetic enough to photodissociate the deuteron (the reaction proceeding from right to left), and thus deuterons accumulate. The following nucleon capture reactions can then build up heavier elements:

$$D^+ + n \rightleftarrows {}^3H^+ + \gamma, \quad D^+ + p \rightleftarrows {}^3He^{++} + \gamma \tag{10.51}$$

and

$${}^3H^+ + p \rightleftarrows {}^4He^{++} + \gamma, \quad {}^3He^{++} + n \rightleftarrows {}^4He^{++} + \gamma. \tag{10.52}$$

These reversible nuclear reactions would **not** go further, to bind into even heavier nuclei because, helium-4 being a particularly tightly bound nucleus, the formation of a nuclear structure involving five nucleons was not energetically favored. Lacking an $A = 5$ stable nucleus, the synthesis of lithium with mass numbers six or seven from stable helium required the much less abundant deuterons or tritium

$${}^4He + D \rightleftarrows {}^6Li + \gamma \quad \text{and} \quad {}^4He + {}^3H \rightleftarrows {}^7Li + \gamma.$$

Big bang nucleosynthesis could not progress further in producing heavier elements ($A > 7$) because there is no stable $A = 8$ element. (If beryllium-8 were formed, it would almost immediately disintegrate into a pair of helium-4s.)

Epoch of primordial nucleosynthesis

But as the universe expanded and cooled, the photons were no longer energetic enough to photodissociate the bound nuclei. This happened, according to a detailed rate calculation, when the thermal energy decreased down to

$$k_B T_{bbn} \simeq 0.7 \text{ MeV}, \tag{10.53}$$

corresponding to an ambient temperature on the order of $T_{bbn} \simeq 10^9$ K, and an age of the universe of $t_{bbn} \simeq O(10^2 \text{ s})$, cf. (10.41) and (10.44). The net effect of the above reactions from (10.50) to (10.52) was to cause all the neutrons to be bound into helium-4 nuclei, because there were more protons than neutrons:

$$2n + 2p \longrightarrow {}^4He^{++} + \gamma. \tag{10.54}$$

We can then conclude that the resultant number density n_{He} for helium-4 must be equal to half of the neutron density $n_{He} = n_n/2$. Using the approximation that the helium mass m_{He} is four times the nucleon mass m_N, and that the number density of hydrogen n_H should equal the proton number density minus the neutron density (i.e. they were the left-over protons after all the other protons had combined with the neutrons to form helium ions), we have

$$y \equiv \left(\frac{^4\mathrm{He}}{\mathrm{H} + {}^4\mathrm{He}} \right)_{\mathrm{mass}} = \frac{n_{\mathrm{He}} m_{\mathrm{He}}}{n_{\mathrm{H}} m_{\mathrm{H}} + n_{\mathrm{He}} m_{\mathrm{He}}}$$

$$= \frac{(n_{\mathrm{n}}/2) \cdot 4m_{\mathrm{N}}}{(n_{\mathrm{p}} - n_{\mathrm{n}}) m_{\mathrm{N}} + (n_{\mathrm{n}}/2) \cdot 4m_{\mathrm{N}}} = \frac{2\lambda}{1 + \lambda}, \tag{10.55}$$

where λ is the neutron to proton ratio: $\lambda = n_{\mathrm{n}}/n_{\mathrm{p}}$. For nonrelativistic nucleons in thermal equilibrium, this ratio is fixed by the Boltzmann distribution, $\exp(-E/k_{\mathrm{B}} T)$:

$$\lambda = \exp\left[-\left(\frac{E_{\mathrm{n}} - E_{\mathrm{p}}}{k_{\mathrm{B}} T_{\mathrm{bbn}}} \right) \right] \simeq \exp\left[-\left(\frac{m_{\mathrm{n}} - m_{\mathrm{p}}}{k_{\mathrm{B}} T_{\mathrm{bbn}}} \right) c^2 \right] = \mathrm{e}^{-1.3/0.7} = \frac{1}{6.4},$$
$$\tag{10.56}$$

where we have used (10.53) and the fact that the neutron is slightly more massive than the proton, with a rest energy difference of $(m_{\mathrm{n}} - m_{\mathrm{p}})c^2 = 1.3$ MeV. After taking into account the fact that some of the neutrons could have decayed into protons via the process $\mathrm{n} \longrightarrow \mathrm{p} + \mathrm{e} + \bar{\nu}$ (but keep in mind that at this stage the free neutron lifetime was still longer than the age of the universe) the neutron to proton ratio λ is reduced somewhat from $\frac{1}{6.4}$ to $\simeq \frac{1}{7}$, yielding a result very close to the observed ratio of 0.24:

$$y = \frac{2\lambda}{1 + \lambda} \simeq \frac{2/7}{8/7} = \frac{1}{4}. \tag{10.57}$$

In summary, once the deuterium was formed by the fusion of protons and neutrons, this chain of fusion reactions proceeded rapidly so that just about all neutrons were bound into helium. Since these reactions were not perfectly efficient, trace amounts of deuterium and helium-3 were left over. (The small leftover tritium would also combine into helium-3.) Formation of nuclei beyond helium progressed slowly; only small amounts of lithium-6 and -7 were synthesized in the big bang.

Spectral lines of $^7\mathrm{Li}$ have been measured in metal-poor stars. Since nuclear elements were produced in stellar thermonuclear reactions, a low abundance of heavy elements indicates that the star was formed from primordial, uncontaminated gas. Thus, at such sites we can assume the elements were produced in big bang nucleosynthesis. Deuterium abundance has been measured in the solar system and in high-redshift clouds of interstellar gas by their absorption of light coming from even more distant quasars.

Again, we must keep in mind that the crude calculations presented here are for illustrative purpose only. They are meant to give us a simple picture of the physics involved in such a cosmological deduction. Realistic calculations often involved the simultaneous inclusion of many reaction rates. In the detailed computations leading to (10.57), one must also use the result of:

1. The assumption that there are only three flavors of light neutrinos (electron-, muon-, and tau-neutrinos), because it changes the effective degrees of freedom in the radiation g^* in (10.38).

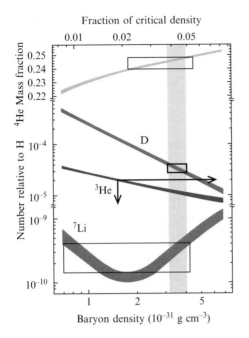

Fraction of critical density

Fig. 10.3 The abundance of light nuclear elements vs. baryon mass density ρ_B of the universe. [Graph from Freedman and Turner (2003).] The curves are big bang nucleosynthesis predictions and the boxes are observational results: the vertical heights represent uncertainties in observation, and the horizontal width the range of ρ_B that theory can accommodate observation. The shaded vertical column represents the value of ρ_B that allows theory and observation to agree for all four elements. Its uncertainty (the width of the column) is basically determined by the deuterium abundance which has both a strong ρ_B dependence and a well-measured value.

2. The observed baryon mass density, ρ_B, because it impacts the rate of cooling of the universe. In particular, deuterium D is very sensitive to ρ_B. Thus, we can use the observed abundance of light elements, in particular deuterium, to determine the baryon density.[16] The best fit, as shown in Fig. 10.3, is at $\rho_B \simeq 4 \times 10^{-31}$ g/cm^3, or as a fraction of the critical density:

$$\Omega_B \simeq 0.042. \tag{10.58}$$

As we already pointed out in Section 9.2.2, when compared to the total mass density $\Omega_M = \Omega_B + \Omega_{DM} \simeq 0.25$, this shows that baryons are only a small part of the matter in the universe. Furthermore, we can obtain an estimate of baryon number density n_B when we divide the baryon energy density $\rho_B c^2 = \Omega_B \cdot \rho_c c^2 = 230$ MeV/m^3 by the energy of each nucleon, which can be taken to be the rest energy of the nucleon (939 MeV) because these particles are nonrelativistic:

$$n_B \simeq 0.25/\text{m}^3. \tag{10.59}$$

[16] As can be seen from (10.50) to (10.52), the production of all light elements passed through deuterium. Thus the observed D-abundance resulted from the remanent intermediate states, hence very sensitive to the reaction and cooling rates.

10.5   Photon decoupling and the CMB

The early universe was a plasma of radiation and matter held together by their mutual interaction. As the universe expanded and cooled, matter had congealed into neutral atoms and the cosmic soup lost its ability to entrap the photons. These free thermal photons survived as the CMB radiation we see today. The uniformly distributed relic photons obey a blackbody spectrum. Their discovery gave strong support to the hot big bang beginning of our universe, as it is difficult to think of any other alternative account for the existence of such physical phenomena on the cosmic scale. Furthermore, its slight temperature fluctuation, the CMB anisotropy, is a picture of the "baby universe." Careful study of this anisotropy has furnished and will continue to provide us with detailed information about the history and composition of the universe. This is a major tool for quantitative cosmology.

10.5.1   The universe became transparent to photons

The epoch when charged nuclear ions and electrons were transformed into neutral atoms is called the **photon-decoupling time**[17] (t_γ). This took place when the thermal energy of photons just dropped below the threshold required to ionize the newly formed atoms. Namely, the dominant reversible reaction during the age of ions,

$$e^- + p^+ \longleftrightarrow H + \gamma, \tag{10.60}$$

ceased to proceed from right to left when the photon energy was less than the ionization energy. All the charged electrons and protons were swept up and bound themselves into stable neutral atoms.

One would naturally expect the temperature at the decoupling time $k_B T_\gamma = O(\text{eV})$ to be comparable to the typical atomic binding energy. In fact, a detailed calculation yields

$$k_B T_\gamma \simeq 0.26 \text{ eV}. \tag{10.61}$$

Dividing out the Boltzmann constant k_B, this energy corresponds to a photon temperature of

$$T_\gamma \simeq 3000 \text{ K}. \tag{10.62}$$

Equation (10.44) can translate this photon decoupling temperature into a cosmic age of

$$t_\gamma \simeq 1.1 \times 10^{13} \text{ s}, \tag{10.63}$$

that is, about 360 000 years after the big bang. The same $\Gamma = H$ calculation also yields the redshift

$$z_\gamma \simeq 1100.$$

That is, the cosmos at the photon decoupling time was about a thousand times smaller in linear dimension and a billion time denser, on average, than it is today, cf. Eqs. (9.50) and (10.24).

Shortly after recombination, the universe became transparent to electromagnetic radiation. Thereafter, the decoupled photons could travel freely through the universe, but they still had the blackbody spectrum which was unchanged as the universe expanded. These relic photons cooled according to the scaling law of $T \propto a^{-1}$. Thus, the big bang cosmology predicted that everywhere in the present universe there should be a sea of primordial photons following a blackbody spectrum.

[17]The photon-decoupling time is also referred to in the literature as the "recombination time." We prefer not to use this terminology as, up to this time, ions and electrons had never been combined. The name has been used because of the analogous situation in the interstellar plasma where such atomic formation is indeed a recombination.

What should the photon temperature be now? From the estimates of $T_\gamma \simeq$ 3000 K and $z_\gamma \simeq 1100$ we can use (10.40) and (9.50) to deduce

$$T_{\gamma,0} = \frac{T_\gamma}{1 + z_\gamma} \simeq 2.7 \text{ K},$$

or a thermal energy $k_B T_{\gamma,0} \simeq 2.5 \times 10^{-3}$ eV. A blackbody spectrum of temperature T has its maximal intensity at the wavelength $\lambda_{max} T \simeq 0.3$ cm K (known as the "Wien displacement constant"). Thus, an estimate $T_{\gamma,0} \simeq 2.7$ K implies a thermal spectrum with the maximal intensity at λ_{max} on the order of millimeters—there should be a relic background radiation in the microwave range.[18]

[18]While this electromagnetic radiation is outside the visible range, we can still "see" it because such a microwave noise constitutes a percentage of the television "snow" between channels.

In summary, photons in the early universe were tightly coupled to ionized matter through Thomson scattering. Such interactions stopped about 360 000 years after the big bang, when the universe had cooled sufficiently to form neutral atoms (mainly hydrogens). Ever since this last scattering time, the photons have traveled freely through the universe, and redshifted to microwave frequencies as the universe expanded. This primordial light should appear today as the CMB thermal radiation with a temperature of about 2.7 K.

10.5.2 The discovery of CMB radiation

The observational discovery of the CMB radiation was one of the great scientific events in the modern era. It made the big bang cosmology much more credible as it is difficult to see how else such thermal radiation could have been produced. The discovery and its interpretation also constitute an interesting story. Gamow (1948), Ralph Alpher and Robert Herman (1948) first predicted that a direct consequence of the big bang model is the presence of a relic background of radiation with a temperature of a few degrees. However their contribution was not widely appreciated and no effort was mounted to detect such a microwave background. Only in 1964 did Robert Dicke at Princeton University lead a research group (including James Peebles, Peter Roll, and David Wilkinson) to detect this CMB. While they were constructing their apparatus, Dicke was contacted by Arno Penzias and Robert W. Wilson at the nearby Bell Lab. Penzias and Wilson had used a horn-shaped microwave antenna in the past year to do astronomical observations of the Galaxy. This "Dicke radiometer" was originally used in a trial satellite communication experiment, and was known to have some "excess noise." Not content to ignore it, Penzias and Wilson made a careful measurement of this background radiation, finding it to be independent of direction, time of the day, or season of the year. While puzzling over the cause of such radiation, they were informed by one of their colleagues of Princeton group's interest in the detection of a cosmic background radiation. (Peebles had given a colloquium on this subject at Johns Hopkins University.) This resulted in the simultaneous publication of two papers: one in which Penzias and Wilson (1965) announced their discovery; the other by the Princeton group (Dicke *et al.* 1965) explaining the cosmological significance of the discovery. At about the same time, a research group (Doroshkevich and Novikov, 1964) around Yakov Zel'dovich in Moscow also recognized the importance of the cosmic background radiation as the relic signature of a big bang beginning of the universe.

Because of microwave absorption by water molecules in the atmosphere, it is desirable to carry out CMB measurements at locations having low humidity and/or at high altitude. Thus, some of the subsequent observations were done with balloon-borne instruments launched in Antarctica (low temperature, low humidity, and high altitude), or even better, to go above the atmosphere in a satellite. This was first accomplished in the early 1990s (Mather *et al.* 1990, and Smoot *et al.* 1990) by the Cosmic Background Explorer satellite (COBE), obtaining results showing that the CMB radiation followed a perfect blackbody spectrum (Fixsen *et al.*, 1996) with a temperature of

$$T_{\gamma,0} = 2.725 \pm 0.002 \text{ K}. \tag{10.64}$$

The COBE observation not only confirmed that the thermal nature of the cosmic radiation was very uniform (the same temperature in every direction), but also discovered the minute anisotropy at the micro-kelvin level. This has been interpreted as resulting from the matter density perturbation, which, through subsequent gravitational clumping, gave rise to the cosmic structure we see today: galaxies, clusters of galaxies, and voids, etc. This will be further discussed in Section 10.5.4, and in Chapter 11.

10.5.3   Photons, neutrinos and the radiation–matter equality time

Knowledge of the CMB's temperature allows us to calculate the photon number density. This reveals that there are about a billion photons to every nucleon in the universe. Such information will enable us to estimate the cosmic time when the universe made the transition from a radiation-dominated to a matter-dominated universe. In this section, we also discuss another cosmic thermal relic: the primordial neutrinos.

The photon to baryon number ratio

Knowing the CMB photon temperature $T_{\gamma,0} = 2.725$ K, we can calculate the relic photon number density via (10.35):

$$n_{\gamma,0} = \frac{2.4}{\pi^2} \left(\frac{k_B T_{\gamma,0}}{hc} \right)^3 \simeq 411/\text{cm}^3. \tag{10.65}$$

That is, there are now in the universe, on average, about 400 photons for every cubic centimeter. Clearly this density is much higher than the baryon matter density obtained in (10.59) from the primordial nucleosynthesis theory, and the observed abundance of light nuclear elements. The baryon and photon number ratio comes out to be

$$\frac{n_B}{n_\gamma} \simeq 0.6 \times 10^{-9}. \tag{10.66}$$

For every proton or neutron there are about 2 billion photons. This also explains why the thermal energy at photon decoupling is as low as 0.26 eV. Considering that the hydrogen ionization energy is 13.6 eV, why was the ionization not shut off until the thermal energy fell so much below this value? This just reflects the fact that there are so many photons for every baryon that the blackbody thermal photons have a broad distribution and the photon number density was very high, $n \sim T^3$. Thus, even though the average photon energy was only 0.26 eV, there was a sufficient number of high energy photons at the tail end of the distribution to bring about a new equilibrium phase.

This ratio (10.66) should hold all the way back to the photon decoupling time, because not only was the number of free photons unchanged, also the baryon number remained the same, since all the interactions in this low energy range respected the law of baryon number conservation. The relevance of this density ratio to the question of matter–antimatter asymmetry of the universe is discussed in Box 10.2 below.

Transition from radiation-dominated to matter-dominated era

The present epoch is matter-dominated $\Omega_M \gg \Omega_R$ The cosmic radiation is now composed of CMB photons[19] and three flavors of neutrinos.[20] Their relative abundance is calculated in Box 10.1: using (10.76) $\Omega_{\nu,0} = 0.68\,\Omega_{\gamma,0}$ so that the radiation density is about 1.68 times larger than the density due to CMB radiation alone. The matter–radiation ratio now can be related to the baryon–photon ratio

$$\frac{\Omega_{M,0}}{\Omega_{R,0}} = \frac{\Omega_{M,0}}{\Omega_{B,0}} \frac{\Omega_{B,0}}{1.68 \times \Omega_{\gamma,0}}$$

$$\simeq \frac{0.25}{0.04} \frac{n_B}{n_\gamma} \frac{m_N c^2}{1.68 \times k_B T_{\gamma,0}} \simeq 8.8 \times 10^3, \qquad (10.67)$$

where the energy density for nonrelativistic baryon matter is given by the product of the number density and rest energy of the nucleon $m_N c^2 = 939$ MeV and the photon energy by $k_B T_{\gamma,0}$. The baryon photon ratio is given by (10.66). We have also used the phenomenological results of (9.26) and (9.22) for $\Omega_{M,0}$ and $\Omega_{B,0}$. It is clear that we now live in a matter-dominated universe as $\Omega_{M,0} \gg \Omega_{R,0}$.

Transition from RDU to MDU Since radiation density scales as $\rho_R \sim a^{-4}$ while matter $\rho_M \sim a^{-3}$, even though $\Omega_{R,0}$ is small in the present epoch, radiation was the dominant contributor in the early universe. The epoch when the universe made this transition from a radiation-dominated to matter-dominated era can be fixed by the condition of $\rho_R(t_{RM}) = \rho_M(t_{RM})$, where t_{RM} is the cosmic age when the radiation and matter densities were equal:

$$1 = \frac{\rho_R}{\rho_M} = \frac{\rho_{R,0}}{\rho_{M,0}} a(t_{RM})^{-1} = \frac{\Omega_{R,0}}{\Omega_{M,0}}(1 + z_{RM}) \simeq \frac{1 + z_{RM}}{8.8 \times 10^3}. \qquad (10.68)$$

Hence the redshift for radiation–matter equality is $z_{RM} \simeq 8800$, which is eight times larger than the photon decoupling time with $z_\gamma \simeq 1100$. This also yields scale factor and temperature ratios of $a(t_\gamma)/a(t_{RM}) = T_{RM}/T_\gamma \simeq 8$, or a radiation thermal energy

$$k_B T_{RM} = 8\,k_B T_\gamma = O(2\text{ eV}). \qquad (10.69)$$

[19]The energy density due to starlight (i.e. all electromagnetic radiation except the microwave background) is much smaller than that of the CMB photons.

[20]Even though we have not measured all the neutrino masses, all indications are that they are very small and we can treat them as relativistic particles with kinetic energy much larger than their rest energy.

Knowing this temperature ratio we can find the "radiation–matter equality time" $t_{RM} \simeq 16\,000$ yrs (Problem 10.9) from the photon decoupling time $t_\gamma \simeq 360\,000$ yrs. From that time on, gravity (less opposed by significant radiation pressure) began to grow, from the tiny lumpiness in matter distribution, the rich cosmic structures we see today.

An estimate of the age of a flat universe   Also, since radiation dominance ceased so long ago ($t_{RM} \ll t_0$) and if the universe is composed of matter and radiation only, we can estimate the age of the universe based on the model of a matter-dominated universe because during the overwhelming part of the universe's history, the dominant energy component has been in the form of nonrelativistic matter. In particular, for a matter-dominated universe with a spatially flat geometry, we have, according to (10.30),

$$(t_0)^{k=0} \simeq (t_0)^{k=0}_{\text{MDU}} = \frac{2}{3} t_H \simeq 9 \text{ Gyr.} \tag{10.70}$$

This value is seen to be significantly less than the age deduced by observation (cf. Section 9.1.3). As we shall see in the next chapter, a flat geometry is indeed favored theoretically and confirmed by observation. This contradiction in the estimate of the cosmic age hinted at the possibility that, besides radiation and matter, there may be some other significant form of energy in the universe.

Box 10.1 Entropy conservation, photon reheating and the neutrino temperature—primordial neutrinos are expected to be colder than the CMB

We have already noted that (10.3), being a linear combination of the Friedmann equations (10.1) and (10.2), has the interpretation of energy conservation, $dE + p\,dV = 0$. This implies, through the second law of thermodynamics $dE + p\,dV = T\,dS$, that the entropy is conserved, $dS = 0$. Holding the volume fixed, a change in entropy is related to that of the energy density $dS = (V/T)du_R$ with the radiation energy density u_R being proportional to T^4, and we can relate entropy to temperature and volume as

$$S = \frac{g^*}{2} a_{\text{SB}} V \int \frac{dT^4}{T} = \frac{2g^*}{3} a_{\text{SB}} V T^3 \tag{10.71}$$

where we have used the Stefan–Boltzmann law (10.36).

Before neutrino decoupling, the radiation particles, photons, neutrinos, antineutrinos, electrons, and positrons, were in thermal equilibrium. Thus photons and neutrinos (as well as antineutrinos) have the same temperature $T_\gamma = T_\nu$. We have already discussed the sequence of cosmic equilibria: first, neutrinos decoupled at $k_B T \simeq 3$ MeV, second, there was photon decoupling at $k_B T = O(\text{eV})$. Whether they were coupled or not, the radiation temperature scaled as $T \propto a^{-1}$. So we would expect the relic photons and neutrinos to have the same temperature now. However, we must take into account the effect of positron disappearance, which happened at $k_B T = O(\text{MeV})$ in between these two epochs of neutrino decoupling and photon decoupling. The annihilation of electrons and positrons into photons would heat up the photons ("**photon reheating**"), rasing the photon temperature over that of neutrinos, which had already decoupled. One can calculate the raised photon temperature $T'_\gamma > T_\gamma$ by the condition of entropy conservation discussed above:

$$S_\gamma + S_{e^-} + S_{e^+} = S'_\gamma. \tag{10.72}$$

With the effective spin degrees of freedom g^* of (10.38) in (10.71), this entropy conservation equation becomes $\left[2 + \frac{7}{8}(2+2)\right] V T^3 = 2V'T'^3$, or

$$\left(\frac{T'_\gamma}{T_\gamma}\right)^3 \frac{V'}{V} = \frac{11}{4}. \tag{10.73}$$

On the other hand, the neutrinos being noninteracting, their entropy was not affected, $S_\nu = S'_\nu$. Namely, $T_\nu^3 V = T'^3_\nu V'$, or

$$\left(\frac{T'_\nu}{T_\nu}\right)^3 \frac{V'}{V} = 1. \tag{10.74}$$

Equations (10.73) and (10.74), together with the thermal equilibrium condition $T_\gamma = T_\nu$ that prevailed before the positron disappearance, lead to

$$T'_\gamma = \left(\frac{11}{4}\right)^{1/3} T'_\nu = 1.4 \times T'_\nu. \tag{10.75}$$

Knowing $T_{\gamma,0} \simeq 2.7\,\text{K}$ the temperature of relic neutrinos and antineutrinos now should be $T_{\nu,0} \simeq 1.9\,\text{K}$. This temperature difference leads to the different photon and neutrino number densities as first stated in Section 10.3.2.

Using (10.75) we can also compare the neutrino and photon contributions to the radiation energy content of the universe via (10.36):

$$\frac{\rho_\nu}{\rho_\gamma} = \frac{g^*_\nu}{g_\gamma} \left(\frac{T'_\nu}{T'_\gamma}\right)^4 = \frac{\frac{21}{4}}{2} \left(\frac{4}{11}\right)^{4/3} = 0.68, \tag{10.76}$$

because the neutrino effective spin degrees of freedom $g^*_\nu = \frac{7}{8}3 \times (1+1)$ must include the three species ("flavors") of neutrinos (electron-, muon- and tau-neutrinos) as well as their antiparticles.

Box 10.2 Cosmological asymmetry between matter and antimatter

The ratio displayed in (10.66) is also a measure of the baryon number asymmetry in the universe. By this we mean that if the universe contained equal numbers of baryons (having baryon number +1) and antibaryons (having baryon number −1) then the net baryon number would vanish, $n_B = 0$. Inthe early universe there was plenty of thermal energy so that antibaryons

(*cont.*)

Box 10.2 (*Continued*)

(particles carrying negative baryon numbers such as antiquarks making up antiprotons, etc.) were present in abundance. In fact there were just about an equal number of particles and antiparticles. The fact that the present universe has only particles and no antiparticles means that there must have been a slight excess of particles. As the universe cooled to the point when particle–antiparticle pairs could no longer be produced by radiation, all antiparticles would disappear through annihilation. (Cf. Section 10.3.2 on positron disappearance.) Thermal equilibrium in the early universe would ensure that photons and quark–antiquark pair numbers should be comparable. From this we can conclude that the population of baryons was only slightly larger than that of antibaryons as indicated, for example, by the quark–antiquark asymmetry ratio:

$$\frac{n_q - n_{\bar{q}}}{n_q + n_{\bar{q}}} \simeq \frac{n_B}{n_\gamma} = O\left(10^{-9}\right). \tag{10.77}$$

Because the universe is observed to be electrically neutral, this statement about baryon number asymmetry can also be extended to electron–positron asymmetry. Namely, (10.77) holds for the entire matter–antimatter asymmetry of the universe.

   This matter–antimatter asymmetry is a puzzle as no cosmological model can generate a net baryon number if all underlying physical processes (such as all the interactions included in the Standard Model of particle physics) conserve baryon number. Thus one had to impose on the standard big bang cosmology an ad hoc asymmetric initial boundary condition (10.66). Why should there be this asymmetry, with this particular value? It would be much more satisfying if starting with a symmetric state (or better, independent of initial conditions) such an asymmetry could be generated by the underlying physical interactions. One of the attractive features of the Standard Model of particle interactions and its natural extensions is that they generally possess precisely the conditions to produce such an excess of matter over antimatter.

10.5.4   CMB temperature fluctuation

After subtracting off the Milky Way foreground radiation, one obtained, in every direction, the same blackbody temperature—the CMB showed a high degree of isotropy. However, such an isotropy is not perfect. One of the major achievements of COBE satellite observations was the detection of a slight variation of temperature: first at the 10^{-3} level associated with the motion of our Local Group of galaxies in the gravitational potential due to neighboring cosmic matter distribution, then at the 10^{-5} level, which, as we shall explain, holds the key to our understanding of the origin of the structure in the universe, how the primordial plasma evolved into stars, galaxies, and clusters of galaxies. Furthermore, such a CMB fluctuation provides us with another means to measure the matter/energy content of the universe.[21]

[21] In this introductory presentation, we only provide a sketchy description of CMB anisotropy. Readers wishing for a more detailed discussion are referred to more advanced texts such as Dodelson (2003) and Weinberg (2008).

The dipole anisotropy

The sensitive Differential Microwave Radiometer (DMR) aboard COBE first revealed the existence of a "dipole anisotropy" in the CMB background. Although each point on the sky has a blackbody spectrum, in one half of the sky the spectrum corresponds to a slightly higher temperature while the other half is slightly lower with respect to the average background temperature

$$\frac{\delta T}{T} \approx 1.237 \times 10^{-3} \cos\theta, \qquad (10.78)$$

where θ is measured from the direction joining the hottest to the coldest spot on the sky. The dipole distortion is a simple Doppler shift, caused by the net motion of the COBE satellite which is 371 km/s relative to the reference frame in which the CMB is isotropic (cf. Problem 10.14). The Doppler effect changes the observed frequency, which in turn changes the energy and temperature of the detected background radiation. The different peculiar motions result from the gravitational attraction as a consequence of uneven distribution of masses in our cosmic neighborhood. The measured value[22] is in fact the vector sum of the orbital motion of the solar system around the galactic center ($\sim$220 km/s), the motion of the Milky Way around the center of mass of the Local Group ($\sim$80 km/s), and the motion of Local Group of galaxies (630 ± 20 km/s) in the general direction of the constellation Hydra. The last, being the peculiar motion of our small galaxy cluster toward the large mass concentration in the neighboring part of the universe, reflects the gravitational attraction by the very massive Virgo Cluster at the center of our Local Supercluster, which is in turn accelerating toward the Hydra–Centaurus Supercluster.

   The peculiar motions mentioned above are measured with respect to the frame in which the CMB is isotropic. The existence of such a CMB rest frame does not contradict special relativity. SR only says that no internal physical measurements can detect absolute motion. Namely, physics laws do not single out an absolute rest frame. It does not say that we cannot compare motion relative to a cosmic structure such as the microwave background. The more relevant question is why constant velocity motion in this CMB rest frame coincides with the Galilean frames of Newton's first law. To the extent that the CMB frame represents the average mass distribution of the universe, this supports **Mach's principle** (cf. Box 1.1). While to a large extent Einstein's GR embodies Mach's principle, there is no definitive explanation of why the CMB rest frame defines the inertial frames for us.

Physical origin and mathematical description of CMB anisotropy

Physical origin of the temperature inhomogeneity   After taking off this 10^{-3} level dipole anisotropy, the background radiation is seen to be isotropic. CMB being a snapshot of our universe, the observed isotropy is direct evidence of our working hypothesis of a homogeneous and isotropic universe. Nevertheless, this isotropy should not be perfect. The observed universe has all sorts of structure; some of the superclusters of galaxies and largest voids have dimensions as large as 100 Mpc across. Such a basic feature of our universe must be reflected in the CMB in the form of small temperature anisotropy.

[22]The quoted number represents the observational result after subtracting out the orbital motion of COBE around the earth ($\sim$8 km/s), and the seasonal motion of earth around the sun ($\sim$30 km/s).

There must be matter density nonuniformity which would have brought about temperature anisotropy through electromagnetic interactions; photons traveling from denser regions were gravitationally redshifted and therefore arrived cooler, while photons from less dense regions did less work and arrived warmer. One of the great achievements of the COBE observation team is the first observation of such an anisotropy, at the level of 10^{-5} (Smoot *et al.*, 1992). Such small temperature variations (≈ 30 μK) coming from different directions was finally detected. This discovery provided the first evidence for a primordial density nonuniformity that, under gravitational attraction, grew into the structures of stars, galaxies, and clusters of galaxies that we observe today.

Dark matter forms the cosmic scaffolding of matter distribution The temperature variation $\delta T / T = O(10^{-5})$ is smaller than expected based on the observed structure of luminous matter. But this "discrepancy" can be resolved by the existence of dark matter. Dark matter has no electromagnetic interaction; its density inhomogeneities cannot be seen through the CMB anisotropy. Thus, it is possible to have sufficient density perturbation (of all the matter, baryonic and dark) to seed the structure we see today even though that part in the baryonic matter is by itself insufficient to bring it about. In fact we expect that the bulk of the structure is formed by the clumping of the dark matter, that forms the gravitational potential (the cosmic scaffolding form in the interval from t_{RM} to t_γ) into which the baryonic matter falls.[23] This extra early growth of density perturbation for nonbaryonic dark matter means that less baryonic inhomogeniety at t_γ is needed to produce the structure seen today. Thus a smaller $\delta T / T$ in the baryonic matter is needed if the bulk of the nonluminous matter is nonbaryonic.

Expansion into spherical harmonics Here we present the basic formalism needed for a description of this CMB anisotropy. The CMB temperature has directional dependence $T(\theta, \phi)$ with an average of

$$\langle T \rangle = \frac{1}{4\pi} \int T(\theta, \phi) \sin\theta \, d\theta \, d\phi = 2.725 \text{ K}. \tag{10.79}$$

The temperature fluctuation

$$\frac{\delta T}{T}(\theta, \phi) \equiv \frac{T(\theta, \phi) - \langle T \rangle}{\langle T \rangle} \tag{10.80}$$

has a root-mean-square value of

$$\left\langle \left(\frac{\delta T}{T} \right)^2 \right\rangle^{1/2} = 1.1 \times 10^{-5}. \tag{10.81}$$

How do we describe such a temperature variation across the celestial sphere? Recall that for a function of one variable, a useful approach is Fourier

[23] Because baryonic matter interacts with radiation, it is difficult to build up its inhomogeneity during the period of $t_{RM} < t < t_\gamma$, as radiation pressure tended to wash it out.

expansion of the function in a series of sine waves with frequencies that are integral multiples of the fundamental wave (with the largest wavelength). Similarly for the dependence on (θ, ϕ) by the temperature fluctuation (think of it as vibration modes on the surface of an elastic sphere), we expand it in terms of spherical harmonics[24]

$$\frac{\delta T}{T}(\theta, \phi) = \sum_{l=0}^{\infty} \sum_{m=-l}^{l} a_{lm} Y_l^m(\theta, \phi). \tag{10.82}$$

[24] The temperature being real, the expansion could equally be written in terms of $a_{lm}^* Y_l^{*m}$.

These basis functions obey the orthonormality condition

$$\int Y_l^{*m} Y_{l'}^{m'} \sin\theta d\theta d\phi = \delta_{ll'} \delta_{mm'} \tag{10.83}$$

and the addition theorem

$$\sum_m Y_l^{*m}(\hat{n}_1) Y_l^m(\hat{n}_2) = \frac{2l+1}{4\pi} P_l(\cos\theta_{12}), \tag{10.84}$$

where $P_l(\cos\theta)$ is the Legendre polynomial, and $\hat{n}_1$ and $\hat{n}_2$ are two unit vectors pointing in directions with an angular separation θ_{12}. Namely, $\hat{n}_1 \cdot \hat{n}_2 = \cos\theta_{12}$. We display a few samples of the spherical harmonics,

$$Y_0^0 = \left(\frac{1}{4\pi}\right)^{1/2}, \quad Y_1^0 = \left(\frac{3}{4\pi}\right)^{1/2} \cos\theta,$$

$$Y_3^{\pm 1} = \mp \left(\frac{21}{64\pi}\right)^{1/2} \sin\theta (5\cos^2\theta - 1) e^{\pm i\phi}.$$

The multipole number "l" represents the number of nodes (locations of zero amplitude) between equator and poles, while "m" is the longitudinal node number. For a given l, there are $2l+1$ values for m $-l, -l+1, \ldots, l-1, l$. The expansion coefficients a_{lm}, much like the individual amplitudes in a Fourier series, are determined by the underlying density perturbation They can be projected out from the temperature fluctuation by (10.83):

$$a_{lm} = \int Y_l^{*m}(\theta, \phi) \frac{\delta T}{T}(\theta, \phi) \sin\theta d\theta d\phi. \tag{10.85}$$

These multipole moments can only be predicted statistically.

Two-point correlation and the power spectrum Cosmological theories predict only statistical information. The most useful statistic is the two-point correlation. Consider two points at $\hat{n}_1$ and $\hat{n}_2$ separated by θ. We define the correlation function

$$C(\theta) \equiv \left\langle \frac{\delta T}{T}(\hat{n}_1) \frac{\delta T}{T}(\hat{n}_2) \right\rangle_{\hat{n}_1 \cdot \hat{n}_2 = \cos\theta}, \tag{10.86}$$

where the angle brackets denote the averaging over an ensemble of realizations of the fluctuation.[25] The inflationary cosmology predicts that the

[25] In principle it means averaging over many universes. Since we have only one universe, this ensemble averaging is carried out by averaging over multiple moments with different m moments, which in theory should be equal because of spherical symmetry.

fluctuation is Gaussian[26] (i.e. maximally random) and is thus independent of the a_{lm}s. Namely, the multipoles a_{lm} are uncorrelated for different values of l and m:

$$\langle a_{lm} \rangle = 0, \qquad \langle a_{lm}^* a_{l'm'} \rangle = C_l \delta_{ll'} \delta_{mm'}, \tag{10.87}$$

which defines the power spectrum C_l as a measure of the relative strength of spherical harmonics in the decomposition of the temperature fluctuations. The lack of m-dependence reflects the rotational symmetry of the underlying cosmological model. When we plug (10.82) into (10.86), the conditions (10.87) and (10.84) simplify the expansion to

$$C(\theta) = \frac{1}{4\pi} \sum_{l=0}^{\infty} (2l + 1) C_l P_l(\cos \theta). \tag{10.88}$$

Namely, the information carried by $C(\theta)$ in the angular space can be represented by C_l in the space of multipole number l. The power spectrum C_l is the focus of experimental comparison with theoretical predictions. From the map of measured temperature fluctuations, one can extract multipole moments by the projection (10.85) and since we do not actually have an ensemble of universes to take the statistical average, this is estimated by averaging over the a_{lm}s with different ms. Such an estimate will be uncertain by an amount inversely proportional to the square-root of the number of samples

$$\left\langle \left(\frac{\delta C_l}{C_l} \right)^2 \right\rangle^{1/2} \propto \sqrt{\frac{1}{2l + 1}}. \tag{10.89}$$

The expression also makes it clear that the variance will be quite significant for low multiple moments when we have only a very small number of samples. This is referred to as the "cosmic variance problem" (cf. Fig. 11.13).

In the next chapter we shall present the basic features of the power spectrum: to show how it can be used to measure the curvature of space, to test different theories of the origin of the cosmic structure that we see today, and to extract many cosmological parameters.

Review questions

1. Describe the relation of the Friedmann equation (10.1) and the Einstein equation, as well as give its Newtonian interpretation. Why can we use nonrelativistic Newtonian theory to interpret the general relativistic equation in cosmology? Also, in what sense is it only quasi-Newtonian?

2. In what sense can the critical density be understood as akin to the more familiar concept of escape velocity?

3. Why do we expect the energy density of radiation to scale as a^{-4}. Why should the energy of the universe be radiation-dominated in its earliest moments?

4. What is the equation of state parameter w for radiation? For nonrelativistic matter? If we know that the universe has a flat geometry, what is the time dependence of the scale factor $a(t)$ in a radiation-dominated universe (RDU)? In a

matter-dominated universe (MDU)? How is the age of the universe t_0 related to the Hubble time t_H in a RDU, and in a MDU? Justify the approximation that the age of our universe is two-thirds of the Hubble time.

5. Draw a schematic diagram showing the behavior of the scale factor $a(t)$ for various values of k in cosmological models (with zero cosmological constant). (It is suggested that all $a(t)$ curves be drawn to meet at the same point $a(t_0)$ with the same slope $\dot{a}(t_0)$.) Also mark the region corresponding to a decelerating universe, an accelerating universe and an empty universe.

6. Give an argument for the scaling behavior of the radiation temperature: $T \backsim a^{-1}$. Show that under such a scaling law, the spectrum distribution of the blackbody radiation is unchanged as the universe expands.

7. What is the condition (called the Gamow condition) for any particular set of interacting particles being in thermal equilibrium during the various epochs of the expanding universe?

8. The cosmic helium synthesis corresponds to the reaction of combining two protons and two neutrons into a helium nucleus and the Boltzmann distribution at thermal energy of the order of MeV yielding a neutron to proton number density ratio of $n_n/n_p \simeq 1/7$. From this how would you estimate the cosmic helium mass fraction?

9. How can one use the theory of big bang nucleosynthesis and the observed abundance of light elements such as deuterium to deduce the baryon number density Ω_B and that the number of neutrino flavors should be three? (The reader is not asked why there are three neutrino flavors, but how the astrophysical observation is only compatible with the neutrino number being three.)

10. What physics process took place around the photon decoupling time t_γ? What are the average thermal energy and temperature at t_γ? Knowing the redshift $z_\gamma \simeq 10^3$, calculate the expected photon temperature now.

11. What is the cosmic time when the universe made that transition from a radiation-dominated to a matter-dominated system. How does it compare to the nucleosynthesis time, and photon decoupling times?

12. Give the argument that relates the matter-antimatter asymmetry in the early universe to the baryon-to-photon ratio now ($\simeq 10^{-9}$).

13. Why would the peculiar motion of our galaxy show up as CMB dipole anisotropy?

14. Besides the dipole anisotropy, how does the CMB temperature anisotropy reflect the origin of cosmic structure?

15. What is the "cosmic variance" of the CMB power spectrum?

Problems

10.1 **Friedmann equations and energy conservation** Show that a linear combination of these two Friedmann equations (10.1) and (10.2) leads to Eq. (10.3).

10.2 **Newtonian interpretation of the second Friedmann equation** Adopting the same approach used in the Newtonian "derivation" of Eq. (10.1) in the text, interpret the second Friedmann equation (10.2) as the $F = ma$ equation of the system.

10.3 **Friedmann equation for a multi-component universe** Show that the Friedmann equation for a multi-component universe may be written as

$$\dot{a}^2 + \frac{kc^2}{R_0^2} = \frac{8\pi G_N}{3} \sum_w \rho_{w,0} a^{-(1+3w)}$$

where w is the equation of state parameter defined in (10.4).

10.4 **The empty universe** A low density universe may be approximated by setting the density function in the Fried-

mann equation to zero,

$$\dot{a}^2 = -\frac{kc^2}{R_0^2}.$$

Besides the uninteresting possibility of $\dot{a} = k = 0$ for a static universe with a Minkowski spacetime, show that the nontrivial solution to this equation is a negatively curved open universe, represented by the straight-line $a(t)$ in Fig. 10.2. Find the Hubble relation between the proper distance and redshift in such a model universe.

10.5 **Hubble plot in a flat matter-dominated universe** We explained at the end of Chapter 9 that the Hubble diagram is usually a plot of the distance modulus vs. redshift. Find this relation for a matter-dominated universe with a flat spatial geometry $k = 0$.

10.6 **Another calculation of photon density** Give a direct estimate of the thermal photon number density from the estimate that at a cosmic era with redshift $z_\gamma \simeq 1100$ the average photon energy was $\bar{u} \simeq 0.26 \, \text{eV}$.

10.7 Distance to a light emitter at redshift z   Given the time dependence of the scale factor as in (10.26) $a(t) = (t/t_0)^x$, use (9.47) to calculate the proper distance $d_p(t)$ between a light emitter (at redshift z) and receiver (at t_0). Check your result with the MDU case worked out in Problem 10.5.

10.8 Scaling behavior of number density and Hubble's constant

(a) Show that the number densities for matter and radiation both scale with the redshift as

$$\frac{n(t)}{n_0} = (1+z)^3 .$$

(b) From the Friedmann equation, show that Hubble's constant $H(t)$ scales as $H^2 = \Omega_{M,0}(1 + z)^3 H_0^2$ in a matter-dominated flat universe, and as $H^2 = \Omega_{R,0}(1+z)^4 H_0^2$ in a radiation-dominated flat universe.

10.9 Radiation and matter equality time   Knowing that the photon decoupling epoch corresponds to a redshift of $z_\gamma = 1100$ and a cosmic time $t_\gamma \simeq 360\,000$ years, convert the radiation-matter equality redshift $z_{RM} \simeq 8800$ from (10.68) to the corresponding cosmic time t_{RM}.

10.10 Density and deceleration parameter   In Problem 9.11 we introduce the deceleration parameter q_0. Use the second Friedmann equation (10.2) and the equation of state

parameter w of (10.4) to show that

$$q_0 = \frac{1}{2} \sum_i \Omega_{i,0} (1 + 3w_i) .$$

In particular in a matter-dominated flat universe $q_0 = +1/2$.

10.11 Temperature and redshift   Knowing how the temperature scales, show that we can also connect $T(z)$ at an epoch to the corresponding redshift z to T_0 at the present era:

$$T = T_0 (1 + z) . \tag{10.90}$$

10.12 Radius of the universe   Show that the radius R_0 of the universe is related to the density parameter Ω_0 and the Hubble constant H_0 by

$$R_0 = \frac{c}{H_0 \sqrt{\Omega_0 - 1}} .$$

10.13 Cosmological limit of neutrino mass   Given that the density ratio of dark matter is $\Omega_{DM} = \Omega_M - \Omega_B \simeq 0.21$, what limit can be obtained for the average mass of neutrinos (average over all neutrino flavors)?

10.14 Temperature dipole anisotropy as Doppler effect   Show that the Doppler effect implies that an observer moving with a nonrelativistic velocity $\mathbf{v}$ through an isotropic CMB would see a temperature dipole anisotropy of

$$\frac{\delta T}{T} (\theta) = \frac{v}{c} \cos \theta$$

where θ is the angle from the direction of the motion.

Inflation and the accelerating universe

- Einstein introduced the cosmological constant in his field equation so as to obtain a static universe solution.
- The cosmological constant is the vacuum energy of the universe: This constant energy density corresponds to a negative pressure, giving rise to a repulsive force that increases with distance. A vacuum-energy dominated universe expands exponentially.
- The inflationary theory of cosmic origin—the universe had experienced a huge expansion at the earliest moment of the big bang—can provide the correct initial conditions for the standard FLRW model of cosmology: solving the flatness, horizon problems, and providing an origin of matter/energy, as well as giving just the right kind of density perturbation for subsequent structure formation.
- The primordial inflation leaves behind a flat universe, which can be compatible with the observed matter density being less than the critical density and a cosmic age greater than 9 Gyr if there remains a small but nonvanishing cosmological constant—a dark energy. This would imply a universe now undergoing an accelerating expansion.
- The measurement of supernovae at high redshift provided direct evidence for an accelerating universe. Such data, together with other observational results, especially the anisotropy of the cosmic microwave background and large structure surveys, gave rise to a concordant cosmological picture of a spatially flat universe $\Omega = \Omega_\Lambda + \Omega_M = 1$, dominated by dark energy $\Omega_\Lambda \simeq 0.75$. Most of the matter $\Omega_M \simeq 0.25$ is exotic dark matter $\Omega_{DM} \simeq 0.21$, compared to the ordinary (baryonic) matter $\Omega_B \simeq 0.04$ The cosmic age $t_0 \simeq 14\,\text{Gyr}$ comes out to be close to the Hubble time.
- The cosmological constant and the cosmic coincidence problems point to the need for new fundamental physics.

As we have discussed in Sections 9.1.3 and 10.2, Newton's and the original Einstein's equations would lead us to expect the expansion of the universe to slow down because of gravitational attraction. In this chapter, we shall see how a modification of the Einstein equation, with the introduction of the **cosmological constant** Λ, allows for the possibility of a gravitational repulsive force that increases with distance. This effect was first discovered by Einstein in his effort of seeking a static solution to the GR field equation. It also allows for the possibility that the universe had undergone an extraordinarily rapid expansion at an early moment (the inflationary epoch). The inflationary scenario of the big bang brings about just the correct initial conditions for the then standard cosmology (the FLRW model of Box. 11.1) and predicts a flat geometry for the universe at large. Finally, a nonvanishing Λ term can account for the recently discovered evidence of an accelerating universe in the present epoch. An accelerating expansion means slower expansion in the past, hence a longer age for the naively expected decelerating universe— long enough to account for the oldest objects observed in the universe. The cosmological constant also provides us with a dark energy that, together with the observed matter content, fulfills the inflationary cosmology's prediction of a flat universe, which requires the mass/energy density of the universe to be equal to the critical density.

11.1   The cosmological constant

Before Hubble's discovery in 1929 of an expanding universe, just about everyone, Einstein included, believed that we lived in a static universe. Recall that the then-observed universe consisted essentially of stars within the Milky Way galaxy. But gravity, whether nonrelativistic or relativistic, is a universal attraction. Hence, theoretically speaking, a static universe is an impossibility. Specifically, as we have demonstrated, the Friedmann cosmological equations (10.1) and (10.2) have solutions corresponding always to a **dynamic** universe— a universe which is either contracting or expanding. Namely, these equations are not compatible with the static condition of an unchanging scale factor $\dot{a} = \ddot{a} = 0$, which would lead to a trivial empty universe,[1] $\rho = p = 0$.

[1] For the Einstein equation without a cosmological constant, a static solution necessarily corresponds to an empty universe. On the other hand, an empty universe is compatible with an expanding universe with negative spatial curvature. See Problem 10.4.

Λ **as a modification of the geometry side**   Recall our brief discussion in Section 6.3.2 of the GR field equation $G_{\mu\nu} = \kappa T_{\mu\nu}$ with $\kappa = -8\pi c^{-4} G_{\mathrm{N}}$. The Einstein tensor $G_{\mu\nu}$ on the left-hand side (LHS) is the curvature of spacetime and $T_{\mu\nu}$ on the right-hand side (RHS), the energy–momentum source term for gravity (the curved spacetime). The goal of obtaining a static universe from general relativity (GR) led Einstein to alter his field equation to make it contain a repulsion component. This could, in principle, balance the usual gravitational attraction to yield a static cosmic solution. Einstein discovered that the geometry side of his field equation could naturally accommodate an additional term. As will be discussed in Section 14.4.3, the simplest term that is mathematically compatible with Einstein's field equation (6.37) is the metric tensor $g_{\mu\nu}$,

$$G_{\mu\nu} - \Lambda g_{\mu\nu} = \kappa T_{\mu\nu}. \tag{11.1}$$

Such a modification will, however, alter its nonrelativistic limit to differ from Newton's equation. In order that this alteration is compatible with known phenomenology, it must have a coefficient Λ so small as to be unimportant in all situations except on truly large cosmic scales. Hence, this additional constant Λ has come to be called the **cosmological constant**.

Λ as a vacuum energy momentum contribution   While we have introduced this term as an additional geometric term, we could just as well move it to the RHS of the equation and view it as an additional source term of gravity. In particular, when the regular energy–momentum is absent $T_{\mu\nu} = 0$ (the vacuum state),

$$G_{\mu\nu} = \Lambda g_{\mu\nu} \equiv \kappa T_{\mu\nu}^{\Lambda}$$

where $T_{\mu\nu}^{\Lambda} = \kappa^{-1}\Lambda g_{\mu\nu} = (-c^4\Lambda/8\pi G_{\mathrm{N}})g_{\mu\nu}$ can be interpreted as the energy–momentum tensor of the vacuum.[2] Just as the $T_{\mu\nu}$ for ordinary radiation and matter depends on two functions of energy density ρ and pressure p, this vacuum-energy–momentum tensor $T_{\mu\nu}^{\Lambda}$ can be similarly parametrized by "**vacuum-energy density**" ρ_Λ and "**vacuum pressure**" p_Λ. As we shall demonstrate in Section 14.4.3 (after we have properly studied the energy–momentum tensor in Section 12.3), these two quantities are related to a positive cosmological constant Λ as follows: the vacuum energy per unit volume,

$$\rho_\Lambda = \frac{\Lambda c^2}{8\pi G_{\mathrm{N}}} > 0, \tag{11.2}$$

is a constant[3] (in space and in time) and the corresponding vacuum pressure,

$$p_\Lambda = -\rho_\Lambda c^2 < 0, \tag{11.3}$$

is negative, corresponding to an equation-of-state parameter $w = -1$ as defined in Eq. (10.4). Such a density and pressure, as we shall presently show, are compatible with basic physics principles, and, most relevant for our cosmological discussion, they give rise to a gravitational repulsion.

Λ as constant energy density and negative pressure   What is a negative pressure? Consider the simple case of a piston chamber filled with ordinary matter and energy, which exerts a positive pressure by pushing out against the piston. If it is filled with this Λ energy, Fig. 11.1, it will exert a negative pressure by **pulling in** the piston. Physically this is sensible because, as its energy per unit volume $\rho_\Lambda c^2$ is a constant, the change in system's energy is strictly proportional to its volume change $dE = \rho_\Lambda c^2 dV$. The system would like to lower its energy by volume-contraction (pulling in the piston). When we increase the volume of the chamber $dV > 0$ (hence its energy $dE > 0$) by pushing out the piston, we have to do positive work to overcome the pulling by the Λ energy. Energy conservation is maintained in such a situation because the negative pressure $p < 0$ is just what is required by the first law of thermodynamics: $dE = -pdV$ when both dE and dV have the same sign. In fact the first law also makes it clear that if the energy density

[2]It is appropriate that the vacuum energy momentum tensor is proportional to the metric $g_{\mu\nu}$, which is Lorentz invariant in local inertial coordinates. This must be the case as such a $T_{\mu\nu}^{\Lambda}$ should not pick out any preferred direction.

[3]In nonrelativistic physics only the relative value of energy is meaningful—the motion of a particle with potential energy $V(x)$ is exactly the same as one with $V(x) + C$, where C is a constant. In GR, since the whole energy–momentum tensor is the source of gravity, the actual value of energy makes a difference.

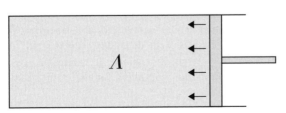

Fig. 11.1 The Λ energy in a chamber has negative pressure and pulls in the piston.

is a constant $dE = \rho c^2 dV$ so that the dV factors cancel from both sides, the pressure must equal the negative of the energy density $p = -\rho c^2$ as shown in (11.3).

11.1.1   Vacuum energy as source of gravitational repulsion

To see that the negative pressure can give rise to a repulsive force, let us first discuss the Newtonian limit of the Einstein equation with a general source, composed of mass density ρ as well as pressure p (as is the case for a cosmology with an ideal fluid as the source). It can be shown (see Box 14.1 for details) that the limiting equation, written in terms of the gravitational potential Φ, is

$$\nabla^2 \Phi = 4\pi G_N \left(\rho + 3\frac{p}{c^2}\right) = 4\pi G_N \left(1 + 3w\right)\rho. \tag{11.4}$$

This informs us that not only mass, but also pressure, can be a source of gravity. For the nonrelativistic matter having a negligible pressure term, we recover the familiar equation (6.36) of Newton.

Explicitly displaying the contributions from ordinary matter and vacuum energy (thus density and pressure each have two parts: $\rho = \rho_M + \rho_\Lambda$ and $p = p_M + p_\Lambda$), the Newton/Poisson equation (11.4) becomes

$$\nabla^2 \Phi = 4\pi G_N \left(\rho_M + 3\frac{p_M}{c^2} + \rho_\Lambda + 3\frac{p_\Lambda}{c^2}\right)$$

$$= 4\pi G_N \rho_M - 8\pi G_N \rho_\Lambda = 4\pi G_N \rho_M - \Lambda c^2, \tag{11.5}$$

where we have used (11.3), $p_\Lambda = -\rho_\Lambda c^2$, and set $p_M = 0$ because $\rho_M c^2 \gg p_M$. For the vacuum-energy dominated case of $\Lambda c^2 \gg 4\pi G_N \rho_M$, the Poisson equation can be solved (after setting the potential to zero at the origin) by

$$\Phi_\Lambda (r) = -\frac{\Lambda c^2}{6} r^2. \tag{11.6}$$

Between any two mass points, this potential corresponds to a repulsive force (per unit mass) that increases with separation r,

$$\vec{g}_\Lambda = -\vec{\nabla} \Phi_\Lambda = +\frac{\Lambda c^2}{3} \vec{r}, \tag{11.7}$$

in contrast to the familiar $-\vec{r}/r^3$ gravitational attraction. With this pervasive repulsion that increases with distance, even a small Λ can have a significant effect on truly large dimensions. It would be possible to counteract the gravitational attraction and allow for the static solution sought by Einstein.

11.1.2 Einstein's static universe

We now consider the Friedmann equations (10.1) and (10.2) with a nonvanishing cosmological constant,

$$\frac{\dot{a}^2 + kc^2/R_0^2}{a^2} = \frac{8\pi G_N}{3}(\rho_M + \rho_\Lambda), \tag{11.8}$$

$$\frac{\ddot{a}}{a} = -\frac{4\pi G_N}{c^2}\left[(p_M + p_\Lambda) + \frac{1}{3}(\rho_M + \rho_\Lambda)c^2\right]. \tag{11.9}$$

The RHS of (11.9) need not necessarily be negative because of the presence of the negative pressure term $p_\Lambda = -\rho_\Lambda c^2$. Consequently, a decelerating universe is no longer the inevitable outcome. For nonrelativistic matter ($p_M = 0$), we have

$$\frac{\ddot{a}}{a} = -\frac{4\pi G_N}{3}(\rho_M - 2\rho_\Lambda). \tag{11.10}$$

The static condition of $\ddot{a} = 0$ now leads to the constraint:

$$\rho_M = 2\rho_\Lambda = \frac{\Lambda c^2}{4\pi G_N}. \tag{11.11}$$

That is, the mass density ρ_M of the universe is fixed by the cosmological constant. The other static condition of $\dot{a} = 0$ implies, through (11.8), the static solution $a = a_0 = 1$

$$\frac{kc^2}{R_0^2} = 8\pi G_N\rho_\Lambda = \Lambda c^2. \tag{11.12}$$

Since the RHS is positive, we must have

$$k = +1. \tag{11.13}$$

Namely, the static universe has a positive curvature (a closed universe) and finite size. The "radius of the universe" is also determined, according to (11.12), by the cosmological constant:

$$R_0 = \frac{1}{\sqrt{\Lambda}}. \tag{11.14}$$

Thus, the basic features of such a static universe, the density and size, are determined by the arbitrary input parameter Λ. Not only is this a rather artificial arrangement, but also the solution is, in fact, unstable. That is, a small variation will cause the universe to deviate from this static point. A slight increase in the separation will cause the gravitational attraction to decrease and repulsion to increase, causing the system to deviate further from the initial point. A slight decrease in the separation will increase the gravitational attraction to cause the separation to decrease further, until the whole system collapses.

Box 11.1 Some historical tidbits of modern cosmology

FLRW cosmology   The Friedmann equations with both ordinary and vacuum energies (11.8) and (11.9) are sometimes call the Friedmann–Lemaître equations. That Einstein's equation had expanding, or contracting, solutions was first pointed out in the early 1920s by A.A. Friedmann. His fundamental contribution to cosmology was hardly noticed by his contemporaries.[4] It was to be rediscovered later by the Belgian civil engineer and priest Georges Lemaître, who published in 1927 his model of cosmology with a contribution coming from both ρ_M and ρ_Λ. More importantly, Lemaître was the first one, having been aware of Hubble's work through his contact with Harvard astronomers (he spent three years studying at Cambridge University and MIT), to show that the linear relation between distance and redshift (Hubble's law) follows from such cosmological considerations. The original derivations by Friedmann and Lemaître were somewhat awkward. Modern presentations have mainly followed the approach initiated by Howard Percy Robertson and Arthur G. Walker. Thus the framework using Einstein's equation for a homogeneous and isotropic universe has come to be known as the FLRW (Friedmann–Lemaître–Robertson–Walker) cosmological model.

Einstein's greatest blunder?   Having missed the chance of predicting an expanding universe before its discovery, Einstein came up with a solution which did not really solve the perceived difficulty. (His static solution is unstable.) It had often been said that later in life Einstein considered the introduction of the cosmological constant to be "the biggest blunder of his life!" This originated from a characterization by George Gamow in his autobiography (Gamow, 1970):

Thus, Einstein's original gravity equation was correct, and changing it was a mistake. Much later, when I was discussing cosmological problems with Einstein, he remarked that the introduction of the cosmological term was the biggest blunder he ever made in his life.

Then Gamow went on to say,

But this blunder, rejected by Einstein, is still sometimes used by cosmologists even today, and the cosmological constant Λ rears its ugly head again and again and again.

What we can conclude for sure is that Gamow himself considered the cosmological constant 'ugly' (because this extra term made the field equation less simple). Generations of cosmologist kept on including it because the quantum vacuum energy gives rise to such a term (cf. Section 11.7.1) and there was no physical principle one could invoke to exclude this term. (If it is not forbidden, it must exist!) In fact the discovery of the cosmological constant as the source of a new cosmic repulsive force must be regarded as one of Einstein's great achievements.[5] Now, as we shall see, the idea of a nonzero cosmological constant was the key in solving a number of

---

[4]On the other hand, the comoslogical model with only ρ_Λ, studied by the Dutch astronomer W. de Sitter soon after Einstein's 1917 paper, was widely discussed.

[5]One can speculate that, if there were regret on Einstein's part, it would be the missed opportunity of predicting the expanding universe before its observational discovery.

fundamental problems in cosmology. That is, Einstein taught us the way to bring about gravitational repulsion. Although the original goal of a static universe solution was misguided, this "tool" of the cosmological constant (a repulsive force) was needed to account for the explosion that was the big bang (inflationary epoch), and was needed to explain how the expansion of the universe could accelerate.

11.2 The inflationary epoch

The standard model of cosmology (the FLRW model) has been very successful in presenting a self-contained picture of the evolution and composition of the universe: how the universe expanded and cooled after the big bang; how the light nuclear elements were formed; after the inclusion of the proper density inhomogeneity, how in an expanding universe matter congealed to form stars, galaxies, and clusters of galaxies. It describes very well the aftermath of the big bang. However, the model says very little about the nature of the big bang itself: how did this "explosion of the space" come about? It assumes that all matter existed from the very beginning. Furthermore, it must assume certain very precise initial conditions (see the flatness and horizon problems discussed later) that just clamor for an explanation.

The inflationary cosmology is an attempt to give an account of this big bang back to an extremely short instant (something like 10^{-36} s) after the $t = 0$ cosmic singularity.[6] During this primordial inflation, the universe had a burst of expansion during which the scale factor increased by more than 30 orders of magnitude, see Fig. 11.2. In this inflationary process, all the matter and energy could have been created virtually from nothing. Afterwards, the universe followed the course of adiabatic expansion and cooling as described

[6]This is to be compared to the even earlier period, comparable to the Planck time $t_{\rm Pl} = O(10^{-43}$ s), when quantum gravity is required for a proper description. See Section 8.5.1.

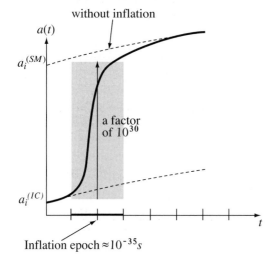

Fig. 11.2 Comparison of the scale factor's time evolution. The standard FLRW model curves are represented by dashed lines; the solid curve is that of the inflation model which coincides with the standard model curve after 10^{-35} s. The intercepts on the a-axis correspond respectively to the initial scales: $a_i^{(SM)}$ in the standard model (without inflation) and $a_i^{(IC)}$ in the inflation cosmology.

by the FLRW cosmology (cf. Chapter 10). Figure 11.2 also makes it clear that in the inflationary scenario, the observable universe originates from an entity some 10^{-30} times smaller than that which would have been the size in the case without inflation.

11.2.1 Initial conditions for the FLRW cosmology

The standard FLRW model requires a number of seemingly unnatural fine-tuned initial conditions. As we shall see, they are just the conditions that would follow from an inflationary epoch. We start the discussion of initial conditions by listing two such theoretical difficulties, two "problems."

The flatness problem

Because of the gravitational attraction among matter and energy, we would expect the expansion of the universe to slow down. This deceleration $\ddot{a}(t) < 0$ means that $\dot{a}(t)$ must be a decreasing function. This is exemplified by the specific case of a radiation-dominated universe $a \sim t^{1/2}$, thus $\dot{a} \sim t^{-1/2}$, or a matter-dominated universe $a \sim t^{2/3}$, and $\dot{a} \sim t^{-1/3}$, as derived in (10.30). Recall that the Friedmann equation can be written in terms of the mass density parameter Ω as in (10.7):

$$1 - \Omega(t) = \frac{-kc^2}{\dot{a}(t)^2 R_0^2}. \tag{11.15}$$

This displays the connection between geometry and matter/energy: if $k = 0$ (a flat geometry), we must have the density ratio $\Omega = 1$ exactly; when $k \neq 0$ for an universe having curvature, then $|1 - \Omega(t)|$ must be **ever-increasing** because the denominator on the RHS is ever decreasing. Thus, the condition for a flat universe $\Omega = 1$ is an **unstable equilibrium point**—if Ω ever deviates from 1, this deviation will increase with time. Or, we may say: gravitational attraction always enhances any initial curvature. In light of this property, it is puzzling that the present mass density Ω_0 has been found observationally (see Section 9.2) to be not too different from the critical density value $(1 - \Omega_0) = O(1)$. This means that Ω must have been extremely close to unity (an extremely flat universe) in the cosmic past. Such a fine-tuned initial condition would require an explanation.

We can make this statement quantitatively. Ever since the radiation–matter equality time $t > t_{RM}$, with $z_{RM} = O(10^4)$, cf. (10.68) the evolution of the universe has been dominated by nonrelativistic matter: $a(t) \sim t^{2/3}$ or $\dot{a} \sim t^{-1/3} \sim a^{-1/2}$. We can then estimate the ratio in (11.15):

$$\frac{1 - \Omega(t_{RM})}{1 - \Omega(t_0)} = \left[\frac{\dot{a}(t_{RM})}{\dot{a}(t_0)}\right]^{-2} = \left[\frac{a_{RM}}{a_0}\right]$$

$$= (1 + z_{RM})^{-1} = O(10^{-4}). \tag{11.16}$$

Successful prediction of light element abundance by primordial nucleosynthesis gave us direct evidence for the validity of the FLRW model of cosmology back to the big bang nucleosynthesis time $t_{bbn} = O(10^2 \, \text{s})$. The time evolution for $t < t_{RM}$ was radiation dominated: $a(t) \sim t^{1/2}$ or $\dot{a} \sim t^{-1/2} \sim a^{-1}$.

This would then imply

$$\frac{1 - \Omega(t_{\text{bbn}})}{1 - \Omega(t_{\text{RM}})} = \left[\frac{\dot{a}(t_{\text{bbn}})}{\dot{a}(t_{\text{RM}})}\right]^{-2} = \left[\frac{a(t_{\text{bbn}})}{a(t_{\text{RM}})}\right]^{2}$$

$$= \left[\frac{k_B T_{\text{bbn}}}{k_B T_{\text{RM}}}\right]^{-2} \simeq O(10^{-11}), \qquad (11.17)$$

where we have used the scaling behavior of the temperature, and (10.53) $k_B T_{\text{bbn}} = O(\text{MeV})$ and (10.69) $k_B T_{\text{RM}} = O(2\,\text{eV})$ to reach the last numerical estimate. Thus, in order to produce a $(\Omega_0 - 1) = O(1)$ now, the combined result of (11.16) and (11.17) tells us that one has to have at the epoch of primordial nucleosynthesis a density ratio equal to unity to an accuracy of one part in 10^{15}. Namely, we must have $\Omega(t_{\text{bbn}}) - 1 = O(10^{-15})$. That the FLRW cosmology requires such an unnatural initial condition constitutes the flatness problem.

The horizon problem

Our universe is observed to be very homogeneous and isotropic. In fact, we can say that it is "too homogeneous and isotropic." Consider two different parts of the universe that are outside of each other's horizons. They are so far apart that no light signal sent from one at the beginning of the universe could have reached the other. Yet they are observed to have similar properties. This suggests their being in thermal contact sometime in the past. How can this be possible?

This horizon problem can be stated most precisely in terms of the observed isotropy of the CMB radiation (up to one part in $100\,000$, after subtracting out the dipole anisotropy due to the peculiar motion of our Galaxy). When pointing our instrument to measure the CMB, we obtain the same blackbody temperature in all directions. However, every two points in the sky with an angular separation on the order of a degree actually correspond to a horizon separation back at the photon-decoupling time t_γ, see (11.31). The age of the universe at the photon decoupling time was about $360\,000$ years, yet the observed isotropy indicates that regions far more than the horizon distance $360\,000$ light-year apart were strongly correlated. This is the horizon problem of the standard FLRW cosmology.

Initial conditions required for the standard cosmic evolution

We have discussed the horizon problem and flatness problem, etc. as the shortcomings of the standard big bang model. Nevertheless, it must be emphasized that they are not contradictions since we could always assume that the universe had just these conditions initially to account for the observed universe today. For example, the horizon problem can be interpreted simply as reflecting the fact that the universe must have been very uniform to begin with. These "problems" should be viewed as informing us of the correct initial conditions for the cosmic evolution after the big bang: "The initial conditions must be **just so**." What we need is a theory of the initial conditions. Putting it in another way, the FLRW model is really a theory for the evolution of the universe **after** the big bang. We now need a theory of the big bang **itself**. A correct theory should have the feature that it would automatically leave behind a universe with just these desired conditions.

11.2.2   The inflation scenario

The initial condition problems can be solved if, in the early moments, the universe had gone through an epoch of extraordinarily rapid expansion. This can solve the flatness problem, as any initial curvature could be stretched flat by the burst of expansion, and can solve the horizon problem if the associated expansion rate could reach superluminal speed. If the expansion rate could be greater than the light speed, then one horizon volume could have been stretched out to such a large volume that corresponded to many horizon volumes after this burst of expansion. This rapid expansion could happen if there existed then a large cosmological constant Λ, which could supply a huge repulsion to the system. The question is, then, what kind of physics can give rise to such a large Λ? In this section, we explain how modern particle physics can suggest a possible mechanism to generate, for a short instant of time, such a large vacuum energy.

False vacuum, slow rollover phase transition and an effective Λ

Here we discuss the possibility of a field system that can give rise to an effective cosmological constant Λ_{eff} that then brings about an explosion of the space.

Hierarchy of particle physics unification and the invention of inflationary cosmology   The inflationary cosmology was invented in 1980 by Alan Guth in his study of the cosmological implications of the grand unified theories (GUTs) of particle interactions. The basic idea of a GUT is that particle interactions possess certain symmetry.[7] As a result, all the fundamental forces (the strong, weak, and electromagnetic interactions, except for gravity) behave similarly at high energy. In fact they are just different aspects of the same (unified) interaction like the different faces of the same die. However, the structure of the theory is such that there is a phase transition at a temperature corresponding to the grand unification energy scale, around 10^{15}–10^{16} GeV. In the energy regime higher than this scale, the system is in a symmetric phase and the unification of particle interactions is manifest (i.e. all interactions behave similarly); when the universe cooled below this scale, the particle symmetry became hidden, showing up as distinctive forces.[8]

Higgs phenomenon in field theory   In quantum field theory, particles are quantum excitations of their associated fields: electrons of the electron field, photons of the electromagnetic field, etc. New fields are postulated to exist, related to yet to be discovered particles. What brings about the above-mentioned spontaneous symmetry breaking and its associated phase transition is the existence of a certain spin-zero field $\phi(x)$, called the Higgs field (its quanta being Higgs particles). Such a field, just like the familiar electromagnetic field, carries energy. What is special about a Higgs field is that it possesses a potential energy density function $V(\phi)$ much like the potential energy function in the ferromagnet example of Section 11.6.

[7] "Particle interaction symmetry" has the same meaning as "symmetry in particle physics" as explained in Chapter 1: physics equations are unchanged under some transformation. However, instead of transformations of space and time coordinates as in relativity, here one considers transformations in some "internal charge space." The mathematical description of symmetry is group theory. An example of a grand unification group is $SU(5)$ and particles form multiplets in this internal charge space. Members of the same multiplet can be transformed into each other: electrons into neutrinos, or into quarks, and the GUT physics equations are covariant under such transformations. After spontaneous symmetry breaking, the interactions possess less symmetry: for example, $SU(5)$ is reduced down to $SU(3) \times SU(2) \times U(1)$, which is the symmetry group of the low energy effective theory known as the Standard Model of quantum chromodynamics and electroweak interactions.

[8] For a discussion of spontaneous symmetry breakdown, that is, hidden symmetry, as illustrated by spontaneous magnetization of a ferromagnet, see Section 11.6 (Appendix C).

Normally one would expect field values to vanish in the vacuum state (the state with the lowest energy). A Higgs field, surprisingly, can have a nonzero vacuum state field permeating throughout space, cf. Fig. 11.3 (a) and (b). The effect of this hidden symmetry can then spread to other particles through their interaction of the Higgs field. For example, a massless particle can gain its mass when propagating in the background of such a Higgs field. Different Higgs fields are posited to exist. Here we are referring to the Higgs particles[9] in GUTs, which may have a mass $O(10^{16} \text{ GeV}/c^2)$.

Slow rollover from a false vacuum gives rise to Λ   In the cosmological context, such a postulated field is simply referred to as the **inflation field**, or **inflation/Higgs field**. In order to have a large Λ_{eff} over an interval long enough to produce the desired initial conditions for the FLRW cosmology, it was suggested that parameters of the unified theory were such that the potential energy function of the inflation field had a very small slope around the $\phi = 0$ origin as in Fig. 11.3(c). As the universe cools, the temperature dependent parameters change so that the potential energy function changes from Fig. 11.15(a) to (b). The prior lowest energy point at zero field value became a local maximum and the system would rollover to the new asymmetric vacuum state where the Higgs field would have a nonvanishing vacuum value. But the parameters are such that this rollover was slow. During this transition, we could regard the system, compared to the true (asymmetric) vacuum state, as having an extra energy density. We say the system (i.e. the universe) was temporarily in a **false vacuum**. Having this vacuum-energy density, which is time and position independent, the universe effectively had a large cosmological constant.

Exponential expansion in a vacuum-energy dominated universe

Let us consider the behavior of the scale factor $a(t)$ in a model with $\Lambda > 0$ when the matter density can be ignored. In such a vacuum-energy dominated situation, the expansion rate $\dot{a}(t)$ is so large, cf. (11.22) that we can always ignore the curvature term in Eq. (11.8):

$$\frac{\dot{a}^2}{a^2} = \frac{8\pi G_N}{3} \rho_\Lambda = \frac{\Lambda c^2}{3}. \tag{11.18}$$

Thus $\dot{a}$ is proportional to the scale factor a itself. Namely, we have the familiar rate equation. It can be solved to yield an exponentially expanding universe (called the **de Sitter universe**):

$$a(t) \equiv a(t_1) e^{(t-t_1)/\Delta\tau} \tag{11.19}$$

with the time constant

$$\Delta\tau = \sqrt{\frac{3}{\Lambda c^2}} = \sqrt{\frac{3}{8\pi G_N \rho_\Lambda}}, \tag{11.20}$$

where we have expressed the cosmological constant in terms of the vacuum-energy density $\rho_\Lambda c^2$ as in (11.2). Physically we can understand this exponential expansion result because the repulsive expansion is self-reinforcing: as the energy density ρ_Λ is a constant, the more the space expands, the greater is the

[9]These Higgs particles should not be confused with the Standard Model's electroweak Higgs particle, thought to have a mass on the order of $10^2 \text{ GeV}/c^2$, which is responsible for giving masses to electrons and quarks as well as the W and Z bosons that mediate weak interactions. Our discussion of the inflation scenario is couched in the language of the grand unified Higgs field. It should be understood the grand unified theories themselves have not been verified experimentally in any detail because its intrinsic energy scale of 10^{16} GeV is so much higher than the highest energy $\approx 10^3$ GeV reachable by our accelerators. On the other hand, we are confident that some version of grand unification is correct, as the simplest GUTs can already explain several puzzles of the Standard Model of particle physics, such as why the strong interaction is strong, the weak interaction weak, and why the quarks and leptons have the charges that they do. Nevertheless, the connection between grand unification and inflation cosmology has remained only as a suggestive possibility. It was our knowledge of the grand unification theory that allowed the construction of a physically viable scenario that could give rise to an inflationary epoch. But what precisely is the inflation field, and what parameters actually govern its behavior remain as topics of theoretical discussion. The remarkable fact is that some reasonable speculation of this type can already lead to the resolution of many cosmological puzzles, and have predictions that have been consistently checked with observation.

Fig. 11.3 Potential energy function of a Higgs field $\phi(x)$ is illustrated by the simple case of $V(\phi) = \alpha\phi^2 + \lambda\phi^4$, possessing a discrete symmetry $V(-\phi) = V(\phi)$. The parameter α has temperature-dependence, for example, $\alpha = \alpha_0(T - T_c)$, with positive constants α_0 and λ. (a) Above the critical temperature ($T > T_c$, hence $\alpha > 0$), we have the normal case of the lowest energy state (the vacuum) being at $\phi_0 = 0$, which is symmetric under $\phi \to -\phi$. (b) Below T_c (hence $\alpha < 0$), the symmetric $V(\phi)$ has the lowest energy at points $\phi_{\pm} = \pm\sqrt{-\alpha/2\lambda}$ while $V(\phi = 0)$ is a local maximum. The choice of the vacuum state being either of the asymmetric ϕ_+ or ϕ_- breaks the symmetry, cf. similar plot in Fig 11.15(b). The dashed box in (b) is displayed in (c) to show that the inflation/Higgs potential $V(\phi)$ has an almost flat portion at the $\phi = 0$ origin for a slow rollover transition. The dot represents the changing location of the system—rolling from a high plateau of the false vacuum towards the true vacuum at the bottom of the trough.

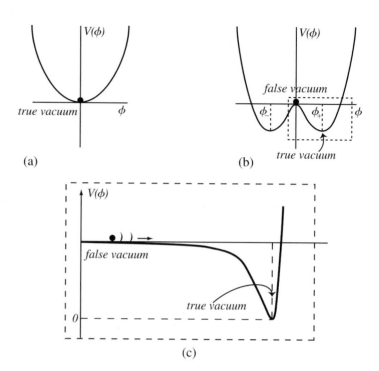

(a)　　(b)

(c)

[10] Because Λ represents a constant energy density, it will be the dominant factor $\rho_\Lambda \gg \rho_M$ at later cosmic time, as $\rho_M \sim a^{-3}$. This dominance means that it is possible for the universe to be geometrically closed ($\Omega > 1$ and $k = +1$), yet does not stop expanding. Namely, with the presence of a cosmological constant, the mass/energy density Ω (hence the geometry) no longer determines the fate of the universe in a simple way. In general, a universe with a nonvanishing Λ, regardless of its geometry, would expand forever. The only exception is when the matter density is so large that the universe starts to contract before ρ_Λ becomes the dominant term.

[11] One can check this estimate of $t_{GU} \simeq 10^{-36}$ s for a thermal energy $E_{GU} = 10^{16}$ GeV with a primordial nucleosynthesis time of $t_{NS} \simeq 10^2$ s when the thermal energy was $E_{NS} = 1$ MeV in this way.

$$\frac{E_{GU}}{E_{NS}} = \frac{kT_{GU}}{kT_{SN}} = \frac{a\,(t_{NS})}{a\,(t_{GU})} = \sqrt{\frac{t_{NS}}{t_{GU}}}$$

because the period between inflation and the nucleosynthesis time was radiation dominated $a \sim \sqrt{t}$. The ratio of $E_{GU}/E_{NS} = 10^{19}$ indeed matches that of $t_{NS}/t_{GU} = 10^{38}$ for $t_{GU} \simeq 10^{-36}$ s.

vacuum energy and negative pressure, causing the space to expand even faster. In fact, we can think of this Λ repulsive force as residing in the space itself, so as the universe expands, the push from this Λ energy increases as well.[10] We note that the total energy was conserved during the inflationary epoch's rapid expansion because of the concomitant creation of the gravitational field, which has a **negative** potential energy (cf. Section 10.3.1).

11.2.3 Inflation and the conditions it left behind

In the previous section we have described how the grand unification Higgs field associated with spontaneous symmetry breaking can serve as the inflation field. A patch of the universe with this "inflation/Higgs matter" might have undergone a slow rollover phase transition and thus lodged temporarily in a false vacuum with a large constant energy density. The resultant effective cosmological constant Λ_{eff} provided the gravitational repulsion to inflate the scale factor exponentially. The a grand unification thermal energy scale is $E_{GU} = O(10^{16}\,\text{GeV})$, that is, a temperature $T_{GU} = O(10^{29}\,\text{K})$, which according to (10.44) corresponds[11] to the cosmic time $t_{GU} \simeq 10^{-36}$ s. The energy density $\rho_{GU}c^2$ can be estimated as follows: in a relativistic quantum system (such as quantum fields) there is the natural energy length scale given by the product of Planck's constant (over 2π) times the velocity of light: $\hbar c = 1.97 \times 10^{-16}\,\text{GeV} \cdot \text{m}$. Using this conversion factor we have the energy density scale for grand unification

$$\rho_{GU}c^2 \simeq \frac{(E_{GU})^4}{(\hbar c)^3} \simeq 10^{100}\,\text{J/m}^3. \tag{11.21}$$

For a vacuum energy density $\rho_\Lambda \approx \rho_{GU}$, the corresponding exponential expansion time constant $\Delta\tau$ of (11.20) had the value $\Delta\tau \simeq 10^{-37}$ s. Namely, the exponential inflationary expansion took place when the universe was $t_{GU} \simeq 10^{-36}$ s old, with an exponential expansion time constant of $\Delta\tau = O(10^{-37}$ s$)$. By a "slow" rollover phase transition we mean that the parameters of the theory are such that inflation might have lasted much longer than 10^{-37} s, for example, 10^{-35} s ($\lesssim 100$ e-fold), expanding the scale factor by more than 30 orders of magnitude, until the system rolled down to the true vacuum, ending the inflation epoch (cf. Fig. 11.3). Afterwards the universe commenced the adiabatic expansion and cooling according to the standard FLRW model until the present epoch.[12] Such dynamics have the attractive property that they would leave behind precisely the features that had to be postulated as the initial conditions for the standard FLRW cosmology.

The horizon and flatness problems solved

With the exponential behavior of the scale factor in (11.19), we can naturally have superluminal ($\dot{a}R_0 > c$) expansion as the rate $\dot{a}(t)$ also grows exponentially.[13] This does not contradict special relativity, which says that an object cannot pass another one faster than c in one fixed frame. Putting it another way, while an object cannot travel faster than the speed of light through space, there is no restriction stipulating that space itself cannot expand faster than c. Having a superluminal expansion rate, this inflationary scenario can solve the horizon problem, because two points that are a large number of horizon lengths apart now (or at the photon decoupling time when the CMB was created) could still be in causal contact before the onset of the inflationary epoch. They started out being thermalized within one horizon volume before the inflation epoch, but became separated by many horizon lengths due to the superluminal expansion.

This inflationary scenario can solve the flatness problem because the space was stretched so much that it became, after the inflationary epoch, a geometrically flat universe to a high degree of accuracy. When this exponential expansion (11.19) is applied to the Friedmann equation (11.15), it yields the ratio

$$\frac{1 - \Omega(t_2)}{1 - \Omega(t_1)} = \left[\frac{\dot{a}(t_2)}{\dot{a}(t_1)}\right]^{-2} = e^{-2(t_2-t_1)/\Delta\tau}. \tag{11.22}$$

Just as the scale factor was inflated by a large ratio, say, $e^{(t_2-t_1)/\Delta\tau} = 10^{30}$, we can have the RHS as small as 10^{-60}. Starting with any reasonable value of $\Omega(t_1)$ we can still have, after the inflation, a $\Omega(t_2) = 1$ to a high accuracy. While the cosmic time evolution in the FLRW model, being determined by gravitational attraction, always enhances the curvature by driving the universe away from $\Omega = 1$ (hence the flatness problem), the accelerating expansion due to the vacuum repulsion always pushes the universe (very rapidly) toward the $\Omega = 1$ point. Thus a firm prediction by the inflationary scenario is that the universe left behind by inflation must have a flat geometry and, according to GR, a density equal to the critical value (11.15)—although it does not specify what components make up such a density.

[12]It had generally been assumed that the effective cosmological constant, associated with the false vacuum, vanished at the end of the inflationary epoch. The general expectation was that the standard FLRW cosmology that followed the inflation epoch was one with no cosmological constant. Part of the rationale was that a straightforward estimate of the cosmological constant, as due to the zero-point energy of a quantum vacuum, yielded such an enormously large Λ (see Section 11.7) that many had assumed that there must be some yet-to-be discovered symmetry argument that would strictly forbid a nonzero cosmological constant. However, as we shall see below, more recent discoveries point to a nonvanishing, but small, Λ. The challenge is now how to explain the presence of such a "dark energy" in the universe.

[13]The Hubble constant, being the ratio of scale change rate per unit scale $\dot{a}/a$, does not change under such an exponential expansion of the scale factor.

The origin of matter/energy and structure in the universe

Besides the flatness and horizon problems, the standard FLRW cosmology requires as initial conditions that all the energy and particles of the universe be present at the very beginning. Furthermore, this hot soup of particles should have just the right amount of **initial density inhomogeneity** (density perturbation) which, through subsequent gravitational clumping, formed the cosmic structure of galaxies, clusters of galaxies, voids, etc. we observe today. One natural possibility is that such a density perturbation resulted from quantum fluctuation of particle fields in a very early universe. However, it is difficult to understand how such microscopic fluctuations can bring forth the astrophysical-sized density nonuniformity required for the subsequent cosmic construction. Remarkably, the inflationary cosmology can provide us with an explanation of the origin of matter/energy, as well as the structure of the universe.

The inflation model suggests that at the beginning of the big bang a patch of the inflation/Higgs matter (smaller than the size of a proton) underwent a phase transition bringing about a huge gravitational repulsion. This is the driving force behind the space-explosion that was the big bang. While this inflation material (the Λ energy) expanded exponentially in size to encompass a space that eventually developed into our presently observed universe, its energy density remained essentially a constant. In this way more and more particle/field energy was "created" during the inflationary epoch. When it ended with the universe reaching the true vacuum, its oscillations at the trough in Fig. 11.3 showed up, according to quantum field theory, as a soup of ordinary particles. According to the inflation theory, the initial potential energy of the inflation/Higgs field (having little kinetic energy) was the origin of our universe's matter content when it was converted into relativistic particles. In short, it is the vacuum energy that drove the inflation that would in the end decay into radiation and matter.

The phenomenon of particle creation in an expanding universe can be qualitatively understood as follows: according to quantum field theory, the quantum fluctuations of the field system can take on the form of the appearance and disappearance of particle–antiparticle pairs in the vacuum. Such energy nonconserving processes are permitted as long as they take place on a sufficiently short time-scale Δt so that the uncertainty relation $\Delta E \Delta t \leq \hbar$ is not violated. In a static space, such "virtual processes" do not create real particles. However, when the space is rapidly expanding, that is, the expansion rate was larger than the annihilation rate, real particles were created.[14] Thus, inflation in conjunction with quantum field theory naturally gives rise to the phenomenon of particle creation. This hot, dense, uniform collection of particles is just the postulated initial state of the standard big bang model. Furthermore, the scale factor had increased by such a large factor that it could stretch the subatomic size fluctuation of a quantum field into astrophysical sized density perturbation to seed the subsequent cosmic structure formation. The resultant density fluctuation was Gaussian (i.e. maximally random) and scale-invariant (i.e. the same fluctuation, of the order of 10^{-5}, in the gravitational potential on all length-scales) as will be discussed in Box 11.2 below.

[14]This way of seeding the structure formation can be viewed as "Hawking radiation from inflation." Recall our discussion in Section 8.5 of Hawking radiation from black holes in which virtual particles are turned into real ones because of a black hole event horizon. Here the production of real particles from quantum fluctuation comes about because of the horizon created by the hyper-accelerating expansion of the universe.

11.3   CMB anisotropy and evidence for a flat universe

As discussed in Section 11.2.3, inflationary cosmology predicts that the space-time geometry of our universe must be flat. This prediction received more direct observational support through detailed measurement of the temperature anisotropy of the CMB radiation.[15]

The CMB is the earliest and largest observable thing in cosmology. Its remarkable uniformity over many horizon lengths reflects its origination from a single pre-inflation horizon volume. Just before the photon decoupling time t_γ, the universe was composed of dark matter and a tightly bound photon–baryon fluid. The inflationary scenario, with its associated phenomenon of particle creation, also generated a small density perturbation on a wide range of distance scales onto this overall homogeneity. Because of gravitational instability, this nonuniform distribution of matter eventually evolved into the cosmic structure we see today. In the early universe up till t_γ, the gravitational clumping of baryons was resisted by photon radiation pressure. This set up acoustic waves of compression and rarefaction with gravity being the driving force and radiation pressure the restoring force. All this took place against a background of dark matter fluctuations, which started to grow right after the radiation–matter equality time because dark matter did not interact with radiation.[16] Such a photon–baryon fluid can be idealized by ignoring the dynamical effects of gravitation and baryons (because the photon number density is much higher than that of baryons). This leads to a sound wave speed

$$c_s \simeq \sqrt{\frac{p}{\rho}} \simeq \frac{c}{\sqrt{3}} \qquad (11.23)$$

as pressure and density being approximated by those for radiation $p \approx \rho c^2/3$. This cosmic sound left an imprint that is still discernible today. The compression and rarefaction was translated through gravitational redshift into a temperature inhomogeneity. By a careful analysis of this wave pattern, we can garner much information about the universe at this early epoch.

11.3.1   Three regions of the angular power spectrum

We shall present only a qualitative discussion of the power spectrum of the temperature anisotropy to give the reader some general idea of how a detailed analysis will allow one to fix a number of important cosmological parameters. From (10.86) and (10.88) for the correlation function, we see that the mean-square temperature anisotropy may be written for large multipole number l as

$$\left\langle \left(\frac{\delta T}{T}\right)^2 \right\rangle = \frac{1}{4\pi} \sum_{l=0}^{\infty} (2l+1)C_l \approx \int \frac{l(l+1)C_l}{2\pi} d(\ln l). \qquad (11.24)$$

$(l(l+1)/2\pi)C_l$ is approximately the power per logarithmic interval, and is the quantity presented in the conventional plot of the power spectrum against a logarithmic multipole number (cf. Figs. 11.4 and 11.13).

[15]The purpose of Section 11.3 is to present the observational evidence for a **flat universe**. As the discussion is somewhat more difficult, some readers may wish to skip it and proceed directly to Section 11.4.

[16]We remind ourselves that the dominant form of matter in the universe is cold dark matter. We can have the somewhat simplified picture: gravitational clumping took place principally among such nonbaryonic dark matter particles after t_{RM}. The baryonic matter inhomogeneity was not amplified until t_γ (when its resistance by radiation pressure disappeared); thereafter it fell into the dominant gravitational potential (the cosmic scaffoldings of Section 10.5.4) formed principally by the dark matter.

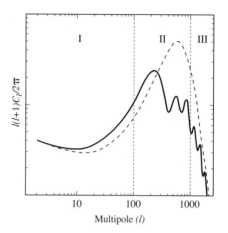

Fig. 11.4 CMB power spectrum as a function of the multipole moments. The solid curve with peaks and troughs (the acoustic peaks) is the prediction by the inflation model (with cold dark matter). The physics corresponding to the three marked regions is discussed in the text. The dashed curve is that by the topological defect model for the origin of the cosmic structure.

On small sections of the sky where curvature can be neglected, the spherical harmonic analysis becomes ordinary Fourier analysis in two dimensions. In this approximation the multipole number l has the interpretation as the Fourier wavenumber. Just as the usual Fourier wavenumber is $k \approx \pi/x$, the multipole moment number is $l \approx \pi/\theta$: large l corresponds to small angular scales with $l \approx 10^2$ corresponding to degree scale separation.

The CMB anisotropy is observed to be adiabatic (all particle species varied together) and this is consistent with the idea that the density fluctuation at t_γ was due to the primordial wrinkles of spacetime left behind by the earlier inflationary epoch. The inflationary scenario left behind density fluctuations that were Gaussian and scale invariant (cf. Box 11.2). Such the initial density perturbation, together with the assumption of a dark matter content dominated by nonrelativistic particles (the "cold dark matter" model) leads to a power spectrum as shown in Fig. 11.4. We can broadly divide it into three regions:

Region I $(l < 10^2)$ This flat portion at large angular scales (the "Sachs–Wolfe plateau") corresponds to oscillations with a period larger than the age of the universe at the photon decoupling time. These waves are essentially frozen in their initial configuration. The flatness of the curve reflects the scale-invariant nature of the initial density perturbation as given by the inflation cosmology (cf. Box 11.2).

Region II $(10^2 < l < 10^3)$ At these smaller angular scales (smaller than the sound horizon), there had been enough time for the photon–baryon fluid to undergo oscillation. The peaks correspond to regions having higher, as well as lower, than average density.[17] The troughs are regions with neutral compression, thus have maximum velocity (recall our knowledge of oscillators). The CMB from such regions underwent a large Doppler shift. In short, here is a snapshot of the acoustic oscillations with modes (fundamental plus harmonics) having different wavelengths and different phases of oscillations. The relative heights of the acoustic peaks are related to cosmological parameters

[17]This is so because the power spectrum is the square of a_{lm} and hence indifferent to their signs.

such as the baryon and the cold dark matter densities. Higher Ω_B enhances the odd-numbered acoustic peaks relative to the even-numbered ones, while higher Ω_{DM} lowers all of the peaks. The positions of the peaks depend on the characteristic length on the surface of last scattering as well as on the spatial curvature of the intervening universe since (see the subsection below).

Region III $(l > 10^3)$ Photon decoupling did not take place instantaneously, i.e. the last scattering surface had a finite thickness.[18] Photons can diffuse out from any over-dense region if it was smaller than the photon's mean free path, which was increasing as the universe expanded. The net effect was an exponential damping of the oscillation amplitude in these sub-arcminute scales.

[18]The transition took about 50 000 years.

Box 11.2 Density fluctuation from inflation is scale-invariant

Inflation produces such a huge expansion that subatomic size quantum fluctuations were stretched to astrophysical dimensions.[19] For fluctuations larger than the sound horizon $\approx c_s H^{-1}$ one can ignore pressure gradients, as the associated sound waves cannot have crossed the perturbation in a Hubble time. The density perturbation without a pressure gradient would evolve like the homogeneous universe (Problem 11.1):

$$\rho a^2 (\Omega^{-1} - 1) = \text{const.} \tag{11.25}$$

where a is the scale factor. With $\Omega = 1 + \Delta\Omega$ and $\rho = \rho_c + \Delta\rho$, the above relation implies, for small $\Delta\Omega$ and $\Delta\rho$,

$$\rho_c a^2 \Delta\Omega = a^2 \Delta\rho = \text{const.} \tag{11.26}$$

We now consider the implication of this scaling behavior for the perturbation in a gravitational potential on a physical distance scale of aL,

$$\Delta\Phi = \frac{G_N \Delta M}{aL} = \frac{4\pi}{3} \frac{G_N \Delta\rho \, (aL)^3}{(aL)} = \left(\frac{4\pi L^2 G_N}{3} \right) a^2 \Delta\rho.$$

Because of (11.26), the gravitational potential perturbation $\Delta\Phi$ over a comoving length L is scale invariant. During the inflationary epoch the scale factor a would change by something like 30 decades; yet we would have the same $\Delta\Phi$ for a huge range[20] of physical distances of aL. Thus, inflationary cosmology makes the strong prediction of a scale-invariant density perturbation—the same fluctuation (of 10^{-5}) on all distance scales. It can be shown that such a density fluctuation, called the Harrison–Zel'dovich spectrum, would produce an angular power spectrum for the CMB anisotropy of the form

$$C_l = \frac{\text{const.}}{l\,(l+1)}.$$

Thus in the plot of $l\,(l+1)\,C_l$ vs. l in Fig 9.4 the power spectrum for the large angle region $(l < 100)$ is a fairly flat curve.

[19]This turns the inevitable quantum effect into the seeds of structure in our universe.

[20]The same level of distortion (the warping of spacetime due to quantum fluctuation) was imprinted on all scales.

In Box 11.2 we have presented the power spectrum as predicted by the inflationary cosmology: a Gaussian density perturbation leading to a random distribution of hot and cold spots on the temperature anisotropy map, and a power spectrum displaying peaks and troughs. It is illuminating to contrast this with an alternative theory of cosmic structure origin, the topological defect model. In this scenario, one posits that as the universe cooled to a thermal energy of 10^{16} GeV, the phase transition that breaks the associated grand unification symmetry also produces defects in the fabric of spacetime—in the form of strings, knots, and domain walls, etc. This introduced the initial density perturbation that seeded the subsequent structure formation. Such a density fluctuation would produce line-like discontinuities in the temperature map and a smooth power spectrum (instead of the wiggly features as predicted by the inflation model), see Fig. 11.4. As we shall discuss in the next subsection, the observed CMB anisotropy favors inflation over this topological defect model for the origin of the cosmic structure.

11.3.2 The primary peak and spatial geometry of the universe

Consider the oscillatory power spectrum in region II of Fig. 11.4. The temperature anisotropy of the CMB is the result of a pattern of density fluctuations on a spherical surface centered on us. It reflects the sound wave spectrum of the photon–baryon fluid at the photon decoupling time, i.e. on the surface of last scattering. There would be standing waves having wavelength $\lambda_n = \lambda_1/n$, with the fundamental wavelength given by the sound horizon,[21] cf. (9.45):

$$\lambda_1 = \int_0^{t_\gamma} \frac{c_s dt}{a(t)} \approx c_s \int_0^{t_\gamma} \frac{dt}{a(t)}. \qquad (11.27)$$

Now such a wavelength on the surface of last scattering would appear as angular anisotropy of scale

$$\alpha_1 \simeq \lambda_1/d(t_\gamma), \qquad (11.28)$$

where $d(t_\gamma)$ is the (proper) radial distance between us now (t_0) and the photon decoupling time (t_γ). Namely, it is the comoving distance a photon would have traveled to reach us (t_0) from the surface of last scattering (also called the angular diameter distance)

$$d(t_\gamma) = c \int_{t_\gamma}^{t_0} \frac{dt}{a(t)}. \qquad (11.29)$$

When evaluating the integrals in (11.27) and (11.29), we shall assume a matter-dominated **flat** universe with time dependence of the scale factor $a(t) \propto t^{2/3}$ as given by (10.30),

$$\int \frac{dt}{a(t)} \propto \int a^{-1/2} da \propto a^{1/2} = (1+z)^{-1/2}. \qquad (11.30)$$

Matter-domination is plausible because the radiation–matter equality time is almost an order of magnitude smaller than the photon decoupling time, that is, according to (10.68) the redshift is $z_{RM} \gg z_\gamma$. Thus the fundamental

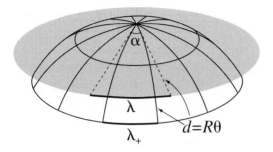

Fig. 11.5 A comparison of subtended lengths in a flat (shaded) vs. positively curved surface. For the same angular diameter distance d, the same angle α subtends a smaller wavelength λ_+ in a closed universe when compared to the corresponding $\lambda = \lambda_+ \left[(R_0/d) \sin (R_0/d) \right]^{-1} > \lambda_+$ in a flat universe.

wavelength corresponds to an angular separation of

$$\alpha_1 \approx \frac{\lambda_1}{d(t_\gamma)} = \frac{c_s(1+z_\gamma)^{-1/2}}{c(1+z_0)^{-1/2} - (1+z_\gamma)^{-1/2}}$$

$$\simeq \frac{(1+z_\gamma)^{-1/2}}{\sqrt{3}} \simeq 0.017 \,\text{rad} \simeq 1°, \tag{11.31}$$

where we have used $z_0 = 0$, $z_\gamma \simeq 1100$ and, as discussed in (11.23), a sound speed $c_s \simeq c/\sqrt{3}$. This fundamental wave angular separation in turn translates into the multipole number

$$l_1 \simeq \frac{\pi}{\alpha_1} \simeq \pi\sqrt{3(1+z_\gamma)} \approx 200. \tag{11.32}$$

Thus, in a flat universe we expect the first peak of the power spectrum to be located at this multipole number.

The above calculation was performed for a flat universe. What would be the result for a spatially curved universe? We will simplify our discussion by the suppression of one dimension and consider a 2D curved surface. In a positive curved closed universe ($k = +1$), light travels along longitudes (Fig. 11.5). A physical separation λ_1 at a fixed latitude, with polar angle θ and a coordinate distance $d = R_0\theta$, subtends an angle

$$\alpha_{1+} = \frac{\lambda_1}{R_0 \sin \theta} = \frac{\lambda_1}{R_0 \sin(d/R_0)} = \frac{\lambda_1}{d}\left(1 + \frac{d^2}{3R_0^2} + \cdots\right) > \frac{\lambda_1}{d}.$$

Namely, at a given scale (λ_1) at a fixed distance (d) the separation angle (α_{1+}) would appear to be larger (than the case of a flat universe). For a negatively curved open universe ($k = -1$), one simply replaces the sine by the hyperbolic sine:

$$\alpha_{1-} = \frac{\lambda_1}{R_0 \sinh(d/R_0)} = \frac{\lambda_1}{d}\left(1 - \frac{d^2}{3R_0^2} + \cdots\right) < \frac{\lambda_1}{d}.$$

At a given scale at a fixed distance, the separation angle would appear to be smaller. With the multipole number being inversely proportional to the separation angular scale, in a universe with spatial curvature the first peak would be shifted away from $l_1 \approx 200$, to a smaller (larger) multipole number for a closed (open) universe.

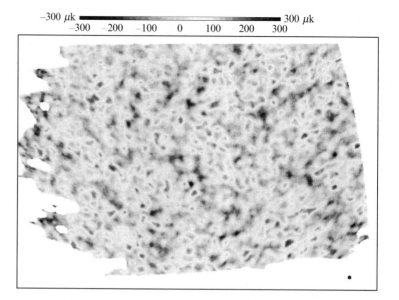

Fig. 11.6 Image of the complex temperature structure of CMB over 2.5% of the sky as captured by the Boomerang balloon-borne detector. It is interpreted as a freeze-frame picture of the sound wave patterns in the universe at the photon decoupling time. The black dot at the lower right-hand corner represents the size of a full moon subtending an angle of about half-a-degree.

Although the COBE satellite mapped the entire sky with high sensitivity discovering the CMB anisotropy at $\delta T/T = O(10^{-5})$, its relatively coarse angular resolution of $O(7°)$ was not able to deduce the geometry of our universe. In the late 1990s a number of high altitude observations, e.g. MAT/TOCO (Miller *et al.,* 1999), and balloon-borne telescopes, Boomerang (de Bernardis *et al.,* 2000), and Maxima-1 (Hanany *et al.,* 2000), had detected CMB fluctuations on smaller sizes. These observations (see Fig. 11.6) produced evidence for a flat universe by finding the characteristic size of the structure to be about a degree wide and a power spectrum peaked at $l \approx 200$, see Fig. 11.4. The $k = 0$ statement is of course equivalent, via the Friedmann equation, to a total density $\Omega_0 = 1$. A careful matching of the power spectrum led to

$$\Omega_0 = 1.03 \pm 0.03. \tag{11.33}$$

In the meantime, another dedicated satellite endeavor, WMAP (Wilkinson Microwave Anisotropy Probe), had reported their results in a series of publications (Bennett, 2003; Hinshaw, 2009). Another influential cosmological project has been the survey of galaxy distributions by SDSS (Sloan Digital Sky Survey). Their high resolution result allowed them to extract many important cosmological parameters: H_0, Ω_0, $\Omega_{M,0}$, Ω_B, and the deceleration parameter q_0, etc. (to be discussed in Section 11.5).

11.4   The accelerating universe in the present epoch

Thus by mid/late-1990s there was definitive evidence that the geometry of the universe is flat as predicted by inflation. Nevertheless, there were several pieces of phenomenology that appeared in direct contradiction to such a picture.

A missing energy problem The Friedmann equation (10.7) requires a flat universe to have a mass/energy density exactly equal to the critical density, $\Omega_0 = 1$. Yet observationally, including both the baryonic and dark matter, we can only find less than a third of this value (radiation energy is negligibly small in the present epoch):

$$\Omega_M = \Omega_B + \Omega_{DM} \simeq 0.25. \tag{11.34}$$

Thus, it appears that to have a flat universe we would have to solve a "missing energy problem."

A cosmic age problem From our discussion of the time evolution of the universe, we learned that the age of a flat universe should be two-third of the Hubble time, see (10.70),

$$(t_0)_{\text{flat}} = \frac{2}{3} t_H \lesssim 9 \, \text{Gyr}, \tag{11.35}$$

which is shorter than the estimated age of old stars. Notably the globular clusters have been deduced to be older than 12 Gyr (cf. Section 9.1.3). Thus, it appears that to have a flat universe we would also have to solve a "cosmic age problem."

Possible resolution with a dark energy A possible resolution of these phenomenological difficulties of a flat universe (hence inflationary cosmology) would be to assume the presence of a **dark energy**. A dark energy is defined as the "negative equation-of-state energy," $w < -1/3$ in Eq. (10.4). It gives rise to a gravitational repulsion, cf. Eq. (11.4). The simplest example of dark energy[22] is Einstein's cosmological constant, with $w = -1$. Such a cosmological constant assumed to be present even after inflation cannot have the immense size as the one it had during the inflation epoch. Rather, the constant dark energy density ρ_Λ should now be about three-quarters of the critical density to provide the required missing energy.

$$\Omega = \Omega_M + \Omega_\Lambda \overset{?}{=} 1, \tag{11.36}$$

where $\Omega_\Lambda \equiv \rho_\Lambda/\rho_c$. A nonvanishing Λ would also provide the repulsion to accelerate the expansion of the universe. In such an accelerating universe the expansion rate in the past must be smaller than the current rate H_0. This means that it would take a longer period (as compared to a decelerating or empty universe) to reach the present era, leading to a longer age $t_0 > 2t_H/3$ even though the geometry is flat. This just might possibly solve the cosmic age problem as well.

[22]One should not confuse dark energy with the energies of neutrinos, WIMPs, etc., which are also "dark," but are counted as parts of the "dark matter" (cf. Section 9.2), as the associated pressure is not negative.

11.4.1 Distant supernovae and the 1998 discovery

In order to obtain observational evidence for any changing expansion rate of the universe (i.e. to measure the curvature of the Hubble curve), one would have to measure great cosmic distances. One needed a distance method that works to over 5 billion light years. Clearly some very bright light sources are required. Since this also means that we must measure objects back in a time interval that is a significant fraction of the age of the universe, the method must be applicable to objects present at the early cosmic era. As it turns out, supernovae are ideally suited for this purpose.

SNe as standard candles and their systematic search

After the suggestion made in the 1970s that type Ia supernovae (SNe Ia) could possibly serve as standard candles, the first SN Ia was discovered in 1988 by a Danish group at redshift $z = 0.3$. At their peaks SNe Ia produce a million times more light than Cepheid variables, the standard candle most commonly used in cosmology (cf. Section 9.4.2). SNe Ia begin as white dwarfs (collapsed old stars sustained by degenerate pressure of their electrons) with mass comparable to the sun. If the white dwarf has a large companion star, which is not uncommon, the dwarf's powerful gravitational attraction will draw matter from its companion. Its mass increases until the "Chandrasekhar limit" $\simeq 1.4\,M_\odot$. As it can no longer be countered by the electron pressure, the gravitational contraction develops and the resultant heating of the interior core triggers the thermonuclear blast that rips it apart, resulting in an SN explosion. The supernova eventually collapses into a neutron star. Because they start with masses in a narrow range, such supernovae have comparable intrinsic brightness. Furthermore, their brightness has a characteristic decline from the maximum which can be used to improve on the calibration of their luminosity (the light-curve shape-analysis), making SNe Ia standardizable candles (Phillips 1993). Supernovae are rare events in a galaxy. The last time a supernova explosion occurred in our Milky Way was about 400 years ago. However, using new technology (large mosaic CCD cameras), astronomers overcame this problem by simultaneously monitoring thousands of galaxies[23] so that on the average some 10–20 supernovae can be observed in a year.

The discovery of an accelerating universe

Because light from distant galaxies was emitted long ago, to measure a star (or a supernova) farther out in distance is to probe the cosmos further back in time. An accelerating expansion means that the expansion rate was smaller in the past. Thus to reach a given redshift (i.e. recession speed) it must be located farther away[24] than expected, see Fig. 11.7. Observationally, the light

[23]Two images of the sky containing thousands of galaxies were taken weeks apart and digitally subtracted; the supernova locations leaped out.

[24]A Hubble curve (as in Fig. 11.7) is a plot of the luminosity distance versus the redshift (measuring recession velocity). A straight Hubble curve means a cosmic expansion that is coasting. This can only happen in an empty universe (cf. Section 9.1.3 and Fig. 10.2). If the expansion is accelerating, the expansion rate H must be smaller in the past ($H < H_0$). From Eq. (9.5): $H\Delta r = z$, we see that, for a given redshift z, the distance Δr to the light-emitting supernova must be larger than that for an empty or decelerating universe.

Fig. 11.7 Hubble diagram: the Hubble curve for an accelerating universe bends upwards. A supernova on this curve at a given redshift would be further out in distance than anticipated.

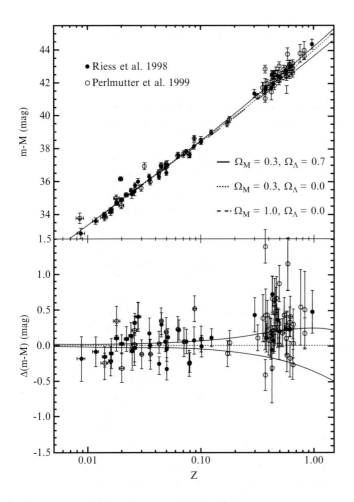

Fig. 11.8 Discovery of an accelerating universe. The Hubble plot showing the data points from Riess *et al.* (1998) and Perlmutter (1999). The horizontal axis is the redshift z; the vertical axes are the luminosity distance expressed in terms of distance modulus (i.e. logarithmic luminosity distance, cf. Box 9.1). In the lower panel $\triangle (m - M)$ is the difference after subtracting out the then expected value for a decelerating universe with $\Omega_M = 0.3$ and $\Omega_\Lambda = 0$. Three curves correspond to models with different matter/energy contents $(\Omega_M, \Omega_\Lambda)$ of the universe. The solid curve for nonvanishing cosmological constant has the best fit of the observational data.

source in an accelerating universe would be measured to be dimmer than expected.

By 1998 two collaborations: the Supernova Cosmological Project, led by Saul Perlmutter of the Lawrence Berkeley National Laboratory (Perlmutter *et al.*, 1999) and the High-z Supernova Search Team, led by Adam Riess of the Astronomy Department at UC Berkeley and Brian Schmidt of the Mount Stromlo and Siding Spring Observatories (Riess *et al.*, 1998), each had accumulated some 50 SNe Ia at high redshifts—z: 0.4–0.7 corresponding to SNe occurring five to eight billion years ago. They made the astonishing discovery that the expansion of the universe was actually accelerating, as indicated by the fact that the measured luminosities were on the average 25% less than anticipated, and the Hubble curve bent upward, Fig. 11.8.

Extracting Ω_M and Ω_Λ from the measured Hubble curve From the Hubble curve plotted in the space of redshift and luminosity distance, one can then extract the mass and dark energy content of the universe. The proper distance d_p from a supernova with a redshift z in the present epoch $a(t_0) = 1$ has been worked out in (9.51). Combined with the result in (9.57), this yields an expression for the luminosity distance:

$$d_L(z) = c(1+z) \int_0^z \frac{dz'}{H(z')}, \tag{11.37}$$

where, using the Friedmann equation (10.1), we can express the epoch-dependent Hubble constant in terms of the scale factor and the density ratios in the present epoch (Problem 11.2), including in particular the cosmological constant density term:

$$H(t) = H_0 \left(\frac{\Omega_{R,0}}{a^4} + \frac{\Omega_{M,0}}{a^3} + \Omega_\Lambda + \frac{1-\Omega_0}{a^2} \right)^{1/2}, \tag{11.38}$$

where $a(t)$ can in turn be replaced by the redshift according to (9.50),

$$H(z) = H_0 \Omega_{R,0}(1+z)^4 + \Omega_{M,0}(1+z)^3 + \Omega_\Lambda + (1-\Omega_0)(1+z)^{2\,1/2}$$

$$\simeq H_0 \Omega_{M,0}(1+z)^3 + \Omega_\Lambda + (1-\Omega_{M,0}-\Omega_\Lambda)(1+z)^{2\,1/2}. \tag{11.39}$$

The resultant Hubble curves $d_L(z)$ in (11.37) with $H(z)$ in the form of (11.39) that best fitted the observation data yields values of $\Omega_{M,0}$ and Ω_Λ as shown in Fig. 11.9. If we further impose the requirement of a flat geometry, $\Omega_{M,0} + \Omega_\Lambda = 1$ as suggested by the CMB data, the favored values from Fig. 11.9 as well as from other supporting evidence obtained later on are

$$\Omega_{M,0} = 0.246 \quad \text{and} \quad \Omega_\Lambda = 0.757 \tag{11.40}$$

suggesting that most of the energy in our universe resides in this mysterious dark energy.[25]

[25] In the present discussion we shall for definiteness assume the dark energy as being the cosmological constant with an equation of state parameter $w = -1$.

The age of universe calculated   These observed values for $\Omega_{M,0}$ and Ω_Λ can also be translated into an age for the flat universe. The Hubble constant being the rate of expansion $H = \dot{a}/a$, we can relate dt to the differential of the scale factor,

$$t_0 = \int_0^{t_0} dt = \int_0^1 \frac{da}{aH}. \tag{11.41}$$

From (11.38) for the scale-dependent Hubble constant, this yields an expression of the age[26] in terms of the density parameters

$$t_0 = t_H \int_0^1 \frac{da}{\Omega_{R,0}a^{-2} + \Omega_{M,0}a^{-1} + \Omega_\Lambda a^2 + (1-\Omega_0)^{1/2}}. \tag{11.42}$$

[26] We can check the limit of (11.42) for a matter-dominated flat universe ($\Omega_{\Lambda,0} = \Omega_{R,0} = 0$ with $\Omega_0 = \Omega_{M,0} = 1$) which yields an age $t_0 = t_H \int_0^1 a^{1/2} da = \frac{2}{3} t_H$, in agreement with the result obtained in (10.30).

The spatially flat universe with negligible amount of radiation energy, $\Omega_0 = \Omega_{M,0} + \Omega_\Lambda = 1$, leads to a simple expression of the age of the universe in terms of the densities

$$\frac{t_0}{t_H} = \int_0^1 \left(\Omega_{M,0}a^{-1} + \Omega_\Lambda a^2 \right)^{-1/2} da$$

$$= \frac{2}{3\sqrt{\Omega_\Lambda}} \ln \frac{\sqrt{\Omega_\Lambda} + \sqrt{\Omega_{M,0} + \Omega_\Lambda}}{\sqrt{\Omega_{M,0}}} = 1.02. \tag{11.43}$$

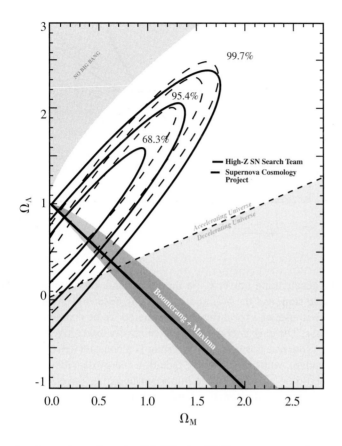

Fig. 11.9 Fitting Ω_Λ and Ω_M to the discovery data as obtained by the High-z SN Search Team and Supernova Cosmology Project. The favored values of Ω_Λ and Ω_M follow from the central values of CMB anisotropy $\Omega_\Lambda + \Omega_M \simeq 1$ (the straight line) and those of the SNe data represented by confidence contours (ellipses) around $\Omega_\Lambda - \Omega_M \simeq 0.5$.

For the density values given in (11.40), the RHS comes very close to unity. Thus the deceleration effect of $\Omega_{M,0}$ and the accelerating effect of Ω_Λ coincidentally cancel each other. The age is very close to that of an empty universe.

$$t_0 = 1.02\, t_{\mathrm{H}} = 13.9\, \mathrm{Gyr}. \tag{11.44}$$

11.4.2   Transition from deceleration to acceleration

Since the immediate observational evidence from these far away supernovae is a smaller-than-anticipated luminosity, one wonders whether there is a more mundane astrophysical explanation. There may be one or a combination of several mundane causes that can mimic the observational effects of an accelerating universe. Maybe this luminosity diminution is brought about not because the supernovae were further away than expected, but due to the absorption by yet-unknown[27] interstellar dust, and/or due to some yet-unknown evolution of supernovae themselves (i.e. supernovae's intrinsic luminosity were smaller in the cosmic past). However, all such scenarios would lead us to expect that the supernovae, at even greater distances (and even further back in time), should have their brightness **continue to diminish**.

[27] The absorption and scattering by ordinary dust shows a characteristic frequency dependence that can in principle be subtracted out. By the unknown dust we refer to any possible "gray dust" that could absorb light in a frequency-independent manner.

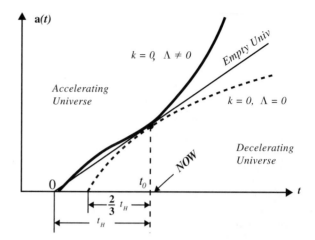

Fig. 11.10 Time evolution of an accelerating universe. It started out in a decelerating phase before taking on the form of an exponential expansion. The transition to an accelerating phase shows up as a "bulge;" this way it has an age longer than the $\Lambda = 0$ flat universe age of $2t_H/3$.

For the accelerating universe, on the other hand, this diminution of luminosity **would stop**, and the brightness would **increase** at even larger distances. This is so because we expect the accelerating epoch be proceeded by a decelerating phase. The dark energy should be relatively insensitive to scale change $\rho_\Lambda \sim a^0(t)$ (the true cosmological constant is a constant density, independent of scale change), while the matter or radiation energy densities, $\rho \sim a^{-3}(t)$ or $a^{-4}(t)$, should be more and more important in earlier times. Thus, the early universe could not be dark energy dominated, and it must be decelerating. This transition from a decelerating to an accelerating phase would show up as a bulge in the Hubble curve, see Fig. 11.10.

The cosmic age at transition   Let us estimate the redshift when the universe made this transition. We define an epoch-dependent **deceleration parameter** which generalizes the q_0 parameter of Problem 9.10,

$$q(t) \equiv \frac{-\ddot{a}(t)}{a(t)H^2(t)}, \tag{11.45}$$

which, through the Friedmann equation, can be related to the density ratios (Problem 10.10)

$$q(t) = \Omega_R(t) + \frac{1}{2}\Omega_M(t) - \Omega_\Lambda$$

$$= \frac{\Omega_{R,0}}{a(t)^4} + \frac{\Omega_{M,0}}{2a(t)^3} - \Omega_\Lambda. \tag{11.46}$$

After dropping the unimportant $\Omega_{R,0}$ and replacing the scale factor by z, we have

$$q(z) \simeq \frac{1}{2}\Omega_{M,0}(1+z)^3 - \Omega_\Lambda. \tag{11.47}$$

The transition from decelerating ($q > 0$) to the accelerating ($q < 0$) phase occurred at redshift z_{tr} when the deceleration parameter vanished $q(z_{tr}) \equiv 0$,

or

$$1 + z_{\text{tr}} = \left(\frac{2\Omega_\Lambda}{\Omega_{\text{M},0}} \right)^{1/3}. \qquad (11.48)$$

The supernovae data translate into a transition redshift of $z_{\text{tr}} \simeq 0.8$, corresponding to a scale factor of $a_{\text{tr}} \simeq 0.56$ and a cosmic time, calculated similarly[28] as in (11.42) and (11.43), of $t_{\text{tr}} = t(a = 0.56) \simeq 7\,\text{Gyr}$—in cosmic terms, the transition took place only recently ("just yesterday")! This reflects the fact that the matter density in the present epoch $\Omega_{\text{M},0}$ happens to be comparable to the dark energy density Ω_Λ.

Discovery of SNe prior to the accelerating phase   Thus, the conclusive evidence for the accelerating universe interpretation of the supernovae data is to observe this bulge structure, which cannot be mimicked by any known astrophysical causes. The 1998 discovery data (z: 0.4–0.7) showed the rise of this bulge, but we need to see the falling part of the Hubble curve. SNe further out ($z > 0.8$) should be still in the decelerating phase; they should be brighter than what is expected of the continuing dimming scenario that a mundane interpretation would have us anticipate. Reassuringly, just such an early decelerating phase had been detected.

After the original discovery of an accelerating universe, researchers had searched for other supernovae at high z. The supernova labeled SN1997ff had been serendipitously recorded by the Hubble Space Telescope, and by other observational means (some intentionally, and some unpremeditated). Through a major effort at data analysis, its properties were deduced in 2001, showing that it is a type Ia SN having a redshift of $z \simeq 1.7$ and, thus an explosion occurring 10 billion years ago, making it by far most distant supernova ever detected. Remarkably, it is brighter by almost a factor of two (see Fig. 11.11) compared to the expectation of continual dimming as a mundane astrophysical explanation would require. This is the bulge feature unique to a Hubble curve for an accelerating universe—the light was emitted so long ago when the

[28]The relation between cosmic time and scale factor of a given epoch is

$$t(a) = t_{\text{H}} \int_0^a \Omega_{\text{M},0}/a' + \Omega_\Lambda a'^2 {}^{-1/2} da'$$

$$= \frac{2t_{\text{H}}}{3\sqrt{\Omega_\Lambda}} \ln \left(\sqrt{\frac{\Omega_\Lambda}{\Omega_{\text{M},0}} a^3} + \sqrt{1 + \frac{\Omega_\Lambda}{\Omega_{\text{M},0}} a^3} \right).$$

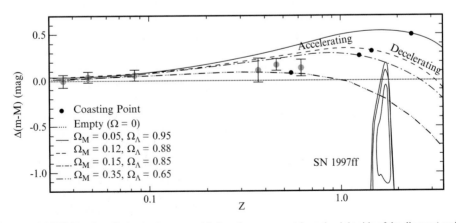

Fig. 11.11 Location of SN1997ff (because of measurement uncertainties, shown as a patch on the right side of the diagram) and other high z SNe are plotted with respect to those for an empty universe (the horizontal line) in a Hubble diagram. The black spots follow an up-turning curve which represents the luminosity and redshift relation showing continuing dimming.

expansion of the universe was still decelerating. Since this first confirmation finding, many more high-z SNe had been observed both from ground-based surveys and with Hubble Space Telescope. These data had provided conclusive evidence for cosmic deceleration that preceded the present epoch of cosmic acceleration (Riess *et al.*, 2004).

11.4.3   Dark energy: Further evidence and the mystery of its origin

After the supernovae discovery, the presence of dark energy $\Omega_\Lambda = 0.75$ was further confirmed by the analysis of the CMB anisotropy power spectrum, as well as of the distribution of galaxies. In the following paragraph we shall discuss the recent result showing that such a dark energy is just the agent needed to explain the observed slow down of galaxies' growth. On the other hand, even though the observational evidence for dark energy is strong, its physical origin remains mysterious. The quantum vacuum energy does have precisely the property of a density being constant with respect to volume changes and hence being a negative pressure, the estimated magnitude of such a vacuum quantum energy is something like 120 orders of magnitude too large. In this section we give a brief summary of these developments; more details can be found in a recent review (Frieman, Turner and Huterer, 2008).

Dark energy stunts the growth of galaxies   We have calculated the cosmic time $t_{tr} \simeq 7\,\mathrm{Gyr}$ when the deceleration phase begin to be replaced by the acceleration phase, and the cosmic time $t_{M\Lambda} \simeq 9.5\,\mathrm{Gyr}$ when the energy content of the universe was just balanced between matter and dark energy (see Problem 11.5). Namely, only in the last 5–7 Gyr or so has the dark energy become the dominant force in the universe that turned the (decelerating) expansion of the universe into an accelerated expansion. Such an effect of having the repulsive gravity of dark energy overcoming the more familiar gravitational attraction also shows up in the slowing down of the growth of the largest conglomeration of matter in the universe, the galaxy clusters. They are relatively easy to find as clusters are filled with hot gas that emits X-rays. A group researcher (Vikhlinin *et al.*, 2009) used the Chandra X-ray satellite telescope to study the intensity and spectra of 86 clusters that had previously been found by the ROSAT (X-ray) All-Sky Survey. A set of 37 clusters at more than 5 gigalight-years away was compared with another set of 49 that are closer than half a gigalight-year. Theoretical models were used to calculate how the numbers of clusters with different masses would change during this span of $\Delta t \simeq 5$ Gyr under different conditions: with different amounts, and values of the equation of state parameter w, and with or without a dark energy. A good fit to the observation data[29] clearly required the presence of dark energy with $w \simeq -1$.

[29]For example, Vikhlinin *et al.* (2009) found only a fifth of the number of the most massive clusters that a universe without dark energy would have.

The problem of interpreting Λ as quantum vacuum energy   The introduction of the cosmological constant in the GR field equation does not explain its physical origin. In the inflation model one postulates that it is the false vacuum energy of an inflation/Higgs field that acts like an effective

Table 11.1 Cosmological parameters deduced from an analysis (Tegmark 2006) based on data collected by WMAP and SDSS. The first column displays the parameter result from an analysis that assumes the cosmological constant as being the dark energy; the second column for a universe assumed to be flat. The equation numbers in the third column refer to part of the text, where such parameters were discussed. The parameter h_0 is the Hubble constant H_0 measured in units of 100 (km/s) /Mpc.

	$\left(\begin{array}{c} DE=\Lambda \\ w=-1 \end{array}\right)$	$\left(\begin{array}{c} FlatU \\ \Omega_0=1 \end{array}\right)$	Parameter description (equation number)
Ω_0	1.003 ± 0.010	1 (fixed)	density parameter (11.33)
Ω_Λ	0.757 ± 0.021	0.757 ± 0.020	dark energy density (11.40)
Ω_M	0.246 ± 0.028	0.243 ± 0.020	matter density (9.23, 11.40)
Ω_B	0.042 ± 0.002	0.042 ± 0.002	baryon density (10.58)
h_0	0.72 ± 0.05	0.72 ± 0.03	present expansion rate (9.7)
t_0	13.9 ± 0.6 Gyr	13.8 ± 0.2 Gyr	age of the universe (11.44)
T_0	2.725 ± 0.001 K	2.725 ± 0.001 K	CMB temperature (10.64)
q_0	-0.63 ± 0.03	-0.57 ± 0.1	deceleration parameter (11.45)
w	-1 (fixed)	-0.94 ± 0.1	dark energy equation of state (10.4)

cosmological constant driving the inflationary expansion. What is the physical origin of the dark energy that brings about the accelerating expansion of the present epoch? A natural candidate is the quantum vacuum energy. The zero point energy of a quantized field automatically has the property of having an energy density that is constant, giving rise to a negative pressure. However as explained in Appendix D (Section 11.7) such a vacuum energy while having the correct property is expected to be way too large to account for the observed $\Omega_\Lambda = O(1)$. What are the other possibilities? One chance is that the dark energy is associated with some yet-unknown scalar field (sometimes referred to as the "quintessence"), somewhat akin to the association of the inflationary expansion to the inflation/Higgs field. Such theories often have an equation-of-state parameter $w \neq w_\Lambda = -1$. However, observational data do not support a dark energy w significantly different from the value of -1 (see Table 11.1).

11.5   The concordant picture

An overall coherent and self-consistent picture of the cosmos has emerged that can account for the geometry and structure of the universe, as well as its evolution onward from a fraction of a second after the big bang. In this section, we first summarize the cosmological parameters and discuss the concordant cosmological model that has emerged. Even though we have a consistent picture, there are still many unsolved problems; we shall mention some of them at the end of this chapter.

Cosmological parameters from CMB and the galaxy distribution

Our previous discussion has concentrated on conceptually and technically simpler approaches in obtaining cosmological parameters—counting and weighing methods, plotting the Hubble curve (including data from high-redshift supernovae), and light nuclear element abundance, etc. These measurements have now been confirmed and hugely improved by the analysis of very different physical phenomena: the CMB temperature anisotropy (in particular as measured by WMAP) in combination with analysis of large-scale structure survey

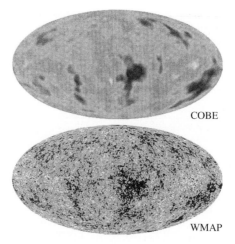

COBE

WMAP

Fig. 11.12 The temperature fluctuation of the CMB is a snap-shot of the baby universe at the photon decoupling time. A comparison of the results by COBE vs. WMAP1 shows the marked improvement in resolution by WMAP. This allowed us to extract many more cosmological parameters from the latest observations.

data (obtained in particular by 2dF and SDSS). We have briefly discussed the CMB anisotropy (cf. Sections 10.5.4 and 11.3): a detailed study of the power spectrum through a spherical harmonics decomposition can be displayed as a curve (relative amplitude vs. angular momentum number) with a series of peaks. The primary peak (i.e. the dominant structure) is at the one degree scale showing that the spatial geometry is flat; the secondary peaks are sensitive to other cosmological parameters such as the baryon contents of the universe, $\Omega_B \simeq 0.04$, etc. WMAP has a much improved angular resolution compared to COBE, Fig. 11.12, The study of the large-scale cosmic distribution of galaxies is beyond the scope of this book, the underlying physics also reflects the relic imprints of primordial acoustic waves as in the case of CMB. The combined results of this array of observations allowed us to extract a large number of cosmological parameters at high accuracy (Table 11.1).

The standard model of cosmology

Cosmology has seen a set of major achievements over the past decade, to the extent that something like a standard model for the origin and development of the universe is now in place: the FLRW cosmology proceeded by an inflationary epoch. Many of the basic cosmological parameters have been deduced in several independent ways, arriving at a consistent set of results. These data are compatible with our universe being infinite and spatially flat, having matter/energy density equal to the critical density, $\Omega_0 = 1$. The largest energy component is consistent with it having Einstein's cosmological constant $\Omega_\Lambda \simeq 0.75$. In the present epoch this dark energy content is comparable in size to the matter density $\Omega_M \simeq 0.25$, which is made up mostly of cold dark matter. Thus this standard model is often call the ΛCDM cosmology model. The expansion of the universe will never stop—in fact having entered the accelerating phase, the expansion will be getting faster and faster.

Still many unsolved problems

Although we have a self-consistent cosmological description, many mysteries remain. We do not really know what makes up the bulk of the dark matter, even though there are plausible candidates as predicted by some yet-to-be-proven particle physics theories. The most important energy component is the mysterious "dark energy," although a natural candidate is the quantum vacuum energy. Such an identification leads to an estimate of its size that is completely off the mark (cf. Section 11.7). If one can show that the quantum vacuum energy must somehow vanish due to some yet-to-be-found symmetry principle, a particular pressing problem is to find out whether this dark energy is time-independent, as is the case of the cosmological constant Λ, or is it more like a Λ_{eff} coming from some quintessence scalar field as in the case of inflation? Despite our lack of understanding of this dark energy, in recent discoveries constitute a remarkable affirmation of the inflationary theory of the big bang. Still, even here question remains as to the true identity of the inflation/Higgs field. We need to find ways to test the existence of such a field in some noncosmological settings.

Besides the basic mystery of dark energy ("the cosmological constant problem") there are other associated puzzles, one of them being the "cosmic coincidence problem:" we have the observational result that in the present epoch the dark energy density is comparable to the matter density, $\Omega_\Lambda \simeq \Omega_M$. Since they scale so differently ($\Omega_M \sim a^{-3}$ vs. $\Omega_\Lambda \sim a^0$) we have $\Omega_M \simeq 1$ in the cosmic past, and $\Omega_\Lambda \simeq 1$ in the future. Thus, the present epoch is very special—the only period when they are comparable.[30] Then the question is why? How do we understand this requirement of fine tuning the initial values in order to have $\Omega_M \simeq \Omega_\Lambda$ now?

A finite dodecahedral universe: A cautionary tale

It cannot be emphasized too much that the recent spectacular advances in cosmology have their foundation in the ever-increasing amount of high precision observational data. Ultimately any cosmological theory will stand or fall, depending on its success in confronting experimental data. In this context we offer the following cautionary tale.

An inspection of the CMB power spectrum in Fig. 11.13 shows that a few data points in the large angle (low l) region tend to be lower than the theoretical curve based on the standard cosmological model outlined above.

[30]Closely related to this is the puzzle of the respective amounts of decelerating matter and the accelerating dark energy so that their effects cancel each other, leaving the age of the universe very close to that of an empty universe (the Hubble time).

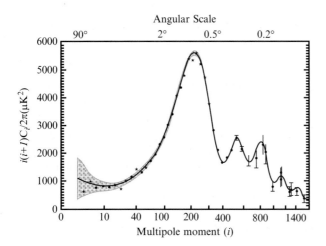

Fig. 11.13 The angular power spectrum of CMB temperature anisotropy. The dots are the first-year data-points from WMAP. The theoretical curve follows from inflationary model (having cold dark matter) with parameters given in Table 11.1. The fan-shaped shaded area at low multiple moments reflects the uncertainty due to cosmic variance, cf. (10.89).

This does not concern most cosmologists because they are still in the shaded area corresponding to the statistical uncertainty called cosmic variance cf. (10.89). Nevertheless, it is possible to interpret these low data points as potential signature of a finite universe. The weakness of the quadrupole ($l = 2$) and octupole term ($l = 3$) can be taken as a lack of temperature correlation on scales greater than $60°$. Maybe the space is not infinite and the broadest waves are missing because space is not big enough to accommodate them. Our discussion above has shown the evidence for the space being locally homogenous and isotropic. However, local geometry constrains, but does not dictate, the shape of the space. Thus, it is possible that the topology of the universe is nontrivial. Luminet *et al.* (2003) constructed just such a model universe based on a finite space with a nontrivial topology (the Poincaré dodecahedral space). It has a positive curvature (closed universe) with $\Omega_0 = 1.013$, which is compatible with observation as of 2003. One of the ways to study the shape, or topology, of the universe is based on the idea that if the universe is finite, light from a distant source will be able to reach us along more than one path. This will produce matching images (e.g. circles) in the CMB anisotropy. A search for such matching circles failed to find such features (Cornish *et al.,* 2004). Thus, this finite universe model may, in the end, be ruled out by observation.

Our purpose in reporting this particular episode in the cosmological study is to remind ourselves of the importance of keeping an open mind of alternative cosmologies. This example showed vividly how drastically different cosmological pictures can be based on cosmological parameters that are not that different from each other. Thus, when looking at a result such as $\Omega_0 = 1.03 \pm 0.03$, as known then in 2003, we should refrain from jumping to the conclusion that data has already shown a $\Omega_0 = 1$ flat universe. This shows the importance of acquiring high precision data, which will ultimately decide which model gives us the true cosmology. On the other hand, while a slight change of one or two parameters may favor different cosmological models, it is the overall theoretical consistency, the ability to account for a whole array of data in cosmology and robust in its cross-checks that ultimately allows us to believe that the current concordant picture has a good chance to survive future experimental tests.

11.6   Appendix C: False vacuum and hidden symmetry

In Section 11.2.2 we discussed the theoretical suggestion that the cosmological inflationary epoch is associated with a "false vacuum" of an inflation/Higgs field. This involves the concept of a "spontaneous breakdown of a symmetry," also described as a "hidden symmetry"—even though a theory is symmetric, its familiar symmetry properties are hidden. Namely, a symmetric theory somehow ends up having asymmetrical solutions. This can happen, as we shall see, when there are "degenerate ground states"—an infinite number of theoretically possible states (related to each other by symmetry transformations)

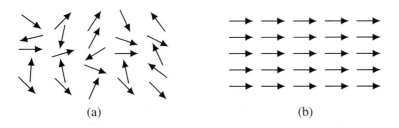

Fig. 11.14 (a) Ground state with zero magnetization $\vec{\mathcal{M}}_0 = 0$ for randomly oriented dipoles, (b) asymmetric ground state with $\vec{\mathcal{M}}_0 \neq 0$.

(a)   (b)

all having the same lowest energy. But the physical vacuum is one of this set, and, by itself, it is not symmetric because it singles out a particular direction in the symmetry space. In this Appendix, we illustrate this phenomenon by the example of the breakdown of rotational symmetry in a ferromagnet near the Curie temperature.

Hidden rotational symmetry in a ferromagnet   A ferromagnet can be thought of as a collection of magnetic dipoles. When it is cooled below a certain critical temperature, the Curie temperature T_c, it undergoes spontaneous magnetization: all its dipoles are aligned in one particular direction (a direction determined not by dipole interactions, but by external boundary conditions). Namely, when $T > T_c$ the ground state has zero **magnetization** $\vec{\mathcal{M}}_0 = 0$ because the dipoles are randomly oriented; but below the critical temperature $T < T_c$, all the dipoles line up, giving arise to a nonzero magnetization $\vec{\mathcal{M}}_0 \neq 0$ (Fig. 11.14). This can happen even though the underlying dynamics of dipole–dipole interaction is rotationally symmetric—no preferred direction is built into the dynamics, that is, the theory has rotation symmetry.

Ginzburg and Landau description of spontaneous symmetry breaking
For a mathematical description we shall follow the phenomenological theory of Ginzburg and Landau. When $T \approx T_c$, the rotationally symmetric free energy $\mathcal{F}(\vec{\mathcal{M}})$ of the system can be expanded in a power series of the magnetization $\vec{\mathcal{M}}$:

$$\mathcal{F}(\vec{\mathcal{M}}) = \left(\nabla_i \vec{\mathcal{M}}\right)^2 + \underbrace{a(T)(\vec{\mathcal{M}} \cdot \vec{\mathcal{M}}) + b(\vec{\mathcal{M}} \cdot \vec{\mathcal{M}})^2}_{V(\vec{\mathcal{M}})}. \tag{11.49}$$

In the potential energy function $V(\vec{\mathcal{M}})$ we have kept the higher order $(\vec{\mathcal{M}} \cdot \vec{\mathcal{M}})^2$ term, with a coefficient $b > 0$ (as required by the positivity of energy at large $\mathcal{M}$), because the coefficient a in front of the leading $(\vec{\mathcal{M}} \cdot \vec{\mathcal{M}})$ term can vanish: $a(T) = \gamma(T - T_c)$. With γ being some positive constant, the temperature-dependent coefficient a is positive when $T > T_c$, negative when $T < T_c$. Since the kinetic energy term $(\nabla_i \vec{\mathcal{M}})^2$ is nonnegative, to obtain the ground state, we need only to minimize the potential energy:

$$\frac{dV}{d\vec{\mathcal{M}}} \propto \vec{\mathcal{M}}\left[a + 2b\left(\vec{\mathcal{M}} \cdot \vec{\mathcal{M}}\right)\right] = 0. \tag{11.50}$$

The solution of this equation gives us the ground state magnetization $\vec{\mathcal{M}}_0$. For $T > T_c$, hence a positive a, we get the usual solution of a zero magnetization $\vec{\mathcal{M}}_0 = 0$ (i.e. randomly oriented dipoles). This situation is shown in the plot of

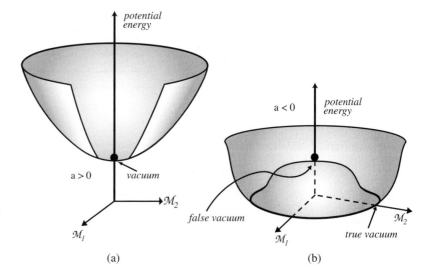

Fig. 11.15 Symmetric potential energy surfaces in the field space: (a) the normal solution, when the ground state is at a symmetric point with $\mathcal{M}_0 = 0$, and (b) the broken symmetry solution, when the energy surface has the shape of a "Mexican hat" with $\mathcal{M} = 0$ being a local maximum and the true ground state being one point in the trough (thus singling out one direction and breaking the rotational symmetry).

[31] We have simplified the display to the case when $\vec{\mathcal{M}}$ is a 2D vector in a plane having two components $\mathcal{M}_1$ and $\mathcal{M}_2$.

$V(\vec{\mathcal{M}})$ of Fig. 11.15(a), where the potential energy surface is clearly symmetric with any rotation (in the 2D plane[31]) around the central axis. However, for subcritical temperature $T < T_c$, the sign change of a brings about a change in the shape of the potential energy surface as in Fig. 11.15(b). The surface remains symmetric with respect to rotation, but the zero magnetization point $\mathcal{M} = 0$ is now a local maximum. There is an infinite number of theoretically possible ground states at the bottom ring of the wine-bottle shaped surface—all having nonzero magnetization $\mathcal{M}_0 = \sqrt{-a/2b}$, but pointing in different directions in the 2D field space. These possible ground states are related to each other by rotations. The physical ground state, picked to be one of them by external conditions, singles out one specific direction, and hence is not rotationally symmetric. Below the Curie temperature, rotational symmetry in the ferromagnet is spontaneously broken and the usual symmetry properties of the underlying dynamics (in this case, rotational symmetry) are not apparent. We say spontaneous symmetry breaking corresponds to a situation of **hidden symmetry**.

Higgs/inflation field   In particle physics we have a system of fields. In particular it is postulated that there are scalar fields (for particles with zero spin) which have potential energy terms displaying the same spontaneous symmetry properties as ferromagnetism near T_c. The magnetization $\vec{\mathcal{M}}$ in (11.49) is replaced, in the case of particle physics, by a scalar field $\phi(x)$. Thus at high energy (i.e. high temperature) the system is in a symmetric phase (normal solution with $a(T) > 0$) and the unification of particle interactions is manifest (cf. Section 11.2.2, see in particular sidenote 6); at lower energy (low temperature) the system enters a broken symmetry phase because of $a(T) < 0$. The ground state of a field system is, by definition, the vacuum. In this hidden symmetry phase we have a nonvanishing scalar field $\phi_0(x) \neq 0$. The relevance to cosmology is as follows: at higher temperature we have a symmetric vacuum. When the universe cools below the critical value, the same state becomes a local maximum and is at a higher energy than, and begins to roll toward, the true vacuum. We say the system (the universe) is temporarily, during the rollover period, in a false vacuum (cf. Fig. 11.3). This semiclassical description indicates the existence of a constant field $\phi_0(x) \neq 0$ permeating everywhere in the universe.

11.7 Appendix D: Quantum vacuum energy as the cosmological constant

We associate the quantum vacuum energy with the dark energy that drives the accelerating universe, while it is the false vacuum energy of a Higgs field that supposedly brings about the primordial inflation. While these two mechanisms may well be related, our presentation assumes that they are separate. In the previous section we discussed the scalar field that can give rise to an inflationary exponential expansion; here we shall concentrate our discussion of the dark energy associated with the accelerating universe. As we shall explain, the zero-point energy of the quantum fields is a natural candidate for such a cosmological constant Λ. However the difficulty of such an association is that the natural size of such a quantum vacuum energy is much too large to account for the observed value of $\Omega_\Lambda \simeq 0.75$. We also briefly note that quantum vacuum energies of boson and fermion fields have opposite signs. Had our universe obeyed supersymmetry exactly with a strict degeneracy of bosonic and fermionic degrees of freedom, their respective contributions to the vacuum energy would exactly cancel, leading to a vanishing vacuum energy and cosmological constant.

Quantum vacuum energy gives rise to a cosmological constant

From the view point of quantum field theory, a vacuum state is not simply "nothingness." The uncertainty principle informs us that the vacuum has an energy because any localization has an associated spread in the momentum value. In fact, quantum field theory pictures the vacuum (defined as the state of lowest energy) as a sea of sizzling activities with constant creation and annihilation of particles. Thus, the cosmological constant Λ, as the energy density of the vacuum, naturally has a non-zero value (Zel'dovich 1968).

The simplest way to see that a quantum vacuum state has energy is to start with the observation that the normal modes of a field are simply a set of harmonic oscillators.[32] Summing over the quantized oscillator energies of all the modes, we have[33] (with the occupation number of the ith state denoted by n_i)

$$E_{\mathrm{b}} = \sum_i \left(\frac{1}{2} + n_i \right) \hbar \omega_i, \qquad \text{with } n_i = 0, 1, 2, 3, \ldots \qquad (11.51)$$

From this we can identify the vacuum energy (also called the zero-point energy) as

$$E_\Lambda = \sum_i \frac{1}{2} \hbar \omega_i. \qquad (11.52)$$

[32]We recall that the Fourier coefficients of, say, an electromagnetic field obey the simple harmonic oscillator equations.

[33]The subscript b stands for "boson;" this bosonic contribution will be compared to that by a fermion field in the last subsection.

At the atomic and subatomic levels, there is abundant empirical evidence for the reality of such a zero-point energy. For macroscopic physics, a notable manifestation of the vacuum energy is the Casimir effect, which has been verified experimentally.

The zero-point energy has the key property of having a density that is unchanged with respect to any volume changes, i.e. an energy density that is constant. The summation of the mode degrees of freedom in Eq. (11.52) involves the enumeration of the phase space volume in units of Planck's constant $\Sigma_i = \int d^3x d^3p \, (2\pi\hbar)^{-3}$, cf. Eq. (11.53). Since the zero-point energy $\hbar\omega_i = E(p)$ has no dependence on position, one obtains a simple volume factor $\int d^3x = V$ so that the corresponding energy per unit volume $E_\Lambda V^{-1}$ is a constant with respect to changes in volume. As explained in Section 11.1, this constant energy density corresponds to a negative pressure ($p = -E/V$) and implies a $-\partial E/\partial x$ force that is attractive, pulling-in the piston in Fig. 11.1. This is the key property of the cosmological constant and is the origin of the Casimir effect—an attractive force between two parallel conducting plates.

Quantum vacuum energy is 10^{120}-fold too large as dark energy

Nevertheless, a fundamental problem exists because the natural size of a quantum vacuum energy is enormous. Here is a simple estimate of the sum in (11.52). The energy of a particle with momentum p is $E(p) = \sqrt{p^2c^2 + m^2c^4}$, see (3.37). From this we can calculate the sum by integrating over the momentum states to obtain the vacuum energy/mass density,

$$\rho_\Lambda c^2 = \frac{E_\Lambda}{V} = \int_0^{E_{Pl}/c} \frac{4\pi p^2 dp}{(2\pi\hbar)^3} \left(\frac{1}{2}\sqrt{p^2c^2 + m^2c^4}\right), \tag{11.53}$$

where $4\pi p^2 dp$ is the usual momentum phase space volume factor. The integral in (11.53) would be a divergent quantity had we carried the integration to its infinity limit. Infinite momentum means zero distance; infinite momentum physics means zero distance scale physics. Since we expect spacetime to be quantized at the Planck scale (cf. quantum gravity in Section 8.5.1), it seems natural that we should cut off the integral at the Planck momentum $p_{Pl} = E_{Pl}/c$ as any GR singularities are expected to be modified at the short distance of the Planck length. The Planck momentum being given in Eq. (8.75),

$$p_{Pl} = \frac{E_{Pl}}{c} = \sqrt{\frac{\hbar c^3}{G_N}} \simeq 10^{19} \text{ GeV}/c. \tag{11.54}$$

the integral (11.53) yields

$$\rho_\Lambda c^2 \cong \frac{1}{16\pi^2} \frac{E_{Pl}^4}{(\hbar c)^3} \simeq \frac{\left(3 \times 10^{27} \text{ eV}\right)^4}{(\hbar c)^3} \tag{11.55}$$

Since the critical density of (9.17) may be written in such natural units[34] as:

$$\rho_c c^2 \cong \frac{\left(2.5 \times 10^{-3} \text{ eV}\right)^4}{(\hbar c)^3},$$

[34] Since the natural "conversion constant" $\hbar c$ has the unit of length times energy, the combination (energy)^4/($\hbar c$)^3 has the correct unit of energy per volume.

we have a quantum vacuum energy density ratio $\Omega_\Lambda \equiv \rho_\Lambda/\rho_c$ that is more than a factor of 10^{120} larger than the observed value of dark energy density,

$$(\Omega_\Lambda)_{qv} \simeq 10^{120} \quad \text{vs.} \quad (\Omega_\Lambda)_{obs} \simeq 0.75.$$

Thus if the observed dark energy originates from quantum vacuum energy, it must involve some mechanism to reduce its enormous natural size down to the critical density size. This "cosmological constant problem," in this context, is the puzzle why and how such a fantastic cancellation takes place —a cancellation of the first 120 significant figures (yet stops at the 121st place)!

Partial cancellation of boson and fermion vacuum energies

We should note that in the above calculation, we have assumed that the field is a boson field (such as the photon and graviton fields), having integer spin and obeying Bose–Einstein statistics. The oscillator's creation and annihilation operators obey commutation relations, leading to symmetric wavefunctions. On the other hand, fermions (such as electrons, quarks, etc.) have half-integer spins, and obey Fermi–Dirac statistics. Their fields have normal modes behaving like **Fermi oscillators**. The corresponding creation and annihilation operators obey anticommutation relations, leading to antisymmetric wavefunctions. Such oscillators have a quantized energy spectrum as (see, for example, Das, 1993)

$$E_f = \sum_i \left(-\frac{1}{2} + n_i\right) \hbar\omega_i, \qquad \text{with } n_i = 0 \text{ or } 1 \text{ only.} \tag{11.56}$$

For a fermion field, the zero point energy is negative! Therefore, there will be a cancellation in the contributions by bosons and fermions.[35] Many of the favored theories to extend the Standard Model of particle physics to the Planck scale incorporate the idea of **supersymmetry**. In such theories the bosonic and fermionic degrees of freedom are equal. In fact, the vacuum energy of systems with exact supersymmetry must vanish (i.e. an exact cancellation). However, we know that in reality supersymmetry cannot be exact[36] because its implication of equal boson and fermion masses $m_f = m_b$ in any supersymmetric multiplet is not observed in nature. If the supersymmetry is broken, we expect only a partial cancellation between bosons and fermions, even though there are equal numbers of bosonic and fermionic degrees of freedom. The first-order fermion and boson contributions in Eq. (11.53) would lead to a result that modifies the boson Eq. (11.55) as

$$\left(\rho_\Lambda c^2\right)_{susy} \cong \frac{1}{16\pi^2} \frac{E_{Pl}^4}{(\hbar c)^3} \left(\frac{\Delta m^2 c^4}{E_{Pl}^2}\right), \tag{11.57}$$

where Δm^2 is the fermion and boson mass difference $m_f^2 - m_b^2$. The fact we have so far not observed any superparticles means that such particles must at least be heavier by $\Delta m^2 \gtrsim (10^2 \text{ GeV}/c)^2$, which can only produce a suppression factor $(\Delta m^2 c^4/E_{Pl}^2) = 10^{-36}$ at the most—thus still some 80 to 90 orders short of the required $O(10^{-120})$. Clearly, something fundamental is missing in our understanding of the physics behind the dark energy.

[35] The vacuum energy is, of course, the sum totaling up the contributions from all quantum fields (gravitons, gauge bosons, leptons, and quarks, etc.).

[36] For example, we do not see a spin-zero particle, a "selectron," having the same properties, and degenerate in mass, as the electron; similarly we have not detected the photon's superpartner, a massless spin-$\frac{1}{2}$ particle called the "photino," etc. A plausible interpretation is that supersymmetry is broken, and the superpartners (selectrons, photinos, etc.) of the known particles (electron, photons, etc.) are much more massive and are yet to be produced and detected in our high energy laboratories.

Review questions

1. Use the first law of thermodynamics to show that the constancy of a system's energy density (even as its volume changes) requires this density to be equal to the negative of its pressure.

2. A vacuum energy dominated system obeys Newton's equation $\nabla^2 \Phi = -\Lambda c^2$, where Λ is a positive constant. What is the gravitational potential $\Phi(r)$ satisfying this equation? From this find the corresponding gravitational field $\mathbf{g}(r) \equiv -\nabla \Phi(r)$.

3. From the Friedmann equation $1 - \Omega(t) = -kc^2/\dot{a}(t)R_0^2$ and the fact that the universe has been matter-dominated since the radiation–matter equality time with redshift $z_{RM} = O(10^4)$, show that the deviation of energy density ratio Ω from unity at t_{RM} must be a factor of 10 000 times smaller than that at the present epoch t_0:

$$[1 - \Omega(t_{RM})] = \left[1 - \Omega(t_0)\right] \times 10^{-4}.$$

Use this result (and its generalization) to explain the flatness problem.

4. What is the horizon problem? Use the result that the angular separation corresponding to one horizon length at the photon decoupling time is about one degree (for a flat universe) to explain this problem.

5. Use a potential energy function diagram to explain the idea of a phase transition in which the system is temporarily in a "false vacuum." How can such a mechanism be used to give rise to an effective cosmological constant?

6. Give a simple physical justification of the rate equation obeyed by the scale factor $\dot{a}(t) \propto a(t)$ in a vacuum energy dominated universe. Explain how the solution $a(t)$ of such a rate equation can explain the flatness and horizon problems.

7. How does the inflationary cosmology explain the origin of mass and energy in the universe as well as the origin of the cosmic structure we see today?

8. The CMB power spectrum can be divided into three regions. What physics corresponds to each region?

9. How can the observed temperature anisotropy of the CMB be used to deduce that the average geometry of the universe is flat?

10. The age of a flat universe without the cosmological constant is estimated to be $\frac{2}{3}t_H \approx 9$ Gyr. Why can an accelerating universe increase this value?

11. What is dark energy? How is it different from dark matter? How is Einstein's cosmological constant related to such energy/matter contents? Do cosmic neutrinos contribute to dark energy?

12. Give two reasons to explain why type Ia Supernovae are ideal "standard candles" for large cosmic scale measurements.

13. Why should the accelerating universe lead us to observe the galaxies, at a given redshift, to be dimmer than expected (in an empty or decelerating universe)?

14. Why is the observation of supernovae with the highest redshifts (> 0.7) to be in the decelerating phase taken to be convincing evidence that the accelerating universe interpretation of SNe data ($z\ 0.2 - 0.7$) is correct?

15. What is the cosmic coincidence problem?

16. What is the standard Λ CDM cosmology? What is the spacetime geometry in this cosmological model? How old is the universe? What is the energy/matter content of the universe?

Problems

11.1 Another form of the expansion equation   Use either the Friedmann equation or its quasi-Newtonian analog to derive (11.25).

11.2 The epoch-dependent Hubble's constant and $a(t)$ Use (10.7) to replace the curvature parameter k in the Friedmann equation (10.1) to show the epoch dependence of Hubble constant through its relation to the density parameters as in (11.38).

11.3 **Luminosity distance and redshift in a flat universe** Knowing the redshift-dependence of the Hubble's constant from Problem 11.2 in a flat universe with negligible $\Omega_{R,0}$, show that the Hubble curve $d_L(z)$ can be used to extract the density parameters Ω_M and Ω_Λ from the simple relation

$$d_L(z) = c(1+z) \int_0^z \frac{c\,dz'}{H_0 \left[\Omega_{M,0} \left(1 + z'\right)^3 + \Omega_\Lambda\right]^{1/2}}.$$

11.4 **Negative Λ and the "big crunch"** Our universe is spatially flat with the dominant component being matter and positive dark energy. Its fate is an unending exponential expansion. Now consider the same flat universe but with a negative dark energy $\Omega_\Lambda = 1 - \Omega_{M,0} < 0$, which provides a gravitational attraction, cf. (11.7). Show that this will slow the expansion down to a standstill when the scale factor reaches $a_{max} = (-\Omega_\Lambda/\Omega_{M,0})^{1/3}$. The subsequent contraction will reach the big crunch $a(t_*) = 0$ at the cosmic time $t_* = \frac{2}{3}\pi t_H(-\Omega_\Lambda)^{-1/2}$.

11.5 **Estimate of matter and dark energy equality time** Closely related to the deceleration/acceleration transition ("inflection") time is the epoch when the matter and dark energy components are equal. Show that the redshift result $z_{M\Lambda}$ obtained in this way is comparable to that of (11.48). Estimate the cosmic time $t_{M\Lambda}$ when the matter-dominated universe changed into our present dark energy-dominated universe.

RELATIVITY
Full Tensor Formulation

Tensors in special relativity

- We introduce the mathematical subject of tensors in a general coordinate system and apply it to the 4D continuum of Minkowski spacetime.
- When a tensor is expanded in terms of a set of basis vectors, the coefficients of expansion are its contravariant components with respect to this basis. Likewise, when a tensor is expanded in terms of inverse (reciprocal) basis vectors, the coefficients of expansion are its covariant components.
- The requirement of metric invariance in Minkowski spacetime leads to a generalized orthogonality condition, from which the Lorentz transformation can be derived.
- When physics equations are written as Minkowski 4-tensor equations, they are automatically unchanged under a coordinate transformation; hence they respect the principle of relativity. Such a formalism is said to be "manifestly covariant."
- The six components of the electric and magnetic fields are elements of the antisymmetric electromagnetic field tensor $F_{\mu\nu}$.
- Maxwell's equations, the Lorentz force law, and the charge conservation equation are presented in their covariant forms.
- The symmetric energy–momentum tensor, $T_{\mu\nu}$, of a field system is introduced and the physical meaning of its components is discussed.

In the Introduction of Chapter 1, we emphasized the coordinate symmetry approach to relativity. The principle of relativity says that physics equations should be covariant under coordinate transformations. To ensure that this principle is **automatically** satisfied, one needs to write physics equations in terms of tensors. Tensors are mathematical objects having definite transformation properties under coordinate transformations. The simplest examples are scalars and vectors. If every term in an equation is a tensor with the same transformation property under a coordinate transformation that has the same tensor property, i.e. transforms in the same way under coordinate transformations, then the relational form of the equation is altered under such transformations. In this and the next two chapters, the full tensor formalism—hence the symmetry viewpoint of relativity—will be presented. In this chapter, we deal with some of the basic properties of tensors in 4D spacetime. This formalism is adequate for global Lorentz transformations, which are the transformations that are relevant for special relativity. In the next chapter, we will discuss the topic of tensor equations that are covariant under the local (position-dependent) transformations of general relativity.

12.1   General coordinate systems

As first noted in Section 3.1.1, **basis vectors** in a general coordinate system (as opposed to a Cartesian coordinate system), are not, in general, mutually orthogonal nor of unit length: $\mathbf{e}_\mu \cdot \mathbf{e}_\nu \equiv g_{\mu\nu} \neq \delta_{\mu\nu}$. Cartesian coordinates can be used for a Euclidean space; we call any other system "general coordinates" (including the coordinates for Minkowski space). This means that, for the generalized coordinate systems, the basis $\{\mathbf{e}_\mu\}$, as well as the corresponding metric $g_{\mu\nu}$, is not its own inverse.[1] For a generalized coordinate system, we can define an inverse basis and an inverse metric, distinct from the regular basis and its metric. Let us denote the **inverse basis** by a set of vectors $\{\mathbf{e}^\mu\}$. Standard notation is to use subscript indices to label regular basis vectors and superscript indices to label inverse basis vectors. The relationship between the regular basis vectors and the inverse basis vectors is expressed as an orthonormality condition through their dot products:

$$\mathbf{e}_\mu \cdot \mathbf{e}^\nu = \delta_\mu^\nu. \tag{12.1}$$

That is, each regular basis vector is orthogonal to all but one of the inverse basis vectors, and vice versa. For the pairs of regular and inverse basis vectors that are not orthogonal, the dot product of the two vectors is one. Also, we have the completeness condition of $\mathbf{e}_\mu \otimes \mathbf{e}^\mu = \mathbf{1}$, where the symbol $\otimes$ stands for "direct product" (also called "outer product"). This corresponds to the component multiplication of $(e_\mu)_\alpha (e^\mu)_\beta = \delta_{\alpha\beta}$. (See Problem 12.1 for an illustrative example.) In a four-dimensional space, there are four basis vectors,[2] $\{\mathbf{e}_\mu\}$, with $\mu = 0, 1, 2, 3$. Each of the basis vectors in turn has four components $(e_\mu)_\alpha$, with $\alpha = 0, 1, 2, 3$. For example, for the Cartesian coordinate system,

$$\mathbf{e}_0 = \begin{pmatrix} 1 \\ 0 \\ 0 \\ 0 \end{pmatrix}, \quad \mathbf{e}_1 = \begin{pmatrix} 0 \\ 1 \\ 0 \\ 0 \end{pmatrix}, \quad \text{etc.} \tag{12.2}$$

The corresponding inverse basis vectors are

$$\mathbf{e}^0 = \begin{pmatrix} 1 & 0 & 0 & 0 \end{pmatrix}, \quad \mathbf{e}^1 = \begin{pmatrix} 0 & 1 & 0 & 0 \end{pmatrix}, \text{ etc.} \tag{12.3}$$

Equation (12.1) can be written out in terms of the vector components $\sum_\alpha (e_\mu)_\alpha (e^\nu)_\alpha = \delta_\mu^\nu$. The dot products of the regular basis vectors with each other and the inverse basis vectors with each other are the metric and inverse metric functions:

$$\mathbf{e}_\mu \cdot \mathbf{e}_\nu \equiv g_{\mu\nu} \qquad \text{metric}$$

$$\mathbf{e}^\mu \cdot \mathbf{e}^\nu \equiv g^{\mu\nu} \qquad \text{inverse metric.} \tag{12.4}$$

[1] Two mathematical objects A and $B \equiv A^{-1}$ are said to be inverse to each other if their product $AB = BA = 1$. Thus, if we have $AA = 1$, we say A is its own inverse. In this sense, if the bases satisfy the relation $\mathbf{e}_\mu \cdot \mathbf{e}_\nu = \delta_{\mu\nu}$, we say the basis is its own inverse, since the Kronecker delta may be regarded as representing the matrix elements of the identity $1_{\mu\nu} = \delta_{\mu\nu}$. The inverse base $\{\mathbf{e}^\mu\}$ are also referred to as the **reciprocal bases** or **dual bases**.

[2] Here we confine ourselves to 4D Minkowski space for which $\mu = 0, 1, 2, 3$. The coordinate frame formed by the set of four basis vectors $\{\mathbf{e}_\mu\}$ is referred to in the literature as a **tetrad** (or, in German, as a **vierbein**). The general case of an n-dimensional space, with indices running from 1 to n, can be similarly worked out.

These metric matrices are inverse to each other,

$$g_{\mu\nu}g^{\nu\lambda} = \delta^{\lambda}_{\mu}, \tag{12.5}$$

as can be shown by the orthonormality condition (12.1) and the completeness condition. Again, we follow the **Einstein summation convention** of omitting the display of the summation symbol; the summation over any pairs of repeated upper and lower indices (called a **contraction**) is understood.

12.1.1 Contravariant and covariant components

Because there are two sets of coordinate basis vectors, $\{\mathbf{e}_{\mu}\}$ and $\{\mathbf{e}^{\mu}\}$, there are two possible expansions for each vector $\mathbf{A}$:

Expansion of $\mathbf{A}$	Projections	Component names	
$\mathbf{A} = A^{\mu}\mathbf{e}_{\mu}$	$A^{\mu} = \mathbf{A} \cdot \mathbf{e}^{\mu}$	**contravariant** components of $\mathbf{A}$	(12.6)
$\mathbf{A} = A_{\mu}\mathbf{e}^{\mu}$	$A_{\mu} = \mathbf{A} \cdot \mathbf{e}_{\mu}$	**covariant** components of $\mathbf{A}$	

Repeated indices are summed in the expansions of the vector $\mathbf{A}$. It is common notation to use A_{μ} or A^{ν} to refer to the vector $\mathbf{A}$ itself, even though A_{μ} and A^{ν} are technically the contravariant and covariant **components**, respectively, of the vector $\mathbf{A}$. In this notation, A_{μ} is a contravariant **vector** and A^{ν} is a covariant **vector**.

It may be helpful to have a way to visualize, in a simple situation, the contravariant and covariant components. For an oblique Cartesian coordinate system in a 2D flat plane (Fig. 12.1), these two types of components correspond to the following projections: the contravariant components are the parallel projections of a vector onto the basis vectors, while the covariant components are the perpendicular projections. In Figure 12.1, the parallel projection of vector $\mathbf{A}$ onto $\mathbf{e}_{1}$, for instance, is obtained by drawing a line from the tip of $\mathbf{A}$ to $\mathbf{e}_{1}$ that is parallel to $\mathbf{e}_{2}$. The perpendicular projection of $\mathbf{A}$ onto $\mathbf{e}_{1}$ is obtained by drawing a line from the tip of $\mathbf{A}$ to $\mathbf{e}_{1}$ that is perpendicular to $\mathbf{e}_{1}$. Note that the distinction between parallel and perpendicular projections of a vector is important only for nonorthogonal bases. For the case in which the basis vectors are orthogonal to each other, the parallel and perpendicular projections of a vector are identical.

One of the principal advantages of introducing these two types of tensor components is the simplicity of the resultant scalar product of two vectors (or of a vector with itself), which, after using (12.1), can always be expressed as

$$\mathbf{A} \cdot \mathbf{B} = \left(A_{\nu}\mathbf{e}^{\nu}\right) \cdot \left(B^{\mu}\mathbf{e}_{\mu}\right) = A_{\nu}\left(\mathbf{e}^{\nu} \cdot \mathbf{e}_{\mu}\right)B^{\mu} = A_{\mu}B^{\mu}. \tag{12.7}$$

We can also write this as the product of a row vector and a column vector:[3]

$$\mathbf{A} \cdot \mathbf{B} = \begin{pmatrix} A_0 & A_1 & A_2 & A_3 \end{pmatrix} \begin{pmatrix} B^0 \\ B^1 \\ B^2 \\ B^3 \end{pmatrix}. \tag{12.8}$$

[3]In our study of vectors and tensors, we sometimes represent them explicitly as matrices, with the pedagogical rationale that a reader is likely be more familiar with such math material. Such a presentation may ease the process of learning to manipulate the abstract tensor components. One should note that there is no rule dictating that a covariant vector should be represented as a row vector and a contravariant vector, as a column vector. In (12.8), for example, we could just as well have written $\mathbf{B} \cdot \mathbf{A}$, which could then be represented as the matrix product of $\mathbf{B}$, the contravariant vector, written as a row vector and $\mathbf{A}$, the covariant vector, written as a column vector. It should also be noted that there is no matrix representation of tensors with more than two indices (unless one fixes all indices except two). In short, there is no general way to represent tensors by matrices and in fact there is no need to do so.

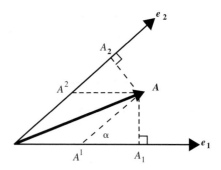

Fig. 12.1 Contravariant components (A^1, A^2) and covariant components (A_1, A_2) of a vector $\mathbf{A}$ in a general coordinate system. For the simple case of basis vectors having unit length, they are seen to be related as $A_1 = (A^1 + A^2 \cos \alpha)$ and $A_2 = (A^1 \cos \alpha + A^2)$.

If we had used the expansions only in terms of the basis $\{\mathbf{e}_\mu\}$, or only the inverse basis $\{\mathbf{e}^\mu\}$, we would have to display the metric tensors in such a contraction:

$$\mathbf{A} \cdot \mathbf{B} = g_{\mu\nu} A^\mu B^\nu \tag{12.9}$$

$$= g^{\mu\nu} A_\mu B_\nu. \tag{12.10}$$

A comparison of (12.7) with (12.9) and (12.10) shows that the contravariant and covariant components of a vector are related to each other through the metric:

$$A_\mu = g_{\mu\nu} A^\nu, \quad A^\mu = g^{\mu\nu} A_\nu. \tag{12.11}$$

We say that tensor indices can be lowered or raised through contractions with a metric tensor or inverse metric tensor. For the case of the vectors being infinitesimal coordinate vectors, $\mathbf{A} = \mathbf{B} = d\mathbf{x}$, the scalar product of (12.9) is just the (infinitesimal) invariant interval $d\mathbf{x} \cdot d\mathbf{x} = ds^2$:

$$ds^2 = g_{\mu\nu} dx^\mu dx^\nu. \tag{12.12}$$

This equation relating the length to the coordinates is often taken as another definition of the metric.

12.1.2   Coordinate transformations

The transformation of contravariant components, under a coordinate (bases) change, has already been presented in Chapter 3 (see especially Section 3.1.2); it may be written as

[4] Any two-indexed quantity can be written as a matrix. We shall follow the convention of letting the first (i.e. the left) index be the row index while the second (the right) the column index. Thus, in Eq. (12.13), for a given row μ we sum over the product of ν column elements of the transformation matrix and ν column elements of the vector. As pointed out in the previous sidenote, there is no simple way for a matrix to distinguish between upper and lower indices. That is, just by looking at a matrix, one will not be able to tell whether it represents $T_{\mu\nu}$ or $T^{\mu\nu}$, or T^μ_ν.

$$\begin{pmatrix} A^0 \\ A^1 \\ A^2 \\ A^3 \end{pmatrix} \longrightarrow \begin{pmatrix} A'^0 \\ A'^1 \\ A'^2 \\ A'^3 \end{pmatrix} = \begin{pmatrix} L^0{}_0 & L^0{}_1 & \cdots & L^0{}_3 \\ L^1{}_0 & L^1{}_1 & & \\ & & & \\ L^3{}_0 & \cdots & & L^3{}_3 \end{pmatrix} \begin{pmatrix} A^0 \\ A^1 \\ A^2 \\ A^3 \end{pmatrix}.$$

A^μ represent the components of the vector $\mathbf{A}$ in the original coordinate system, while A'^μ are those with respect to the transformed system. This relation can be written in a more compact notation[4] as (see Box 2.1 and Section 3.1.2)

$$A^\mu \longrightarrow A'^\mu = [\mathbf{L}]^\mu{}_\nu A^\nu. \tag{12.13}$$

It is important to keep in mind that contravariant and covariant components transform differently under a coordinate transformation. In fact the transformation of covariant components is[5]

$$A_\mu \longrightarrow A'_\mu = A_\nu \left[\mathbf{L}^{-1}\right]^\nu_{\ \mu}. \tag{12.14}$$

where $\mathbf{L}^{-1}\mathbf{L} = \mathbf{1}$, so that the contraction between a contravariant and a covariant vector corresponds to a scalar, invariant under the coordinate transformation:

$$A'_\mu B'^\mu = A_\nu \left[\mathbf{L}^{-1}\right]^\nu_{\ \mu} [\mathbf{L}]^\mu_{\ \lambda} B^\lambda = A_\nu [\mathbf{1}]^\nu_{\ \lambda} B^\lambda = A_\lambda B^\lambda. \tag{12.15}$$

The transformation property of general tensor components can be illustrated by an example using a third-rank tensor:

$$T^\lambda_{\mu\nu} \longrightarrow T'^\lambda_{\mu\nu} = [\mathbf{L}]^\lambda_{\ \gamma} T^\gamma_{\alpha\beta} \left[\mathbf{L}^{-1}\right]^\alpha_{\ \mu} \left[\mathbf{L}^{-1}\right]^\beta_{\ \nu}. \tag{12.16}$$

For each superscript index we have an $\mathbf{L}$ factor, and for each subscript index an $\mathbf{L}^{-1}$ factor. The tensor components $T^\lambda_{\mu\nu}$ must transform in this way so that, after using $\mathbf{L}^{-1}\mathbf{L} = \mathbf{1}$, any contraction of a pair of upper and lower indices reduces the rank of the tensor[6] by two. For example,

$$T^\mu_{\mu\nu} \longrightarrow T'^\mu_{\mu\nu} = [\mathbf{L}]^\mu_{\ \gamma} T^\gamma_{\alpha\beta} \left[\mathbf{L}^{-1}\right]^\alpha_{\ \mu} \left[\mathbf{L}^{-1}\right]^\beta_{\ \nu} = T^\alpha_{\alpha\beta} \left[\mathbf{L}^{-1}\right]^\beta_{\ \nu}.$$

The tensor components $T^\mu_{\mu\nu}$ transform like the components of a covariant vector. We note that the metric $g_{\mu\nu}$, having two covariant indices, must transform as

$$g'_{\mu\nu} = g_{\alpha\beta} \left[\mathbf{L}^{-1}\right]^\alpha_{\ \mu} \left[\mathbf{L}^{-1}\right]^\beta_{\ \nu}, \tag{12.17}$$

which can be written in matrix form as[7]

$$[\mathbf{g}'] = \left[\mathbf{L}^{-1}\right]^\mathsf{T} [\mathbf{g}] \left[\mathbf{L}^{-1}\right] = \left[\mathbf{L}^{-1}\right] [\mathbf{g}] \left[\mathbf{L}^{-1}\right]^\mathsf{T}. \tag{12.18}$$

The matrix $\mathbf{L}^{-1\mathsf{T}}$ is the transpose of $\mathbf{L}^{-1}$; this comes about because in (12.17) we need to interchange the row and column (i.e. the first and second) indices α and μ of the first $\mathbf{L}^{-1}$ matrix in order to have the proper summation of the products of column/row elements in a matrix multiplication.

In a flat space, such as the Minkowski space of SR, we can always use a coordinate system such that the metric is position independent, $\mathbf{g} = \eta$, as in (3.15). We can also require the coordinate transformation that leaves the metric invariant. Keeping the same metric $\eta_{\mu\nu}$ means we require coordinate transformations not to change the geometry, not to take us out of Minkowski spacetime. This is consistent with our conception of spacetime in special relativity as being a fixed stage on which physical processes take place without effecting the spacetime background. With $\mathbf{g}' = \mathbf{g} = \eta$, Eq. (12.18) becomes, after noting that $\eta = \eta^{-1}$,

$$[\eta] = \mathbf{L}^\mathsf{T} [\eta] \mathbf{L} = \mathbf{L} [\eta] \mathbf{L}^\mathsf{T}. \tag{12.19}$$

[5] We remind ourselves that mathematical objects with indices, being the components of tensors and matrices, are ordinary numbers; they are communitive for example, $A_\nu \mathbf{L}^\nu_\mu = \mathbf{L}^\nu_\mu A_\nu$, even though their corresponding matrices are not generally communitive, e.g. $AL \neq LA$. This means that we can move them (the components) around with ease.

[6] A tensor of rank (p, q) is one with p upper indices and q lower indices. Thus, $T^\lambda_{\mu\nu}$ is a component of the tensor with the rank of $(1, 2)$. A contravariant vector is a $(1, 0)$ tensor; a scalar is the $(0, 0)$ tensor, etc.

[7] For the case of the inverse metric $g^{\mu\nu}$, we would be working with $\mathbf{L}$ instead of $\mathbf{L}^{-1}$. Also, all the matrices under discussion, $\mathbf{L}^{-1}$ and $\mathbf{g}$, are square matrices.

The relation can be regarded as the **generalized orthogonality condition** as it reduces to the familiar orthogonality property of $\mathbf{L}^T\mathbf{L} = \mathbf{L}\mathbf{L}^T = \mathbf{1}$ for the Euclidean space with $\mathbf{g}$ replaced by $\mathbf{1}$. (In Problem 12.8, we have used such conditions to derive the explicit form of the Lorentz transformation.)

While the tensor formalism in a flat and a curved space is basically same, there is, however, the crucial difference that in a curved space the basis vectors are necessarily position-dependent, $\{\mathbf{e}_i(x)\}$. Consequently, the metric tensors $g_{\mu\nu} = \mathbf{e}_\mu \cdot \mathbf{e}_\nu$ must also be position-dependent, and they transform nontrivially under coordinate transformations as shown in (12.17), $g'_{\mu\nu} \neq g_{\mu\nu}$. In fact curved spacetime of general relativity is dynamical; the constantly changing geometry reflects the changing distribution of mass and energy.

12.1.3   Position and del operators in Minkowski spacetime

As discussed in Section 3.1.2, if physical quantities are represented by 4-vectors and 4-tensors, the resultant physics equations are automatically covariant under a Lorentz transformation (which is the global coordinate transformation in Minkowski spacetime)—these equations are, manifestly, relativistic. The position-time components x^μ are the contravariant components of the **position 4-vector**:

$$x^\mu = \begin{pmatrix} x^0 \\ x^1 \\ x^2 \\ x^3 \end{pmatrix} = \begin{pmatrix} ct \\ x \\ y \\ z \end{pmatrix}. \tag{12.20}$$

For Minkowski spacetime, the spacetime interval

$$s^2 = -(x^0)^2 + (x^1)^2 + (x^2)^2 + (x^3)^2 = \eta_{\mu\nu} x^\mu x^\nu \tag{12.21}$$

has the same value with respect to every inertial observer. This equation can be interpreted as defining the metric for the flat Minkowski space:

$$g_{\mu\nu} = \text{diag}(-1, 1, 1, 1) \equiv \eta_{\mu\nu}. \tag{12.22}$$

Under a Lorentz transformation, the new components are related to the original ones by

$$x^\mu \rightarrow x'^\mu = [\mathbf{L}]^\mu{}_\nu x^\nu. \tag{12.23}$$

Specifically, for a boost with a relative velocity v in the $+x$ direction, we have, as in Eq. (3.16),

$$\begin{pmatrix} ct' \\ x' \\ y' \\ z' \end{pmatrix} = \begin{pmatrix} \gamma & -\beta\gamma & 0 & 0 \\ -\beta\gamma & \gamma & 0 & 0 \\ 0 & 0 & 1 & 0 \\ 0 & 0 & 0 & 1 \end{pmatrix} \begin{pmatrix} ct \\ x \\ y \\ z \end{pmatrix}, \tag{12.24}$$

where β and γ are defined in (2.12). Because the metric is given by (12.22) the position vector (in contrast to the contravariant position vector) has the covariant components as given by (12.11):

$$x_\mu = \eta_{\mu\nu} x^\nu = (-ct \ x \ y \ z). \tag{12.25}$$

We say that the position 4-vector **x** is naturally contravariant as its covariant vector has an "unnatural" minus sign in the zeroth component. The contraction between the contravariant and covariant components of the position 4-vector yields the invariant interval s^2, and is related to the proper time τ:

$$x_\mu x^\mu = -c^2 t^2 + x^2 + y^2 + z^2 = s^2 = -c^2 \tau^2. \tag{12.26}$$

The gradient operator as the prototype covariant vector

From calculations in Chapter 2 (see, in particular, Problem 2.3) we know that the 4-gradient operator transforms according to an inverse Lorentz transformation

$$\frac{\partial}{\partial x^\mu} \longrightarrow \frac{\partial}{\partial x'^\mu} = \left[\mathbf{L}^{-1}\right]^\nu_\mu \frac{\partial}{\partial x^\nu}. \tag{12.27}$$

For a boost in the $+x$ direction, we have

$$\left[\mathbf{L}^{-1}\right] = \begin{pmatrix} \gamma & \beta\gamma & 0 & 0 \\ \beta\gamma & \gamma & 0 & 0 \\ 0 & 0 & 1 & 0 \\ 0 & 0 & 0 & 1 \end{pmatrix}, \tag{12.28}$$

which is the inverse of **L** in (12.24), and is obtained by the replacement of $(\beta \to -\beta)$. Equation (12.14) makes it clear that the 4-gradient operator is a covariant 4-vector. We shall often use the notation ∂_μ to represent this covariant del operator:

$$\partial_\mu \equiv \frac{\partial}{\partial x^\mu} = \left(\frac{1}{c}\frac{\partial}{\partial t} \ \ \frac{\partial}{\partial x} \ \ \frac{\partial}{\partial y} \ \ \frac{\partial}{\partial z} \right). \tag{12.29}$$

We will use the notation ∂^μ to represent the corresponding contravariant gradient-operator:

$$\partial^\mu = \eta^{\mu\nu}\partial_\nu = \left(-\frac{1}{c}\frac{\partial}{\partial t} \ \ \frac{\partial}{\partial x} \ \ \frac{\partial}{\partial y} \ \ \frac{\partial}{\partial z} \right). \tag{12.30}$$

We see that while the position vector **x** is naturally contravariant, the corresponding derivative operator is naturally covariant. A contraction of the two operators in (12.29) and (12.30) leads to the Lorentz-invariant 4-Laplacian (usually called the **D'Alembertian**) operator:

$$\Box \equiv \partial^\mu \partial_\mu = -\frac{1}{c^2}\frac{\partial^2}{\partial t^2} + \vec{\nabla}^2, \tag{12.31}$$

with the 3-Laplacian operator being $\vec{\nabla}^2 = \partial^2/\partial x^2 + \partial^2/\partial y^2 + \partial^2/\partial z^2$. Thus the relativistic wave equation has the form of $\Box \psi = -c^{-2}\partial^2\psi/\partial t^2 + \vec{\nabla}^2\psi = 0$.

12.2 Manifestly covariant formalism for electromagnetism

In Section 3.2 we have already introduced the velocity 4-vector U^μ, the momentum 4-vector p^μ, and the covariant force K^μ. These are the basic elements that are needed in order to construct equations of relativistic mechanics that are covariant under the Lorentz transformation. They are more general than those of Newtonian mechanics because they satisfy the requirement for (special) relativity symmetry and reduce to the corresponding Newtonian version in the low velocity limit. On the other hand, as shown in Chapter 2 (see, in particular, Box 2.2), the electromagnetic theory is already Lorentz symmetric. In this section we will present the equations of electromagnetism in their "manifestly covariant form." We will write the equations in a form that makes it obvious that these relations do not change under a Lorentz transformation. This is in contrast to the more familiar 3-vector form (2.14) and (2.15) of Maxwell's equations, for which the demonstration of Lorentz covariance requires rather tedious calculations as seen in Problem 2.4.

12.2.1 The electromagnetic field tensor

Relativity unifies space and time: space and time coordinates become components of a common vector (12.20) in the covariant formalism; the spatial and time coordinates can be transformed into each other (12.24) when viewed in different inertial frames. Relativity also makes clear the unification of electricity and magnetism, as the $\vec{E}$ and $\vec{B}$ fields can be transformed into each other by Lorentz transformations (2.16). The components of the $\vec{E}$ and $\vec{B}$ fields must be components belonging to the same Minkowski 4-tensor. The six components, E_i and B_i ($i = 1, 2, 3$), are independent elements of a common antisymmetric 4-field tensor, the **electromagnetic field strength tensor,**

$$F_{\mu\nu} = -F_{\nu\mu}, \tag{12.32}$$

with the assignment of

$$F_{0i} = -F_{i0} = -E_i \qquad F_{ij} = \varepsilon_{ijk} B_k, \tag{12.33}$$

where ε_{ijk} is the totally antisymmetric Levi-Civita symbol with $\varepsilon_{123} = 1$, and all diagonal elements vanish. Writing out (12.33) explicitly, we have the covariant form of the electromagnetic field tensor:

$$F_{\mu\nu} = \begin{pmatrix} 0 & -E_1 & -E_2 & -E_3 \\ E_1 & 0 & B_3 & -B_2 \\ E_2 & -B_3 & 0 & B_1 \\ E_3 & B_2 & -B_1 & 0 \end{pmatrix}. \tag{12.34}$$

Using the metric tensor $\eta^{\mu\nu}$ to raise the indices, we can also write this field tensor in contravariant form:

$$F^{\mu\nu} = \eta^{\mu\lambda} F_{\lambda\rho} \eta^{\rho\nu} = \begin{pmatrix} 0 & E_1 & E_2 & E_3 \\ -E_1 & 0 & B_3 & -B_2 \\ -E_2 & -B_3 & 0 & B_1 \\ -E_3 & B_2 & -B_1 & 0 \end{pmatrix}. \tag{12.35}$$

It is also useful to define the **dual field strength tensor**,

$$\tilde{F}_{\mu\nu} \equiv -\frac{1}{2}\varepsilon_{\mu\nu\lambda\rho} F^{\lambda\rho}, \tag{12.36}$$

where $\varepsilon_{\mu\nu\lambda\rho}$ is the four-dimensional Levi-Civita symbol[8] with $\varepsilon_{0ijk} \equiv \varepsilon_{ijk}$ and thus $\varepsilon_{0123} = 1$. The elements of the dual field tensor are

$$\tilde{F}_{0i} = -\tilde{F}_{i0} = -B_i \quad \tilde{F}_{ij} = -\varepsilon_{ijk} E_k, \tag{12.37}$$

or explicitly,

$$\tilde{F}_{\mu\nu} = \begin{pmatrix} 0 & -B_1 & -B_2 & -B_3 \\ B_1 & 0 & -E_3 & E_2 \\ B_2 & E_3 & 0 & -E_1 \\ B_3 & -E_2 & E_1 & 0 \end{pmatrix}. \tag{12.38}$$

[8]The Levi-Civita symbol in an n-dimensional space is a quantity with n indices. Thus, in a 3D space, we have ε_{ijk}, with the indices running from 1 to 3 (or we could choose for them to run from 0 to 2, for example); in a 4D space, we have $\varepsilon_{\mu\nu\lambda\rho}$, with the indices running from 1 to 4 (or from 0 to 3, which is useful for relativity). The Levi-Civita symbol is totally antisymmetric: an interchange of any two indices results in a minus sign: $\varepsilon_{ijk} = -\varepsilon_{jik} = \varepsilon_{jki}$ *etc.* The symbol vanishes whenever any two indices are equal. For instance, ε_{122} must be zero because an interchange of the last two indices must result in a minus sign ($\varepsilon_{122} = -\varepsilon_{122}$), and the only way this can be true is if $\varepsilon_{122} = 0$. All the nonzero elements can be obtained by permutation of indices from $\varepsilon_{12} = \varepsilon_{123} = \varepsilon_{0123} \equiv 1$. One use of the Levi-Civita symbol is to express the cross-product of vectors, $(\vec{A} \times \vec{B})_i = \varepsilon_{ijk} A_j B_k$. For further discussion of the Levi-Civita symbol, see Section 13.3.

Box 12.1 Lorentz transformation of the electromagnetic fields

With respect to moving observers, the electric and magnetic fields transform into each other. That they are components of a 4-tensor means that they must transform according[9] to (12.16):

$$F^{\mu\nu} \longrightarrow F'^{\mu\nu} = [\mathbf{L}]^{\mu}_{\ \lambda} [\mathbf{L}]^{\nu}_{\ \rho} F^{\lambda\rho}. \tag{12.39}$$

The explicit form of this transformation under a boost [see (12.24)] is given by

$$\begin{pmatrix} 0 & E'_1 & E'_2 & E'_3 \\ -E'_1 & 0 & B'_3 & -B'_2 \\ -E'_2 & -B'_3 & 0 & B'_1 \\ -E'_3 & B'_2 & -B'_1 & 0 \end{pmatrix} \tag{12.40}$$

$$= \begin{pmatrix} \gamma & -\beta\gamma & 0 & 0 \\ -\beta\gamma & \gamma & 0 & 0 \\ 0 & 0 & 1 & 0 \\ 0 & 0 & 0 & 1 \end{pmatrix} \begin{pmatrix} 0 & E_1 & E_2 & E_3 \\ -E_1 & 0 & B_3 & -B_2 \\ -E_2 & -B_3 & 0 & B_1 \\ -E_3 & B_2 & -B_1 & 0 \end{pmatrix} \begin{pmatrix} \gamma & -\beta\gamma & 0 & 0 \\ -\beta\gamma & \gamma & 0 & 0 \\ 0 & 0 & 1 & 0 \\ 0 & 0 & 0 & 1 \end{pmatrix}.$$

One can easily check that the relation given by the tensor transformation (12.40) is just the Lorentz transformation $(\vec{E}, \vec{B}) \to (\vec{E}', \vec{B}')$ as shown in (2.16).

The dual tensor $\tilde{F}^{\mu\nu}$ has the same Lorentz transformation property as the tensor $F^{\mu\nu}$ itself. From this we deduce that there are *two* Lorentz invariant combinations of the electromagnetic fields (constructed from $F^{\mu\nu}$ and $\tilde{F}_{\mu\nu}$

(*cont.*)

[9]In (12.16) we have illustrated the tensor transformation by that of a mixed tensor with one contravariant superscript index and two covariant subsript indices. The rule is that for each contravariant index we have an $\mathbf{L}$ factor, and for each covariant index an $\mathbf{L}^{-1}$ factor. For the present case, we have two superscript indices, hence two $\mathbf{L}$ factors.

[10]Similar to (12.39), have $\tilde{F}'^{\mu\nu} = \mathbf{L}^{\mu}_{\lambda}\mathbf{L}^{\nu}_{\rho}\tilde{F}^{\lambda\rho}$ and $\tilde{F}'_{\mu\nu} = \tilde{F}_{\lambda\rho}\mathbf{L}^{-1\lambda}_{\mu}\mathbf{L}^{-1\rho}_{\nu}$. Also we have $F^{\mu\nu}F_{\mu\nu} = \tilde{F}^{\mu\nu}\tilde{F}_{\mu\nu}$. The combination $\vec{E}^2 + \vec{B}^2$ is not a Lorentz scalar. This combination is the energy density of the electromagnetic field and it transforms as a component of the field's energy–momentum tensor, which we shall discuss in Box 12.4.

Box 12.1 (*Continued*)

with all indices contracted):[10]

$$F^{\mu\nu}F_{\mu\nu} = 2\left(\vec{E}^2 - \vec{B}^2\right) \tag{12.41}$$

and

$$F^{\mu\nu}\tilde{F}_{\mu\nu} = 4\left(\vec{E}\cdot\vec{B}\right). \tag{12.42}$$

We now use the field tensor $F_{\mu\nu}$ to write the equations of electromagnetism in a form that clearly displays their Lorentz covariance.

Lorentz force law

Using (12.33) one can easily show that the electromagnetic force felt by a charge q moving with velocity $\vec{v}$, the Lorentz force,

$$\vec{F} = q\left(\vec{E} + \frac{1}{c}\vec{v}\times\vec{B}\right) \tag{12.43}$$

is just the $\mu = i$ ($i = 1, 2, 3$) part of the covariant force law

$$K^{\mu} = \frac{q}{c}F^{\mu\nu}U_{\nu}, \tag{12.44}$$

where $K^{\mu} = mdU^{\mu}/d\tau$, with U^{μ} being the 4-velocity of the charged particle, and is the covariant force discussed in Box 3.4. The interpretation of the $\mu = 0$ component is left as an exercise (Problem 12.9).

Inhomogeneous Maxwell's equations

Gauss's and Ampere's laws in (2.15),

$$\vec{\nabla}\cdot\vec{E} = \rho, \qquad \vec{\nabla}\times\vec{B} - \frac{1}{c}\frac{\partial\vec{E}}{\partial t} = \frac{1}{c}\vec{j}, \tag{12.45}$$

may be written as one covariant Maxwell equation:

$$\partial_{\mu}F^{\mu\nu} = -\frac{1}{c}j^{\nu}, \tag{12.46}$$

[11]To show that j^{μ} is a *bona fide* 4-vector, we recall the relation between charge-current-density and charge-density as $\vec{j} = \rho\vec{v}$, where $\vec{v}$ is the velocity field. Thus, (12.47) may be written as $j^{\mu} = \rho(c, \vec{v})$. If we replace the density ρ by the rest-frame density ρ' (which is a Lorentz scalar) through the relation $\rho = \gamma\rho'$ (reflecting the usual Lorentz length/volume contraction), we can relate j^{μ} to the 4-velocity field (3.31), $j^{\mu} = \rho'\gamma(c, \mathbf{v}) = \rho'U^{\mu}$. Since ρ' is a scalar and U^{μ} is a 4-vector, this shows explicitly that j^{μ} is also a 4-vector.

where the electromagnetic current 4-vector[11] is given as

$$j^{\mu} = \left(j^0, \vec{j}\right) = \left(c\rho, \vec{j}\right). \tag{12.47}$$

The reader is invited to check that the four equations[12] in (12.45) are just the four equations corresponding to different values of the index ν in Eq. (12.46).

Homogeneous Maxwell's equations

Faraday's and Gauss's laws for magnetism in (2.14),

$$\vec{\nabla}\times\vec{E} + \frac{1}{c}\frac{\partial\vec{B}}{\partial t} = 0, \quad \vec{\nabla}\cdot\vec{B} = 0 \tag{12.48}$$

[12]Beside the one equation for Gauss' law, we have three equations, corresponding to the three components of the vector equation, of Ampere's law.

correspond to the "Bianchi identities" of the electromagnetic (EM) field tensor,

$$\partial_{\mu}F_{\nu\lambda} + \partial_{\lambda}F_{\mu\nu} + \partial_{\nu}F_{\lambda\mu} = 0. \tag{12.49}$$

We note that this equation can be written (Problem 12.11) in an alternative form as

$$\partial_\mu \tilde{F}^{\mu\nu} = 0. \tag{12.50}$$

Electromagnetic duality   Looking at Eqs. (12.46) and (12.50), it is clear that Maxwell's equations in free space ($j^\mu = 0$) are invariant under the **duality transformation** of $F_{\mu\nu} \to \tilde{F}_{\mu\nu}$, which can be viewed as a rotation by 90° in the plane spanned by the perpendicular E and B axes: $E \to B$ and $B \to -E$. This is just a mathematical way to formalize the symmetry between electricity and magnetism that we all notice in Maxwell's theory. If $j^\mu \neq 0$, this duality symmetry can be restored by the introduction in Eq. (12.50) of the magnetic current $j_M^\mu = (c\rho_M, \vec{j}_M)$, where ρ_M is the magnetic charge (monopole) density, and $\vec{j}_M$ the magnetic charge current density.

Box 12.2   EM potential and gauge symmetry

The covariant formalism not only provides a simplified description in terms of the field tensor $F^{\mu\nu}$, but also a more compact notation for the electromagnetic potentials. The homogeneous Maxwell equation (12.50), because $\tilde{F}_{\mu\nu} \equiv -\varepsilon_{\mu\nu\lambda\rho} F^{\lambda\rho}/2$ can be written as $\epsilon^{\mu\nu\lambda\rho} \partial_\mu F_{\lambda\rho} = 0$, can be solved by expressing the field tensor in terms of the electromagnetic 4-potential $A^\mu = (\phi, \vec{A})$ with ϕ being the familiar scalar potential and $\vec{A}$ the vector potential

$$F_{\lambda\rho} = \partial_\lambda A_\rho - \partial_\rho A_\lambda. \tag{12.51}$$

Identifying the various components with the $\vec{E}$ and $\vec{B}$ fields, we see that (12.51) is just the well-known relation between EM fields and the scalar and vector potentials

$$\vec{B} = \vec{\nabla} \times \vec{A} \quad \vec{E} = -\vec{\nabla}\phi - \frac{\partial}{\partial t}\vec{A}. \tag{12.52}$$

Instead of $F^{\mu\nu}$ (i.e. $\vec{E}$ and $\vec{B}$), one can describe electromagnetic fields using A^μ (i.e. ϕ and $\vec{A}$) potentials. The dynamics of electromagnetism is then determined by the inhomogeneous Maxwell equations[13](12.46) after the replacement of $F^{\mu\nu}$ by A^μ as in (12.51). This simplifies the description of electrodynamics by reducing the number of dynamical variables from six in $F^{\mu\nu}$ down to four in A^μ. However the correspondence between $F_{\mu\nu}$ and A_μ is not unique in the sense that different sets of potentials A^μ lead to the same field strength $F^{\mu\nu}$. This is so because of the relation (12.51), hence Maxwell's equations are invariant under the **gauge transformation**:

$$A_\mu \longrightarrow A'_\mu = A_\mu - \partial_\mu \chi, \tag{12.53}$$

where $\chi(x)$ is an arbitrary spacetime-dependent scalar function (called the gauge function). A change of the potential according to (12.53) will not alter the electromagnetic description given by the $\vec{E}$ and $\vec{B}$ fields.

---

[13]In a sense the homogenous Maxwell equations, in contrast to the inhomogeneous equations, may be regarded as only providing the "boundary conditions"—they just tell us that the EM potentials exist.

12.2.2   Electric charge conservation

When the 4-del operator ∂_ν of (12.29) is applied to the inhomogeneous Maxwell equation (12.46), the LHS vanishes because the combination $\partial_\nu \partial_\mu$ is symmetric under an exchange of the (μ, ν) indices, while $F^{\mu\nu}$ is antisymmetric as in (12.32). This implies that the RHS must also vanish:

$$\partial_\nu j^\nu = 0. \tag{12.54}$$

To investigate the physical meaning of this 4-divergence equation, we display its components as shown in (12.47), noting that $\partial_\mu = (c^{-1}\partial_t, \vec{\nabla})$,

$$\frac{\partial \rho}{\partial t} + \vec{\nabla}\cdot\vec{j} = 0. \tag{12.55}$$

This is the familiar "equation of continuity." To show more explicitly that this represents a conservation statement, we integrate both terms in this equation over the volume (and moving one term to the RHS),

$$\frac{d}{dt}\int_V \rho \, dV = -\int_V \vec{\nabla}\cdot\vec{j}\,dV = -\oint_S \vec{j}\cdot d\vec{\sigma} \tag{12.56}$$

where we have used the divergence theorem to arrive at the last integral (over the closed surface S that covers the volume V). This expression clearly shows the physical interpretation of this equation as a statement of electric charge conservation: the RHS shows the inflow of electric charge across the surface S (flux) resulting in an increase of charge in the volume V, as expressed on the LHS. As a general rule, the expression of any conservation law in a system of continuous medium (e.g. a field system) is in the form of a continuity equation (12.55), or more directly as a vanishing 4-divergence condition like (12.54).

Physical interpretation of the current components   Let us also review the physical meaning of the various elements of the electromagnetic current $j^\mu = (c\rho, \vec{j})$, where ρ is the electric charge density (the electric charge per volume) and $\vec{j}$ the current density (charge flow per unit time and per cross-sectional area, i.e. charge flux). More explicitly, we can display their physical meaning as

$$\rho = \frac{\Delta q}{\Delta x \Delta y \Delta z}, \quad j^1 = \frac{\Delta q}{\Delta t \Delta y \Delta z}, \text{ etc.,} \tag{12.57}$$

where $\Delta x \Delta y \Delta z$ in the expression for ρ is an infinitesimal 3-volume, and $\Delta y \Delta z$ in j^1 is the surface area perpendicular to the current direction (x^1). We can express the meaning of all four components more compactly:

$$j^\mu = \frac{c\Delta q}{\Delta S^\mu}, \tag{12.58}$$

where ΔS^μ is the Minkowski volume with x^μ held fixed (i.e. it is a Minkowski 3-surface). For example, $\Delta S^0 = (\Delta x^1 \Delta x^2 \Delta x^3)$, $\Delta S^1 = (\Delta x^0 \Delta x^2 \Delta x^3) = c(\Delta t \Delta y \Delta z)$, etc.

12.3 Energy–momentum tensors

Electric charge conservation can be expressed as a 4-divergence condition (12.54). Other conservation laws can all be written similarly. For example, instead of electric charge, if we consider the case of mass, we can define in a manner similar to (12.58) a mass-current 4-vector $j^\mu = c\Delta m/\Delta S^\mu$ so that

$$j^\mu = (c\rho, \; \vec{j}) = (c \times \text{mass density}, \; \text{mass 3-current-density}) \qquad (12.59)$$

and (12.54) becomes a statement of mass conservation.

In the same manner, we can use Eq. (12.58) to consider the 4-current for energy and momentum. Recall, from Section 3.2.2, that energy and 3-momentum form a momentum 4-vector $p^\mu = (p^0, \vec{p}) = (E/c, \vec{p})$, with $(p^i/E) = (v^i/c^2)$ of (3.40). This suggests that we can define, for a field system, the 4-current for this momentum 4-vector, called the **energy–momentum tensor**, by

$$T^{\mu\nu} = \frac{c\Delta p^\mu}{\Delta S^\nu}. \qquad (12.60)$$

The conservation of energy and momentum can then be expressed as

$$\partial_\mu T^{\mu\nu} = 0. \qquad (12.61)$$

We are particularly interested in this quantity because, energy and momentum being the source of gravity, $T^{\mu\nu}$ enters directly into the the relativistic field equation of gravity.[14]

Component meaning of the energy–momentum tensor   Here we examine the physical meaning the components of the tensor $T^{\mu\nu}$. In the process we will also show that $T^{\mu\nu}$ is a symmetric tensor:

$$T^{\mu\nu} = T^{\nu\mu}. \qquad (12.62)$$

We shall use (12.60) to investigate the meaning of each element. For $\mu = 0$ and $\nu = 0$,

$$T^{00} = \frac{c\Delta p^0}{\Delta S^0} = \frac{\Delta E}{\Delta x \Delta y \Delta z}, \qquad (12.63)$$

showing that T^{00} is the energy density of the system. From

$$T^{01} = \frac{c\Delta p^0}{\Delta S^1} = \frac{\Delta E}{c\,(\Delta t \Delta y \Delta z)}, \qquad (12.64)$$

we see that T^{0i} is the ith-component of the 3-current density for energy (divided by c), i.e. the energy flux. On the other hand, from

$$T^{10} = \frac{c\Delta p^1}{\Delta S^0} = \frac{c\Delta p^1}{\Delta x \Delta y \Delta z} \qquad (12.65)$$

we see that T^{i0} is the ith-component of the 3-momentum density (multiplied by c). In fact, it is not too difficult to deduce that the momentum density is equal to the energy flux: $T^{i0} = T^{0i}$. This can be verified by a comparison of

[14]This Einstein equation, first displayed in Section 6.3.2, will be discussed in detail in Section 14.2.

(12.65) and (12.64), with the help of the momentum–energy ratio mentioned above, $(\Delta p^i / \Delta E) = (v_i / c^2)$:

$$\frac{c \Delta p^i}{\Delta x^i} = \frac{\Delta E}{c \Delta t}. \tag{12.66}$$

Thus we have, for example,

$$T^{10} = c \frac{\Delta p^1}{\Delta x \Delta y \Delta z} = \frac{\Delta E}{c \Delta t \Delta y \Delta z} = T^{01}. \tag{12.67}$$

T^{ij} can similarly be interpreted as the jth component of the ith momentum current density. In particular, the diagonal $i = j$ momentum flux T^{ii} has the simple interpretation as the pressure:

$$\text{momentum-flux} = \frac{\text{momentum}}{(\text{area})_\perp \Delta t} = \frac{\text{normal force}}{\text{area}} = \text{pressure}, \tag{12.68}$$

where we have used the fact that the rate of momentum change is force. The off-diagonal terms (T^{ij} for $i \neq j$) would involve (parallel) shear forces. Also, just as we showed that $T^{i0} = T^{0i}$, it is straightforward to show that $T^{ij} = T^{ji}$. Thus we have a symmetric energy–momentum tensor (for all its indices) as shown in (12.62). Energy and momentum conservation for an isolated system is then expressed as in Eq. (12.61). If the system is subject to some external force field, the RHS will then be a force-density field,

$$\partial_\mu T^{\mu\nu} = \phi^\nu. \tag{12.69}$$

This is analogous to the single particle case, in which the change of momentum is related to a force: $dp^\mu / d\tau = K^\mu$; and, in the absence of any external force, we have momentum conservation $dp^\mu / d\tau = 0$.

A swarm of noninteracting particles

Let us construct the energy–momentum tensor for the simplest example of a continuous medium. The simplest such system is a swarm of noninteracting particles, or a "cloud of dust." We choose first to describe the swarm of particles using a coordinate system in which each spatial position label is carried by the particles themselves. In such **comoving frames**, all particles have fixed spatial coordinates at all times; thus, the particles are viewed in this comoving coordinate system as effectively at rest ($v = 0$, $\gamma = 1$), the 4-velocity field taking on the simple form of $U^\mu = (c, \vec{0})$. The only nonvanishing energy–momentum tensor term[15] is the rest energy density T^{00} of (12.63):

$$T^{\mu\nu} = \begin{pmatrix} \rho' c^2 & 0 & 0 & 0 \\ 0 & 0 & 0 & 0 \\ 0 & 0 & 0 & 0 \\ 0 & 0 & 0 & 0 \end{pmatrix}, \tag{12.70}$$

where ρ' is the mass density in this comoving (i.e. rest) coordinate system. But in this reference frame we can also write the above tensor in terms of the 4-velocity field $U^\mu = (c, \vec{0})$ as

$$T^{\mu\nu} = \rho' U^\mu U^\nu. \tag{12.71}$$

[15] In the effective rest frame, $\Delta p^i = 0$. Thus, momentum density and energy flux all vanish, $T^{i0} = T^{0i} = 0$. Since the particles are noninteracting, there are no normal or shear forces; hence, $T^{ij} = 0$.

Even though we have arrived at this expression for $T^{\mu\nu}$ in a particular (comoving) coordinate frame, because this equation is a proper tensor equation (i.e. every term has the same tensor transformation property), the equation does not change under coordinate transformations. Consequently, this expression is valid in **every** inertial frame, and hence is the general expression of the energy–momentum tensor for this system of noninteracting particles.

Ideal fluid

We now consider the case of an **ideal fluid**, in which fluid elements interact only through a normal (perpendicular) force. There is no shear force (thus $T^{ij} = 0$ for $i \neq j$). This implies that in this system there is pressure but no viscosity. So, according to (12.68), to obtain the $T^{\mu\nu}$ of an ideal fluid, all we need to do is to add pressure p at the diagonal (i, i) positions in (12.70):

$$T^{\mu\nu} = \begin{pmatrix} \rho'c^2 & 0 & 0 & 0 \\ 0 & p & 0 & 0 \\ 0 & 0 & p & 0 \\ 0 & 0 & 0 & p \end{pmatrix} = \left(\rho' + \frac{p}{c^2}\right) U^\mu U^\nu + p\eta^{\mu\nu}, \qquad (12.72)$$

where $\eta^{\mu\nu}$ is the Minkowski space metric (3.13). We are again using the comoving coordinate system. The equality of $T^{11} = T^{22} = T^{33} = p$ expresses the isotropy property of the ideal fluid; the pressure applied to a given portion of the fluid is transmitted equally in all directions and is everywhere perpendicular to the surface on which it acts. Also, because the given volume element is at rest in the chosen reference frame, its momentum density also vanishes, $T^{0i} = T^{i0} = 0$. Similarly to the case of a cloud of dust as discussed above, the proper tensor expression on the RHS of (12.72) means that this expression is valid for all coordinate frames. The nonrelativistic limit of an ideal fluid energy–momentum tensor and its relation to the Euler equation for fluid mechanics are considered in Box 12.3.

Box 12.3 Nonrelativistic limit and the Euler equation

It is instructive to consider the nonrelativistic limit ($\gamma \to 1$) of the energy–momentum tensor of an ideal fluid. In this limit, with $U^\mu = \gamma(c, v^i) \simeq (c, v^i)$, since the rest energy dominates over the pressure (which results from particle momenta), $\rho c^2 \gg p$, the tensor in (12.72) takes on the form of

$$T^{\mu\nu} \stackrel{\text{NR}}{=} \begin{pmatrix} \rho c^2 & \rho c v^i \\ \rho c v_j & \rho v^i v^j + p\delta^{ij} \end{pmatrix}. \qquad (12.73)$$

In this way of representing a 4×4 matrix, the lower right element is actually a 3×3 matrix, etc. Let us now examine the conservation law $\partial_\mu T^{\mu\nu} = 0$ for this nonrelativistic system:

(*cont.*)

Box 12.3 (*Continued*)

The $\nu = 0$ part is

$$\partial_\mu T^{\mu 0} = \partial_0 T^{00} + \partial_i T^{i0} = c(\partial \rho / \partial t + \vec{\nabla} \cdot \vec{j}) = 0, \qquad (12.74)$$

which is just the continuity equation, expressing mass conservation. We have used the expression for mass current density $\vec{j} = \rho \vec{v}$.

The $\nu = j$ part is

$$\partial_\mu T^{\mu j} = \partial_0 T^{0j} + \partial_i T^{ij} = \partial_t \rho v^j + \partial_i (\rho v^i v^j + p \delta^{ij}) = 0. \qquad (12.75)$$

After using the continuity equation (12.74) with $\vec{j} = \rho \vec{v}$, this equation can be written in the form known as the **Euler equation** of fluid mechanics:

$$\rho \left[\frac{\partial \vec{v}}{\partial t} + (\vec{v} \cdot \vec{\nabla}) \vec{v} \right] = -\vec{\nabla} p. \qquad (12.76)$$

What is the physical significance of the Euler equation? First, let us look at the meaning of the terms in the square brackets. The velocity field $\vec{v}(t, \vec{r})$ depends on time as well as spatial position:

$$\frac{d\vec{v}}{dt} = \frac{\partial \vec{v}}{\partial t} + \frac{dx}{dt} \frac{\partial \vec{v}}{\partial x} + \frac{dy}{dt} \frac{\partial \vec{v}}{\partial y} + \frac{dz}{dt} \frac{\partial \vec{v}}{\partial z} = \frac{\partial \vec{v}}{\partial t} + (\vec{v} \cdot \vec{\nabla}) \vec{v}. \qquad (12.77)$$

Thus, the square bracket of (12.76) is just an expression for the **total derivative** (also called the **substantial derivative**). To see the meaning of the RHS of (12.76), we perform a volume integration:

$$-\int_V \vec{\nabla} p \, dV = -\oint_S p \, d\vec{\sigma}. \qquad (12.78)$$

This equation is a kind divergence theorem; to make this explicit, one can treat the pressure as a vector $\vec{p} = p\hat{n}$ with $\hat{n}$ being a unit vector normal to the surface. Pressure p being the force per unit area, the RHS is just the total force acting on a closed surface; this equation then tells us that $-\vec{\nabla} p \, dV$ is the force acting on a fluid element having a volume dV and, thus, $-\vec{\nabla} p$ must be the force per unit volume. We then have the interpretation of the Euler equation (12.76) as the "per-volume $\vec{F} = m\vec{a}$ equation" for fluid mechanics:

$$\rho \frac{d\vec{v}}{dt} = -\vec{\nabla} p, \qquad (12.79)$$

where $d\vec{v}/dt$ represents the rate of change of velocity of a fluid element as it moves about in space.

Electromagnetic field system

The electromagnetic field is a physical system carrying energy and momentum. In Box 12.4 we discuss the energy–momentum tensor $T_{\mu\nu}$ of such a field system.

Box 12.4 $T^{\mu\nu}$ **of the electromagnetic field**

A field is a physical system and it carries energy and momentum. It can be shown[16] that the energy–momentum tensor for an electromagnetic field is

$$T^{\mu\nu}_{\text{field}} = \frac{1}{2}\eta_{\alpha\beta}(F^{\mu\alpha}F^{\nu\beta} + \tilde{F}^{\mu\alpha}\tilde{F}^{\nu\beta}) = \eta_{\alpha\beta}F^{\mu\alpha}F^{\nu\beta} - \frac{1}{4}\eta^{\mu\nu}F^{\alpha\beta}F_{\alpha\beta},$$
(12.80)

where we have used the identity

$$\eta_{\alpha\beta}(F^{\mu\alpha}F^{\nu\beta} - \tilde{F}^{\mu\alpha}\tilde{F}^{\nu\beta}) = \frac{1}{2}\eta^{\mu\nu}F^{\alpha\beta}F_{\alpha\beta}$$
(12.81)

to go from the first expression to the second. This relation can be proven by summing over the Levi-Civita symbols appearing in the definitions of the dual fields $\tilde{F}^{\mu\alpha}$ and $\tilde{F}^{\nu\beta}$, or by direct multiplication of the field tensor matrices of (12.34) and (12.38). From the component expression of the field tensor, one can easily check (Problem 12.14) that $T^{00} = \frac{1}{2}(\vec{E}^2 + \vec{B}^2)$ and $T^{0i} = (\vec{E} \times \vec{B})_i$, which are the familiar EM expressions for the energy density and the energy current density (the Poynting vector), respectively.

For the simpler case of free space, for which $j^\mu = 0$, we expect to have conservation of the field energy–momentum $\partial_\mu T^{\mu\nu}_{\text{field}} = 0$. This can be checked as follows:

$$\partial_\mu T^{\mu\nu}_{\text{field}} = \eta_{\alpha\beta}F^{\mu\alpha}\left(\partial_\mu F^{\nu\beta}\right) - \frac{1}{2}\eta^{\mu\nu}F_{\alpha\beta}\left(\partial_\mu F^{\alpha\beta}\right)$$
(12.82)

$$= F_{\alpha\beta}\left(\partial^\alpha F^{\nu\beta}\right) - \frac{1}{2}F_{\alpha\beta}\left(\partial^\nu F^{\alpha\beta}\right)$$

$$= F_{\alpha\beta}\left(\partial^\alpha F^{\nu\beta}\right) + \frac{1}{2}F_{\alpha\beta}\left(\partial^\beta F^{\nu\alpha} + \partial^\alpha F^{\beta\nu}\right)$$

$$= \frac{1}{2}F_{\alpha\beta}\left(\partial^\alpha F^{\nu\beta} + \partial^\beta F^{\nu\alpha}\right) = 0$$

where on the first line we have used Maxwell's equation (12.46) for $j^\mu = 0$, to reach the second line we have relabeled some dummy indices, to reach the third line we have used the Bianchi identity. (12.49), i.e. the homogeneous Maxwell equation, to reach the fourth line we have used the antisymmetric property of $F^{\nu\beta} = -F^{\beta\nu}$, and the last equality follows from the fact that the antisymmetric $F_{\alpha\beta}$ is contracted with the combination in parentheses, which is symmetric in (α, β).

For the case in which there are charged particles in space, so that $j^\mu \neq 0$, energy and momentum are stored in the field as well as in the motion of the charged particles,

$$T^{\mu\nu} = T^{\mu\nu}_{\text{field}} + T^{\mu\nu}_{\text{charge}},$$
(12.83)

where $T^{\mu\nu}_{\text{charge}} = \rho'_{\text{mass}}U^\mu U^\nu$ with ρ'_{mass} being the proper mass density of the charged particles, as shown in (12.71). It can be shown (Problem 12.15) that neither $T^{\mu\nu}_{\text{field}}$ nor $T^{\mu\nu}_{\text{charge}}$ are conserved, but their divergences mutually cancel so that $\partial_\mu T^{\mu\nu} = 0$. Thus, for the system as a whole, energy and momentum are conserved.

[16]This is usually obtained by relating the energy–momentum tensor to the variations of the Lagrangian density of the field system.

Review questions

1. What are the covariant and contravariant components of a vector? Why do we need these two kinds of vector (tensor) components? Operationally how can they be related to each other?

2. Why do we say that the position 4-vector x^μ is naturally contravariant and that the gradient operator ∂_μ is naturally covariant?

3. Contravariant and covariant vectors transform differently. How are their transformations related?

4. Display the coordinate transformation for a mixed tensor $T^\mu_\nu \to T'^\mu_\nu$.

5. What is the Lorentz transformation of coordinates $(t, \vec{x})$ and of the differential operators $(\partial_t, \vec{\nabla})$ for a boost $\vec{v} = v\hat{x}$ of the reference frame?

6. What does one mean by saying that the inhomogeneous Maxwell equation $\partial_\mu F^{\mu\nu} = -j^\nu/c$ is manifestly covariant? Show that this equation also includes the statement of electric charge conservation.

7. Use Minkowski index notation (i.e. Greek indices such as $\mu = 0, 1, 2, 3$) to express in a compact form the meaning of the components of the current 4-vector $j^\mu = (j^0, \vec{j})$.

8. Use the answer to the previous question to define the energy–momentum tensor $T^{\mu\nu}$, and state the meaning of its various components.

9. Write out the elements of $T^{\mu\nu}$ for an ideal fluid in the rest frame of a fluid element (the comoving frame).

Problems

12.1 Basis and inverse-basis vectors: a simple exercise The basis vectors for a two-dimensional space are given explicitly as

$$\mathbf{e}_1 = a \begin{pmatrix} 1 \\ 0 \end{pmatrix} \quad \text{and} \quad \mathbf{e}_2 = b \begin{pmatrix} \cos\theta \\ \sin\theta \end{pmatrix}.$$

(a) Find the inverse basis vectors $\{\mathbf{e}^i\}$ so that $\mathbf{e}_i \cdot \mathbf{e}^j = \delta_{ij}$.

(b) Write out the metric matrices g_{ij} and g^{ij} and check their inverse relationship: $\sum_j g_{ij} g^{jk} = \delta_{ik}$.

(c) Show that the sum of outer products, $\sum_i \mathbf{e}_i \otimes \mathbf{e}^i$, is the identity matrix. This is an expression of the completeness condition.

12.2 Perpendicular vs. parallel projections In a coordinate system with nondiagonal unit base vectors: $e_1^2 = e_2^2 = 1$, and $e_1 \cdot e_2 = \cos\alpha$, as shown in Fig. 12.1, use the matrix form (3.8) for the metric to check (geometrically) the relations (3.8) between the perpendicular and parallel projections as drawn in Fig 12.1:

$$V_1 = g_{11} V^1 + g_{12} V^2 = V^1 + (\cos\alpha) V^2. \qquad (12.84)$$

12.3 Coordinate transformations and permutation symmetry If a tensor has some symmetry properties, for example, $T_{\mu\nu} \pm T_{\nu\mu} = 0$, then after a coordinate transformation, the transformed tensor still has the same properties; in this case, $T'_{\mu\nu} \pm T'_{\nu\mu} = 0$.

12.4 Transformations: coordinates vs. basis vectors The reason that A^μ are called the contravariant components and A_μ the covariant components is that they transform "oppositely" and "in the same way" as the basis vectors $\mathbf{e}_\mu$. From the definitions given in (12.6), explain why there must be such relations.

12.5 $g_{\mu\nu}$ is a tensor We have called $g_{\mu\nu} = \mathbf{e}_\mu \cdot \mathbf{e}_\nu$ a tensor.

(a) Demonstrate that this metric definition does imply the requisite transformation property.

(b) Use the role played by metrics in the contractions $A^\mu B^\nu g_{\mu\nu}$ or $A_\mu B_\nu g^{\mu\nu}$ to confirm these transformation properties.

[(b) is an equivalent way to define the metric as $g_{\mu\nu} = \mathbf{e}_\mu \cdot \mathbf{e}_\nu$.]

12.6 The quotient theorem This theorem states that in a tensor equation such as $A_{\mu\nu} = C_{\mu\lambda\rho} B_\nu^{\lambda\rho}$, if we know that $A_{\mu\nu}$ and $B_\nu^{\lambda\rho}$ are tensors, then the coefficients $C_{\mu\lambda\rho}$ also form a tensor. Show that the proof that $g^{\mu\nu}$ is a set of tensor components in Problem 12.5(b) is an illustration of the quotient theorem.

12.7 Transformation of the metric The transformation matrix **L** taking polar coordinates in the 2D plane to Cartesian coordinates can be obtained from (5.34), and from this demonstrate that the metric transforms as $\mathbf{g'} = \mathbf{L}^{-1\mathsf{T}}\mathbf{g}\mathbf{L}^{-1}$.

12.8 Generalized orthogonality condition and the boost transformation Consider the relation between two inertial frames connected by a boost (with velocity v) in the $+x$ direction. We have effectively a two-dimensional problem:

$$\begin{pmatrix} ct' \\ x' \end{pmatrix} = \begin{pmatrix} a & b \\ c & d \end{pmatrix} \begin{pmatrix} ct \\ x \end{pmatrix}$$

where the effective 2×2 boost matrix **L** with real elements (a, b, c, d) must satisfy the generalized orthogonal condition $\mathbf{L}\eta\mathbf{L}^{\mathsf{T}} = \eta$ so that the length $x^2 - c^2t^2$ is an invariant. Show that this condition fixes (in a way entirely similar to Problem 3.4) the explicit form of the Lorentz transformation to be

$$\begin{pmatrix} a & b \\ c & d \end{pmatrix} = [\mathbf{L}] = \begin{pmatrix} \cosh\psi & \sinh\psi \\ \sinh\psi & \cosh\psi \end{pmatrix},$$

where $\tanh\psi = v/c$.

12.9 Covariant Lorentz force law

(a) From Box 3.4 we have identified the spatial part of the covariant force $\vec{K} = \gamma\vec{F}$ with the relativistic force $\vec{F} = d\vec{p}/dt$ where $\vec{p}$ is the relativistic momentum $\gamma m\vec{v}$. Justify the nonrelativistic identification $\vec{F}$ in Eq. (12.43) with $m\vec{a}$.

(b) Check that the $\mu = i$ components of (12.44) correspond to the familiar Lorentz force law of (12.43).

(c) Check that the $\mu = 0$ component does have the interpretation correct as the time component of

a covariant force $K^0 = \gamma\vec{F}\cdot\vec{v}/c$ as required by (3.53).

12.10 Manifestly covariant Maxwell equations Use the identification of field tensor elements $F_{\mu\nu}$ and $\tilde{F}_{\mu\nu}$ as given in (12.33) and (12.37), to check that the components of the covariant equations (12.46) and (12.50) are just the Maxwell equations in the familiar form of (12.45) and (12.48).

12.11 Homogeneous Maxwell equations Explicitly demonstrate that the two forms of the homogeneous equations (12.49) and (12.50) are equivalent. *Suggestion*: Derive Eq. (12.49) from Eq. (12.50) starting with the $\nu = 0$ component of the equation. The proof of the converse statement, from (12.50) to (12.49), is more straightforward.

12.12 Electromagnetic potentials Verify the solution (12.51) of the homogeneous Maxwell equations, by substituting it into (12.49).

12.13 $T^{\mu\nu}$ for a swarm of dust Use the explicit form of $\rho'U^\mu U^\nu$ in (12.71) for $T^{\mu\nu}$ to check the physical meaning of the elements of the energy–momentum tensor as discussed in Eqs. (12.63)–(12.68).

12.14 $T^{\mu\nu}$ for the electromagnetic field Check the physical meaning of the elements of the energy–momentum tensor (12.80) for the electromagnetic field as given in Eqs. from (12.63)–(12.68).

12.15 $T^{\mu\nu}$ for a system of EM fields and charges Because fields and particles can exchange energy and momenta between them, energy and momentum are conserved only for the combined system. Show that neither $T^{\mu\nu}_{\text{field}}$ of (12.80) nor $T^{\mu\nu}_{\text{charge}}$ of (12.83) is conserved, but their divergences mutually cancel so that $\partial_\mu T^{\mu\nu} = 0$. Thus, for the system as a whole, energy and momentum are conserved.

12.16 Radiation pressure and energy density A system of radiation can be treated as a system of EM fields, or as an ideal fluid of photons with energy density ρc^2 and pressure p. Derive the pressure and density relation, $p = \rho c^2/3$. **Hint**: First examine the trace ($\eta_{\mu\nu}T^{\mu\nu}$) of the energy–momentum tensor for the EM field (12.80).

13

Tensors in general relativity

- While the tensors used in general relativity are basically the same as those in special relativity, differentiation of tensor components in a curved space must be handled with extra care.
- By adding another term (related to Christoffel symbols) to the ordinary derivative operator, we can form a "covariant derivative;" such a differentiation operation does not spoil the tensor property.
- The relation between Christoffel symbols and first derivatives of metric functions is reestablished.
- Using the concept of parallel transport, the geometric meaning of covariant differentiation is further clarified.
- The curvature tensor for an n-dimensional space is derived by the parallel transport of a vector around a closed path.
- Symmetry and contraction properties of the Riemann curvature tensor are considered. We find just the desired tensor needed for the GR field equation.

In Chapters 4 and 6, we have discussed that, because of the equivalence principle, the equation of motion for a test particle in a gravitational field (4.8) is totally independent of any properties of the test particle. This led Einstein to the idea of gravity being the structure of a curved spacetime. In the absence of gravity, spacetime is flat. The central point of EP that gravity can be transformed away locally fits snugly in the Riemannian geometrical description of a curved space as being locally flat. General relativity requires that physics equations be covariant under any general coordinate transformation that leaves invariant the infinitesimal length

$$ds^2 = g_{\mu\nu}dx^\mu dx^\nu. \tag{13.1}$$

Just as special relativity requires physics equations to be tensor equations with respect to Lorentz transformations, GR equations must be tensor equations with respect to general coordinate transformations. In this way, the principle of general relativity can be fulfilled.

Physics equations usually involve differentiation. While tensors in GR are basically the same as SR tensors, the derivative operators in a curved space require considerable care. In contrast to the case of flat space, basis vectors in a curved space must change from position to position. This implies that general coordinate transformations must necessarily be position dependent. As a consequence, ordinary derivatives of tensors, except for scalars, are no longer tensors. Nevertheless it can be shown that one can construct "covariant differentiation operations" so that they result in tensor derivatives. We demonstrate this by formal manipulation (Section 13.1) as well as by a more geometric introduction (Section 13.2). This geometric concept of parallel transport will also be employed to generalize the Gaussian curvature of a 2D space to the Riemann curvature tensor for a curved space of arbitrary dimensions. We conclude this chapter with a study of the symmetry and contraction properties of the Riemann tensor, which will be needed when we study the GR field equation, the Einstein equation, in the next chapter.

13.1 Derivatives in a curved space

The tensors used in general relativity are basically the same as those in special relativity, except when differentiation is involved. This difference reflects the fact that coordinate transformations in a curved space are necessarily position dependent. One finds that differentiation of a tensor results in a quantity which is no longer a tensor. This poses a serious problem as relativistic equations must be tensor equations. To overcome this, we introduce in this section the "covariant derivative" which does not spoil the tensor properties, and allows us to have relativistic physics equations.

13.1.1 General coordinate transformations

The coordinate transformations in special relativity (the Lorentz transformations) are position-independent "global transformations." The rotation angles and boost velocity are the same for every spacetime point; we rotate the same amount of angle and boost with the same velocity everywhere. In GR we must deal with position-dependent "local transformations," the general coordinate transformation. This position dependence is related to the fact that in a curved space the basis vectors $\{\mathbf{e}_\mu\}$ must necessarily change from point to point, leading to position-dependent metric functions:

$$g_{\mu\nu} \equiv \left[\mathbf{e}_\mu(x) \cdot \mathbf{e}_\nu(x)\right] = g_{\mu\nu}(x). \tag{13.2}$$

That the metric in a curved space is always position dependent immediately leads to the conclusion that a general coordinate transformation $\mathbf{L}$ must also be position dependent (i.e. nonlinear):[1]

$$\partial[\mathbf{L}] \neq 0. \tag{13.3}$$

A position-dependent transformation is also called a **local** transformation. Namely, a transformation performed at each location may well be different. The demand that physics equations be covariant under such a local transformation is much more severe. As we shall see in Chapter 14, it not only imposes restriction on the form of the physic equations but also requires the presence

[1] The metric $\mathbf{g}$ is a rank-2 tensor and thus transforms, cf. (12.16), as $\mathbf{g}' = \mathbf{L}^{-1}\mathbf{L}^{-1}\mathbf{g}$. If we differentiate both sides of this relation, we get

$$\partial[\mathbf{g}'] = 2\left[\mathbf{L}^{-1}\right][\mathbf{g}]\left(\partial\left[\mathbf{L}^{-1}\right]\right)$$
$$+ \left[\mathbf{L}^{-1}\right]\left[\mathbf{L}^{-1}\right](\partial[\mathbf{g}]).$$

For a flat space, one can always work with a coordinate system having a position-independent metric, $\mathbf{g}' = \mathbf{g} = \eta$ with $\partial\eta = 0$. The above relation then shows that the transformation matrix must also be position independent, $\partial\mathbf{L} = 0$. In a curved space the metric must be position dependent, $\partial\mathbf{g} \neq 0$, implying that the transformation also has an x-dependence, $\partial\mathbf{L} \neq 0$.

of a "force field" (in our case, the gravitational field), which is introduced via the covariant derivatives.

Coordinate transformation as a matrix of partial derivatives

The coordinate transformations in special relativity (the Lorentz transformations) leave invariant the separation[2] $s^2 = g_{\mu\nu} x^\mu x^\nu$. In a curved space the bases and metric necessarily vary from point to point. General transformations in such a space are not expected to have such a finite invariant separation. However, since a curved space is locally flat, it will be possible to demand the coordinate transformation

$$dx'^\mu = [\mathbf{L}]^\mu_\nu \, dx^\nu \tag{13.4}$$

that leaves invariant the infinitesimal length. Equation (13.1) defines the metric for a given coordinate system. Let us now recall the (chain-rule) differentiation relation:

$$dx'^\mu = \frac{\partial x'^\mu}{\partial x^\nu} dx^\nu. \tag{13.5}$$

A comparison of (13.4) and (13.5) suggests that the coordinate transformation can be written as a matrix of partial derivatives:

$$[\mathbf{L}]^\mu_\nu = \frac{\partial x'^\mu}{\partial x^\nu}. \tag{13.6}$$

Therefore, the transformation (12.13) for a contravariant vector is

$$A^\mu \longrightarrow A'^\mu = \frac{\partial x'^\mu}{\partial x^\nu} A^\nu. \tag{13.7}$$

Writing out this equation more explicitly, we have

$$\begin{pmatrix} A'^0 \\ A'^1 \\ A'^2 \\ A'^3 \end{pmatrix} = \begin{pmatrix} \frac{\partial x'^0}{\partial x^0} & \frac{\partial x'^0}{\partial x^1} & \frac{\partial x'^0}{\partial x^2} & \frac{\partial x'^0}{\partial x^3} \\ \frac{\partial x'^1}{\partial x^0} & \frac{\partial x'^1}{\partial x^1} & \frac{\partial x'^1}{\partial x^2} & \frac{\partial x'^1}{\partial x^3} \\ \frac{\partial x'^2}{\partial x^0} & \frac{\partial x'^2}{\partial x^1} & \frac{\partial x'^2}{\partial x^2} & \frac{\partial x'^2}{\partial x^3} \\ \frac{\partial x'^3}{\partial x^0} & \frac{\partial x'^3}{\partial x^1} & \frac{\partial x'^3}{\partial x^2} & \frac{\partial x'^3}{\partial x^3} \end{pmatrix} \begin{pmatrix} A^0 \\ A^1 \\ A^2 \\ A^3 \end{pmatrix}. \tag{13.8}$$

This way of writing a transformation also has the advantage of preventing us from mis-identifying the transformation $\mathbf{L}^\mu_\nu$ as a tensor. From now on, we shall always adopt this practice.[3] Similarly, from the transformation property of the del operator discussed in the previous chapter, Eq. (12.27),

$$\frac{\partial}{\partial x'^\mu} = \left[\mathbf{L}^{-1}\right]^\nu_\mu \frac{\partial}{\partial x^\nu}, \tag{13.9}$$

the chain rule of differentiation leads to the identification

$$\left[\mathbf{L}^{-1}\right]^\nu_\mu = \frac{\partial x^\nu}{\partial x'^\mu}. \tag{13.10}$$

From (13.6) and (13.10), we see clearly that the identity $\mathbf{L}\mathbf{L}^{-1} = \mathbf{1}$ is satisfied because $(\partial x^\lambda/\partial x'^\mu)(\partial x'^\mu/\partial x^\nu) = \delta^\lambda_\nu$. For covariant components of a vector,

[2]This naturally includes the infinitesimal separation of $ds^2 = g_{\mu\nu} dx^\mu dx^\nu$ also.

[3]This notation is also applicable to the global transformation of SR discussed in previous chapters. As an instructive exercise, one can show that the elements of the Lorentz transformation matrix (12.24) can be recovered from partial differentiation of the Lorentz boost formulae. Namely, the familiar Lorentz transformation (12.24) can also be written as a matrix of partial differentiation. For example, from $t' = \gamma(t - vx/c^2)$, or $x'^0 = \gamma(x^0 - \beta x^1)$ we have $\partial x'^0/\partial x^1 = -\gamma\beta$, etc.

we have the transformation (12.14),

$$A_\mu \longrightarrow A'_\mu = \frac{\partial x^\nu}{\partial x'^\mu} A_\nu. \tag{13.11}$$

In general a tensor with contravariant and covariant components, $T^{\mu\nu\,\cdots}_{\lambda\,\cdots}$, transforms as a direct product of contravariant and covariant vectors $T^{\mu\nu\,\cdots}_{\lambda\,\cdots} \sim A^\mu B^\nu \ldots C_\lambda \ldots$. For example, the simplest mixed tensor has the transformation

$$T^\mu_\nu \longrightarrow T'^\mu_\nu = \frac{\partial x^\lambda}{\partial x'^\nu} \frac{\partial x'^\mu}{\partial x^\rho} T^\rho_\lambda. \tag{13.12}$$

Because we have the expansion of $\mathbf{A} = A^\mu \mathbf{e}_\mu = A_\mu \mathbf{e}^\mu$, with the vector $\mathbf{A}$ being coordinate independent, the transformations of the expansion coefficients A^μ and A_μ must be "cancelled out" by those of the corresponding bases:

$$\mathbf{e}'_\mu = \frac{\partial x^\nu}{\partial x'^\mu} \mathbf{e}_\nu, \quad \text{and} \quad \mathbf{e}'^\nu = \frac{\partial x'^\nu}{\partial x^\rho} \mathbf{e}^\rho. \tag{13.13}$$

This is the reason that $\{V_\mu\}$ are called the **co**variant components: they transform in the same way as the basis vectors, while the **contra**variant components transform oppositely.

In summary we are interested in a general coordinate transformation $x'^\mu = x'^\mu(x)$ that leaves the infinitesimal length ds^2 of (13.1) invariant. This is sometimes described as a general reparametrization. Under such a transformation $\mathbf{L} = \partial x'/\partial x$, the metric tensor transforms, as shown in (13.12), as

$$g'_{\mu\nu} = g_{\alpha\beta} \left[\mathbf{L}^{-1} \right]^\alpha_{\ \mu} \left[\mathbf{L}^{-1} \right]^\beta_{\ \nu}, \tag{13.14}$$

or in matrix language $\mathbf{g}' = \mathbf{L}^{-1\mathsf{T}} \mathbf{g} \mathbf{L}^{-1}$. We are interested in the general reparametrizations that are "smooth" so that operations such as $\partial x'/\partial x$ exist. Also, in the small and flat space, we should still have the residual transformation that leaves the metric invariant: $\mathbf{g}' = \mathbf{g} = \eta$. That is, the metric remains Minkowskian. Recall from Chapters 12 and 3, the generalized orthogonality condition $\eta = \mathbf{L}_0^{-1\mathsf{T}} \eta \mathbf{L}_0^{-1}$ informs us that $\mathbf{L}_0$ must be the Lorentz transformation. That is, in a local small flat space the transformation must be reducible to a Lorentz transformation. In this sense the general coordinate transformation we will study can be regarded as a "local Lorentz transformation"—an independent Lorentz transformation at every spacetime point.

We remind ourselves that a description in Riemannian geometry is through distance measurements (Chapter 5). Coordinates in Riemannian geometry have no intrinsic meaning (see, e.g. Box 7.1) as they are related to distance measurements only through the metric. We are allowed to make all sorts of coordinate changes so long as the metric changes correspondingly. Such a transformation will change the terms in the physics equations set up in a curved space. These equations will automatically retain their forms (i.e. covariant), hence obey the principle of relativity, if they are tensor equations. That is, every term in the equation transforms in the same way under a coordinate transformations—each is a proper tensor and transforms according to the rules as given in (13.12).

13.1.2   Covariant differentiation

The above discussion would seem to imply that there is no fundamental difference between tensors in flat and in curved space. But as we shall demonstrate below, this is not so when differentiation is involved.

Ordinary derivatives of tensor components are not tensors

In a curved space, the derivative $\partial_\nu A^\mu$ is a nontensor. Namely, even though we have A^μ and ∂_ν being good vectors, as indicated by (13.7), (13.9) and (13.10)

$$\partial_\mu \longrightarrow \partial'_\mu = \frac{\partial x^\lambda}{\partial x'^\mu} \partial_\lambda, \tag{13.15}$$

their combination $\partial_\nu A^\mu$ still does not transform properly,

$$\partial_\nu A^\mu \longrightarrow \partial'_\nu A'^\mu \neq \frac{\partial x^\lambda}{\partial x'^\nu} \frac{\partial x'^\mu}{\partial x^\rho} \partial_\lambda A^\rho, \tag{13.16}$$

as required by (13.12). We can find the full expression for $\partial'_\nu A'^\mu$ by differentiating $\partial'_\nu \equiv (\partial/\partial x'^\nu)$ on both sides of (13.7)

$$\partial'_\nu A'^\mu = \frac{\partial}{\partial x'^\nu} \left(\frac{\partial x'^\mu}{\partial x^\rho} A^\rho \right)$$

$$= \frac{\partial x^\lambda}{\partial x'^\nu} \frac{\partial x'^\mu}{\partial x^\rho} \left(\partial_\lambda A^\rho \right) + \frac{\partial^2 x'^\mu}{\partial x'^\nu \partial x^\rho} A^\rho, \tag{13.17}$$

where (13.15) has been used. Compared to the RHS of (13.16), there is an extra term

$$\frac{\partial}{\partial x'^\nu} \left(\frac{\partial x'^\mu}{\partial x^\rho} \right) \neq 0, \tag{13.18}$$

which is (13.3) with the transformation written in terms of partial derivatives. Thus $\partial_\nu A^\mu$ not being a tensor is related to the position-dependent nature of the transformation, which in turn reflects (as discussed at the beginning of this subsection) the position-dependence of the metric. Thus the root problem lies in the "moving bases," $\mathbf{e}^\mu = \mathbf{e}^\mu(x)$, of the curved space. More explicitly, because the tensor components are the projections of the tensor onto the basis vectors $A^\mu = \mathbf{e}^\mu \cdot \mathbf{A}$, the moving bases $\partial_\nu \mathbf{e}^\mu \neq 0$ produce an extra (second) term in the derivative:

$$\partial_\nu A^\mu = \mathbf{e}^\mu \cdot (\partial_\nu \mathbf{A}) + \mathbf{A} \cdot \left(\partial_\nu \mathbf{e}^\mu \right). \tag{13.19}$$

The properties of the two terms on the RHS will be studied separately below.

Covariant derivatives as expansion coefficients of $\partial_\nu \mathbf{A}$

In order for an equation to be manifestly relativistic we must have it as a tensor equation such that the equation is unchanged under coordinate transformations. Thus, we seek a **covariant derivative** D_ν to be used in covariant physics equations. Such a differentiation is constructed so that when acting on tensor components it still yields a tensor.

$$D_\nu A^\mu \longrightarrow D'_\nu A'^\mu = \frac{\partial x^\lambda}{\partial x'^\nu} \frac{\partial x'^\mu}{\partial x^\rho} D_\lambda A^\rho. \tag{13.20}$$

As will be demonstrated below, the first term on the RHS of (13.19) is just this desired covariant derivative term.

We have suggested that the difficulty with the differentiation of vector components is due to their coordinate dependence. By this reasoning, derivatives of a scalar function Φ should not have this complication—because a scalar tensor does not depend on the bases: $\Phi' = \Phi$,

$$\partial_\mu \Phi \longrightarrow \partial'_\mu \Phi' = \frac{\partial x^\lambda}{\partial x'^\mu} \partial_\lambda \Phi. \tag{13.21}$$

Similarly, the derivatives of the vector $\mathbf{A}$ itself (not its components) transform properly because $\mathbf{A}$ is coordinate independent,

$$\partial_\mu \mathbf{A} \longrightarrow \partial'_\mu \mathbf{A} = \frac{\partial x^\lambda}{\partial x'^\mu} \partial_\lambda \mathbf{A}. \tag{13.22}$$

Both (13.21) and (13.22) merely reflect the transformation of the del operator (13.15). If we dot both sides of (13.22) by the two sides of the transformation (13.13) of the inverse basis vectors, $\mathbf{e}'^\nu = (\partial x'^\nu / \partial x^\rho) \mathbf{e}^\rho$, we obtain

$$\mathbf{e}'^\nu \cdot \partial'_\mu \mathbf{A} = \frac{\partial x^\lambda}{\partial x'^\mu} \frac{\partial x'^\nu}{\partial x^\rho} \mathbf{e}^\rho \cdot \partial_\lambda \mathbf{A}. \tag{13.23}$$

This shows that $\mathbf{e}^\nu \cdot \partial_\mu \mathbf{A}$ is a proper mixed tensor[4] as required by (13.12), and can be the covariant derivative we have been looking for:

$$D_\mu A^\nu = \mathbf{e}^\nu \cdot \partial_\mu \mathbf{A}. \tag{13.24}$$

This relation implies that $D_\mu A^\nu$ can be viewed as the projection of the vectors[5] $\partial_\mu \mathbf{A}$ along the direction of $\mathbf{e}^\nu$; we can then interpret $D_\mu A^\nu$ as the coefficient of expansion of $\partial_\mu \mathbf{A}$ in terms of the basis vectors:

$$\partial_\mu \mathbf{A} = \left(D_\mu A^\nu \right) \mathbf{e}_\nu \tag{13.25}$$

with the repeated indices ν summed over.

Christoffel symbols as expansion coefficients of $\partial_\nu \mathbf{e}^\mu$

On the other hand, we do not have a similarly simple transformation relation like (13.22) when the coordinate independent $\mathbf{A}$ is replaced by one of the coordinate basis vectors ($\mathbf{e}_\mu$), which by definition change under coordinate transformations. Thus, an expansion of $\partial_\nu \mathbf{e}^\mu$ in a manner similar to (13.25):

$$\partial_\nu \mathbf{e}^\mu = -\Gamma^\mu_{\nu\lambda} \mathbf{e}^\lambda \quad \text{or} \quad \mathbf{A} \cdot \left(\partial_\nu \mathbf{e}^\mu \right) = -\Gamma^\mu_{\nu\lambda} A^\lambda, \tag{13.26}$$

does not have coefficients $(-\Gamma^\mu_{\nu\lambda})$ that are tensors. Anticipating the result, we have here used the same notation for these expansion coefficients as the **Christoffel symbols** introduced in Chapter 6, cf. (6.10).

Plugging (13.24) and (13.26) into (13.19), we find

$$D_\nu A^\mu = \partial_\nu A^\mu + \Gamma^\mu_{\nu\lambda} A^\lambda. \tag{13.27}$$

Thus, in order to produce the covariant derivative, the ordinary derivative $\partial_\nu A^\mu$ must be supplemented by another term. This second term directly reflects the position dependence of the basis vectors, as in (13.26). Even though both $\partial_\nu A^\mu$ and $\Gamma^\mu_{\nu\lambda} A^\lambda$ do not have the correct transformation properties, the unwanted

---

[4] We can reach the same conclusion by applying the quotient theorem to Eq. (13.25), with the observaton that since both $\partial_\mu \mathbf{A}$ and $\mathbf{e}_\nu$ are "good tensors," so must be their quotient $(D_\mu A^\nu)$.

[5] We are treating $\partial_\mu \mathbf{A}$ as a set of vectors, each being labeled by an index μ, and $D_\mu A^\nu$ is a projection of $\partial_\mu \mathbf{A}$ in the same way that $A^\nu = \mathbf{e}^\nu \cdot \mathbf{A}$ is a projection of the vector $\mathbf{A}$.

terms produced from their respective transformations (13.17) cancel each other so that their sum $D_\nu A^\mu$ is a good tensor. Further insight about the structure of the covariant derivative can be gleaned by invoking the basic geometric concept of parallel displacement of a vector, to be presented in Section 13.2.

Compared to the contravariant vector A^μ of (13.27), the covariant derivative for a covariant vector A_μ takes on the form (Problem 13.1) of

$$D_\nu A_\mu = \partial_\nu A_\mu - \Gamma^\lambda_{\nu\mu} A_\lambda. \tag{13.28}$$

A mixed tensor such as T^μ_ν, transforming in the same way as the direct product $A^\mu B_\nu$, will have a covariant derivative

$$D_\nu T^\rho_\mu = \partial_\nu T^\rho_\mu - \Gamma^\lambda_{\nu\mu} T^\rho_\lambda + \Gamma^\rho_{\nu\sigma} T^\sigma_\mu. \tag{13.29}$$

There should be a set of Christoffel symbols for each index of the tensor—a set of $(+\Gamma T)$ for a contravariant index, a $(-\Gamma T)$ for a covariant index, *etc.* A specific example is the covariant differentiation of the (covariant) metric tensor $g_{\mu\nu}$:

$$D_\lambda g_{\mu\nu} = \partial_\lambda g_{\mu\nu} - \Gamma^\rho_{\lambda\mu} g_{\rho\nu} - \Gamma^\rho_{\lambda\nu} g_{\mu\rho}. \tag{13.30}$$

13.1.3 Christoffel symbols and the metric tensor

We have introduced the Christoffel symbols $\Gamma^\mu_{\nu\lambda}$ as the coefficients of expansion for $\partial_\nu e^\mu$ as in (13.26). In this section we shall relate such $\Gamma^\mu_{\nu\lambda}$ to the first derivative of the metric tensor. This will justify the identification with the symbols first defined in (6.10). To derive this relation, we need to point out[6] (Problem 13.3) an important feature of $\Gamma^\mu_{\nu\lambda}$, as defined by (13.26) and (13.30):

$$\Gamma^\mu_{\nu\lambda} = \Gamma^\mu_{\lambda\nu} \tag{13.31}$$

i.e. Christoffel symbols are symmetric with respect to the interchange of the two lower indices.

[6]This is valid in the framework of Riemannian geometry where a curved space is locally flat. However, GR coupled to a fermionic source has been formulated in a way (as in the Einstein–Cartan theory) involving the difference $\Gamma^\mu_{\nu\lambda} - \Gamma^\mu_{\lambda\nu}$. This antisymmetric tensor is called the **torsion**.

The metric tensor is covariantly constant

While the metric tensor is position dependent, $\partial g \neq 0$, it is a constant with respect to covariant differentiation, $D g = 0$ (we say, $g_{\mu\nu}$ is **covariantly constant**):

$$D_\lambda g_{\mu\nu} = 0. \tag{13.32}$$

One way to prove this is to use the expression of the metric in terms of the basis vectors: $g_{\mu\nu} = \mathbf{e}_\mu \cdot \mathbf{e}_\nu$, and apply the definition of the connection, $\partial_\nu \mathbf{e}_\mu = +\Gamma^\rho_{\mu\nu} \mathbf{e}_\rho$, as given in (13.26):

$$\partial_\lambda (\mathbf{e}_\mu \cdot \mathbf{e}_\nu) = (\partial_\lambda \mathbf{e}_\mu) \cdot \mathbf{e}_\nu + \mathbf{e}_\mu \cdot (\partial_\lambda \mathbf{e}_\nu)$$
$$= \Gamma^\rho_{\lambda\mu} \mathbf{e}_\rho \cdot \mathbf{e}_\nu + \Gamma^\rho_{\lambda\nu} \mathbf{e}_\mu \cdot \mathbf{e}_\rho. \tag{13.33}$$

Written in terms of the metric tensors, this relation becomes

$$\partial_\lambda g_{\mu\nu} - \Gamma^\rho_{\lambda\mu} g_{\rho\nu} - \Gamma^\rho_{\lambda\nu} g_{\mu\rho} = D_\lambda g_{\mu\nu} = 0, \tag{13.34}$$

where we have applied the definition of the covariant derivative of a covariant tensor $g_{\mu\nu}$ as in (13.30). That the metric tensor is covariantly constant is also the key ingredient in the proof of the "flatness theorem" first discussed in Section 5.2.3, and proven in Box 13.1. As we shall discuss (see Section 14.4.3), this key property allowed Einstein to introduce his "cosmological constant term" in the general relativistic field equation.

Christoffel symbols as the metric tensor derivative

In the above discussion we have used the definition (13.26) of Christoffel symbols as the coefficients of expansion of the derivative $\partial_\nu \mathbf{e}^\mu$. Here we shall derive an expression for Christoffel symbols, as the first derivatives of the metric tensor, which agrees with the definition first introduced in (6.10). We start by using several versions of (13.34) with their indices permuted cyclically:

$$D_\lambda g_{\mu\nu} = \partial_\lambda g_{\mu\nu} - \Gamma^\rho_{\lambda\mu} g_{\rho\nu} - \Gamma^\rho_{\lambda\nu} g_{\mu\rho} = 0$$

$$D_\nu g_{\lambda\mu} = \partial_\nu g_{\lambda\mu} - \Gamma^\rho_{\nu\lambda} g_{\rho\mu} - \Gamma^\rho_{\nu\mu} g_{\lambda\rho} = 0 \qquad (13.35)$$

$$-D_\mu g_{\nu\lambda} = -\partial_\mu g_{\nu\lambda} + \Gamma^\rho_{\mu\nu} g_{\rho\lambda} + \Gamma^\rho_{\mu\lambda} g_{\nu\rho} = 0.$$

Summing over these three equations and using the symmetry property of (13.31), we obtain:

$$\partial_\lambda g_{\mu\nu} + \partial_\nu g_{\lambda\mu} - \partial_\mu g_{\nu\lambda} - 2\Gamma^\rho_{\lambda\nu} g_{\mu\rho} = 0 \qquad (13.36)$$

or, in its equivalent form,

$$\Gamma^\lambda_{\mu\nu} = \frac{1}{2} g^{\lambda\rho} \left[\partial_\nu g_{\mu\rho} + \partial_\mu g_{\nu\rho} - \partial_\rho g_{\mu\nu} \right]. \qquad (13.37)$$

This relation showing $\Gamma^\mu_{\nu\lambda}$ as the first derivative of the metric tensor is called "the fundamental theorem of Riemannian geometry." It is just the definition stated previously in (6.10). From now on we shall often use this intrinsic geometric description of the Christoffel symbols (13.37) rather than (13.26). The symmetry property of (13.31) is explicitly displayed in (13.37).

Box 13.1 A proof of the flatness theorem

The flatness theorem, as first stated in Section 5.2.3, asserts that at any point P one can always make a coordinate transformation $x^\mu \to \bar{x}^\mu$ and $g^{\mu\nu} \to \bar{g}^{\mu\nu}$ where the metric tensor $\bar{g}^{\mu\nu}$ is a constant, up to a second-order correction (i.e. the first-order terms vanish):

$$\bar{g}^{\mu\nu}(\bar{x}) = \bar{g}^{\mu\nu}(0) + b^{\mu\nu\lambda\rho} \bar{x}_\lambda \bar{x}_\rho + \cdots, \qquad (13.38)$$

where for simplicity we have taken the point P to be at the origin of the coordinate system and the position vector $\bar{x}^\mu$ is assumed to be infinitesimally

(*cont.*)

Box 13.1 (*Continued*)

small. The coefficients $b^{\mu\nu\lambda\rho}$ are of course related to the second derivatives of the metric at P. We shall prove this result by explicit construction. Namely, we display a coordinate transformation

$$\frac{\partial x^{\mu}}{\partial \bar{x}^{\nu}} = \delta^{\mu}_{\nu} - \Gamma^{\mu}_{\nu\lambda}\bar{x}^{\lambda} \tag{13.39}$$

that will be shown leading to the result of (13.38).

Proof Substituting (13.39), as well as the power series expansion $g^{\mu\nu}(x) = g^{\mu\nu}(0) + \partial_{\lambda}g^{\mu\nu}x^{\lambda} + \cdots$, into the metric transformation equation (13.14)

$$\bar{g}_{\mu\nu}(\bar{x}) = \frac{\partial x^{\lambda}}{\partial \bar{x}^{\mu}}\frac{\partial x^{\rho}}{\partial \bar{x}^{\nu}}g_{\lambda\rho}(x), \tag{13.40}$$

we have

$$\bar{g}_{\mu\nu}(\bar{x}) = \left(\delta^{\lambda}_{\mu} - \Gamma^{\lambda}_{\mu\alpha}\bar{x}^{\alpha}\right)\left(\delta^{\rho}_{\nu} - \Gamma^{\rho}_{\nu\beta}\bar{x}^{\beta}\right)\left(g_{\lambda\rho}(0) + \partial_{\gamma}g_{\lambda\rho}x^{\gamma} + \cdots\right)$$

$$= g_{\mu\nu}(0) - \left[\Gamma^{\lambda}_{\mu\alpha}g_{\lambda\nu}(0) + \Gamma^{\lambda}_{\alpha\nu}g_{\mu\lambda}(0) - \partial_{\alpha}g_{\mu\nu}\right]x^{\alpha} + \cdots$$

The coefficient of x^{α} (square bracket) vanishes because, according to (13.34), the metric is covariantly constant. Thus the transformation in (13.39) indeed has the claimed property of leading to a metric with a vanishing first derivative as shown in (13.38). ■

This proves the flatness theorem, and the assertion that $\{\bar{x}^{\mu}\}$, the local Euclidean frame (LEF), always exists. While the constant tensor can be diagonalized to the principal axes (with length adjusted correctly) so that $\bar{g}^{\mu\nu}(0)$ becomes the standard flat space metric $\eta^{\mu\nu}$, it is apparent that the second derivatives of $\bar{g}_{\mu\nu}$, related to the intrinsic curvature of the space, cannot be eliminated by adjusting the coordinate system.

Now we have shown that the transformation in (13.39) can perform the task of changing any coordinates to one which is explicitly flat in an infinitessimal region around a given point. How did one find this transformation in the first place? One can motivate the result (13.39) by comparing it to the covariant derivative (13.27) for the case of $A^{\mu} = x^{\mu}$. The covariant term $(Dx^{\mu}/d\bar{x}^{\nu})$, being valid in every coordinate system (including the frame of $\{x^{\mu}\} = \{\bar{x}^{\mu}\}$), is identified with the identity matrix δ^{μ}_{ν}. Its difference with the coordinate transformation $(\partial x^{\mu}/\partial \bar{x}^{\nu})$ must then be the Christoffel symbols as dictated by (13.27).

Just as the covariant constancy of the metric tensor is the key ingredient in the proof that the LEF exists (Box 13.1), we also have the reverse statement that the existence of an LEF proves that the metric tensor must be covariantly constant: since the first derivative of the metric, hence the Christoffel symbols, vanish in LEF, $0 = \partial_{\mu}g_{\nu\lambda} = D_{\mu}g_{\nu\lambda}$. The last expression being convariant is valid in every frame of reference.

13.2 Parallel transport

Parallel transport is a fundamental notion in differential geometry. It illuminates the idea of covariant differentiation, and the associated Christoffel symbols. Furthermore, using this operation, we can present another view of the geodesic as the "straightest possible curve"—a geodesic line as the curve traced out by the parallel transport of its tangent vector. In Section 13.3 we shall derive the Riemann curvature tensor by way of parallel transporting a vector around a closed path.

13.2.1 Component changes under parallel transport

Equation (13.27) follows from (13.19) and expresses the relation between ordinary and covariant derivatives. Writing $DA^\mu = (D_\nu A^\mu)dx^\nu$ and $dA^\mu = (\partial_\nu A^\mu)dx^\nu$, (13.19) becomes

$$dA^\mu = DA^\mu - \Gamma^\mu_{\nu\lambda}A^\nu dx^\lambda. \tag{13.41}$$

We will show that the Christoffel symbols in the derivative of tensor components reflect the effects of parallel transport of a vector by a distance of dx. First, what is parallel transport? Why does one need to perform such an operation? Recall the definition of differentiation for the case of a scalar function $\Phi(x)$,

$$\frac{d\Phi(x)}{dx} = \lim_{\Delta x \to 0} \frac{\Phi(x+\Delta x) - \Phi(x)}{\Delta x}. \tag{13.42}$$

It is the difference of the functional values **at two different positions**. For the coordinate-independent scalar function $\Phi(x)$, this issue of two locations does not introduce any complication. This is not the case for vector components. The differential dA^μ on the LHS (13.41) is the difference $A^\mu(x+\Delta x) - A^\mu(x) \equiv A^\mu_{(2)} - A^\mu_{(1)}$, being the vector components $A^\mu = \mathbf{e}^\mu \cdot \mathbf{A}$, evaluated at two different positions (1) and (2), separated by dx. There are two sources for their difference: the change of the vector itself, $\mathbf{A}_{(2)} \neq \mathbf{A}_{(1)}$, and a coordinate change $\mathbf{e}^\mu_{(2)} \neq \mathbf{e}^\mu_{(1)}$, corresponding to the two terms on the RHS of (13.19). Thus the total change is the sum of two terms

$$dA^\mu = \left[\Delta A^\mu\right]_{\text{total}} = \left[\Delta A^\mu\right]_{\text{true}} + \left[\Delta A^\mu\right]_{\text{coord}}, \tag{13.43}$$

as indicated by (13.19). One term representing the change of the vector itself may be called the "true change,"

$$\left[\Delta A^\mu\right]_{\text{true}} = \mathbf{e}^\mu \cdot d\mathbf{A} = DA^\mu, \tag{13.44}$$

and another term represents the projection of $\mathbf{A}$ onto the coordinate (basis vector) change between the two points separated by dx. This change is expected to be proportional to the vector component A^ν and to the separation dx^λ with the proportional constants in (13.45) being identified with the Christoffel symbols:

$$\left[\Delta A^\mu\right]_{\text{coord}} = \mathbf{A} \cdot d\mathbf{e}^\mu = -\Gamma^\mu_{\nu\lambda}A^\nu dx^\lambda. \tag{13.45}$$

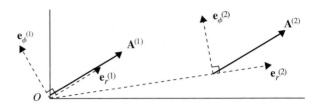

Fig. 13.1 Parallel transport of a vector **A** in a flat plane with polar coordinates: from position-1 at the origin $\mathbf{A}^{(1)}$ to another position-2, $\mathbf{A}^{(2)}$. The differences of the basis vectors at these two positions $(\mathbf{e}_\phi^{(1)}, \mathbf{e}_r^{(1)}) \neq (\mathbf{e}_\phi^{(2)}, \mathbf{e}_r^{(2)})$ bring about component changes. In particular, $A_\phi^{(1)} = 0 \neq A_\phi^{(2)}$, and $A_r^{(1)} = \{A_\phi^{(2)2} + A_r^{(2)2}\}^{1/2}$.

This discussion motivates us to introduce the geometric concept of **parallel transport**. It is the process of moving a tensor without changing the tensor itself. The only change of the tensor components under parallel displacement is due to coordinate changes, $dA^\mu = \Delta A^\mu{}_{\text{coord}}$. In a flat space with a Cartesian coordinate system, this is trivial as there is no coordinate change from point-to-point. But in a flat space with a curvilinear coordinate system, such as polar coordinates, this parallel transport itself induces component changes as shown in Fig. 13.1. For the vector example being discussed here, we have $\Delta A^\mu{}_{\text{true}} = \mathbf{e}^\mu \cdot d\mathbf{A} = DA^\mu = 0$. Thus the mathematical expression for a **parallel transport of vector components** is

$$DA^\mu = dA^\mu + \Gamma^\mu_{\nu\lambda} A^\nu dx^\lambda = 0. \tag{13.46}$$

Recall that we have shown the metric tensor as being covariantly constant, $D_\mu g_{\nu\lambda} = 0$. We now understand covariant constancy to mean the change of tensor components due to a coordinate change only. But a change of the metric, by definition, is a pure coordinate change. Hence, it must have a vanishing covariant derivative.

13.2.2   The geodesic as the straightest possible curve

The process of parallel transporting a vector A^μ along a curve $x^\mu(\sigma)$ can be expressed, according to (13.46), as

$$\frac{DA^\mu}{d\sigma} = \frac{dA^\mu}{d\sigma} + \Gamma^\mu_{\nu\lambda} A^\nu \frac{dx^\lambda}{d\sigma} = 0. \tag{13.47}$$

From this we can define the geodesic line, as the straightest possible curve, by the condition of it being the line constructed by parallel transport of its tangent vector. See Fig. 13.2(a) for an illustration of such an operation in flat space. In this way the condition can be formulated by setting $A^\mu = dx^\mu/d\sigma$ in (13.47):

$$\frac{D}{d\sigma}\left(\frac{dx^\mu}{d\sigma}\right) = 0, \tag{13.48}$$

or, more explicitly,

$$\frac{d}{d\sigma}\frac{dx^\mu}{d\sigma} + \Gamma^\mu_{\nu\lambda}\frac{dx^\nu}{d\sigma}\frac{dx^\lambda}{d\sigma} = 0. \tag{13.49}$$

This agrees with the geodesic equation as shown in Eq. (6.9).

Fig. 13.2 Straight line as the geodesic in a flat plane: (a) As a curve traced out by parallel transport of its tangents. (b) When a vector is parallel transported along a straight line, the angle between them is unchanged.

Example   When a vector A_μ is parallel transported along a geodesic, we can show that the angle subtended by the vector and the geodesic (i.e. the tangent of the geodesic) is unchanged, see Fig. 13.2(b). Namely, we need to show

$$\frac{D}{d\sigma}\left(A_\mu \frac{dx^\mu}{d\sigma}\right) = 0. \tag{13.50}$$

The proof is straightforward:

$$\frac{D}{d\sigma}\left(A_\mu \frac{dx^\mu}{d\sigma}\right) = \frac{DA_\mu}{d\sigma}\left(\frac{dx^\mu}{d\sigma}\right) + A_\mu \frac{D}{d\sigma}\left(\frac{dx^\mu}{d\sigma}\right). \tag{13.51}$$

The RHS indeed vanishes: the first term is zero because we parallel transport the vector, (13.47); the second term is zero because the curve is a geodesic satisfying Eq. (13.48).

13.3   Riemannian curvature tensor

Curvature measures how much a curved space is curved because it measures the amount of deviation of any geometric relations from their corresponding Euclidean equalities. We have already proven in Section 5.3.3 a particular relation showing that for a 2D curved surface the angular excess ϵ (the sum of the interior angles over its Euclidean value) of an infinitesimal polygon is proportional to the Gaussian curvature K at this location:

$$\epsilon = K\sigma \tag{13.52}$$

where σ is the area of the polygon. In the following, this relation (13.52) for a 2D curvature K will be generalized to that of an n-dimensional curved space.

Angular excess ϵ and directional change of a vector

How can an angular excess be measured in general? To implement this, we first use the concept of parallel transport to cast this relation (13.52) in a form that allows for such an n-dimensional generalization. It can be shown that the angular excess ϵ is related to the directional change of a vector after it has been parallel transported around the perimeter of the polygon. The simplest example is a spherical triangle with three 90° interior angles. In Fig. 13.3 we see that a parallel transported vector changes its direction by 90°, which is the angular excess of this triangle. The generalization to an arbitrary triangle, hence to any polygon, is assigned as an exercise (Problem 13.6).

Recall the definition of an angle being the ratio of arc length to the radius, Fig. 13.4(a). Hence, the directional angular change $d\theta$ can be written as the ratio of the change of a vector to its magnitude $(dA)/A$. In this way we can

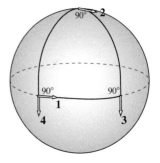

Fig. 13.3 A triangle with all interior angles being 90° on a spherical surface. The parallel transport of a vector around this triangle (vector-1, clockwise to vector-2, to -3, and finally back to the starting point as vector-4) leads to a directional change of the vector by 90° (the angular difference between vector-4 and vector-1).

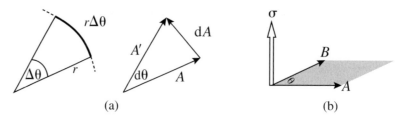

Fig. 13.4 (a) The directional change of a vector can be expressed as a fractional change of the vector: $d\theta = (dA)/A$. (b) The area of parallelogram as the cross-product of its two sides, $\vec{\sigma} = \vec{A} \times \vec{B}$.

relate the angular excess ϵ to the change of a vector after a transport: $\epsilon A = dA$. Substituting this into (13.52), we obtain

$$dA = KA\sigma. \tag{13.53}$$

Namely, the change of a vector after a round-trip parallel transport is proportional to the vector itself and the area of the closed path. The coefficient of proportionality is identified as the curvature.

The area tensor

We will use this relation (13.53) to seek out the curvature for a higher dimensional curved space. To do so we need to write the 2D equation (13.53) in the proper index form in order to generalize it to an n-dimensional space. Recall the 2D area spanned by two vectors $\vec{a}$ and $\vec{b}$ can be calculated as a vector product, Fig 13.4(b): $\vec{\sigma} = \vec{a} \times \vec{b}$. Or, using the antisymmetric Levi-Civita symbol in the index notation,[7] we can write this as

$$\sigma_k = \varepsilon_{ijk} a^i b^j. \tag{13.54}$$

The area vector $\vec{\sigma}$ has magnitude $(ab \sin\theta)$ and direction given by the right-hand rule.

But (13.54) is not a convenient form to use in a higher dimensional space: (i) It refers to the embedding space with a 3-valued index $i = 1, 2, 3$, even though the parallelogram resides in a 2D space, say $i = 1, 2$. (ii) For different dimensions we would need to use an area tensor with a different number of indices because the number of indices in the Levi-Civita symbol increases with the dimension, for example, for four dimensions: ε_{ijkl}, etc. We will instead use a two-index object σ^{ij} to represent the area:[8]

$$\sigma^{ij} \equiv \varepsilon^{ijk}\sigma_k = \varepsilon^{ijk}\varepsilon_{mnk} a^m b^n = \left(a^i b^j - a^j b^i\right). \tag{13.55}$$

Furthermore, since the index $i = 3$ is irrelevant (being outside the 2D space), we can write this entirely with the two-dimensional indices ($i = 1, 2$) without any reference to the embedding space. For an area in an n-dimensional space, we can represent the area spanned by a^λ and b^ρ by the antisymmetric combination,

$$\sigma^{\lambda\rho} = \left(a^\lambda b^\rho - b^\lambda a^\rho\right), \tag{13.56}$$

with the indices ranging within the space: $\mu = 1, 2, \ldots, n$.

[7] Levi-Civita symbols are discussed in sidenote 8 of Chapter 12.

[8] We have the identity $\epsilon^{ijk}\epsilon_{mnk} = \delta^i_m \delta^j_n - \delta^i_n \delta^j_m$.

13.3.1   The curvature tensor in an *n*-dimensional space

With the area written as a tensor of (13.56), Eq. (13.53) suggests that we can represent the change dA^μ of a vector due to a parallel transport around a parallelogram spanned by two vectors a^λ and b^ρ by a tensor equation,

$$dA^\mu = R^\mu_{\ \nu\lambda\rho} A^\nu a^\lambda b^\rho. \tag{13.57}$$

Namely, the change is proportional to the vector A^ν itself and to the two vectors $(a^\lambda b^\rho)$ spanning the parallelogram. The coefficient of proportionality $R^\mu_{\ \nu\lambda\rho}$ is a quantity with four indices (antisymmetric in λ and ρ) and we shall take this to be the definition of the curvature (called the **Riemann curvature tensor**) of this *n*-dimensional space.[9] Explicit calculation in Box 13.2 of the parallel transport of a vector around an infinitesimal parallelogram leads to the expression:

$$R^\mu_{\ \lambda\alpha\beta} = \partial_\alpha \Gamma^\mu_{\lambda\beta} - \partial_\beta \Gamma^\mu_{\lambda\alpha} + \Gamma^\mu_{\nu\alpha}\Gamma^\nu_{\lambda\beta} - \Gamma^\mu_{\nu\beta}\Gamma^\nu_{\lambda\alpha}. \tag{13.58}$$

The Christoffel symbol Γ being a first derivative, the Riemann curvature $R = d\Gamma + \Gamma\Gamma$ is then a nonlinear second derivative function of the metric, $\partial^2 g + (\partial g)^2$.

[9]We can plausibly expect this coefficient $R^\mu_{\ \nu\lambda\rho}$ to be a tensor because the differential dA^μ (being taken at the same position), a^λ, b^ρ and A^ν being tensors, the quotient theorem tells us that $R^\mu_{\ \nu\lambda\rho}$ should be a good tensor of rank 4 (i.e. a tensor with four indices).

Box 13.2 $R^\mu_{\lambda\alpha\beta}$ from parallel transporting a vector around a closed path

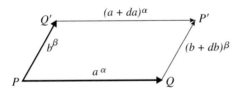

To fix the form of the Riemann tensor as in (13.57), we carry out the operation of parallel transporting a vector around an infinitesimal parallelogram $(PQP'Q')$ spanned by two infinitesimal vectors, a^α and b^β in Fig. 13.5. Recall that parallel transport of a vector $DA^\mu = 0$ means that the total vectorial change is due entirely to a coordinate change, see (13.46):

$$dA^\mu = -\Gamma^\mu_{\nu\lambda} A^\nu dx^\lambda. \tag{13.59}$$

The opposite sides of the parallelogram in Fig. 13.5, $(a + da)^\alpha$ and $(b + db)^\beta$, are obtained by parallel transport of a^α and b^β, respectively. Since $da^\mu = b^\mu$ and $db^\mu = a^\mu$, the expression for parallel transport of (13.59) gives:

$$(a + da)^\alpha = a^\alpha - \Gamma^\alpha_{\mu\nu} a^\mu b^\nu,$$

$$(b + db)^\beta = b^\beta - \Gamma^\beta_{\mu\nu} a^\mu b^\nu. \tag{13.60}$$

(cont.)

Fig. 13.5 The parallelogram $PQP'Q'$ is spanned by two vectors a^α and b^β. The opposite sides $(a + da)^\alpha$ and $(b + db)^\beta$ are obtained by parallel transport of a^α and b^β, respectively.

Box 13.2 (*Continued*)

Using again (13.59), we now calculate the change of a vector A^μ due to parallel transport from $P \to Q \to P'$:

$$dA^\mu_{PQP'} = dA^\mu_{PQ} + dA^\mu_{QP'} \tag{13.61}$$

$$= -\left(\Gamma^\mu_{\nu\alpha} A^\nu\right)_P a^\alpha - \left(\Gamma^\mu_{\nu\beta} A^\nu\right)_Q (b + db)^\beta .$$

The subscripts P and Q on the last line denote the respective positions where these functions are to be evaluated. Since eventually we shall compare all quantities at one position, say P, we will Taylor expand the quantities $(\ldots)_Q$ around the point P:

$$\left(\Gamma^\mu_{\nu\beta}\right)_Q = \left(\Gamma^\mu_{\nu\beta}\right)_P + a^\alpha \left(\partial_\alpha \Gamma^\mu_{\nu\beta}\right)_P \tag{13.62}$$

$$\left(A^\nu\right)_Q = \left(A^\nu\right)_P + a^\alpha \left(\partial_\alpha A^\nu\right)_P = \left(A^\nu\right)_P - a^\alpha \left(\Gamma^\nu_{\lambda\alpha} A^\lambda\right)_P$$

where we have used (13.59) to reach the last expression. From now on we shall drop the subscript P. Substitute into (13.61) the expansions of (13.60) and (13.62):

$$dA^\mu_{PQP'} = -\Gamma^\mu_{\nu\alpha} A^\nu a^\alpha - \left(\Gamma^\mu_{\nu\beta} + a^\alpha \partial_\alpha \Gamma^\mu_{\nu\beta}\right) \tag{13.63}$$

$$\times \left(A^\nu - a^\alpha \Gamma^\nu_{\lambda\alpha} A^\lambda\right) \left(b^\beta - \Gamma^\beta_{\rho\sigma} a^\rho b^\sigma\right) .$$

Multiplying it out and keeping terms up to $O(ab)$, we have

$$dA^\mu_{PQP'} = -\Gamma^\mu_{\nu\alpha} A^\nu a^\alpha - \Gamma^\mu_{\nu\beta} A^\nu b^\beta + A^\nu \Gamma^\mu_{\nu\beta} \Gamma^\beta_{\rho\sigma} a^\rho b^\sigma$$

$$-\partial_\alpha \Gamma^\mu_{\lambda\beta} A^\lambda a^\alpha b^\beta + \Gamma^\mu_{\nu\beta} \Gamma^\nu_{\lambda\alpha} A^\lambda a^\alpha b^\beta . \tag{13.64}$$

The vectorial change due to parallel transport along the other sides, $P \to Q' \to P'$, can be obtained from the above equation simply by the interchange of $a \leftrightarrow b$:

$$dA^\mu_{PQ'P'} = -\Gamma^\mu_{\nu\alpha} A^\nu b^\alpha - \Gamma^\mu_{\nu\beta} A^\nu a^\beta + A^\nu \Gamma^\mu_{\nu\beta} \Gamma^\beta_{\rho\sigma} a^\rho b^\sigma$$

$$-\partial_\alpha \Gamma^\mu_{\lambda\beta} A^\lambda a^\alpha b^\beta + \Gamma^\mu_{\nu\beta} \Gamma^\nu_{\lambda\alpha} A^\lambda a^\alpha b^\beta . \tag{13.65}$$

For a round-trip parallel transport from P back to P, the vectorial change dA^μ corresponds to the difference of the above two equations (which results in the cancellation of the first three terms on the RHS):

$$dA^\mu = dA^\mu_{PQ'P'} - dA^\mu_{PQP'} \tag{13.66}$$

$$= \left[\partial_\alpha \Gamma^\mu_{\lambda\beta} - \partial_\beta \Gamma^\mu_{\lambda\alpha} + \Gamma^\mu_{\nu\alpha} \Gamma^\nu_{\lambda\beta} - \Gamma^\mu_{\nu\beta} \Gamma^\nu_{\lambda\alpha}\right] A^\lambda a^\alpha b^\beta .$$

We conclude, after a comparison (13.66) with (13.57), that the sought-after Riemann curvature tensor in terms of Christoffel symbols is just the quoted result of (13.58).

Since this expression (13.58) for the curvature is in terms of $\Gamma^{\mu}_{\lambda\beta}$ which are not tensor components, we still need to show that $R^{\mu}_{\lambda\alpha\beta}$ has the proper tensor transformation property. This can be accomplished by writing (Problem 13.7) the curvature as the commutator of covariant derivatives:

$$\left[D_{\alpha}, D_{\beta}\right] A^{\mu} = R^{\mu}_{\lambda\alpha\beta} A^{\lambda}. \tag{13.67}$$

Since the covariant differentials $D_{\alpha}s$, together with $A^{\nu}s$, are good vectors, then according to the quotient theorem, $R^{\mu}_{\lambda\alpha\beta}$ must be a tensor.[10]

In a flat space, one can always find a coordinate system so that the metric is position independent. Namely, not only do the first derivatives vanish, $\partial g = 0$ (as in the local Euclidean frame of any curved space), but the second derivatives of the metric are $\partial^2 g = 0$ likewise. In such a coordinate frame, $R^{\mu}_{\lambda\alpha\beta} \propto \partial^2 g + (\partial g)^2 = 0$. Since it is a good tensor, if it vanishes in one coordinate, it vanishes for all coordinates.[11] In fact we can also show that this is a sufficient condition for a space to be flat, i.e. $R^{\mu}_{\lambda\alpha\beta} = 0$ implies a flat space.

[10] This is in agreement with our expectation (see sidenote 9) based on a heuristic argument.

[11] This is exactly the same as in the simpler 2D situation with the Gaussian curvature K.

13.3.2   Symmetries and contractions of the curvature tensor

We discuss the symmetries of the Riemann curvature tensor, and count its independent components. We note that the Riemann curvature tensor with all lower indices

$$R_{\mu\nu\alpha\beta} = g_{\mu\lambda} R^{\lambda}_{\nu\alpha\beta} \tag{13.68}$$

has the following symmetry features (Problem 13.8):

- It is **antisymmetric** with respect to the interchange of the first and second indices, and that of the third and fourth indices, respectively:

$$R_{\mu\nu\alpha\beta} = -R_{\nu\mu\alpha\beta} \tag{13.69}$$

$$R_{\mu\nu\alpha\beta} = -R_{\mu\nu\beta\alpha}. \tag{13.70}$$

- It is **symmetric** with respect to the interchange of the pair made up of the first and second indices with the pair of third and fourth indices:

$$R_{\mu\nu\alpha\beta} = +R_{\alpha\beta\mu\nu}. \tag{13.71}$$

- It also has the **cyclic** symmetry:

$$R_{\mu\nu\alpha\beta} + R_{\mu\beta\nu\alpha} + R_{\mu\alpha\beta\nu} = 0. \tag{13.72}$$

Knowing its symmetry properties, we find (Problem 13.9) the number of independent components of a curvature tensor in an n-dimensional space:

$$N_{(n)} = \frac{1}{12}n^2 \left(n^2 - 1\right). \tag{13.73}$$

For various n-dimensions this gives the following numbers:

- **Line:** $N_{(1)} = 0$. It is not possible for a one-dimensional inhabitant to see any curvature.

- **Surface:** $N_{(2)} = 1$. One can check (Problem 13.11) that the one independent element R_{1212}, according to (13.58), is related to the Gaussian curvature K of (5.35), so that Eq. (13.57) reduces to (13.53), by $K = -R_{1212}/g$, where $g = g_{11}g_{22} - g_{12}^2$ is the determinant of the 2D metric tensor. In fact, in the 2D space the Gaussian curvature K is simply related to the Ricci scalar R of (13.76) by $R = -2K$.
- **Spacetime:** $N_{(4)} = 20$. There are 20 independent components in the curvature tensor for a 4D curved spacetime.
- **Metric's second derivatives:** It can be shown (Problem 13.10) that the number in (13.73) for the curvature just matches the number of independent second derivatives of the metric tensor. Thus, we can indeed claim the Riemann tensor as the nonlinear second derivative of the metric tensor.

Contractions of the curvature tensor

We show how the covariantly constant **Einstein tensor**, which appears in the GR field equation (the Einstein equation), arises from contractions of the Riemann tensor.

Ricci tensor $R_{\mu\nu}$ This is the Riemann curvature tensor with the first and third indices contracted,[12]

$$R_{\mu\nu} \equiv g^{\alpha\beta} R_{\alpha\mu\beta\nu} = R^{\beta}_{\ \mu\beta\nu} \tag{13.74}$$

which is a symmetric tensor,

$$R_{\mu\nu} = R_{\nu\mu}. \tag{13.75}$$

It is straightforward to show that, because of the symmetry relations discussed above, contractions of the curvature tensor are essentially unique.[13]

Ricci scalar R   This is the Riemann curvature tensor contracted twice,

$$R \equiv g^{\alpha\beta} R_{\alpha\beta} = R^{\beta}_{\ \beta}. \tag{13.76}$$

Bianchi identities and the Einstein tensor

There is a set of constraints[14] on the curvature tensor, called the **Bianchi identities:**

$$D_\lambda R_{\mu\nu\alpha\beta} + D_\nu R_{\lambda\mu\alpha\beta} + D_\mu R_{\nu\lambda\alpha\beta} = 0, \tag{13.77}$$

which can be proven by following the steps outlined in Problem 13.12.

We now perform contractions on these Bianchi identities. Contracting with $g^{\mu\alpha}$ and use the fact that the metric tensor is covariantly constant, $D_\lambda g^{\alpha\beta} = 0$. Hence this metric contraction can be pushed right through the covariant differentiation:

$$D_\lambda R_{\nu\beta} - D_\nu R_{\lambda\beta} + D_\mu g^{\mu\alpha} R_{\nu\lambda\alpha\beta} = 0. \tag{13.78}$$

Contracting another time with $g^{\nu\beta}$,

$$D_\lambda R - D_\nu g^{\nu\beta} R_{\lambda\beta} - D_\mu g^{\mu\alpha} R_{\lambda\alpha} = 0. \tag{13.79}$$

[12]Note, in effect we have made a choice for sign convention in the definition of the Ricci tensor. For other sign conventions in our presentation see further comments in the next chapter when we present the GR field equation.

[13]Contracting different pairs of indices will always result in $\kappa R_{\mu\nu}$ with $\kappa = \pm 1, 0$.

[14]We note its resemblance to the homogeneous Maxwell equation as displayed in (12.49). There is a close analogy between the curvature tensor $R_{\mu\nu\alpha\beta}$ and the electromagnetic field tensor $F_{\mu\nu}$. It can easily be shown that, just as (13.67), we have $[D_\mu, D_\nu] = -ieF_{\mu\nu}$, where $D_\mu = \partial_\mu - ieA_\mu$ is the gauge covariant derivative with A_μ being the 4-potential (EM gauge field). In fact, the structure of electromagnetism can best be understood through its basic property of gauge symmetry, which has a deep Riemannian geometric interpretation.

At the last two terms, the metric just raises the indices,

$$D_\lambda R - D_\nu R_\lambda^\nu - D_\mu R_\lambda^\mu = D_\lambda R - 2D_\nu R_\lambda^\nu = 0. \tag{13.80}$$

Pushing through yet another $g^{\mu\lambda}$ in order to raise the λ index at the last term,

$$D_\lambda \left(R g^{\mu\lambda} - 2R^{\mu\lambda} \right) = 0. \tag{13.81}$$

We see that the combination,

$$G^{\mu\nu} = R^{\mu\nu} - \frac{1}{2} R g^{\mu\nu} \tag{13.82}$$

is covariantly constant (i.e. divergence free with respect to covariant differentiation),

$$D_\mu G^{\mu\nu} = 0. \tag{13.83}$$

To summarize, $G^{\mu\nu}$, called the **Einstein tensor**, is a covariantly constant, rank-2, symmetric tensor involving the second derivatives of the metric $\partial^2 g$ as well as the quadratic in ∂g.

$$
\boxed{
\begin{array}{c}
G^{\mu\nu} \text{ has the property:} \\[6pt]
\hline \\[-4pt]
\text{conserved (covariantly constant)} \\
\text{symmetric rank-2 tensor} \\
\partial\Gamma,\ \Gamma^2 \frown \left(\partial^2 g\right),\ (\partial g)^2.
\end{array}
}
\tag{13.84}
$$

As we shall see in the next chapter, this is just the sought-after mathematical quantity in the field equation of general relativity.

Review questions

1. What is the fundamental difference between the coordinate transformations in a curved space and those in flat space (e.g. Lorentz transformations in flat Minkowski space)?

2. Writing the coordinate transformation as a partial derivative matrix, give the transform law for a contravariant vector $A^\mu \rightarrow A'^\mu$, as well as that for a mixed tensor $T_\nu^\mu \rightarrow T'^\mu_\nu$.

3. From the transformation of $A_\mu \rightarrow A'_\mu$ in the answer to the previous question, work out the coordinate transformation of the derivatives $\partial_\mu A_\nu$. Why do we say that $\partial_\mu A_\nu$ is not a tensor? How do the covariant derivatives $D_\mu A_\nu$ transform? Why is it important to have differentiations that result in tensors?

4. What is the basic reason why $\partial_\mu A_\nu$ is not a tensor?

5. Write out the covariant derivative $D_\mu T_\nu^{\lambda\rho}$ (in terms of the connection symbols) of a general tensor $T_\nu^{\lambda\rho}$.

6. The relation between the Christoffel symbol and the metric tensor is called "the fundamental theorem of Riemannian geometry." Write out this relation.

7. As the Christoffel symbols $\Gamma_{\alpha\beta}^\mu$ are not components of a tensor, how do we know $R_{\lambda\alpha\beta}^\mu = \partial_\alpha \Gamma_{\lambda\beta}^\mu - \partial_\beta \Gamma_{\lambda\alpha}^\mu + \Gamma_{\nu\alpha}^\mu \Gamma_{\lambda\beta}^\nu - \Gamma_{\nu\beta}^\mu \Gamma_{\lambda\alpha}^\nu$ is really a tensor?

8. What are the two basic properties of the Einstein tensor $G^{\mu\nu} = R^{\mu\nu} - \frac{1}{2} R g^{\mu\nu}$?

9. What is "the flatness theorem?" Use this theorem to show that the metric tensor is covariantly constant, $D_\mu g_{\nu\lambda} = 0$.

Problems

13.1 Covariant derivative for covariant components Given that the covariant derivatives for contravariant components have the form of Eq. (13.27), show that the covariant derivatives for covariant components are $D_\nu A_\mu = \partial_\nu A_\mu - \Gamma^\lambda_{\nu\mu} A_\lambda$. Hint: $A_\mu A^\mu$ is an invariant.

13.2 Christoffel symbols in polar coordinates for a flat plane Even in a flat space, one can have moving bases. The polar coordinate system (r, θ) on a plane surface is such an example.

(a) Work out their respective (moving) basis vectors $(\mathbf{e}_r, \mathbf{e}_\theta)$ and $(\mathbf{e}^r, \mathbf{e}^\theta)$ in terms of the (fixed) Cartesian bases $(\mathbf{i}, \mathbf{j})$.

(b) Calculate the Christoffel symbols through their definition of $\partial_\nu \mathbf{e}^\mu = -\Gamma^\mu_{\nu\lambda} \mathbf{e}^\lambda$ given in (13.26).

(c) Calculate the divergence in a polar coordinate system: work out $D_\mu A^\mu = \partial_\mu A^\mu + \Gamma^\mu_{\mu\nu} A^\nu$ in terms of the component fields (A^r, A^θ).

(d) Calculate the covariant Laplacian $D_\mu D^\mu \Phi(x)$ in a polar coordinate system.

(e) Use the Christoffel symbols obtained in (b) to show that the metric tensors are constant with respect to covariant differentiation.

(f) Use the fundamental theorem of Riemannian geometry (13.37) to calculate a few $\Gamma^\mu_{\nu\lambda}$ to check the results obtained in (b).

(g) Use the explicit form of the Christoffel symbols calculated in (b) to show that the only independent component for the curvature tensor in 2D space vanishes, $R_{1212} = 0$, as expected for a flat space.

13.3 Symmetry property of Christoffel symbols Show that the symmetry property $\Gamma^\mu_{\nu\lambda} = \Gamma^\mu_{\lambda\nu}$ follows from the "torsion-free" statement of $[D_\mu, D_\nu]\Phi = 0$ where Φ is some scalar function.

13.4 The metric is covariantly constant: by explicit calculation We have already provided two proofs of $D_\lambda g_{\mu\nu} = 0$ one given in Eqs. (13.33) and (13.34); another at the end of Section 13.2.1. You are asked to perform an explicit calculation of $D_\lambda g_{\mu\nu}$ using the expression of the Christoffel symbols as the first derivatives of the metric as given by (13.37).

13.5 $D_\nu V_\lambda$ is a good tensor: another proof Use Eq. (13.50) and the geodesic Eq. (13.49) to prove that

$$(D_\nu V_\lambda) \frac{dx^\nu}{d\sigma} \frac{dx^\lambda}{d\sigma} = 0.$$

This is another way to see, via the quotient theorem, that $D_\nu V_\mu$ is a good tensor.

13.6 Parallel transport and the angular excess Prove that the directional change of a vector, after being parallel transported around the perimeter of an arbitrary triangle on a spherical surface, is equal to the angular excess of the triangle. This result then holds for any spherical polygon, since any polygon can always be divided into triangles. This in turn implies that such a relation is valid for any infinitesimal closed geodesic path in a general 2D space.

13.7 Riemann curvature tensor as the commutator of covariant derivatives To show that $R^\mu_{\lambda\alpha\beta}$ is indeed a tensor, we can perform the following calculation: Take the double derivative $D_\alpha D_\beta A^\mu = D_\alpha(\partial_\beta A^\mu + \Gamma^\mu_{\beta\lambda} A^\lambda) =$ $\ldots$ as well as that in the reverse order $D_\beta D_\alpha A^\mu = D_\beta(\partial_\alpha A^\mu + \Gamma^\mu_{\alpha\lambda} A^\lambda) = \ldots$ Show that their difference is just the expression for the Riemann tensor as given by Eq (13.58):

$$\left[D_\alpha, D_\beta\right] A^\mu = R^\mu_{\lambda\alpha\beta} A^\lambda \quad \text{or}$$

$$\left[D_\alpha, D_\beta\right] A_\mu = -R^\lambda_{\mu\alpha\beta} A_\lambda. \tag{13.85}$$

This result for a vector A_μ can easily be generalized to tensors of higher rank, for example,

$$\left[D_\mu, D_\nu\right] T_{\alpha\beta} = -R^\gamma_{\alpha\mu\nu} T_{\gamma\beta} - R^\gamma_{\beta\mu\nu} T_{\alpha\gamma}, \tag{13.86}$$

that is, a Riemann tensor factor for each tensor index.

13.8 Symmetries of $R_{\mu\nu\alpha\beta}$ Since the symmetry properties are not changed by coordinate transformations, one can choose a particular coordinate frame to prove these symmetry relations, and once proven in one frame, we can then claim their validity in all frames. An obvious choice is the locally Euclidean frame (with $\Gamma = 0$, $\partial \Gamma \neq 0$) where the curvature takes on a simpler form, $R_{\mu\nu\alpha\beta} = g_{\mu\lambda}(\partial_\alpha \Gamma^\lambda_{\nu\beta} - \partial_\beta \Gamma^\lambda_{\nu\alpha})$, and the symmetry properties are easy to inspect. In this way, check the validity of the symmetry properties as shown in Eqs. (13.69) to (13.72).

13.9 Counting independent elements of Riemann tensor The Riemann curvature tensor has the symmetry properties of (13.69) to (13.72). Show that the number of independent components of a curvature tensor in an n-dimensional space is $N_{(n)} = \frac{1}{12} n^2 (n^2 - 1)$.

13.10 Counting a metric's independent second derivatives In this problem you are asked to show that the number $N_{(n)}$

found in the previous problem of the Riemann tensor just matches that of the independent second derivatives of the metric tensor.

(a) Calculate the number of independents elements in $g_{\mu\nu}$, $\partial_\alpha g_{\mu\nu}$ and $\partial_\alpha \partial_\beta g_{\mu\nu}$, taking into consideration only the symmetry properties of these tensors. Collectively call these numbers $A_{(n)}$. First give the result $A_{(4)}$ in a four-dimensional space, and then record in particular the number for $\partial_\alpha \partial_\beta g_{\mu\nu}$ in a general n-dimensional space.

(b) The number $A_{(n)}$ obtained in (a) for the independent elements in $g_{\mu\nu}$, $\partial_\alpha g_{\mu\nu}$ and $\partial_\alpha \partial_\beta g_{\mu\nu}$ is an overcount, in the sense that some of them can be eliminated by coordinate transformations. If we are interested in the number of "truly independent elements" that reflects the property of the space itself (rather than the coordinate system), we should subtract out the elements that can be transformed away. Now calculate the number of elements $B_{(4)}$ that can be transformed away. *Suggestion:* The transformation of the metric $g_{\mu\nu}$ is given in (12.17). The relevant transformation matrix can be written as the shorthand: $(\partial x^\beta / \partial x'^\alpha) \equiv (\partial_\alpha x_\beta)$. We see that transformations of tensor derivatives depends on the derivatives of the transformation matrices. For example, the first derivative $\partial_\alpha g_{\mu\nu}$ transformation is determined by the first derivative of the number of transformations $\partial_\gamma (\partial_\alpha x_\beta)$, and $\partial_\alpha \partial_\beta g_{\mu\nu}$ by $\partial_\gamma \partial_\delta (\partial_\alpha x_\beta)$, etc. The number of parameters in these transformations (and their derivatives) should be the number of elements that can be transformed away by coordinate transformations. It will be useful to compare the proof (Box 13.1) of the flatness theorem by way of power series expansions.

(c) The difference $N_{(4)} = A_{(4)} - B_{(4)}$ obtained in (a) and (b) should correspond to the number of independent elements in $g_{\mu\nu}$, $\partial_\alpha g_{\mu\nu}$ and $\partial_\alpha \partial_\beta g_{\mu\nu}$. Do these counts make physical sense? Give your interpretation for each case.

(d) Calculate the independent elements of $\partial_\alpha \partial_\beta g_{\mu\nu}$ in the n-dimensional space. You should find a result $N_{(n)}$ that matches that obtained in Problem 13.9 for the Riemann tensor.

13.11 Reducing the Riemann tensor to Gaussian curvature For a 2D space, the curvature tensor has only one independent element.

(a) Show that it is just the Gaussian curvature of (5.35) with the identification of

$$K = -R_{1212}/g$$

g being the determinant of the metric tensor.

(b) Demonstrate that the Ricci scalar of (13.76) of a 2D space is simply twice the Gaussian curvature, $R = -2K$.

(c) With the result in (a), show that Eq. (13.57) reduces to the 2D result of (13.52): $\epsilon = K\sigma$.

13.12 Bianchi identities Prove the Bianchi identity (13.77) by the following steps:

(i) Prove the Jacobi identity for double commutators:

$$[D_\lambda, [D_\mu, D_\nu]] + [D_\nu, [D_\lambda, D_\mu]]$$
$$+ [D_\mu, [D_\nu, D_\lambda]] = 0$$

(ii) The three double commutators can be expressed in terms of the three terms on the LHS of (13.77) when one makes use of the commutator results of Problem 13.7 and the cyclic symmetry property of the curvature tensor shown in (13.72).

13.13 The Ricci tensor is symmetric From the definition of $R_{\mu\nu} \equiv g^{\alpha\beta} R_{\alpha\mu\beta\nu}$, show that $R_{\mu\nu} = R_{\nu\mu}$.

13.14 Contraction of Christoffel symbols Show that

$$\Gamma^\mu_{\mu\alpha} = \frac{1}{\sqrt{-g}} \frac{\partial}{\partial x^\alpha} \sqrt{-g}$$

where g is the determinant of the matrix $g_{\mu\nu}$.

13.15 Contraction of the Riemann tensor Recall that the Ricci tensor is obtained by contracting the first and third indices of the Riemann tensor. This result shows that all contractions of the Riemann tensor, based on its symmetry properties, are either the same as the Ricci tensor or are zero. For example, $R^\mu_{\mu\alpha\beta} = g^{\mu\nu} R_{\nu\mu\alpha\beta} = 0$ because these two terms have opposite symmetry properties in the interchange of (μ, ν) indices. Here you are asked to show that $R^\mu_{\mu\alpha\beta} = 0$ by explicit calculation, using the relation obtained in Problem 13.14.

14

GR as a geometric theory of gravity – II

- The mathematical realization of the equivalence principle (EP) is the principle of general covariance. General relativity (GR) equations must be covariant with respect to general coordinate transformations.
- To go from special relativity (SR) to GR equations, one replaces ordinary by covariant derivatives. The SR equation of motion $d^2x^\mu/d\tau^2 = 0$ turns into $D^2x^\mu/d\tau^2 = 0$, which is the geodesic equation.
- The Einstein equation, as the relativistic gravitation field equation, relates the energy momentum tensor to the Einstein curvature tensor.
- We solve the Einstein equation in the space exterior to a spherical source to obtain the Schwarzschild metric as the solution.
- The solutions of the Einstein equation that satisfy the cosmological principle must have a space with constant curvature. This is the Robertson–Walker spacetime.
- The relation of the cosmological Friedmann equations to the Einstein field equation is explicated.
- The compatibility of the cosmological-constant term with the mathematical structure of the Einstein equation and the interpretation of this term as the vacuum energy tensor are discussed.

In Chapter 6 we presented arguments for a geometric theory of gravity. The gravitational field is identified with the warped spacetime described by the metric function $g_{\mu\nu}(x)$. After one accepts that spacetime can be curved and the Riemannian geometry is the appropriate mathematics to describe such a space, we can now use the tensor calculus learned in Chapters 12 and 13 to write down the physics equations satisfying the principle of general relativity (GR). In Section 14.1 we present the principle of general covariance, which guides us to the GR equations in a curved spacetime. A proper derivation of the geodesic equation as the GR equation of motion will be presented, and we can finally write down the GR field equation, the Einstein equation. Its connection to the Newton/Poisson equation is discussed. Finally, we show how to obtain the Schwarzschild metric as the solution to the Einstein equation with a spherical source. In Section 14.4, the geometric formalisms used in cosmology as discussed in Chapters 9–11 are shown as solutions of the Einstein equation compatible with the cosmological principle.

14.1   The principle of general covariance

According to the strong principle of equivalence, gravity can always be transformed away locally. As first discussed in Section 6.1.2, Einstein suggested an elegant formulation of the new theory of gravity based on a curved spacetime. In this way EP is a fundamental built-in feature. Local flatness (a metric structure of spacetime) means that SR (theory of flat spacetime with no gravity) is an automatic property of the new theory. Gravity is not a force but the structure of spacetime. A particle just follows geodesics in such a curved spacetime. The physical laws, or the field equation for the relativistic potential, the metric function $g_{\mu\nu}(x)$, must have the same form no matter what generalized coordinates are used to locate, i.e. label, worldpoints (events) in spacetime. One expresses this by the requirement that the physics equations must satisfy the **principle of general covariance**. This is a two-part statement:

1. Physics equations must be covariant under the general coordinate transformations which leave the infinitesimal spacetime line element ds^2 invariant.
2. Physics equations should reduce to the correct special relativistic form in local inertial frames.[1] Namely, we must have the correct SR equations in the free-fall frames, in which gravity is transformed away. Additionally, gravitational equations reduce to Newtonian equations in the limit of low-velocity particles in a weak and static field.

[1] Cf. the discussion in Section 13.1.1, especially below Eq. (13.14), on general coordinate transformations that are reducible to the Lorentz transformation in the small empty space.

14.1.1   The minimal substitution rule

The principle of general covariance provides us with a well-defined path to go from SR equations, valid in the local inertial frames with no gravity, to GR equations that are valid in every coordinate system in the curved spacetime—curved because of the presence of gravity. Such GR equations must be covariant under general local transformations. The key feature of a general covariance transformation, in contrast to the (Lorentz) transformation in a flat spacetime, is its spacetime dependence. The tensor formalism in a curved spacetime differs from that for a flat spacetime of SR in its derivatives. To go from an SR equation to the corresponding GR equation is simple: we need to replace the ordinary derivatives ∂ in SR equations by covariant derivatives D:

$$\partial \longrightarrow D \,(= \partial + \Gamma)\,. \tag{14.1}$$

This is known as the **minimal coupling** because we are assuming the absence on the RHS of terms such as the Riemann tensor $R^{\mu}_{\nu\lambda\rho}$, which vanishes in the flat spacetime limit. Since Christoffel symbols Γ are the derivatives of the metric, hence the derivatives of the gravitational potential (i.e. they represent the gravitational field strength), the introduction of covariant derivatives naturally brings the gravitational field into the physics equations. In this way we can, for example, find the electromagnetic equations in the presence of a gravitational field. In Table 14.1, we show how GR equations arise from the SR results.

Table 14.1 SR electromagnetic equations in flat spacetime vs. GR equations in a curved spacetime.

SR equations			GR equations
Lorentz force law in flat spacetime			in curved spacetime
Eq. (12.44) $\frac{dU^\mu}{d\tau} = \frac{q}{c} F^{\mu\nu} U_\nu$		$\longrightarrow$	$\frac{DU^\mu}{d\tau} = \frac{q}{c} F^{\mu\nu} U_\nu$
Maxwell's equations in flat spacetime			in curved spacetime
Eq. (12.50) $\partial_\mu \tilde{F}^{\mu\nu} = 0$		$\longrightarrow$	$D_\mu \tilde{F}^{\mu\nu} = 0$
Eq. (12.46) $\partial_\mu F^{\mu\nu} = -\frac{1}{c} j^\nu$		$\longrightarrow$	$D_\mu F^{\mu\nu} = -\frac{1}{c} j^\nu$

This discussion of introducing gravitational coupling in GR illustrates how a local symmetry can dictate the form of dynamics—in this case, the precise way the Christoffel symbol $\Gamma^\lambda_{\mu u}$ (gravitational field) enters into physics equations. For example, in the last line of Table 14.1, starting from the familiar special relativistic equation (12.46), we have the set of GR equations in curved spacetime,

$$\partial_\mu F^{\mu\nu} + \Gamma^\mu_{\mu\lambda} F^{\lambda\nu} + \Gamma^\nu_{\mu\lambda} F^{\mu\lambda} = -\frac{1}{c} j^\nu, \tag{14.2}$$

which are interpreted as Gauss's and Ampere's laws in the presence of a gravitational field.

14.1.2   Geodesic equation from SR equation of motion

Now that we have the GR equations for electromagnetism, what about the GR equations for gravitation? In Table 14.1, we have already written down the gravitational equation of motion: the equation that allows us to find the motion of a test charge in the presence of electromagnetic, as well as gravitational, fields. Just concentrating on the gravitational part, hence setting the EM field tensor $F_{\mu\nu} = 0$, we have the equation of motion for a particle in a gravitational field:

$$\frac{DU^\mu}{d\tau} = 0, \tag{14.3}$$

where U^μ is the 4-velocity of the test particle, and τ is the proper time. In fact, we should think of its derivation more directly as the generalization from the special relativistic equation of motion for a free particle:

$$\frac{dU^\mu}{d\tau} = 0, \tag{14.4}$$

which simply states that in the absence of an external force the test particle follows a trajectory of constant velocity.

We now demonstrate that Eq. (14.3) is just the geodesic Eq. (6.9). Using the explicit form of the covariant differentiation (13.47), the above equation can be written as

$$\frac{DU^\mu}{d\tau} = \frac{dU^\mu}{d\tau} + \Gamma^\mu_{\nu\lambda} U^\nu \frac{dx^\lambda}{d\tau} = 0. \tag{14.5}$$

Plugging in the expression of the 4-velocity in terms of the position vector[2]

$$U^\mu = \frac{dx^\mu}{d\tau}, \tag{14.6}$$

we immediately obtain an equation

$$\frac{d^2 x^\mu}{d\tau^2} + \Gamma^\mu_{\nu\lambda} \frac{dx^\nu}{d\tau} \frac{dx^\lambda}{d\tau} = 0, \tag{14.7}$$

which is recognized as the geodesic equation (6.9). This supports our heuristic argument—"gravity is not regarded as a force; a test body will move freely in a curved spacetime representing the gravitational field"—used in Section 6.2 to suggest that the GR equation of motion should be the geodesic equation.

[2] For the 4-velocity, we have $U^\mu = Dx^\mu / d\tau = dx^\mu / d\tau$ because $dx^\mu / d\tau$ is already a "good vector" as can been seen from the fact that $(ds/d\tau)^2 = g_{\mu\nu}(dx^\mu / d\tau)(dx^\nu / d\tau)$ is a scalar.

14.2   Einstein field equation

The procedure of going from SR to GR equations, illustrated in the above derivation of the GR equation of motion, fails for the case of the GR field equation because there is no SR equation for the gravitational field. Thus, for a gravitational field equations we need a fresh start—for guidance we need to go back to the Newtonian theory of gravitation, see Section 4.1. We have to answer the question: "What should be the relativistic generalization of the Newtonian field equation?"

$$\nabla^2 \Phi = 4\pi G_N \rho \rightarrow ?, \tag{14.8}$$

where G_N is Newton's constant, and ρ is the mass density function, just as the geodesic equation is the GR generalization of the Newtonian equation of motion:

$$\left[\frac{d^2 \vec{r}}{dt^2} = -\vec{\nabla}\Phi \right] \rightarrow \left[\frac{d^2 x^\mu}{d\tau^2} + \Gamma^\mu_{\nu\lambda} \frac{dx^\nu}{d\tau} \frac{dx^\lambda}{d\tau} = 0 \right]. \tag{14.9}$$

14.2.1   Finding the relativistic gravitational field equation

We have already learned that the metric tensor is the relativistic generalization of the gravitational potential (Section 6.1) and mass density is the (0,0) component of the energy–momentum tensor (Section 12.3):

$$\left(1 + \frac{2\Phi(x)}{c^2} \right) \rightarrow g_{00}(x) \quad \text{and} \quad \rho(x) \rightarrow T_{00}(x). \tag{14.10}$$

The GR field equation, being the relativistic generalization of the Newtonian field equation (14.8), should have the structure, written out in operator form, as

$$\hat{O} g = \kappa T, \tag{14.11}$$

namely, some differential operator $\hat{O}$ acting on the metric g to yield the energy–momentum tensor T with κ being the "conversion factor" proportional to Newton's constant G_N that allows us to relate the energy density and the spacetime curvature. Since we expect $\hat{O}g$ to have the Newtonian limit of $\nabla^2 \Phi$, the operator $\hat{O}$ must be a second derivative operator. Besides the $\partial^2 g$ terms, we also expect it to contain nonlinear terms of the type of $(\partial g)^2$. This is suggested by the fact that energy, just like mass, is a source of gravitational fields, and gravitational fields themselves hold energy—just as electromagnetic fields hold energy, with density being quadratic in the fields $(\vec{E}^2 + \vec{B}^2)$. That is, the gravitational field energy density must be quadratic in the gravitational field strength, $(\partial g)^2$. In terms of Christoffel symbols $\Gamma \sim \partial g$, we anticipate $\hat{O}g$ to contain not only $\partial \Gamma$ but also Γ^2 terms as well. Furthermore, because the right-hand side (RHS) is a symmetric tensor of rank 2 which is covariantly constant, $D_\mu T^{\mu\nu} = 0$ (reflecting energy–momentum conservation), $\hat{O}g$ on the LHS must have these properties also. The basic properties that the LHS of the field equation must have, in order to match those of $T^{\mu\nu}$ on the RHS, are summarized below:

$$
\boxed{\;\left[\hat{O}g\right] \text{ must have the property:}\;}
$$

conserved (covariantly constant)
symmetric rank-2 tensor
$(\partial^2 g), (\partial g)^2 \leadsto \partial \Gamma, \Gamma^2.$ (14.12)

There is only one such second rank tensor: the Einstein tensor $G_{\mu\nu}$, see Section 13.3.2, Eq. (13.84). Thus Einstein proposed the GR field equation to be

$$
G_{\mu\nu} = \kappa T_{\mu\nu}, \tag{14.13}
$$

where the proportional constant κ will be determined when we compare this field equation with that in the Newtonian theory. Writing out the Einstein equation in terms of the Ricci scalar and tensor we have

$$
R_{\mu\nu} - \frac{1}{2} R g_{\mu\nu} = \kappa T_{\mu\nu}. \tag{14.14}
$$

This field equation can be written in an alternative form. Taking the trace of the above equation, we have

$$
-R = \kappa T, \tag{14.15}
$$

where T is the trace of the energy–momentum tensor, $T = g^{\mu\nu} T_{\mu\nu}$. In this way we can rewrite the field equation in an equivalent form by replacing $R g_{\mu\nu}$ by $-\kappa T g_{\mu\nu}$:

$$
R_{\mu\nu} = \kappa \left(T_{\mu\nu} - \frac{1}{2} T g_{\mu\nu} \right). \tag{14.16}
$$

14.2.2   Newtonian limit of the Einstein equation

Here we shall show that the familiar Newtonian field equation (14.8) is simply the leading approximation to the Einstein equation (14.16) in the Newtonian limit (defined in Section 6.2.1) as being for a nonrelativistic source particle producing a weak and static gravitational field.

- **Nonrelativistic velocity**   In the nonrelativistic regime of small v/c, the rest energy density term T_{00} being dominant, we shall concentrate on the 00-component of (14.16), as other terms are down by $O(v/c)$:

$$R_{00} = k \left(T_{00} - \frac{1}{2} T g_{00} \right) \tag{14.17}$$

with

$$T = g^{\mu\nu} T_{\mu\nu} \simeq g^{00} T_{00} = \frac{1}{g_{00}} T_{00}. \tag{14.18}$$

Equation (14.17) becomes

$$R_{00} = \frac{1}{2} \kappa T_{00}. \tag{14.19}$$

To recover the Newtonian field equation, we need to show that $R_{00} \to \nabla^2 g_{00}$. From the definition of the Ricci tensor (in terms of the Riemann–Christoffel tensor), we have

$$R_{00} = g^{\mu\nu} R_{\mu 0 \nu 0} = g^{ij} R_{i0j0} \tag{14.20}$$

where $i = 1,2,3$ and in reaching the last equality we have used the fact that the tensor components such as R_{0000} and R_{i000} all vanish because of symmetry properties of the curvature tensor, $R_{\mu\nu\lambda\rho} = -R_{\mu\nu\rho\lambda}$, etc.

- **Weak field limit**   The Newtonian limit also corresponds to the weak field limit, $g_{\mu\nu} = \eta_{\mu\nu} + h_{\mu\nu}$ with $h_{\mu\nu}$ being small. Since $\partial g = \partial h$, we will keep as few powers of ∂g as possible, i.e. keep $\partial\partial g$ terms rather than $(\partial g)^2$s, etc.

$$R_{\mu\nu\alpha\beta} = \frac{1}{2} \left(\partial_\mu \partial_\alpha g_{\nu\beta} - \partial_\nu \partial_\alpha g_{\mu\beta} + \partial_\nu \partial_\beta g_{\mu\alpha} - \partial_\mu \partial_\beta g_{\nu\alpha} \right). \tag{14.21}$$

Substituting this into (14.20) we have

$$R_{00} = g^{ij} R_{i0j0} = \frac{g^{ij}}{2} \left(\partial_i \partial_j g_{00} - \partial_0 \partial_j g_{i0} + \partial_0 \partial_0 g_{ij} - \partial_i \partial_0 g_{0j} \right). \tag{14.22}$$

- **Static limit**   The Newtonian limit also corresponds to a static situation we can drop in (14.22) all terms having a time derivative ∂_0 factor,

$$R_{00} = \frac{1}{2} \nabla^2 g_{00}. \tag{14.23}$$

After using the relation (6.20) between g_{00} and the Newtonian potential Φ and $T_{00} = \rho c^2$ as given in (12.63), Eq. (14.19) becomes

$$-\frac{1}{2} \nabla^2 \left(1 + 2\frac{\Phi}{c^2} \right) = \frac{1}{2} \kappa \rho c^2,$$

or

$$\nabla^2 \Phi = -\frac{1}{2}\kappa\rho c^4. \tag{14.24}$$

Thus we see that the Einstein equation indeed has the correct Newtonian limit of $\nabla^2 \Phi = 4\pi G_N \rho$ when we make the identification of

$$\kappa = -\frac{8\pi G_N}{c^4}. \tag{14.25}$$

The Einstein equation

Putting this value of (14.25) into the field equation (14.14) we have the Einstein equation[3]

$$R_{\mu\nu} - \frac{1}{2}Rg_{\mu\nu} = -\frac{8\pi G_N}{c^4}T_{\mu\nu}, \tag{14.26}$$

or, written in its equivalent form as (14.16):

$$R_{\mu\nu} = -\frac{8\pi G_N}{c^4}\left(T_{\mu\nu} - \frac{1}{2}Tg_{\mu\nu}\right). \tag{14.27}$$

Given the source distribution $T_{\mu\nu}$, one can solve this set of 10 coupled nonlinear partial differential equations for the spacetime metric $g_{\mu\nu}(x)$. We reiterate the central point: spacetime geometry in GR is not a fixed quantity, but is dynamically determined by the mass/energy distribution—in contrast to the fixed SR's spacetime. In general they are extremely difficult to solve. However, for the spherically symmetric situation (in the three spatial dimensions), an analytic solution can be obtained. We shall study this solution in the following section.

[3]Beware of various sign conventions $S = \pm 1$ used in the literature:

$$\eta_{\mu\nu} = S1 \times \mathrm{diag}(-1, 1, 1, 1),$$

$$R^{\mu}_{\lambda\alpha\beta} = S2 \times (\partial_\alpha \Gamma^\mu_{\lambda\beta} - \partial_\beta \Gamma^\mu_{\lambda\alpha},$$

$$+ \Gamma^\mu_{\nu\alpha}\Gamma^\nu_{\lambda\beta} - \Gamma^\mu_{\nu\beta}\Gamma^\nu_{\lambda\alpha})$$

$$G_{\mu\nu} = S3 \times \frac{8\pi G}{c^4}T_{\mu\nu}.$$

Thus our convention is $S1, S2, S3 = (+ + -)$. The sign in the Einstein equation $S3$ is related to the sign convention in the definition of the Ricci tensor $R_{\mu\nu} = R^\alpha_{\mu\alpha\nu}$.

Box 14.1 Newtonian limit for the general source having mass density and pressure

In certain situations, with cosmology being the notable example, we consider the source of gravity as being a plasma having mass density and pressure. This is unlike the more familiar nonrelativistic matter for which we can drop the negligibly small pressure term. However, if the corresponding matter density is particularly low, or just comparable to the pressure contribution, we need to work out the Newtonian limit for a general source with an energy–momentum tensor of the ideal fluid as in (12.72). As shown in (14.17), the dominant term in this limit is the 00-component of the Einstein equation (14.16),

$$R_{00} = \kappa\left(T_{00} - \frac{1}{2}Tg_{00}\right) = \frac{\kappa}{2}(T_{00} + T_{11} + T_{22} + T_{33})$$

$$= \frac{\kappa}{2}\left(\rho c^2 + 3p\right). \tag{14.28}$$

The trace (T) of the energy–momentum tensor, to leading order in the Newtonian limit, has been calculated by using the flat spacetime metric: $T = \eta^{\mu\nu}T_{\mu\nu}$ with $\eta^{\mu\nu} = \mathrm{diag}(-1, 1, 1, 1)$. From this and the expression of

R_{00} in terms of the gravitational potential of (14.23), we obtain the quasi-Newtonian equation for the gravitational potential Φ, as first displayed in (11.4) when we were discussing the cosmological constant as a vacuum pressure,

$$\nabla^2 \Phi = 4\pi G_N \left(\rho + 3\frac{p}{c^2} \right).$$ (14.29)

This makes it clear that not only mass but also pressure can be a source of gravitational field.

14.3   The Schwarzschild exterior solution

We now solve the Einstein equation for a spherically symmetric (nonrotating) source with total mass M. The solution is the metric function $g_{\mu\nu}(x)$ for the spacetime geometry outside the source, and is called the Schwarzschild exterior solution. In Section 7.1.1 we have shown that a spatially spherical symmetric metric tensor (7.12) has only two unknown scalar functions:

$$ds^2 = g_{00}(r, t)c^2 dt^2 + g_{rr}(r, t)dr^2 + r^2(d\theta^2 + \sin^2\theta d\phi^2).$$ (14.30)

Here we shall use the Einstein equation to solve for g_{00} and g_{rr}. The first step involves expressing the Ricci tensor elements $R_{\mu\nu}$ in terms of these metric elements.

The spherical symmetric Christoffel symbols

We begin by calculating the connection symbols based on the spherically symmetric form of (14.30). It will be convenient to introduce the notation:

$$g_{00} = \frac{1}{g^{00}} \equiv -e^\nu, \quad g_{rr} = \frac{1}{g^{rr}} \equiv e^\rho$$ (14.31)

so that the unknown metric functions are now $\nu(r, t)$ and $\rho(r, t)$. Here we state the result (see Box 14.2 for comments on the calculational procedure):

$$\Gamma^0_{00} = -\frac{\dot\nu}{2}, \quad \Gamma^0_{rr} = \frac{\dot\rho}{2}e^{\rho-\nu}, \quad \Gamma^0_{0r} = \frac{\nu'}{2},$$

$$\Gamma^r_{00} = \frac{\nu'}{2}e^{\nu-\rho}, \quad \Gamma^r_{rr} = \frac{\rho'}{2}, \quad \Gamma^r_{0r} = \frac{\dot\rho}{2},$$ (14.32)

$$\Gamma^r_{\theta\theta} = -re^{-\rho}, \quad \Gamma^r_{\phi\phi} = -r\sin^2\theta e^{-\rho}, \quad \Gamma^\theta_{\phi\phi} = -\sin\theta\cos\theta,$$

$$\Gamma^\theta_{r\theta} = r^{-1}, \quad \Gamma^\phi_{\phi\theta} = \cot\theta, \quad \Gamma^\phi_{r\phi} = r^{-1},$$

where a dot denotes differentiation with respect to the coordinate time $x^0 = ct$, while the prime is differentiation with respect to the radial coordinate r: for example,

$$\dot\nu = \frac{1}{c}\frac{\partial\nu}{\partial t} \qquad \nu' = \frac{\partial\nu}{\partial r}.$$ (14.33)

Box 14.2 $\Gamma^{\mu}_{\nu\lambda}$ **via the Euler–Lagrange equation**

In principle, we can obtain the result in (14.33) by differentiating the metric tensor as in (13.37). A more efficient procedure will be through the interpretation of the geodesic equation (with σ being the curve parameter)

$$\frac{d^2x^{\mu}}{d\sigma^2} + \Gamma^{\mu}_{\nu\lambda}\frac{dx^{\nu}}{d\sigma}\frac{dx^{\lambda}}{d\sigma} = 0 \tag{14.34}$$

as the Euler–Lagrange equation

$$\frac{d}{d\sigma}\frac{\partial L}{\partial \dot{x}^{\mu}} - \frac{\partial L}{\partial x^{\mu}} = 0 \tag{14.35}$$

with the Lagrangian being (see Section 5.2.1 for more details)

$$L = g_{\mu\nu}\frac{dx^{\mu}}{d\sigma}\frac{dx^{\nu}}{d\sigma} \tag{14.36}$$

$$= -e^{\nu}\left(\frac{dx^0}{d\sigma}\right)^2 + e^{\rho}\left(\frac{dr}{d\sigma}\right)^2 + r^2\left(\frac{d\theta}{d\sigma}\right)^2 + r^2\sin^2\theta\left(\frac{d\phi}{d\sigma}\right)^2.$$

Once the geodesic equation is written out this way as in (14.35), we can then extract the value of $\Gamma^{\mu}_{\nu\lambda}$ by comparing it to (14.34). For example, because we have

$$\frac{\partial L}{\partial x^0} = -\dot{\nu}e^{\nu}\left(\frac{dx^0}{d\sigma}\right)^2 + \dot{\rho}e^{\rho}\left(\frac{dr}{d\sigma}\right)^2, \quad \text{and} \quad \frac{\partial L}{\partial \dot{x}^0} = -2e^{\nu}\left(\frac{dx^0}{d\sigma}\right)$$

the $\mu = 0$ component of the Euler–Lagrange equation (14.35) reads as

$$\frac{d}{d\sigma}\left[-2e^{\nu}\left(\frac{dx^0}{d\sigma}\right)\right] - \left[-\dot{\nu}e^{\nu}\left(\frac{dx^0}{d\sigma}\right)^2 + \dot{\rho}e^{\rho}\left(\frac{dr}{d\sigma}\right)^2\right] = 0$$

or

$$-e^{\nu}\left[\frac{d^2x^0}{d\sigma^2} + \nu'\frac{dr}{d\sigma}\frac{dx^0}{d\sigma} - \frac{\dot{\nu}}{2}\left(\frac{dx^0}{d\sigma}\right)^2 + \frac{\dot{\rho}}{2}e^{\rho-\nu}\left(\frac{dr}{d\sigma}\right)^2\right] = 0.$$

This is to be compared to the $\mu = 0$ component of (14.34), which with only the nonvanishing $(dx^{\nu}/d\sigma)(dx^{\lambda}/d\sigma)$ factors displayed, has the form:

$$\frac{d^2x^0}{d\sigma^2} + 2\Gamma^0_{r0}\frac{dr}{d\sigma}\frac{dx^0}{d\sigma} + \Gamma^0_{00}\left(\frac{dx^0}{d\sigma}\right)^2 + \Gamma^0_{rr}\left(\frac{dr}{d\sigma}\right)^2 = 0.$$

Hence we can extract the result

$$\Gamma^0_{r0} = \frac{\nu'}{2}, \quad \Gamma^0_{00} = -\frac{\dot{\nu}}{2}, \quad \Gamma^0_{rr} = \frac{\dot{\rho}}{2}e^{\rho-\nu}, \tag{14.37}$$

as displayed in (14.32).

The spherically symmetric curvature

From the Christoffel symbols we then use (13.58) to calculate the curvature tensor $R^\alpha_{\mu\beta\nu}$ from which we can contract the indices $R^\alpha_{\mu\alpha\nu}$ to form the Ricci tensor:

$$R_{00} = -\left(\frac{v''}{2} + \frac{v'^2}{4} - \frac{v'\rho'}{4} + \frac{v'}{r}\right)e^{v-\rho} + \left(\frac{\ddot{\rho}}{2} + \frac{\dot{\rho}^2}{4} - \frac{\dot{v}\dot{\rho}}{4}\right),$$

$$R_{rr} = \left(\frac{v''}{2} + \frac{v'^2}{4} - \frac{v'\rho'}{4} - \frac{\rho'}{r}\right) - \left(\frac{\ddot{\rho}}{2} + \frac{\dot{\rho}^2}{4} - \frac{\dot{v}\dot{\rho}}{4}\right)e^{\rho-v},$$

$$R_{0r} = -\frac{\dot{\rho}}{r}, \qquad\qquad\qquad\qquad (14.38)$$

$$R_{\theta\theta} = \left[1 + \frac{r}{2}\left(v' - \rho'\right)\right]e^{-\rho} - 1,$$

$$R_{\phi\phi} = \sin^2\theta\, R_{\theta\theta}.$$

So far we have only discussed the restriction that spherical symmetry places on the solution, and have not sought the actual solution to the Einstein field equation. This we shall do in the following section.

The Einstein equation for the spacetime exterior to the source

Here we wish to find the metric in the region outside a spherically symmetric source. Because the energy–momentum tensor $T_{\mu\nu}$ vanishes in the exterior, the Einstein field equation becomes

$$R_{\mu\nu} = 0. \qquad\qquad (14.39)$$

Do not be deceived by the superficially simple form of this equation.[4] Keep in mind that the Ricci tensor is a set of a second-order nonlinear differential operators acting on the metric functions, as displayed in (14.39). In this spherical symmetrical case with only two nontrivial scalar functions $v(r, t)$ and $\rho(r, t)$, we expect this to represent two coupled partial differential equations.

Isotropic metric is time independent

Before getting the solution for the two unknown metric functions $g_{00}(r, t) \equiv -\exp v(r, t)$ and $g_{rr}(r, t) \equiv \exp\rho(r, t)$, we first point out that the metric must necessarily be time-independent (the Birkhoff theorem, see Box 14.3)

$$v(r, t) = v(r) \quad\text{and}\quad \rho(r, t) = \rho(r). \qquad (14.40)$$

After substituting in this condition that all t-derivative terms vanish $\dot{v} = \dot{\rho} = \ddot{\rho} = 0$, the Einstein vacuum relations in (14.38) yield three component equations:

the $R_{00} = 0$ equation:

$$\frac{v''}{2} + \frac{v'^2}{4} - \frac{v'\rho'}{4} + \frac{v'}{r} = 0, \qquad (14.41)$$

[4]One should keep in mind that a vanishing Ricci tensor $R_{\mu\nu} = 0$ does not imply a vanishing Riemann tensor $R_{\mu\nu\alpha\beta} = 0$. Namely, an empty space ($T_{\mu\nu} = 0$) does not need to be flat, even though a flat space $R_{\mu\nu\alpha\beta} = 0$ must have a vanishing Ricci tensor. (It may be helpful to compare the situation to the case of a matrix having a vanishing trace. This certainly does not require the entire matrix to vanish.)

the $e^{\rho-\nu} R_{00} + R_{rr} = 0$ equation:

$$\nu' + \rho' = 0, \tag{14.42}$$

the $R_{\theta\theta} = 0$ equation:

$$\left[1 + \frac{r}{2}(\nu' - \rho')\right] e^{-\rho} - 1 = 0. \tag{14.43}$$

Actually one of these three equations is redundant. It can be shown that the solution to two equations, for example, (14.42) and (14.43), automatically satisfies the remaining equation (14.41).

Box 14.3 The Birkhoff theorem

Theorem   Every spherically symmetric vacuum solution to $R_{\mu\nu} = 0$ is static. That is, $\dot{\nu} = \dot{\rho} = 0$.

Proof   That ρ has no time dependence follows simply from the equation $R_{0r} = -\dot{\rho}/r = 0$ in (14.39). That ν has no time dependence can be demonstrated as follows. Because ρ and, hence also, ρ' have no t-dependence, the Einstein equation

$$R_{\theta\theta} = \left[1 + \frac{r}{2}(\nu' - \rho')\right] e^{-\rho} - 1 = 0, \tag{14.44}$$

implies that ν' is also time independent (as there is no time dependence in the entire equation). The statement

$$\nu' \equiv \frac{d\nu}{dr} = f(r) \tag{14.45}$$

means that the function ν must depend on the variables r and t, separately:

$$\nu(r, t) = \nu(r) + n(t). \tag{14.46}$$

The appearance of $\nu(r)$ and $n(t)$ in the infinitesimal interval ds^2 has a form so that a possible time-dependence $n(t)$ can be absorbed in a new time variable $\tilde{t}$:

$$-e^{\nu(r)} e^{n(t)} c^2 dt^2 \equiv -e^{\nu(r)} c^2 d\tilde{t}^2.$$

In terms of these coordinates, the metric functions are time independent. This completes our proof of the Birkhoff theorem.   ∎

- Recall the simple physical argument for the Newtonian analog of the Birkhoff theorem, given at the end of Section 7.1.
- Historically, Schwarzschild obtained his solution by explicitly assuming a static spherical source. Only several years later did Birkhoff prove his theorem showing that the solution Schwarzschild obtained was actually valid for an exploding, collapsing, or pulsating spherical star.

Solving the Einstein equation

We now carry out the solution to (14.42) and (14.43). After an integration over r of (14.42), we obtain the equality

$$v(r) = -\rho(r), \tag{14.47}$$

where we have set the integration constant to zero by a choice of new time coordinates in exactly the same manner as done in the proof of the Birkhoff theorem (Box 14.3). Because v and ρ are exponents of the metric scalar functions (14.31), this relation (14.47) translates into

$$-g_{00} = \frac{1}{g_{rr}}, \tag{14.48}$$

as first quoted in (7.17). Equation (14.42) also allows us to rewrite (14.43) as

$$(1 - r\rho')e^{-\rho} - 1 = 0. \tag{14.49}$$

We can simplify this equation by introducing a new variable:

$$\lambda(r) \equiv e^{-\rho(r)}, \quad \frac{d\lambda}{dr} = -\rho' e^{-\rho}$$

so that (14.49) becomes

$$\frac{d\lambda}{dr} + \frac{\lambda}{r} = \frac{1}{r}, \tag{14.50}$$

which has the general solution $\lambda(r) = \lambda_0(r) + \lambda_1$ where λ_0 is the solution to the homogeneous equation

$$\frac{d\lambda_0}{dr} = -\frac{\lambda_0}{r}. \tag{14.51}$$

This can be solved by straightforward integration, $\ln \lambda_0 = -\ln r + c_0$. It implies that the product of $\lambda_0 r$ is a constant, which we label

$$\lambda_0 r \equiv -r^*. \tag{14.52}$$

Combining this with a particular solution of $\lambda_1 = 1$, we have the general solution of

$$\lambda = 1 - \frac{r^*}{r} = \frac{1}{g_{rr}} = -g_{00}, \tag{14.53}$$

where we have used (7.17) and noted that the λ function is just g_{rr}^{-1}. From this solution we have the **Schwarzschild metric** in the (ct, r, θ, ϕ) coordinate system

$$g_{\mu\nu} = \mathrm{diag}\left[\left(-1 + \frac{r^*}{r}\right), \left(1 - \frac{r^*}{r}\right)^{-1}, r^2, r^2 \sin^2\theta\right], \tag{14.54}$$

which is quoted in (7.18). The parameter r^* is then related to Newton's constant and source mass $r^* = 2G_N M/c^2$ through the relation between the metric element and gravitational potential in the Newtonian limit:

$$g_{00} = -\left(1 + \frac{2\Phi}{c^2}\right) = -1 + \frac{2G_N M}{c^2 r} = -1 + \frac{r^*}{r}. \tag{14.55}$$

This derivation is rather formal and lengthy. To get an intuitive understanding by connecting it to the familiar Newtonian result of an $1/r$ gravitational potential, the reader is urged to study Chapter 15 where the linearized Einstein equation is presented in Section 15.1. That equation has a form similar to the standard wave equation. Its solution for a spherically symmetric source immediately leads to a metric that approximates (14.54) up to a correction of $O(r^{*2})$. (See Problem 15.2.)

This Schwarzschild solution (14.54) to the Einstein field equation must be considered as a main achievement of GR in the field of astrophysics. It is an exact solution which corresponds historically to Newton's treatment of the $1/r^2$ force law (14.8) in classical gravitational theory. Numerous GR applications, from the bending of a light-ray to black holes, are based on this solution (see Chapters 7 and 8).

As nonlinear equations are very difficult to solve, it is astonishing that Karl Schwarzschild, the Director of the Potsdam Observatory, discovered these exact solutions[5] only two months after Einstein's final formulation of GR at the end of November 1915. At this time Schwarzschild was already in the German army on the Russian front. Tragically by the summer of 1916 he died there (of an illness)—one of the countless victims of the First World War.

[5]There is also the Schwarzschild interior solution for the Einstein field equation with $T_{\mu\nu} \neq 0$ (e.g. that for an ideal fluid). Such solutions are relevant for the discussion of gravitational collapse.

14.4   The Einstein equation for cosmology

Cosmological study must be carried out in the framework of GR. The dynamical spacetime of general relativity can naturally accommodate some of the basic observational features such as the expanding universe. The basic equation is the Einstein equation. In Section 14.4.1 we find the solution of Einstein's equation that is compatible with a 3D space being homogeneous and isotropic as required by the cosmological principle. This solution is the Robertson–Walker metric presented in Chapter 9. In Box 14.4 we show that the Einstein equation with a Robertson–Walker metric leads to the Friedmann equations discussed in Chapter 10. Finally in Section 14.4.2 we show how the Einstein equation can be modified by the addition of the cosmological constant term. The physical implications of such a Λ term have been studied in Chapter 11.

14.4.1   Solution for a homogeneous and isotropic 3D space

The cosmological principle gives us a picture of the universe as a system of a "cosmic fluid." It is convenient to pick the coordinate time t to be the proper time of each fluid element. In this comoving coordinate system, the 4D metric has the form (as discussed in Section 9.3) of $g_{\mu\nu} = \mathrm{diag}(-1, g_{ij})$ so that the spacetime line element is

$$ds^2 = -c^2 dt^2 + dl^2 \tag{14.56}$$

with

$$dl^2 = g_{ij} x^i x^j = R(t)^2 d\hat{l}^2, \tag{14.57}$$

where $R(t)$ is the dimensionful scale factor, equal to $a(t)R_0$. The 3D line element $d\hat{l}^2$ is then dimensionless. Previously we argued that the requirement of a homogeneous and isotropic space means that the space must have constant curvature. Then we used the result obtained in Section 5.3.2 for a constant curvature 3D space (derived heuristically from the result of 2D surfaces of constant curvature):

$$d\hat{l}^2 = \frac{d\xi^2}{1 - k\xi^2} + \xi^2 d\theta^2 + \xi^2 \sin^2\theta d\phi^2 \tag{14.58}$$

with ξ being the dimensionless radial distance. This is the Robertson–Walker metric. Here in this subsection, we shall use the intermediate steps of Section 14.3 (in arriving at the Schwarzschild solution) to provide another derivation of this result (14.58). The purpose is to make it clear that such a metric is indeed the solution of the Einstein equation for a homogeneous and isotropic space.[6]

Homogeneity and isotropy mean that the space must be spherically symmetric with respect to every point in that space. We can work out the metric that satisfies this requirement as follows:

- **Spherically symmetric with respect to the origin** This means that the metric for the three-dimensional space (ξ, θ, ϕ) should have the form as discussed Section 7.1.1 and Section 14.3.1. Keeping only the spatial part of (14.30), we have

$$d\hat{l}^2 = \hat{g}_{\xi\xi}d\xi^2 + \xi^2\left(d\theta^2 + \sin^2\theta d\phi^2\right). \tag{14.59}$$

Birkhoff's theorem (Box 14.3) then implies that the metric element $\hat{g}_{\xi\xi}$ is independent of the coordinate time.[7] We will also follow the previous notation of $\hat{g}_{\xi\xi} \equiv e^{\rho(\xi)}$ as shown in (14.31).
- **Spherically symmetric with respect to every point** To broaden from the spherical symmetry with respect to one point (the origin) as discussed in Section 14.3.1 to that with respect to every point (as required by homogeneity and isotropy), we demand that the Ricci scalar for this 3D space, which in general is a function of ξ, is a constant; with some foresight we set it equal to $-6k$,

$$R^{(3)} \equiv -6k. \tag{14.60}$$

This is just the mathematical statement expressing our expectation that the space should be one with constant curvature. We can look up the expression for the Ricci tensor in (14.38), and after setting $\nu = \dot{\nu} = \nu' = \nu'' = \dot{\rho} = \ddot{\rho} = 0$, we obtain the Ricci tensor elements $R^{(3)}_{ij}$ for the three-dimensional space:

$$R^{(3)}_{\xi\xi} = -\frac{1}{\xi}\frac{d\rho}{d\xi}, \quad R^{(3)}_{\theta\theta} = \left(1 - \frac{\xi}{2}\frac{d\rho}{d\xi}\right)e^{-\rho} - 1$$

$$R^{(3)}_{\phi\phi} = \sin^2\theta R^{(3)}_{\theta\theta} \tag{14.61}$$

[6]It should nevertheless be emphasized that the Robertson–Walker metric follows from the symmetry of 3D space, rather from any specific property of gravity as encoded in the Einstein equation. A more rigorous derivation would involve the mathematics of symmetric spaces, Killing vectors, and isometry. See Chapter 13 of S. Weinberg, *Gravitation and Cosmology*, Wiley, 1972.

[7]This shows the consistency of our assumption that the reduced metric $\hat{g}_{ij}$, after factoring out the scale factor $a^2(t)$, does not change with time.

which is to be contracted with the inverse metric $\hat{g}^{ij}$ of (14.59),

$$\hat{g}^{\xi\xi} = e^{-\rho(\xi)}, \qquad \hat{g}^{\theta\theta} = \xi^{-2}, \qquad \hat{g}^{\phi\phi} = 1/\left(\xi^2 \sin^2\theta\right), \qquad (14.62)$$

to obtain the Ricci scalar:

$$R^{(3)} = \sum_i R_{ii}^{(3)} \hat{g}^{ii} = R_{\xi\xi}^{(3)} \hat{g}^{\xi\xi} + 2R_{\theta\theta}^{(3)} \hat{g}^{\theta\theta}$$

$$= -\frac{e^{-\rho}}{\xi}\frac{d\rho}{d\xi} + \frac{2}{\xi^2}\left[\left(1 - \frac{\xi}{2}\frac{d\rho}{d\xi}\right)e^{-\rho} - 1\right].$$

Setting it to $-6k$ as in (14.60)

$$\frac{2}{\xi^2}\frac{d}{d\xi}\left(\xi e^{-\rho} - \xi\right) = -6k. \qquad (14.63)$$

We can solve this differential equation by straightforward integration

$$d\left(\xi e^{-\rho} - \xi\right) = -3k\xi^2 d\xi$$

$$\left(1 - e^{-\rho}\right)\xi = k\xi^3 + A \qquad (14.64)$$

where the integration constant is $A = 0$, as can be seen in the $\xi = 0$ limit. We obtain the desired solution

$$\hat{g}_{\xi\xi} = e^{\rho(\xi)} = \frac{1}{1 - k\xi^2}. \qquad (14.65)$$

Plugging this into (14.59), we have the dimensionless separation in 3D space as given by (14.58), confirming the heuristic results of (5.55) and (9.38).

Box 14.4 Friedmann equations

Here we shall explicate the exact relation between the Einstein and Friedmann equations used in Chapter 10. In the Einstein equation $G_{\mu\nu} = \kappa T_{\mu\nu}$ (with $\kappa = -8\pi G_N/c^4$) for the homogeneous and isotropic universe, the LHS is determined by the Robertson–Walker metric with its two parameters: the curvature constant k and the scale factor $a(t)$. We still need to specify the energy–momentum tensor on the RHS, which must be compatible with the cosmological principle. The simplest plausible choice is to take the cosmic fluid as an ideal fluid as discussed in Section 12.3. In special relativity, we have already shown in (12.72) that

$$T_{\mu\nu} = pg_{\mu\nu} + \left(\rho + \frac{p}{c^2}\right)U_\mu U_\nu, \qquad (14.66)$$

where p is the pressure, ρ is the mass density, and U^μ is the 4-velocity field of the fluid. Since there is no derivative, the same form also holds for GR. In the cosmic rest frame (the comoving coordinates) in which each fluid element (galaxy) carries its own position label, all the fluid elements are at rest $U^\mu = (c, 0)$. In such a frame with a metric given by $g_{\mu\nu} = \text{diag}(-1, g_{ij})$

as in (9.32), the energy–momentum takes on the particularly simple form

$$T_{\mu\nu} = \begin{pmatrix} \rho c^2 & 0 \\ 0 & p g_{ij} \end{pmatrix}. \tag{14.67}$$

The cosmological Friedmann equations are just the Einstein equation with Robertson–Walker metric and with an ideal fluid energy–momentum tensor.

1. The $G_{00} = -8\pi G_N \rho/c^2$ equation can then be written (again after a long calculation) in terms of the Robertson–Walker metric elements $a(t)$ and k. We have the first Friedmann equation (10.1),

$$\frac{\dot{a}^2(t)}{a^2(t)} + \frac{kc^2}{R_0^2 a^2(t)} = \frac{8\pi G_N}{3} \rho. \tag{14.68}$$

2. From the equation $G_{ij} = -8\pi G_N p g_{ij}/c^4$, we have the second Friedmann equation (10.2),

$$\frac{\ddot{a}(t)}{a(t)} = -\frac{4\pi G_N}{c^2} \left(p + \frac{1}{3}\rho c^2 \right). \tag{14.69}$$

As we have shown in Chapter 10 these Friedmann equations, because of the cosmological principle, have a simple Newtonian interpretation. Nevertheless, they must be understood in the context of GR as they still involve geometric concepts like curvature, etc. The proper view is that they are Einstein equations applied to cosmology.

14.4.2 Einstein equation with a cosmological constant term

Einstein's desideratum for a static universe led him to modify his original field equation for GR. Given the strong theoretical arguments (see Section 14.2) used in arriving at (14.13) and its success in describing gravitational phenomena (at least up to the solar system), how can we go about making such a modification? The possibility is that there is some gravitational feature which is too small to be observed for systems at sub-cosmic scales, but becomes important only on truly large dimensions. Still, whatever we add to the Einstein equation, it must be compatible with its tensor structure—a symmetric rank-2 tensor that is covariantly constant (i.e. its covariant derivative vanishes). Recall that the Einstein tensor $G_{\mu\nu}$, being a nonlinear second-order derivative of the metric, is such a tensor. But the metric tensor $g_{\mu\nu}$ itself is also symmetric, rank-2, and covariantly constant, see (13.32). Thus it is mathematically consistent to include such a term on the LHS of the field equation (with $\kappa = -8\pi G_N/c^4$):

$$G_{\mu\nu} - \Lambda g_{\mu\nu} = \kappa T_{\mu\nu}. \tag{14.70}$$

Λ is some unknown constant coefficient. The addition will alter the Newtonian limit of the field equation as discussed in Section 14.2.2, and leads to a nonrelativistic equation different from the Newton/Poisson equation as shown in (14.8). This difficulty can, however, be circumvented by assuming that Λ is of such a small size as to be unimportant except for cosmological applications.[8] Hence Λ is called the **cosmological constant**.

While it is more straightforward to see, from a mathematical viewpoint, how the geometric side of Einstein's equation can be modified by this addition, the physical interpretation of this new term can be more readily gleaned if we move it to the energy–momentum side:

$$G_{\mu\nu} = \kappa(T_{\mu\nu} + \kappa^{-1}\Lambda g_{\mu\nu}) = \kappa(T_{\mu\nu} + T^{\Lambda}_{\mu\nu}), \tag{14.71}$$

where $T^{\Lambda}_{\mu\nu} = \kappa^{-1}\Lambda g_{\mu\nu}$ can be called the "vacuum energy–momentum tensor." Since the metric tensor is covariantly constant, the conservation of vacuum energy–momentum is satisfied. In the absence of an ordinary mass/energy distribution $T_{\mu\nu} = 0$ (hence, the vacuum), the source term $T^{\Lambda}_{\mu\nu}$ can still bring about a gravitational field in the form of a nontrivial spacetime curvature.[9]

In the cosmic rest frame (the comoving coordinates) with the velocity field being $U^{\mu} = (c, 0, 0, 0)$ and the metric $g_{\mu\nu} = \text{diag}(-1, g_{ij})$ of (9.32), this vacuum energy–momentum tensor can be written in a form analogous to the conventional ideal fluid stress tensor (14.67):

$$T^{\Lambda}_{\mu\nu} = \frac{\Lambda}{\kappa}\begin{pmatrix} -1 & 0 \\ 0 & g_{ij} \end{pmatrix} \equiv \begin{pmatrix} \rho_{\Lambda}c^2 & 0 \\ 0 & p_{\Lambda}g_{ij} \end{pmatrix}. \tag{14.72}$$

This implies a constant vacuum energy density,

$$\rho_{\Lambda} = -\frac{\Lambda}{\kappa c^2} = \frac{\Lambda c^2}{8\pi G_{\text{N}}}, \tag{14.73}$$

which is the result quoted in Chapter 11, see (11.2). If we take $\Lambda > 0$, so that $\rho_{\Lambda} > 0$, it implies a **negative** vacuum pressure:

$$p_{\Lambda} = -\rho_{\Lambda}c^2 < 0. \tag{14.74}$$

Thus the cosmological constant corresponds to an energy density which is constant in time and in space. No matter how we change the volume, this energy density is unchanged. As we have discussed in Chapter 11, such negative pressure is the source of gravitational repulsion which can drive the inflationary epoch of the big bang, and can give rise to a universe undergoing an accelerated expansion.

Review questions

1. What is the **principle of general covariance**?

2. Since SR equations are valid only in the absence of gravity, turning SR into GR equations implies the introduction of a gravitational field into the relativistic equations. If the physics equation is known in the special relativistic limit, how does one turn such an SR equation into a general

relativistic one? Also discuss the difference of coordinate symmetries involved in SR and in GR.

3. How can one "deduce" the GR equation of motion from that of SR?

4. Why did Einstein expect the relativistic gravitational field equation to have the form of $\hat{O}g = \kappa T$ with the LHS

being a covariantly constant symmetric tensor of rank-2 involving $(\partial^2 g)$ as well as $(\partial g)^2$ terms?

5. Write out the two equivalent versions of the Einstein field equation, with the coupling expressed in terms of Newton's constant. (A coupling measures the interaction strength, which typically appears as the coefficient constant in front of the source term.)

6. How can we use the result of a metric for a spherical symmetric space to derive that for a space that is homogeneous and isotropic?

7. What is the relation between the Friedmann equations and the Einstein equation?

8. What are the mathematical properties of the cosmological constant term that allow it to be added to the Einstein equation?

9. Write out the Einstein equation with the Λ term. Explain why such a term can be interpreted as the vacuum energy–momentum source of gravity.

Problems

14.1 Another derivation of the geodesic equation Starting from the no-force condition $(d\mathbf{p})^\mu = Dp^\mu = 0$ in a curved spacetime with 4-momentum $p^\mu = mU^\mu$, we can arrive at the geodesic equation (14.7) by using the covariant derivative expression of (13.46).

14.2 Vacuum Einstein equations

(a) Show that a vanishing Einstein tensor implies a vanishing Ricci tensor.

(b) The three equations (14.41)–(14.43) are redundant. Show that the solution to two equations, e.g. Eqs. (14.42) and (14.43), automatically satisfies the remaining equation (14.41).

14.3 Friedmann equations and energy conservation Show that the energy conservation statement (10.3) that results from the linear combination of these two Friedmann equations (14.68) and (14.69) can also be derived directly from the energy–momentum conservation equation of $D_\mu T^{\mu\nu} = 0$, where $T^{\mu\nu}$ is the energy–momentum of the cosmic fluid.

14.4 The equation of geodesic deviation We have derived an expression for the curvature (13.58) on pure geometric considerations. A more physical approach would be to seek the GR generalization of tidal forces as discussed in Section 6.3. Following exactly the same steps used to derive the Newtonian deviation equation in Box 6.3, one can obtain its GR version, called the equation of geodesic deviation,

$$\frac{D^2 s^\mu}{D\tau^2} = -R^\mu{}_{\alpha\nu\beta} s^\nu \frac{dx^\alpha}{d\tau} \frac{dx^\beta}{d\tau}, \qquad (14.75)$$

where s^ν is the coordinate separation between the two particles undergoing their respective geodesic motion. Namely, the tensor of the gravitational potential's second derivatives in Eq. (6.32) is replaced by the Riemann

curvature tensor (13.58). This derivation requires a careful discussion of the second derivative along a geodesic curve as discussed in (13.47).

14.5 From geodesic deviation to NR tidal forces The equation of geodesic deviation (14.75) reduces the to Newtonian deviation equation (6.32) in the Newtonian limit. In the NR limit of a slow moving particle with 4-velocity $dx^\alpha/d\tau \simeq (c, 0, 0, 0)$, the GR equation (14.75) is reduced to

$$\frac{d^2 s^i}{dt^2} = -c^2 R^i{}_{0j0} s^j$$

We have also set $s^0 = 0$ because we are comparing the two particle's acceleration at the same time. Thus (6.32) can be recovered if we have in the Newtonian limit the relation

$$R^i{}_{0j0} = \frac{1}{c^2} \frac{\partial^2 \Phi}{\partial x^i \partial x^j}.$$

You are asked to prove this limit expression for the Riemann curvature.

14.6 Relativistic spin precession Consider a test body following a circular orbit (radius R) in the equatorial plane ($\theta = \pi/2$) in the Schwarzschild geometry. From Problem (7.7) we learnt that its 4-velocity has components $U^\mu = (U^t, 0, 0, U^\phi)$ with $U^\phi = \Omega U^t$ where the orbit frequency $\Omega = d\phi/dt$ obeys Kepler's third law (we generally set $c = 1$)

$$\Omega = \frac{G_N M}{R^3} = \frac{r^*}{2R^3} \quad \text{and} \quad U^t = \left(1 - \frac{3r^*}{2R}\right)^{-1/2}. \qquad (14.76)$$

If this test body also carries a intrinsic angular momentum (spin) S^μ, we need to work out the relativistic problem of spin precession. Just as the orbital geodesic equation corresponds to the parallel transport of the velocity vector $DU^{\mu}/d\tau = 0$, the geodesic equation for spin, the "gyroscope equation," is $DS^{\mu}/d\tau = 0$, with the constraint that the spin and velocity 4-vectors are orthogonal $S^{\mu}U_{\mu} = 0$ (since this invariant clearly vanishes in the instantaneous rest frame of the test body). Suppose the spin initially points in the radial direction

$$S^{\mu}(0) = \left(0, S_0^r, 0, 0\right).$$ (14.77)

Show that the solutions of the gyroscope equation yield time-dependence of the spin components:

$$S^r(t) = S_0^r \cos \Omega' t$$ (14.78)

$$S^{\phi}(t) = -\frac{\Omega}{R\Omega'} S_0^r \sin \Omega' t$$ (14.79)

with $S^{\theta}(t) = 0$ while the time component $S^t(t)$ is directly proportional to $S^r(t)$. The spin rotation frequency Ω' differs from the orbital frequency Ω

$$\Omega' = \frac{\Omega}{U^t} = \left(1 - \frac{3r^*}{2R}\right)^{1/2} \Omega.$$ (14.80)

Thus after one period $t = T = 2\pi/\Omega$, while the test body returns to the initial position, its spin orientation has changed

$$S^r(T) = S_0^r \cos\left(2\pi\Omega'/\Omega\right) \neq S_0^r$$ (14.81)

$$S^{\phi}(T) = -\frac{\Omega}{R\Omega'} S_0^r \sin\left(2\pi\Omega'/\Omega\right) \neq 0.$$ (14.82)

To return to the original orientation, it needs to rotate an extra $\Delta\phi$

$$2\pi\frac{\Omega'}{\Omega} + \Delta\phi = 2\pi.$$

Namely, the spin direction has precessed an amount of

$$\Delta\phi = 2\pi\left(1 - \frac{\Omega'}{\Omega}\right) = 2\pi\left[1 - \left(1 - \frac{3r^*}{2R}\right)^{1/2}\right]$$

$$\simeq \frac{3\pi}{2}\frac{r^*}{R} = \frac{3\pi}{4c^2}\frac{G_N M}{R}.$$ (14.83)

This is the relativistic result for spin precession. It is often referred to as the **geodetic** (i.e. geodesic) **precession**. Since it was first calculated by Willem de Sitter, this is also sometimes referred to as the **de Sitter precession**. For the corresponding two-body problem, it is useful to think of this as resulting from spin-orbit coupling. If we had a rotating gravitational source giving rise to a Kerr geometry (as discussed in Section 8.4) there is yet another spin precession known as **Lens-Thirring precession**. The corresponding two-body phenomenon may then be thought of as resulting from spin–spin coupling. Since this is closely connected to the "dragging of frame" feature of the Kerr geometry, it is sometimes incorrectly stated that while Lens–Thirring precession represents frame-dragging, the geodetic precession calculated in this problem does not (as it is done in the Schwarzschild spacetime). The correct statement should actually be that the geodetic precession in a two-body system also represents frame-dragging as we can always make a coordinate change and work in the instantaneous rest frame of the spinning body, which gives rise to a Kerr geometry.

Linearized theory and gravitational waves

15

- In the weak-field limit Einstein's equation can be linearized and it takes on form of the familiar wave equation.
- Gravitational waves may be viewed as ripples of curvature propagating in a background of flat spacetime.
- The strategy of detecting such tidal forces by a gravitational wave interferometer is outlined.
- The rate of energy loss due to the quadrupole radiation by a circulating binary system is calculated, and found to be in excellent agreement with the observed orbit decay rate of the Hulse–Taylor binary pulsar.

Newton's theory of gravitation is a static theory. The Newtonian field due to a source is established instantaneously. Thus, while the field has nontrivial dependence on the spatial coordinates, it does not depend on time. Einstein's theory, being relativistic, treats space and time on an equal footing. Just like Maxwell's theory, it has the feature that a field propagates outward from the source with a finite speed. In this chapter we study the case of a weak gravitational field. This approximation linearizes the Einstein theory. In this limit, a gravitational waves may be viewed as small curvature ripples (the metric field) propagating in a background of flat spacetime. It is a transverse wave, having two independent polarization states, traveling at the speed of light.

Because gravitational interaction is so weak, any significant emission of gravitational radiation can come only from a strong field region involving dynamics that directly reflects GR physics. Once gravitational waves are emitted, they will not scatter and they propagate out undisturbed from the inner core of an imploding star, from the arena of black hole formation, and from the earliest moments of the universe, etc. That is, they come from regions which are usually obscured in electromagnetic, even neutrino astronomy: gravitational waves can provide us with a new window into astrophysical phenomena.

These ripples of curvature can be detected as tidal forces. We provide an outline of the detection strategy using gravitational wave interferometers, which can measure the minute compression and elongation of orthogonal lengths that are caused by the passage of such a wave. In the final section, we present the indirect, but convincing, evidence for the existence of gravitational waves as predicted by general relativity (GR). This came from the observation, spanning more than 25 years, of orbital motion of the relativistic Hulse–Taylor binary pulsar system (PSR 1913+16). Even though the binary pair is 5 kpc away from us, the basic parameters of the system can be deduced by carefully monitoring the radio pulses emitted by the pulsar, which effectively acted as an accurate and stable clock. From this record we can verify a number of GR effects. In particular the orbital period is observed to decrease. According to GR, this is brought about by the gravitational wave quadrupole radiation from the system. The observed orbital rate decrease is in splendid agreement with the prediction by Einstein's theory.

15.1   Linearized theory of a metric field

Even though the production of gravitational waves usually involves strong field situations, but, because of the weakness of the gravitational interaction, the produced gravitational waves are only tiny displacements of the flat spacetime metric. Thus it is entirely adequate for the description of a gravity wave to restrict ourselves to the situation of a weak gravitation field. In this limit,[1] the metric is almost Minkowskian $\eta_{\mu\nu} = \text{diag}(-1, 1, 1, 1)$:

[1] Contrasting this to the Newtonian limit of nonrelativistic motion in a weak static field, here we remove the restriction of slow motion and allow for a time-dependent field.

$$g_{\mu\nu} = \eta_{\mu\nu} + h_{\mu\nu} \equiv g_{\mu\nu}^{(1)} \tag{15.1}$$

where the metric perturbation $|h_{\mu\nu}| \ll 1$ everywhere in spacetime. Thus we will keep only first-order terms in $h_{\mu\nu}$, and denote the relevant quantities with a superscript $^{(1)}$. The idea is that slightly curved coordinate systems exist and they are suitable coordinates to use in the weak field situation. We can still make coordinate transformations among such systems—from one slightly curved one to another. In particular we can make a "background Lorentz transformation." Distinguishing the indices, $\{\mu\}$ vs. $\{\mu'\}$, to indicate the pre-transformed and transformed coordinates, we have

$$x^\mu \to x^{\mu'} = [\mathbf{L}]^{\mu'}_{\ \nu} x^\nu \tag{15.2}$$

where $\mathbf{L}$ is the position-independent Lorentz transformation of special relativity, see (12.13) and (12.17). The key property of such transformations is that they keep the Minkowski metric invariant, see (12.19),

$$\left[\mathbf{L}^{-1}\right]^\mu_{\ \alpha'} \left[\mathbf{L}^{-1}\right]^\nu_{\ \beta'} \eta_{\mu\nu} = \eta_{\alpha'\beta'} \tag{15.3}$$

This leads to the transformation of the full metric as

$$\left[\mathbf{L}^{-1}\right]^\mu_{\ \alpha'} \left[\mathbf{L}^{-1}\right]^\nu_{\ \beta'} g_{\mu\nu}^{(1)} = \eta_{\alpha'\beta'} + \left[\mathbf{L}^{-1}\right]^\mu_{\ \alpha'} \left[\mathbf{L}^{-1}\right]^\nu_{\ \beta'} h_{\mu\nu} = g_{\alpha'\beta'}^{(1)}. \tag{15.4}$$

Thus

$$h_{\alpha'\beta'} = \left[\mathbf{L}^{-1}\right]^\mu_{\ \alpha'} \left[\mathbf{L}^{-1}\right]^\nu_{\ \beta'} h_{\mu\nu}. \tag{15.5}$$

Namely, $h_{\mu\nu}$ is just a Lorentz tensor. Thus this part of the metric can be taken as a tensor defined on a flat Minkowski spacetime. Since the nontrivial physics is contained in $h_{\mu\nu}$, we can have the convenient picture of a weak gravitational field as being described by this symmetric field $h_{\mu\nu}$ in a flat spacetime.[2]

Dropping higher order terms of $h_{\mu\nu}$, we have the Riemann curvature tensor

$$R^{(1)}_{\alpha\mu\beta\nu} = \frac{1}{2}\left(\partial_\alpha\partial_\nu h_{\mu\beta} + \partial_\mu\partial_\beta h_{\alpha\nu} - \partial_\alpha\partial_\beta h_{\mu\nu} - \partial_\mu\partial_\nu h_{\alpha\beta}\right), \tag{15.6}$$

the Ricci tensor

$$R^{(1)}_{\mu\nu} = \eta^{\alpha\beta}R^{(1)}_{\alpha\mu\beta\nu}$$

$$= \frac{1}{2}\left(\partial_\alpha\partial_\nu h^\alpha_{\ \mu} + \partial_\mu\partial_\alpha h^\alpha_{\ \nu} - \Box h_{\mu\nu} - \partial_\mu\partial_\nu h\right), \tag{15.7}$$

and the Ricci scalar

$$R^{(1)} = \partial_\mu\partial_\nu h^{\mu\nu} - \Box h, \tag{15.8}$$

where $\Box = \partial_\mu\partial^\mu$ and $h = h^\mu_\mu$ is the trace. Clearly the resultant Einstein tensor

$$G^{(1)}_{\mu\nu} = R^{(1)}_{\mu\nu} - \frac{1}{2}R^{(1)}\eta_{\mu\nu} \tag{15.9}$$

is also linear in $h_{\mu\nu}$, and so is the Einstein equation:

$$G^{(1)}_{\mu\nu} = -\frac{8\pi G_N}{c^4}T^{(0)}_{\mu\nu}. \tag{15.10}$$

For a spacetime being slightly curved the left hand side (LHS) is of order $h_{\mu\nu}$; this means that the energy–momentum tensor must also be small, $T^{(0)}_{\mu\nu} = O(h_{\mu\nu})$. Thus, its conservation condition $D^\mu T_{\mu\nu} = 0$ can be simply expressed in terms of ordinary derivatives

$$\partial^\mu T^{(0)}_{\mu\nu} = 0 \tag{15.11}$$

as the difference between D_μ and ∂_μ is of the order of $h_{\mu\nu}$.

15.1.1 The coordinate change called a gauge transformation

In the following, we shall make coordinate transformations so that the linearized Einstein equation (15.10) can be written more compactly in terms of $h_{\mu\nu}$. This class of coordinate transformations (within the slightly curved spacetime) is called, collectively, **gauge transformations** because of their close resemblance to the electromagnetic gauge transformations. Consider a small shift of the position vector:

$$x^{\mu'} = x^\mu + \chi^\mu(x) \tag{15.12}$$

where $\chi^\mu(x)$ are four arbitrary small functions. Collectively they are called the "vector gauge function" (as opposed to the scalar gauge function in electromagnetic gauge transformations see Box 12.3). Clearly this is not a tensor equation, as indices do not match on the two sides. (Our notation indicates the relation of the position vector as labeled by the transformed

[2] Eventually in a quantum description, $h_{\mu\nu}$ is a field for the spin-2 gravitons, and the perturbative description of gravitational interactions as due to the exchanges of massless gravitons.

and pre-transformed coordinates.) The transformation matrix elements (for the contravariant components) can be obtained by differentiating (15.12):

$$\frac{\partial x^{\mu'}}{\partial x^\alpha} = \delta^\mu{}_\alpha + \partial_\alpha \chi^\mu. \tag{15.13}$$

The smallness of the shift $\chi \ll x$ means

$$\left| \partial_\mu \chi^\nu \right| \ll 1. \tag{15.14}$$

This implies an inverse transformation of

$$\frac{\partial x^\mu}{\partial x^{\alpha'}} = \delta^\mu_\alpha - \partial_\alpha \chi^\mu + O\left(|\partial \chi|^2 \right). \tag{15.15}$$

Apply it to the metric tensor:[3]

$$
\begin{aligned}
g^{(1)}_{\alpha'\beta'} &= \frac{\partial x^\mu}{\partial x^{\alpha'}} \frac{\partial x^\nu}{\partial x^{\beta'}} g^{(1)}_{\mu\nu} \\
&= \delta^\mu_\alpha \delta^\nu_\beta g^{(1)}_{\mu\nu} - \partial_\alpha \chi^\mu \eta_{\mu\beta} - \partial_\beta \chi^\nu \eta_{\nu\alpha} \\
&= g^{(1)}_{\alpha\beta} - \partial_\alpha \chi_\beta - \partial_\beta \chi_\alpha
\end{aligned}
\tag{15.16}
$$

where $\chi_\alpha = \chi^\mu \eta_{\mu\alpha}$. Expressing both sides in term of $h_{\alpha\beta}$, we have the gauge transformation of the perturbation field

$$h_{\alpha'\beta'} = h_{\alpha\beta} - \partial_\alpha \chi_\beta - \partial_\beta \chi_\alpha \tag{15.17}$$

which closely resembles the transformation (12.53) for the electromagnetic 4-vector potential $A_\alpha(x)$.

15.1.2   The wave equation in the Lorentz gauge

Just as in electromagnetism, one can streamline some calculations by an appropriate choice of gauge conditions. Here this means that a particular choice of coordinates can simplify the field equation formalism for gravitational waves. We are interested in the coordinate system (Problem 15.1) for which the **Lorentz gauge** (also known as the **harmonic gauge**) condition holds:

$$\partial^\mu \bar{h}_{\mu\nu} = 0 \tag{15.18}$$

where $\bar{h}_{\mu\nu}$ is the **trace reversed perturbation**:

$$\bar{h}_{\mu\nu} = h_{\mu\nu} - \frac{h}{2} \eta_{\mu\nu} \tag{15.19}$$

with a trace of opposite sign, $\bar{h}^\mu{}_\mu \equiv \bar{h} = -h$. From (15.18) and (15.19), we have the **Lorentz gauge** relation $\partial^\mu h_{\mu\nu} = \frac{1}{2} \partial_\nu h$, which implies, in (15.7) and (15.8), a simplified Ricci tensor $R^{(1)}_{\mu\nu} = -\frac{1}{2} \Box h_{\mu\nu}$, and Ricci scalar $R^{(1)} = -\frac{1}{2} \Box h$. This turns the linearized Einstein equation (15.10) into the form of a standard wave equation:[4]

$$\Box \bar{h}_{\mu\nu} = \frac{16\pi G_N}{c^4} T^{(0)}_{\mu\nu}. \tag{15.20}$$

[3]We note the index structure of this equation. While the first equality in (15.16) represents the standard general coordinate transformation (with primed indices on both sides of the equation), the primed indices disappear in the subsequent right-hand-sides because of the gauge transformation of (15.12).

[4]For a discussion of the Schwarzschild exterior solution for this linearized GR field equation, see Problem 15.2.

Table 15.1 Analog between the electromagnetic and linearized gravitational field theory.

	Electromagnetism	**Linearized gravity**
Source	j^{μ}	$T^{\mu\nu}$
Conservation law	$\partial_{\mu} j^{\mu} = 0$	$\partial_{\mu} T^{\mu\nu} = 0$
Field	A_{μ}	$h_{\mu\nu}$
Gauge transformation	$A_{\mu} \to A_{\mu} - \partial_{\mu}\chi$	$h_{\mu\nu} \to h_{\mu\nu} - \partial_{\mu}\chi_{\nu} - \partial_{\nu}\chi_{\mu}$
Preferred gauge (Lorentz gauge)	$\partial^{\mu} A_{\mu} = 0$	$\partial^{\mu} \bar{h}_{\mu\nu} = 0$ $\bar{h}_{\mu\nu} = h_{\mu\nu} - \frac{1}{2}h\eta_{\mu\nu}$
Field equation in the preferred gauge	$\Box A_{\mu} = \frac{4\pi}{c} j_{\mu}$	$\Box \bar{h}_{\mu\nu} = \frac{16\pi G_{N}}{c^4} T_{\mu\nu}$

One can also view this as the equation for the metric field with the energy–momentum tensor being the source of the field. Its retarded solution, expressed as a spatial integral over the source, is

$$\bar{h}_{\mu\nu}(\mathbf{x}, t) = \frac{4G_{\mathrm{N}}}{c^4} \int d^3\mathbf{x}' \frac{T^{(0)}_{\mu\nu}\left(\mathbf{x}', t - |\mathbf{x} - \mathbf{x}'|/c\right)}{|\mathbf{x} - \mathbf{x}'|}, \tag{15.21}$$

which is certainly compatible with the gauge condition $\partial^{\mu} \bar{h}_{\mu\nu} = 0$ because of the energy momentum conservation (15.11).

To reiterate, in this linear approximation of the Einstein theory, the metric perturbation $h_{\mu\nu}$ may be regarded as the symmetric field of gravity waves propagating in the background of a flat spacetime. A comparison of the linearized Einstein theory with the familiar electromagnetic equations can be instructive. Such an analog is presented in Table 15.1.

15.2   Plane waves and the polarization tensor

We shall first consider the propagation of a gravitational wave in vacuum. Such ripples in the metric can always be regarded as a superposition of plane waves. A gravity wave has two independent polarization states. Their explicit form will be displayed in a particular coordinate system, the transverse-traceless (T T) gauge.

Plane waves

The linearized Einstein equation in vacuum, (15.20) with $T^{(0)}_{\mu\nu} = 0$, is $\Box \bar{h}_{\mu\nu} = 0$. Because the trace $\bar{h} = -h$ satisfies the same wave equation, we also have, from applying the $\Box$ operator to (15.19),

$$\Box h_{\mu\nu} = 0. \tag{15.22}$$

Consider the plane wave solution in the form of

$$h_{\mu\nu}(x) = \epsilon_{\mu\nu} e^{ik_{\alpha}x^{\alpha}} \tag{15.23}$$

where $\epsilon_{\mu\nu}$, the polarization tensor of the gravitational wave, is a set of constants forming a symmetric tensor, $\epsilon_{\mu\nu} = \epsilon_{\nu\mu}$, and k^α is the 4-wavevector $k^\alpha = (\omega/c, \vec{k})$. Substituting (15.23) into (15.22), we obtain $k^2 \epsilon_{\mu\nu} e^{ikx} = 0$; thus the wavevector must be a null-vector

$$k^2 = k_\alpha k^\alpha = -\frac{\omega^2}{c^2} + \vec{k}^2 = 0. \tag{15.24}$$

Gravitational waves propagate at the same speed $\omega/|\vec{k}| = c$ as electromagnetic waves. Furthermore, because the wave equation (15.22) is valid only in the coordinates satisfying the Lorentz gauge condition (15.18), the polarization tensor must be "transverse"

$$k^\mu \epsilon_{\mu\nu} = 0. \tag{15.25}$$

The transverse-traceless gauge

There is still some residual gauge freedom left: one can make further coordinate gauge transformations as long as the transverse condition (15.25) is not violated. This requires that the associated gauge vector function χ_μ be constrained by the condition:

$$\Box \chi_\mu = 0. \tag{15.26}$$

Such coordinate freedom can be used to simplify the polarization tensor (see Problem 15.1): one can pick $\epsilon_{\mu\nu}$ to be traceless

$$\epsilon_\mu^{\ \mu} = 0, \tag{15.27}$$

as well as

$$\epsilon_{\mu 0} = \epsilon_{0\mu} = 0. \tag{15.28}$$

This particular choice of coordinates is called the "transverse-traceless gauge," which is a subset of the coordinates satisfying the Lorentz gauge condition.

The 4×4 symmetric polarization matrix $\epsilon_{\mu\nu}$ has 10 independent elements. Equations (15.25), (15.27), and (15.28) which superficially represent nine conditions actually fix only eight parameters because part of the transversality condition (15.25), $k^\mu \epsilon_{\mu 0} = 0$, is trivially satisfied by (15.28). Thus $\epsilon_{\mu\nu}$ has only two independent elements. Namely, the gravitational wave has two independent polarization states. Let us display them. Consider a wave propagating in the z direction $k^\alpha = (\omega, 0, 0, \omega)/c$, the transversality condition (15.25), together with (15.28), implies that $\omega \epsilon_{3\nu} = 0$, or $\epsilon_{3\nu} = \epsilon_{\nu 3} = 0$. Together with the conditions (15.27) and (15.28), the metric perturbation has the form

$$h_{\mu\nu}(z,t) = \begin{pmatrix} 0 & 0 & 0 & 0 \\ 0 & h_+ & h_\times & 0 \\ 0 & h_\times & -h_+ & 0 \\ 0 & 0 & 0 & 0 \end{pmatrix} e^{i\omega(z-ct)/c}. \tag{15.29}$$

The two polarization states can be taken to be

$$\epsilon^{\mu\nu}_{(+)} = h_+ \begin{pmatrix} 0 & 0 & 0 & 0 \\ 0 & 1 & 0 & 0 \\ 0 & 0 & -1 & 0 \\ 0 & 0 & 0 & 0 \end{pmatrix} \quad \text{and} \quad \epsilon^{\mu\nu}_{(\times)} = h_\times \begin{pmatrix} 0 & 0 & 0 & 0 \\ 0 & 0 & 1 & 0 \\ 0 & 1 & 0 & 0 \\ 0 & 0 & 0 & 0 \end{pmatrix}$$

with h_+ and $h_\times$ being the respective "plus" and "cross" amplitudes.

15.3   Detection of gravitational waves

The coordinate-independent feature of any gravitational field is its tidal effect. Thus, the detection of gravitational waves involves the recording of minute changes in the relative positions of a set of test particles. In this section, we shall first deduce the oscillatory pattern of such displacements, then briefly describe the principle underlying the gravitational wave interferometer as a detector of such ripples in spacetime.

15.3.1   Effect of gravitational waves on test particles

Consider a free particle before its encounter with a gravitational wave. It is at rest with a 4-velocity $U^\mu = (c, 0, 0, 0)$. The effect of the gravitational wave on this test particle is determined by the geodesic equation $dU^\mu/d\tau + \Gamma^\mu_{\nu\lambda} U^\nu U^\lambda = 0$. Since only U^0 is non-vanishing at the beginning, it reduces to an expression for the initial acceleration of $(dU^\mu/d\tau)_0 = -c^2 \Gamma^\mu_{00}$, which vanishes because the Christoffel symbols are $\Gamma^\mu_{00} = 0$ in the TT gauge.[5] The particle is stationary with respect to the chosen coordinate system—the TT gauge coordinate labels stay attached to the particle. Thus one cannot discover any gravitational field effect on a single particle. This is compatible with our expectation, based on the equivalence principle (EP), that gravity can always be transformed away at a point by an appropriate choice of coordinates. We need to examine the relative motion of at least two particles in order to detect the oncoming change in the curvature of spacetime.

Consider the effect of a gravitational wave with "plus-polarization" $\epsilon^{\mu\nu}_{(+)}$ on two test particles at rest: one at the origin and other located at an infinitesimal distance ξ away on the x axis, hence at an infinitesimally small separation $dx^\mu = (0, \xi, 0, 0)$. Using the expression in (15.29), this translates into a proper separation of

$$ds = \sqrt{g_{\mu\nu}dx^\mu dx^\nu} = \sqrt{g_{11}}\xi \simeq \left[\eta_{11} + \frac{1}{2}h_{11}\right]\xi$$

$$= \left[1 + \frac{1}{2}h + \sin\omega\,(t - z/c)\right]\xi \tag{15.30}$$

showing that the proper distance does change with time. Similarly for two particles separated along the y axis, $dx^\mu = (0, 0, \xi, 0)$, the effect of the

[5] The connection $\Gamma^\mu_{00} = \eta^{\mu\nu}(\partial_0 h_{\nu0} + \partial_0 h_{0\nu} - \partial_\nu h_{00})/2 = 0$ because the metric perturbation $h_{\mu\nu}$ has, in the TT gauge, polarization components of $\epsilon_{\nu0} = \epsilon_{0\nu} = \epsilon_{00} = 0$. The vanishing of the initial acceleration means that the particle will be at rest a moment later. Repeating the same argument for later moments, we find the particle at rest for all times. In this way we conclude $dU^\mu/d\tau = 0$.

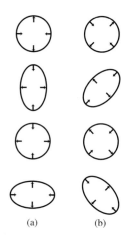

(a)   (b)

Fig. 15.1 Tidal force effects on a circle of test particles due to gravitational waves in (a) the plus-polarization, and (b) the cross-polarization states.

gravitational wave is to alter the separation according to

$$ds = \left[1 - \frac{1}{2}h_+ \sin \omega \, (t - z/c) \right] \xi. \tag{15.31}$$

Thus, the separation along the x direction is elongated while along the y direction it is compressed. There is no change in the longitudinal separation along the z direction. Just like the electromagnetic waves, gravitational radiation is a transverse field. To better exhibit this pattern of relative displacement we illustrate in Fig. 15.1(a) the effect of a plus-polarized wave, but instead of impinging on two particles as discussed above, acting on a set of test particles when the second particle is replaced by a circle of particles with the first test particle at the center. The outcome that generalizes (15.30) and (15.31) is shown through the wave's one cycle of oscillation.

The effect of a wave with cross-polarization $\epsilon_{(\times)}^{\mu\nu}$ on two particles with differential intervals of $dx^\mu = (0, 1, \pm 1, 0)\xi/\sqrt{2}$ alters the proper separation as $ds = 1 \pm \frac{1}{2}h \times \sin \omega(t - z/c)\xi$. The generalization to a circle of particles through one cycle of oscillation is shown in Fig. 15.1(b), which is just a 45° rotation of the plus-polarized wave result of Fig. 15.1(a). While the two independent polarization directions of an electromagnetic wave are at 90° from each other, those of a gravity wave are at 45°. This is related to the feature that, in the dual description of the wave as streaming particles, the associated particles of these waves have different intrinsic angular momenta: the photon has spin 1 while the graviton has spin 2. It is also instructive to compare the tidal force effects on such test-particles' relative displacement in response to an oncoming oscillatory gravitational field to that of a static gravitational field as discussed in Section 6.3.1.

15.3.2   Gravitational wave interferometers

A gravitational wave can be thought of as a propagating metric, affecting distance measurements. Thus, as a wave passes through, the separation s between two test masses changes with time. Gravitational interaction is very weak. The longitudinal and transverse separation of test particles discussed above is expected to be tiny. Before any detailed calculation (such as the one given in Section 15.4), it is useful to have some idea of the size of the expected gravitational wave signal. Here we give an estimate of the fractional change of separation, called the **strain** $\sigma = (\delta s)/s$, by a "hand-waving" argument.

The separation between two test masses are is by the equation of geodesic deviation (Problems 14.4 and 14.5), but we shall estimate it by using the simpler Newtonian deviation equation of (6.32), which expresses the acceleration per unit separation by the second derivative of the gravitational potential. We assume that the relativistic effect can be included by a multiplicative factor. The Newtonian potential for a spherical source is $\Phi = -G_N M r^{-1}$. A gravitational wave propagating in the z direction is a disturbance in the gravitational field:

$$\delta\Phi = -\psi \frac{G_N M}{r} \sin(kz - \omega t), \tag{15.32}$$

where $k = \omega/c$. A dimensionless factor of ψ has been inserted to represent the relativistic correction. The second derivative can be approximated by

$$\frac{\partial^2}{\partial z^2}\delta\Phi = \psi\frac{G_N M}{rc^2}\omega^2 \sin(kz - \omega t),\qquad(15.33)$$

where we have dropped subleading terms coming from differentiation of the r^{-1} factor. This being the acceleration as given in (6.32), the strain amplitude (for the time interval ω^{-1}) is then given by

$$\sigma = \frac{\delta s}{s} = \psi\frac{G_N M}{rc^2}.\qquad(15.34)$$

A similar approximation of the radiation formula (15.61) to be derived below suggests the relativistic correction factor ψ as being the nonrelativistic velocity squared $(v/c)^2$ of the source. The first generation of gravitational wave interferometers have been set up with the aim of detecting gravitational wave emission by neutron stars from the richest source of galaxies in our neighboring part of the universe, the Virgo Cluster, at $r \approx 15$ Mpc distance away. Thus, even for a sizable $\psi = O(10^{-1})$ from a solar mass source $M = M_\odot$ the expected strain is only $\sigma = O(10^{-21})$. For two test masses separated by a distance of 10 km the gravity wave induced separation is still one hundredth of a nuclear size dimension. This shows that spacetime is a very stiff medium, as a large amount of energy can still bring about a tiny disturbance in the spacetime metric. This fact poses great challenges to experimental observation of gravitational waves.

The above discussion makes it clear that one needs to design sensitive detectors to measure the minute length changes between test masses over long distances. Several detectors have been constructed based on the Michelson interferometer configuration (Fig. 15.2). The test masses are mirrors suspended to isolate them from external perturbation forces. Light from a laser source is divided into the two arms by a beam splitter. The light entering into an arm of length L is reflected back and forth in a Fabry–Perot cavity for n times so that the optical length is greatly increased and the storage time is $n(L/c) = \Delta t_n$. The return light beams from the two arms are combined after they pass through the beam splitter again. By choosing the path length properly, the optical electric field can be made to vanish (destructive interference) at the photodetector. Once adjusted this way, a stretch in one arm and a compression

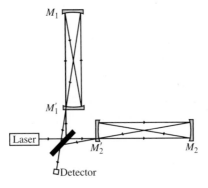

Fig. 15.2 Schematic diagram for gravitational wave Michelson interferometer. The four mirrors $M_{1,2}$, $M'_{1,2}$ and the beam splitter mirror are freely suspended. The two arms are optical cavities that increase the optical paths by many factors. A minute length change of the two arms, one expands and the other contracts, will show up as changes in fringe pattern of the detected light.

Fig. 15.3 LIGO Hanford Observatory in Washington state.

in the other, when induced by the polarization of a passing gravitational wave, will change the optical field at the photodetector in proportion to the product of the field times the wave amplitude. Such an interferometer should be uniformly sensitive to wave frequencies less than $\frac{1}{4}\Delta t_n^{-1}$ (and a loss of sensitivity to higher frequencies). The basic principle to achieve high sensitivity is based on the idea that most of the perturbing noise forces are independent of the baseline lengths while the gravitational-wave displacement grows with the baseline.

The Laser Interferometer Gravitational Observatory (LIGO) comprises of two sites: one at the Hanford Reservation in Central Washington (Fig. 15.3) housing two interferometer one 2 km- and another 4 km-long arms, while the other site is at Livingston Parish, Louisiana. The three interferometers are being operated in coincidence so that the signal can be confirmed by data from all three sites. Other gravitational wave interferometers in operation are the French/Italian VIRGO project, the German/Scottish GEO project, and the Japanese TAMA project. Furthermore, study is underway both at the European Space Agency and NASA for the launching of three spacecraft placed in solar orbit with one AU radius, trailing the earth by $20°$. The spacecraft are located at the corners of an equilateral triangle with sides 5×10^6 km long. The Laser Interferometer Space Antenna (LISA) consists of single-pass interferometers, set up to observe a gravitational wave at low frequencies (from 10^{-5} to 1 Hz). This spectrum range is expected to include signals from several interesting interactions of black holes at cosmological distances.[6]

[6]Besides planning and building ever large-scale gravitational wave detectors, a major effort by the theoretical community in relativity is involved in the difficult task of calculating wave shapes in various strong gravity situations (e.g. neutron-star/neutron-star collision, black hole mergers, etc.) to guide the detection and comparison of theory with experimental observations.

15.4   Emission of gravitational waves

Although, as of this writing, there has not been any generally accepted proof for a direct detection of gravitational wave, there is nevertheless convincing, albeit indirect, evidence for the existence of such waves as predicted by Einstein's theory. Just as any shaking of electric charges produces electromagnetic waves, a shaking of masses will result in the generation of gravitational wave, which carries away energy. In this section we present the relativistic binary pulsar system showing that the decrease of its orbital period due to gravitational wave radiation is in excellent agreement with what is predicted by general relativity. Since such a calculation is somewhat lengthy, we provide an outline of the steps required for its derivation in Box 15.1.

Box 15.1 The steps for calculating the energy loss by a radiating source

- The rate of energy loss dE/dt due to gravitational radiation is the area integral over the radiation flux. This flux (energy per unit area and per unit time) should be the energy density (energy per unit volume) of the radiation gravitational field times the radiation velocity, $f = ct_{00}$, where t_{00} is the $(0, 0)$ component of the energy–momentum tensor $t_{\mu\nu}$ of the radiation field.

- Just as the energy–momentum tensor of an electromagnetic field can be expressed directly in terms of the field itself (as in Box 12.5), $t_{\mu\nu}$ can be expressed in terms of the gravitational (perturbation) field $h_{\mu\nu}$. In particular, the energy density term will be shown in (15.50) as relating to the time derivative of the spatial components of the perturbation in the TT gauge as

$$t_{00} = \frac{c^2}{32\pi G_N} \left\langle \left(\partial_t h_{ij}^{TT}\right) \left(\partial_t h_{ij}^{TT}\right) \right\rangle, \tag{15.35}$$

  where $\langle ... \rangle$ represents the averaging over many wavelengths.
- The gravitational field is related to the energy–momentum tensor $T_{\mu\nu}$ of the source through the Einstein field equation. In our case we can find the metric perturbation $h_{\mu\nu}$ by solving the wave equation in terms of the source distribution. The leading term for h_{ij} is given in (15.53):

$$h_{ij} = \frac{2G_N}{c^4 r} \left(\partial_t^2 I_{ij}\right) \tag{15.36}$$

  where $I_{ij} = \int d^3\mathbf{x} \rho(\mathbf{x}) x_i x_j$, with $\rho(\mathbf{x})$ being the mass density of the source. Namely, it is quadrupole radiation.
- After averaging over all directions, one finds in (15.61) that the radiation power is related to the "reduced quadrupole moments $\tilde{I}_{ij}$" of (15.57) as

$$\frac{dE}{dt} = \frac{G_N}{5c^5} \left\langle \left(\partial_t^3 \tilde{I}_{ij}^{TT}\right) \left(\partial_t^3 \tilde{I}_{ij}^{TT}\right) \right\rangle. \tag{15.37}$$

Thus, one can start the calculation of the gravitational radiation loss due to orbiting binary stars by computing, from its equation of motion, the quadrupole moments of such a system. This will be carried out in Section 15.4.3.

15.4.1 Energy flux in linearized gravitational waves

In the linearized Einstein theory, gravitational waves are regarded as small curvature ripples propagating in a background of flat spacetime. But gravity waves, just like electromagnetic waves, carry energy and momentum; they will in turn produce additional curvature in the background spacetime. Thus we should have a slightly curved background and (15.1) should be generalized to

$$g_{\mu\nu} = g_{\mu\nu}^{(b)} + h_{\mu\nu} \tag{15.38}$$

where $g_{\mu\nu}^{(b)} = \eta_{\mu\nu} + O(h^2)$ is the background metric. The Ricci tensor can similarly be decomposed as

$$R_{\mu\nu} = R_{\mu\nu}^{(b)} + R_{\mu\nu}^{(1)} + R_{\mu\nu}^{(2)} + \cdots$$

where $R_{\mu\nu}^{(n)} = O(h^n)$ with $n = 1, 2, \ldots$ Thus, the background curvature $R_{\mu\nu}^{(b)}$ should be of the same order as $R_{\mu\nu}^{(2)}$. In free space the Einstein equation being

$R_{\mu\nu} = 0$, terms corresponding to different orders of metric perturbation on the RHS must vanish separately: $R^{(1)}_{\mu\nu} = 0$ as well as

$$R^{(b)}_{\mu\nu} + R^{(2)}_{\mu\nu} = 0. \tag{15.39}$$

The energy–momentum tensor $t_{\mu\nu}$ carried by the gravity wave provides the slight curvature of the background spacetime. It must therefore be related, at this order, to the background Ricci tensor by way of the Einstein equation (14.26):

$$R^{(b)}_{\mu\nu} - \frac{1}{2}\eta_{\mu\nu}R^{(b)} = -\frac{8\pi G_N}{c^4}t_{\mu\nu}.$$

That is, $t_{\mu\nu}$ is fixed by $R^{(b)}_{\mu\nu}$, which in turn is related to $R^{(2)}_{\mu\nu}$ by way of (15.39). This allows us to calculate $t_{\mu\nu}$ through the second-order Ricci tensor and scalar

$$t_{\mu\nu} = \frac{c^4}{8\pi G_N}\left(R^{(2)}_{\mu\nu} - \frac{1}{2}\eta_{\mu\nu}R^{(2)}\right). \tag{15.40}$$

Before carrying out the calculation of $t_{\mu\nu}$, we should clarify one point: the concept of **local** energy of a gravitational field does not exist. Namely, one cannot specify the gravitational energy at any single point in space. This is so because the energy, being a coordinate-independent function of field, one can always, according to the EP, find a coordinate (the local inertial frame) where the gravity field vanishes locally. Saying it in another way, just as in electromagnetism, we expect the energy density to be proportional to the square of the potential's first derivative. But, according to the flatness theorem, the first derivative of the metric vanishes in the local inertial frame. Thus we cannot speak of gravity's local energy. Nevertheless, one can associate an effective energy–momentum tensor with the gravitational field of finite volume. Specifically, we can average over a spatial volume that is much larger than the wavelength of the relevant gravitational waves to obtain

$$t_{\mu\nu} = \frac{c^4}{8\pi G_N}\left[\left\langle R^{(2)}_{\mu\nu}\right\rangle - \frac{1}{2}\eta_{\mu\nu}\left\langle R^{(2)}\right\rangle\right] \tag{15.41}$$

where $\langle\ldots\rangle$ stands for the average over many wave cycles.

Let us calculate the energy flux carried by a linearly polarized plane wave, say the h_+ state, propagating in the z direction. The metric and its inverse, accurate up to first order in the perturbation, in the TT gauge can be written as

$$g_{\mu\nu} = \begin{pmatrix} -1 & 0 & 0 & 0 \\ 0 & 1+\tilde{h}_+ & 0 & 0 \\ 0 & 0 & 1-\tilde{h}_+ & 0 \\ 0 & 0 & 0 & 1 \end{pmatrix} \text{ and } g^{\mu\nu} = \begin{pmatrix} -1 & 0 & 0 & 0 \\ 0 & 1-\tilde{h}_+ & 0 & 0 \\ 0 & 0 & 1+\tilde{h}_+ & 0 \\ 0 & 0 & 0 & 1 \end{pmatrix}$$

$$\tag{15.42}$$

where

$$\tilde{h}_+ = h_+ \cos\left[\omega\left(t - z/c\right)\right]. \tag{15.43}$$

To obtain the energy–momentum tensor of the gravity wave by way of $R^{(2)}_{\mu\nu}$ as in (15.40), we need first to calculate the Christoffel symbols by differentiating the metric of (15.42). It can be shown (Problem 15.4-a) that the nonvanishing elements are

$$\Gamma^1_{10} = \Gamma^1_{01} = \Gamma^0_{11} = \frac{1}{2}\left(\partial_0\tilde{h}_+ - \tilde{h}_+\partial_0\tilde{h}_+\right) \tag{15.44}$$

and, similarly,

$$\Gamma^1_{13} = \Gamma^1_{31} = -\Gamma^3_{11} = -\frac{1}{2}(\partial_0\tilde{h}_+ - \tilde{h}_+\partial_0\tilde{h}_+). \tag{15.45}$$

The Riemann tensor has the structure of $(\partial\Gamma + \Gamma\Gamma)$. Since we are only interested in an $O(h^2)$ calculation, the above $\tilde{h}_+\partial_0\tilde{h}_+$ factor in the Christoffel symbols can only enter in the $\partial\Gamma$ terms, leading to the time-averaged term of $\langle\tilde{h}_+\partial_0\tilde{h}_+\rangle \propto \langle\sin 2\omega(t - z/c)\rangle = 0$. Hence we will drop the $\tilde{h}_+\partial_0\tilde{h}_+$ terms in (15.44) and (15.45), and calculate the (averaged) curvature tensor in (13.58) by dropping the $\langle\partial\Gamma\rangle$ factors,

$$\left\langle R^{(2)}_{\mu\nu}\right\rangle = \left\langle\Gamma^\alpha_{\alpha\lambda}\Gamma^\lambda_{\mu\nu} - \Gamma^\alpha_{\mu\lambda}\Gamma^\lambda_{\alpha\nu}\right\rangle. \tag{15.46}$$

A straightforward calculation (Problem 15.4-b) shows that

$$R^{(2)}_{11} = R^{(2)}_{22} = 0 \quad\text{and}\quad R^{(2)}_{00} = R^{(2)}_{33} = \frac{1}{2}\left(\partial_0\tilde{h}_+\right)^2 \tag{15.47}$$

leading to a vanishing Ricci scalar

$$R^{(2)} = \eta^{\mu\nu}R^{(2)}_{\mu\nu} = -R^{(2)}_{00} + R^{(2)}_{11} + R^{(2)}_{22} + R^{(2)}_{33} = 0. \tag{15.48}$$

In particular, the effective energy density of the gravitational plane wave in the plus polarization state, as given by (15.41) and (15.47), yields the first term on the RHS of the following relation:

$$t_{00} = \frac{c^4}{16\pi G_N}\left\langle\left(\partial_0\tilde{h}_+\right)^2 + \left(\partial_0\tilde{h}_\times\right)^2\right\rangle \tag{15.49}$$

where we have also added, the second term on the RHS, the corresponding contribution from the cross-polarization state. If we choose to write the transverse traceless metric perturbation as $\tilde{h}_+ \equiv h^{TT}_{11} = -h^{TT}_{22}$ and $\tilde{h}_\times \equiv h^{TT}_{12} = h^{TT}_{21}$ and $h^{TT}_{3i} = 0$ (with $i = 1, 2, 3$), we then have

$$\left\langle\left(\partial_0\tilde{h}_+\right)^2 + \left(\partial_0\tilde{h}_\times\right)^2\right\rangle = \left\langle\left(\partial_t h^{TT}_{ij}\right)\left(\partial_t h^{TT}_{ij}\right)\right\rangle / \left(2c^2\right).$$

For a wave travelling at the speed c the energy flux being related to the density by $f = ct_{00}$, hence can be expressed it in terms of the metric perturbation as

$$f = \frac{c^3}{32\pi G_N}\left\langle\left(\partial_t h^{TT}_{ij}\right)\left(\partial_t h^{TT}_{ij}\right)\right\rangle \tag{15.50}$$

with repeated indices summed over. It is useful to recall the counterpart in the more familiar electromagnetism. The EM flux is given by the field energy density (multiplied by c) which is proportional to the square of the field, or the square of the time derivatives of the (vector) potential. Equation (15.50) shows that a gravitational wave is just the same, with the proportionality constant built out of c and G_N. One can easily check that c^3/G_N has just the right units (energy times time per unit area). It is a large quantity, again reflecting the stiffness of spacetime—a tiny disturbance in the metric corresponds to a large energy flux.

15.4.2   Energy loss due to gravitational radiation emission

In the previous subsection we have expressed the energy flux of a gravitational wave in terms of the metric perturbation $h_{ij} = g_{ij} - g_{ij}^{(b)}$. Here we will relate h_{ij} to the source of a gravitational wave by way of the linearized Einstein equation (15.21).

Calculate the wave amplitude due to quadrupole moments

We shall be working in the long-wavelength limit for a field-point far away from the source. Let D be the dimension of the source. This limit corresponds to

$$r \gg D \qquad \text{large distance from source}$$

$$\lambda \gg D \qquad \text{long wavelength.}$$

In such a limit we can approximate the integral over the energy–momentum source in (15.21) as

$$\int d^3\mathbf{x}' \frac{T_{\mu\nu}\left(\mathbf{x}', t - |\mathbf{x} - \mathbf{x}'|/c\right)}{|\mathbf{x} - \mathbf{x}'|} \longrightarrow \frac{1}{r} \int d^3\mathbf{x}' T_{\mu\nu}\left(\mathbf{x}', t - \frac{r}{c}\right) \quad (15.51)$$

[7] The long-wavelength approximation means small (D/λ) and small ωt as $t \sim D/c$ and $\omega \sim c/\lambda$.

because in the long wave limit[7] the harmonic source $T_{\mu\nu} \propto \cos \omega t - 2\pi |\mathbf{x} - \mathbf{x}'|/\lambda$ will not change much when integrated over the source. To calculate the energy flux through (15.50) we have, from (15.21) and (15.51),

$$h_{ij}(\mathbf{x}, t) = \frac{4G_N}{c^4 r} \int d^3\mathbf{x}' T_{ij}\left(\mathbf{x}', t - \frac{r}{c}\right), \quad (15.52)$$

where we have not distinguished between h_{ij} and $\bar{h}_{ij}$ as they are the same in the TT gauge.

To calculate $\int d^3\mathbf{x}' T_{ij}(\mathbf{x}')$ we find it convenient to convert it into a second mass moment[8] by way of the energy–momentum conservation relation $\partial_\mu T^{\mu\nu} = 0$, and express the integral as the time derivative of the moment integral of the (0, 0) component of the energy–momentum tensor:

$$h_{ij}(\mathbf{x}, t) = \frac{2G_N}{c^4 r} \frac{\partial^2}{\partial t^2} I_{ij}\left(t - \frac{r}{c}\right) \quad (15.53)$$

where I_{ij} is the second mass moment, after making the Newtonian approximation of the energy density as $T_{00} = \rho c^2$ with $\rho(\mathbf{x})$ being the mass density, given by

$$I_{ij} = \int d^3\mathbf{x}\, \rho(\mathbf{x})\, x_i x_j. \quad (15.54)$$

[8] Differentiating the conservation conditions $\partial_0 \partial_\mu T^{\mu 0} = 0$ leads to

$$\frac{\partial^2 T^{00}}{c^2 \partial t^2} = -\frac{\partial^2 T^{i0}}{c \partial t \partial x^i} = -\frac{\partial}{\partial x^i} \frac{\partial T^{0i}}{c \partial t}.$$

We can apply the conservation relation $\partial_0 T^{0i} + \partial_j T^{ij} = 0$ one more time to get

$$\frac{\partial^2 T^{00}}{c^2 \partial t^2} = +\frac{\partial^2 T^{ij}}{\partial x^i \partial x^j}.$$

Multiply both sides by $x^k x^l$ and integrate over the source volume

$$\frac{\partial^2}{c^2 \partial t^2} \int d^3\mathbf{x} T^{00} x^k x^l = \int d^3\mathbf{x} \frac{\partial^2 T^{ij}}{\partial x^i \partial x^j} x^k x^l$$

$$= 2 \int d^3\mathbf{x} T^{kl}.$$

To reach the last equality we have performed two integrations-by-parts and discarded the surface terms because the source dimension is finite.

We have already explained that, just as in the electromagnetic case, there is no monopole radiation (Birkhoff's theorem). But unlike electromagnetism, there is also no gravitational dipole radiation because the second-order time derivative of the dipole moment

$$\partial_t^2 \mathbf{d} = \int d^3 \mathbf{x} \rho\,(\mathbf{x})\,\partial_t \mathbf{v} = 0 \qquad (15.55)$$

is the total force on the system. It vanishes for an isolated system (reflecting momentum conservation). Thus the leading gravitational radiation must be quadrupole radiation, as shown in (15.53) and (15.54).

Summing over the flux in all directions in the TT gauge

The energy flux we need to calculate is, according to (15.50), directly related to the metric perturbation h_{ij} in the traceless-transverse gauge, while the result (15.53) shows that h_{ij} is given by the quadrupole moment I_{ij}. To have the mass moment with the same traceless and transverse structure as the metric perturbation, h_{ij}^{TT}, we must apply the traceless-transverse projection operator onto the mass moment of (15.54). Consider a plane wave propagating in an arbitrary direction, specified by the unit vector $\vec{n} = \vec{r}/r$. The projection operator that imposes the transversality condition is

$$\Pi_{ij} = \delta_{ij} - n_i n_j, \qquad (15.56)$$

clearly satisfying the condition $n_i \Pi_{ij} = 0$. As it turns out, the algebra will be simplified if we work with the "reduced mass moment" by subtracting out a term proportional to the trace $I = \delta_{ij} I_{ij}$:

$$\tilde{I}_{ij} = I_{ij} - \frac{1}{3}\delta_{ij} I, \qquad (15.57)$$

which is traceless, $\delta_{ij}\tilde{I}_{ij} = 0$. However, one finds that the resultant projection $\Pi_{il}\Pi_{jl}\tilde{I}_{kl}$ is still not traceless. It is not too difficult to find (Problem 15.5) the traceless-transverse reduced mass moment to be

$$\tilde{I}_{ij}^{\mathrm{TT}} = \Pi_{ik}\Pi_{jl}\tilde{I}_{kl} - \frac{1}{2}\Pi_{ij}\Pi_{kl}\tilde{I}_{kl}. \qquad (15.58)$$

It is then straightforward to find (Problem 15.6) that

$$\tilde{I}_{ij}^{\mathrm{TT}}\tilde{I}_{ij}^{\mathrm{TT}} = \frac{1}{2}\left[2\tilde{I}_{ij}\tilde{I}_{ij} - 4\tilde{I}_{ik}\tilde{I}_{il}n_k n_l + \tilde{I}_{ij}\tilde{I}_{kl}n_i n_j n_k n_l\right]. \qquad (15.59)$$

To calculate the total power emitted by the source, we need to integrate over the flux for a wave propagating out in **all** directions. We obtain[9]

$$\int \frac{1}{2}\left[2\tilde{I}_{ij}\tilde{I}_{ij} - 4\tilde{I}_{ik}\tilde{I}_{il}n_k n_l + \tilde{I}_{ij}\tilde{I}_{kl}n_i n_j n_k n_l\right]d\Omega$$

$$= 2\pi\left(2 - \frac{4}{3} + \frac{2}{15}\right)\tilde{I}_{ij}\tilde{I}_{ij} = \frac{8\pi}{5}\tilde{I}_{ij}\tilde{I}_{ij}. \qquad (15.60)$$

Integrating the flux (15.50), with the wave amplitude h_{ij} given by (15.53), over all directions by using the result of (15.60) we arrive at the expression for the total luminosity

$$\frac{dE}{dt} = \int f \cdot r^2 d\Omega = \frac{G_{\mathrm{N}}}{5c^5}\left\langle\left(\partial_t^3 \tilde{I}_{ij}^{\mathrm{TT}}\right)\left(\partial_t^3 \tilde{I}_{ij}^{\mathrm{TT}}\right)\right\rangle. \qquad (15.61)$$

[9]For this we need to use the formulas

$$\int d\Omega = 4\pi$$

$$\int n_k n_l d\Omega = \frac{4\pi}{3}\delta_{kl}$$

$$\int n_i n_j n_k n_l d\Omega = \frac{4\pi}{15}\left(\delta_{ij}\delta_{kl}\right.$$

$$\left. + \delta_{kj}\delta_{il} + \delta_{ik}\delta_{jl}\right).$$

These integration results are easy to understand: the only available symmetric tensor that is invariant under rotation is the Kronecker delta δ_{ij}. After fixing the tensor structure of the integrals, the coefficients in front, $4\pi/3$ and $4\pi/15$, can be obtained by contracting the indices on both sides and using the relation $\delta_{ij}\delta_{ij} = 3$.

Fig. 15.4 The Arecibo Radio telescope. (Courtesy of the NAIC–Arecibo Observatory, a facility of the NSF.)

Let us recapitulate: the energy carried away by a gravitational wave must be proportional to the square of the time-derivative of the wave amplitude (recall the Poynting vector), which is the second derivative of the quadrupole moment, cf. (15.53). The energy flux falls off like r^{-2}. To get the total luminosity by integrating over a sphere of radius r, the dependence of radial distance disappears. The factor of $G_N c^{-5}$ must be present on dimensional grounds. The detailed calculation fixes the proportional constant of $1/5$ and we have the gravitational wave luminosity in the quadrupole approximation displayed above.

15.4.3 Hulse–Taylor binary pulsar

A radio survey, using the Arecibo Radio Telescope in Puerto Rico (Fig. 15.4), for pulsars in our galaxy made by Russel Hulse and Joseph Taylor discovered the unusual system PSR 1913 + 16. Observations made since 1974 allowed them to check GR to great precision including the verification of the existence of gravitational waves as predicted in Einstein's theory.

From the small changes in the arrival times of the pulses recorded in the past decades a wealth of properties of this binary system can be extracted. This is achieved by modeling the orbit dynamics and expressing these in terms of the arrival time of the pulse. Different physical phenomena (such as bending of the light, periastron advance, etc.) are related to the pulse time through different combinations of system parameters. In this way the masses and separation of the stars and the inclination and eccentricity of their orbit can all be deduced (see Table 15.2). It is interesting to note that these two neutron star have just the masses $1.4M_\odot$ of the Chandrasekhar limit (first mentioned in Section 8.3.1).

In this section we shall demonstrate that from these numbers, without any adjustable parameters, we can compute the decrease (decay) of the orbital period due to gravitational radiation by the orbiting binary system. Instead of a full-scale GR calculation, we shall consider the simplified case of two equal mass stars in a circular orbit (Fig. 15.5), as all essential features of gravitational radiation and orbit decay can be easily computed. At the end we then quote the exact GR expression for the pulsar and its companion $M_p \neq M_c$ in an orbit

Table 15.2 Parameters of the Hulse–Taylor binary pulsar system as compiled by Weisberg and Taylor (2003).

Pulsar mass	$M_p = 1.4408 \pm 0.0003\,M_\odot$
Companion mass	$M_c = 1.3873 \pm 0.0003\,M_\odot$
Eccentricity	$e = 0.6171338 \pm 0.000004$
Binary orbit period	$P_b = 0.322997462727\,\text{d}$
Orbit decay rate	$\dot{P}_b = (-2.4211 \pm 0.0014) \times 10^{-12}\,\text{s/s}$

with high eccentricity as a straightforward modification of the result obtained by our simplified calculation.

Energy loss due to gravitational radiation

Let us first concentrate on the instantaneous position of one of the binary stars as shown in Fig. 15.5:

$$x_1(t) = R\cos\omega_b t, \qquad x_2(t) = R\sin\omega_b t, \qquad x_3(t) = 0.$$

From this we can calculate the second mass moment according to (15.54),

$$I_{11} = 2MR^2 \cos^2\omega_b t$$

$$I_{22} = 2MR^2 \sin^2\omega_b t$$

$$I_{12} = 2MR^2 \sin\omega_b t \cos\omega_b t$$

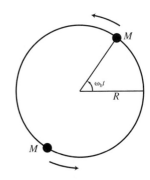

Fig. 15.5 A binary of two equal masses circulating each other in a circular orbit with angular frequency ω_b.

leading to the traceless reduced moment as defined in (15.57),

$$\tilde{I}_{ab} = I_{ab} - \frac{1}{2}\delta_{ab}I = I_{ab} - MR^2\delta_{ab}$$

so that

$$\tilde{I}_{11} = MR^2 \cos 2\omega_b t$$

$$\tilde{I}_{22} = -MR^2 \cos 2\omega_b t$$

$$\tilde{I}_{12} = MR^2 \sin 2\omega_b t.$$

The quadrupole formula (15.61) for luminosity involves time derivatives. For the simple sinusoidal dependence given above, each derivative just brings down a factor of $2\omega_b$; together with the averages $\langle\sin^2\rangle = \langle\cos^2\rangle = 1/2$, we obtain the rate of energy loss due to gravitational radiation:

$$\frac{dE}{dt} = \frac{G_N}{5c^5}(2\omega_b)^6\left(\tilde{I}_{11}{}^2 + \tilde{I}_{22}{}^2 + 2\,\tilde{I}_{12}{}^2\right) = \frac{128G_N}{5c^5}\omega_b^6 M^2 R^4. \qquad (15.62)$$

From energy loss to orbital decay

Energy loss leads to orbital decay, namely the decrease in orbital period P_b of the binary system. We start the calculation of this orbital period change through the relation $(dP_b)/P_b \propto -(dE)/E$. Again we shall only work out the simpler situation of a binary pair of equal mass M separated by $2R$ in circular motion. The total energy being

$$E = MV^2 - \frac{G_N M^2}{2R} \qquad (15.63)$$

with velocity determined by the Newtonian equation of motion $MV^2/R = G_N M^2/(2R)^2$ satisfies

$$V^2 = \frac{G_N M}{4R} \qquad (15.64)$$

so that the total energy of the binary system (15.63) comes out to be

$$E = -\frac{G_N M^2}{4R}. \qquad (15.65)$$

We wish to have an expression of the energy in terms of the orbital period by replacing R using (15.64)

$$R = \frac{G_N M}{4V^2} = \frac{G_N M}{4} \left(\frac{2\pi R}{P_b} \right)^{-2} \quad \text{or} \quad R^3 = \frac{G_N M}{16\pi^2} P_b^2. \qquad (15.66)$$

Plugging this back into (15.65), we have

$$E = -M \left(\frac{\pi M G_N}{2} \right)^{2/3} P_b^{-2/3}. \qquad (15.67)$$

Through the relation, $dE/E = -\frac{2}{3} dP_b/P_b$, the rate of period decrease $\dot{P}_b \equiv dP_b/dt$ can be related to the energy loss rate

$$\dot{P}_b = -\frac{3P_b}{2E} \left(\frac{dE}{dt} \right). \qquad (15.68)$$

Substituting in the expression (15.67) for E in the denominator, (15.62) for (dE/dt) where the wave frequency is given by the orbit frequency $\omega_b = 2\pi/P_b$ and where R is given by (15.66), we obtain the expression for orbital decay rate in this simplified case of two equal masses in circular orbit:

$$\dot{P}_b = -\frac{48\pi}{5c^5} \left(\frac{4\pi G_N M}{P_b} \right)^{5/3}. \qquad (15.69)$$

That the orbit for the Hulse–Taylor binary, rather than circular, is elliptical with high eccentricity can be taken into account (Peters and Mathews 1963) with the result involving a multiplicative factor of

$$\frac{1 + (73/24)e^2 + (37/96)e^4}{(1 - e^2)^{7/2}} = 11.85681, \qquad (15.70)$$

where we have use the observed binary orbit eccentricity as given in Table 15.2. That the pulsar and its companion have slightly different masses, $M_p \neq M_c$, means we need to make the replacement $(2M)^{5/3} \longrightarrow 4 M_p M_c (M_p + M_c)^{-1/3}$. The exact GR prediction is found to be

$$\dot{P}_{b\,\text{GR}} = \frac{-192\pi}{5c^5} \frac{1 + (73/24)e^2 + (37/96)e^4}{(1 - e^2)^{7/2}} \left(\frac{2\pi G_N}{P_b} \right)^{5/3} \frac{M_p M_c}{(M_p + M_c)^{1/3}}$$

$$= -(2.40247 \pm 0.00002) \times 10^{-12} \text{ s/s}. \qquad (15.71)$$

This is to be compared to the observed value corrected for the galactic acceleration of the binary system and the sun, which also cause a change of orbit

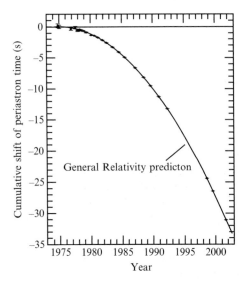

Fig. 15.6 Gravitational radiation damping causes orbital decay of the Hulse–Taylor binary pulsar. Plotted here is the accumulating shift in the epoch of periastron (Weisberg and Taylor, 2003). The parabola is the GR prediction, and observations are depicted by data points. In most cases the measurement uncertainties are smaller than the line widths. The data gap in the 1990s reflects the downtime when the Arecibo Observatory was being upgraded.

period $\dot{P}_{b\,gal} = -(0.0125 \pm 0.0050) \times 10^{-12}$ s/s. From the measured values given in Table 15.2, we then have

$$\dot{P}_{b\,corrected} = \dot{P}_{b\,observed} - \dot{P}_{b\,gal}$$

$$= -(2.4086 \pm 0.0052) \times 10^{-12} \text{ s/s}, \qquad (15.72)$$

in excellent agreement with the theoretical prediction shown in (15.71). This result (Fig. 15.6) provides strong confirmation of the existence of gravitational radiation as predicted by Einstein's theory of general relativity.

With the confirmation of the existence of gravitational radiation according to Einstein's general theory of relativity. The next stage will be the detection of gravitational waves through interferometer observations to confirm the expected wave kinematics, and tests of various strong field situations. But just like all pioneering efforts of fresh ways to observe the universe, gravitational wave observatories will surely discover new phenomena that will deepen and challenge our understanding of astronomy, gravitation and cosmology.

Review questions

1. Give a qualitative discussion showing why one would expect gravitational waves from Einstein's GR theory of gravitation but not from Newton's theory.

2. Why is it important to have a gravitational wave observatory?

3. What approximation is made to have the linearized theory of general relativity? In this framework, how should we view the propagation of gravitational waves?

4. What are the differences and similarities between electromagnetic and gravitational waves?

5. What is a gauge transformation in the linearized theory? What is the Lorentz gauge? Can we make further gauge transformations within the Lorentz gauge?

6. Consider a set of test particles, all of them lying in a circle except one at the center. When a gravitational wave with the + polarization passes through them what will be the relative displacements of these particle going through one period of the wave? How would the relative displacement be different if the polarization is of × type?

7. Give a qualitative argument showing that the wave strain is of the order $\psi G_N M / r c^2$ where $\psi \epsilon$ is a relativistic correction factor typically less than unity. Such a strain would be $O(10^{-21})$ when the wave is generated by a

solar mass source in the Virgo Cluster ($r \approx 15$ Mpc) from us.

8. Using what you know of the Poynting vector as the energy flux of an EM wave, guess the form of energy flux in terms of the gravitational wave amplitude. What should be the proportionality constant (up to some numerical constant that can only derived by detailed calculation)?

9. The leading term in gravitational radiation is quadrupole. Why is there no monopole and dipole radiation?

10. Evidence for gravitational waves is obtained by the study of the Hulse–Taylor binary pulsar system. What is being observed? Which results show strong evidence for the existence of gravitation waves as predicted by GR?

Problems

15.1 Gauge transformations

(a) Show that the gauge transformation for the trace-reversed perturbation $\bar{h}_{\mu\nu}$

$$\bar{h}'_{\alpha\beta} = \bar{h}_{\alpha\beta} - \partial_\alpha \chi_\beta - \partial_\beta \chi_\alpha + \eta_{\alpha\beta}(\partial\chi)$$

follows from (15.17) and (15.19).

(b) Demonstrate the existence of the Lorentz gauge by showing that, starting with an arbitrary coordinate system where $\partial^\mu \bar{h}_{\mu\nu} \neq 0$, one can always find a new system such that $\partial^\mu \bar{h}'_{\mu\nu} = 0$ with a gauge vector function χ_μ being the solution to the inhomogeneous wave equation $\Box \chi_\nu = \partial^\mu \bar{h}_{\mu\nu}$. This also means that one can make further coordinate transformations within the Lorentz gauge, as long as the associated gauge vector function satisfies the wave equation

$$\Box \chi_\nu = 0. \tag{15.73}$$

(c) The solution to Eq (15.73) may be written as $\chi_\nu = X_\nu e^{ikx}$ where k^α is a null-vector. Show that the four constants X_ν can be chosen so that the polarization tensor in the metric perturbation $h_{\mu\nu}(x)$ is traceless $\epsilon^\mu{}_\mu = 0$ and every zeroth component vanishes $\epsilon_{\mu 0} = 0$.

(d) From the results discussed above, show that there are two independent elements in the polarization tensor $\epsilon_{\mu\nu}$.

15.2 The Schwarzschild solution   Obtain the exterior solution to the linearized Einstein equation (15.20) for a spherical source. Since it is of the form of a standard wave equation, the spherically symmetric solution should have the same form as the $1/r$ potential:

$$\bar{h}_{\mu\nu} = \frac{C_{\mu\nu}}{r}, \tag{15.74}$$

where $C_{\mu\nu}$ is a constant tensor. Show that the Lorentz gauge condition (15.18) implies that every element of this tensor vanishes except C_{00}, which is fixed by its Newtonian limit to be $C_{00} = 2r^*$, twice the Schwarzschild radius. From this you are asked to show that the resultant metric may be written in a spherical coordinate system as

$$g_{\mu\nu} = \text{diag}\left[\left(-1 + \frac{r^*}{r}\right), \left(1 + \frac{r^*}{r}\right), r^2, r^2 \sin^2\theta\right], \tag{15.75}$$

which approximates the Schwarzschild metric of (14.54) up to a correction $O(r^{*2})$. **Beware** of the fact that the linearized theory developed in this chapter, hence Eq. (15.20), assumes a flat space metric in the Cartesian coordinate $\eta_{\mu\nu} = \text{diag}(-1, 1, 1, 1)$. This is the coordinate system you should use to start your discussion; only at later stage should you change it to a spherical one—in order to compare your result with (14.54).

15.3 Wave effect via the deviation equation   As we have shown in Section 15.3.1, a gravitational wave can only

be detected through the tidal effect. Since the equation of geodesic deviation (Problem 14.4) is an efficient description of the tidal force, show that the results of (15.30) and (15.31) can be obtained by using this equation.

15.4 **$\Gamma^{\mu}_{\nu\lambda}$ and $R^{(2)}_{\mu\nu}$ in the TT gauge** Show that the Christoffel symbols of (15.44) and (15.45), as well as the second-order Ricci tensor (15.47), are obtained in the TT gauge with the metric given in (15.42).

15.5 **Trace calculation of $\tilde{I}^{\mathrm{TT}}_{ij}$** We claimed that $\tilde{I}^{\mathrm{TT}}_{ij}$, as defined by (15.58), is traceless. Prove this by explicit calculation.

15.6 **Derive the relation (15.59)** Carry out the contraction of the two $\tilde{I}^{\mathrm{TT}}_{ij}$, defined by (15.58), to show

$$\tilde{I}^{\mathrm{TT}}_{ij} \tilde{I}^{\mathrm{TT}}_{ij} = \frac{1}{2} 2\tilde{I}_{ij} \tilde{I}_{ij} - 4\tilde{I}_{ik} \tilde{I}_{il} n_k n_l + \tilde{I}_{ij} \tilde{I}_{kl} n_i n_j n_k n_l. \tag{15.76}$$

ENDNOTES

Part
V

Answer keys to review questions

Here we provide brief answers to the review questions presented at the end of each chapter.

Chapter 1: Introduction and overview

1. Relativity is the coordinate symmetry. SR is the one with respect to inertial frames; GR, to general coordinate frames.
2. Covariance of the physics equations (i.e. physics is not changed) under symmetry transformations. Not able to detect a particular physical feature means physics is unchanged. It is a symmetry. If one cannot detect the effect after changing the orientation it means the physics equation is covariant under rotation—thus rotationally symmetric. Similar statements can be made for the coordinate symmetry of relativity.
3. A tensor is a mathematical object having definite transformation properties under coordinate transformation. A tensor equation, having the same transformation for every term, maintains the same form and is thus symmetric under coordinate changes.
4. (a) The frames in which Newton's first law holds. (b) The frames moving with constant velocity with respect to the fixed stars. (c) The frames in which gravity is absent.
5. Equations of Newtonian physics are covariant under Galilean transformations; electrodynamics, under Lorentz transformations. Galilean transformations are the Lorentz transformations with low relative velocity.
6. The coordinate transformations in GR are necessarily position dependent.
7. See Section 1.2.1.
8. In GR the gravitational field is the curved spacetime. The Einstein equation is the GR field equation. The geodesic equation is the GR equation of motion.
9. In Newtonian physics, space is the arena in which physical events take place. In GR, space simply reflects the relationship among physical events taking place in the world and has no independent existence.

Chapter 2: Special relativity: the basics

1. See Eqs. (2.11) and (2.8).
2. See Eqs. (2.19) to (2.22).
3. See first paragraph in Box 2.2: A static charge produces an electric field according to Coulomb/Gauss law. When it appears, to a moving observer, as a moving charge, it still has an electric field according to Coulomb's law, but because it is moving, it also gives rise to a magnetic field according to Ampere's law.

4. Einstein clarified the meaning of time measurement when the signal transmission could not be instantaneous and showed that $\Delta t' \neq \Delta t$ was physically meaningful.

5. See Section 2.2.1.

6. In this way, the key feature of SR of absolute light speed can be directly incorporated.

7. "Relativity of simultaneity" means that it is possible for two events to be viewed as taking place at the same time in one frame but to be viewed as taking place at different times in another frame. It is counter-intuitive because our intuition is based on nonrelativistic physics ($v \ll c$); in our everyday lives, we generally deal with velocities that are very low compared to the speed of light, and so the effect of relativity of simultaneity is unobservably small to us, being $O(v/c)$.

8. Space and time should be treated symmetrically: time is treated as a coordinate, on the same footing as the space coordinates—in contrast to the situation in Newtonian physics where space and time are treated very differently.

9. See Section 2.2.3.

10. See Section 2.2.3.

11. See Section 2.2.1.

12. See Section 2.3.2. The quantity Δs is absolute because all observers can agree on the value of time as measured in a given clock's rest frame (the proper time) as there is only one rest frame for this clock.

Chapter 3: Special relativity: the geometric formulation

1. (a) $\mathbf{e}_i \cdot \mathbf{e}_j \equiv g_{ij}$. (b) $\Delta s^2 = g_{ij} \Delta x^i \Delta x^j$; in particular, in the Minkowski spacetime, $\Delta s^2 = \eta_{\mu\nu} \Delta x^\mu \Delta x^\nu$, with $\eta_{\mu\nu} = \text{diag}(-1, 1, 1, 1)$. (c) The diagonal elements are squared lengths of basis vectors and the off-diagonal elements the deviation of pairs of basis vectors from orthogonality. (d) $g_{ij} = \delta_{ij}$, which are the matrix elements of an $n \times n$ identity matrix.

2. (a) Light velocity c is absolute. See Box 3.1. (b) $\Delta s^2 = -c^2 \Delta \tau^2$, with τ being the proper time.

3. $x^\mu = (x^0, x^1, x^2, x^3) = (ct, x, y, z)$.

4. All 4-vectors, by definition, transforms as $A^\mu \to A'^\mu = \mathbf{L}^\mu_\nu A^\nu$. See discussion just before (3.18).

5. With all indices contracted, it is a Lorentz invariant $\eta_{\mu\nu} A^\mu B^\nu = \eta_{\mu\nu} A'^\mu B'^\nu$. See Eq. (3.10) and the discussion that follows.

6. See Box 3.2.

7. See the discussion just before Eq. (3.27). $U^\mu = dx^\mu/d\tau = \gamma dx^\mu/dt$. $\eta_{\mu\nu} U^\mu U^\nu = -c^2$ for particles with mass, $\eta_{\mu\nu} U^\mu U^\nu = 0$ for massless particles.

8. $E = \gamma mc^2 \xrightarrow{\text{NR}} mc^2 + \frac{1}{2}mv^2$ and $\vec{p} = \gamma m\vec{v} \xrightarrow{\text{NR}} m\vec{v}$ 4-momentum $p^\mu = (p^0, \vec{p}) = (E/c, \vec{p})$. For particles with mass, $E^2 - (pc)^2 = (mc^2)^2$. For massless particles, $E = pc$; since they always move with $v = c$, there is no nonrelativistic limit.

9. See Fig. 3.2 and Fig. 3.3.

10. See Fig. 3.5 and the related discussion.

11. See Fig. 3.9(a).

12. See Fig. 3.9(b) and the discussion in the related subsection. Since A is not inside the forward lightcone of B (i.e. the separation between A and B is spacelike); it is not causally related to B. In order for B to influence A, a signal traveling faster than light would be required.

13. For a moving object, the two ends of the object must be measured simultaneously, $t_1 = t_2$, by an observer who is moving with respect to the object; thus $\Delta t = 0$ in (3.54). For a clock at rest, there is no displacement: $x_1' = x_2'$, thus $\Delta x' = 0$ in (2.40).

Chapter 4: The principle of equivalence

1. Newton's field equation is (4.6). The equation of motion is (4.8), which is totally independent of any properties of the test particle.
2. Equations (4.9) and (4.10). Experimental evidence: $(\ddot{\mathbf{r}})_A = (\ddot{\mathbf{r}})_B$ for any two objects A and B.
3. EP is the statement that the physics in a freely falling frame in a gravitational field is indistinguishable from the physics in an inertial frame without gravity. Weak EP is EP restricted only to Newtonian mechanics; strong EP is EP applied to all physics including electrodynamics.
4. See Fig. 4.3.
5. (a) See the derivation of Eq. (4.22); (b) Eq. (4.52).
6. (a) See the derivation of Eq. (4.33); (b) Eq. (4.36).
7. See derivation of Eq. (4.38).
8. See Eq. (4.40). The speed of light is absolute as long as it is measured with respect to the proper time of the observer. It may vary if measured by the coordinate time, for example by an observer far away from the gravitational source. An example of the physical manifestations of changing light speed according to some observer is light deflection in a gravitational field.

Chapter 5: Metric description of a curved space

1. An "intrinsic geometric description" is one that an inhabitant living within the space can perform without referring to any embedding.
2. The metric elements can be found by distance measurement once the coordinates have been fixed. See Eq. (4.16).
3. The geodesic equation is the Euler–Lagrange equation resulting from extremization of the length integral $\int g_{ab}\,\dot{x}^a\,\dot{x}^b\,d\sigma$.
4. The metric is an intrinsic quantity because it can be determined through intrinsic operations (cf. Question 2). Other intrinsic geometric quantities, such as angle, geodesic curves, etc. can then be derived from the metric function.
5. A flat surface can also have a position-dependent metric. Example: polar coordinates on a flat plane.
6. Transformation in a curved space must be position dependent.
7. A small region of any curved space can be approximated by a flat space: we can always find a coordinate transformation at a given point so that the new metric is a constant up to second-order correction.
8. K vanishes only for a flat surface, independent of coordinate choices. It measures the deviation from Euclidean geometric relations.
9. 2-sphere, 2-pseudosphere, and flat plane.
10. In a 4D Euclidean space $W^2 + X^2 + Y^2 + Z^2 = R^2$ describes a 3-sphere; in a 4D Minkowski space $-W^2 + X^2 + Y^2 + Z^2 = -R^2$, a 3-pseudosphere.
11. The circumference of a circle is related to its radius r as $S = 2\pi r - \pi r^3 K/3$ for small r. The angular excess per unit area is simply the curvature $\epsilon/\sigma = K$.
12. Cf. (5.40) and Fig. 5.7(a).

Chapter 6: GR as a geometric theory of gravity – I

1. In a "geometric theory of physics" the physical results can be directly attributed to the underlying geometry of space and time. Example: that latitudinal distances decrease as they approach either of the poles, see Fig. 5.2, reflects the geometry of the spherical surface (rather than the physics of ruler lengths). See Section 6.1.

2. Time interval changes because the spacetime metric is position dependent, see (6.3), with the metric element being directly related to the gravitational potential, $g_{00} = -(1 + 2\Phi/c^2)$.

3. See Section 6.1.1. The relation between circumference and radius deviates from the Euclidean $S = 2\pi R$.

4. See Section 6.1.2. Curved spacetime is the gravitational field. EP is automatically incorporated in a description of gravity as the curved spacetime.

5. Space and time are not predetermined to have some fixed structure; rather, they are dynamically determined by the distribution of mass and energy. Space and time are described by a metric function, which is the solution to the GR field equation.

6. Gravity, being the structure of spacetime, is not regarded as a force (bring about acceleration). Thus a test body will move freely in such a curved spacetime, with the equation of motion identified with the geodesic equation.

7. Newtonian limit corresponds to $v \ll c$ particles in a weak and static gravitational field. See Eqs. (6.23) and (6.25) for a geometric derivation of the gravitational redshift.

8. The tidal forces are the relative gravitational forces on neighboring test particles. They are the second derivatives of gravitational potential, thus falls off as r^{-3}. The extra power in the denominator and $r_s \gg r_m$ so that solar tidal force can be smaller than lunar ones even though the leading gravitational force due to the sun is larger. Relativistic gravitational potential being the metric, the tidal forces must be second derivatives of the metric, hence the curvature.

9. GR field equation has the structure of the curvature being proportional to the energy–momentum tensor $G_{\mu\nu} = \kappa T_{\mu\nu}$ with the proportionality constant $\kappa \sim G_N$. The curvature and energy density having different measurement units, Newton's is the basic "conversion factor" in GR. In relativity, space and time are treated on equal footing; the speed c is the conversion factor between space and time. In quantum theory, we have wave-particle duality; Planck's constant is the conversion factor between these two descriptions: converting wave frequency to particle energy $E = \hbar\omega$, or converting wavelength to particle momentum $p = h/\lambda$.

10. See Box 6.3.

Chapter 7: Spherically symmetric spacetime – GR tests

1. See Eq. (7.12). Space is curved because $d\rho = \sqrt{g_{rr}}dr$ and time, because $d\tau = \sqrt{-g_{00}}dt$.

2. Newtonian gravity for a spherical source is the same as if all the mass were concentrated at the center. Thus, time dependence of the source does not show up in the exterior field, and there is no monopole radiation.

3. The Schwarzschild solution yields $g_{00} = -1 + r^*/r = -1 + 2G_N M/(rc^2)$, which should be $-(1 + 2\Phi/c^2)$. Thus, $\Phi = -G_N M/r$.

4. See Eqs. (7.28) and (7.29).

5. When the source, lensing mass, and observer are perfectly aligned, the resultant azimuthal symmetry leads to an "Einstein ring." When the lensing mass is not huge, the separations among multiple images are small, and the images cannot be resolved. This results in the overlap of images and an enhancement of the brightness.

6. The $1/r$ Newtonian gravitational potential has the distinct property of yielding a closed elliptical orbit. The GR correction can be thought of as adding a perturbing $1/r^3$ term in the potential. This will bring about an open orbit. Since it is a small perturbation, it can be thought of as bringing about a closed elliptical orbit with a precessing axis (the perihelion).

7. $K^{\mu}_{(\phi)} = (0, 0, 0, 1)$. The metric's independence of ϕ means $\partial L/\partial \phi = 0$, with $L = g_{\mu\nu}\dot{x}^{\mu}\dot{x}^{\nu}$. This can be translated, through the geodesic equation (7.40), into a conservation statement of $d(\partial L/\partial\dot{\phi})/d\tau = 0$. The constant of motion can then be written as $\partial L/\partial\dot{\phi} \sim g_{\mu\phi}\dot{x}^{\mu} = g_{\mu\nu}\dot{x}^{\mu}K^{\nu}_{(\phi)}$.

8. While $dt/d\lambda$ and $d\phi/d\lambda$ are constants of motion, what we need here is $dr/d\lambda$, which can be obtained in terms of (κ, j) through the equation $L = g_{\mu\nu}\dot{x}^{\mu}\dot{x}^{\nu} = 0$. The integrand $d\phi/dr$ can then be found from the ratio $(d\phi/d\lambda)/(dr/d\lambda) \sim j(dr/d\lambda)^{-1}$.

Chapter 8: Black holes

1. Physical measurement at $r = r^*$ is not singular. A coordinate singularity is present only in certain coordinate systems. Schwarzschild geometry is singular at $r = r^*$ in the Schwarzschild coordinates, but this singularity disappears in coordinates such as the Eddington–Finkelstein system.

2. An event horizon is an one-way surface across which particles and light can traverse only in one direction. As the Schwarzschild surface $r = r^*$ of the Schwarzschild geometry does not allow particles and light to move outward to the $r > r^*$ region, this is a black hole.

3. The proper time is the time measured by an observer on the surface of a collapsing star as the stellar radius decreases beyond $r = r^*$. The coordinate time is the time measured by a far-away observer where the spacetime approaches the flat geometry. Infinite coordinate time means that, to the far-away observer, it would take an infinite amount of time for a particle or light to come over the horizon. Alternatively, we can express this light coming over the horizon as suffering an infinite redshift.

4. A three-dimensional space is spanned by three independent basis vectors. It is said to be null 3-surface, if one of these basis vectors is light-like. Thus, a light signal can propagate on this 3-surface which forms the edges of lightcones. That is, for all points on this surface, their lightcones lie completely on one side of (with their edges touching) the null surface. Since the (time-like) worldlines of all particles must be contained within the lightcone, such geodesics can only traverse this null surface only in one direction. This one-way surface must be an event horizon. See Fig. 8.5.

5. The ingoing light geodesics in the advanced E-F system (the outgoing lights in the retarded E-F system) are straight $45°$ lines so that lightcones tip over smoothly inward (outward) across the $r = r^*$ surfaces. The $r = 0$ is a future singularity in the advanced coordinate case, and a past singularity in the retarded coordinate case. In the first case, all particles and lights are "absorbed" by the event horizon, hence a black hole, while the second case, nothing can enter the event horizon, it is a white hole.

6. The surface defined by the condition $g_{tt} = 0$ is the Stationary Limit Surface (also called Infinite Redshift Surface). Particles in the region inside this surface can never be stationary and a light signal sent from that region would have its wavelength stretched to infinite length. The surface defined by $g_{rr} = \infty$ is an event horizon.

7. A Kruskal diagram is divided into four quadrants by the two 45° straight-lines (representing the $r = r^*$ event horizons) passing through the origin. One quadrants contains the future $r = 0$ singularity (the black hole) while another contains the past $r = 0$ singularity (the white hole). Each of the remaining two quadrants contains an asymptotically flat region. These two regions are connected by a wormhole, which is a structure in the $r \leq r^*$ region where $g_{tt} \geq 0$. Since in this region the r coordinate is time-like, the wormhole must be a dynamical entity that evolves with time.

8. The effective potential energy leads to an expression for the radial force $F = -\partial_r V$:

$$F_{\text{eff}} = \frac{A}{r^2} - \frac{2Bl^2}{r^3} + \frac{3Cl^2}{r^4}$$

The C term is absent in Newtonian limit so that in that case the angular momentum barrier (centrifugal force) is the dominant term in the small r region that pushed the particle away. In the GR description with $C \neq 0$, the last term for a compact source can always pull-in the particle towards the center.

9. Particles can fall into very tight orbits around a compact gravitational sources such as black hole (before finally spiral in). In the process tremendous amount of energy can be released because the gravitational (binding) energy of a particle in such orbits (with radius comparable to r^*) is huge.

10. AGN (active galactic nuclei), quasars, long-duration gamma ray bursts, and some X-ray binaries.

11. Although white holes and wormholes are also possible solutions of GR field equations, nature may not have the physical conditions to realize them. Also, quantum gravity is required to understand the origin of white holes and worm-holes. Thus, it is possible that the theory of quantum gravity (when we finally have such a theory) just may tell us that white holes and wormholes are not in principle realizable.

Chapter 9: The homogeneous and isotropic universe

1. $v = H_0 r$ is linear if H_0 is independent of v and r. This means that every galaxy sees all other galaxies as rushing away according to Hubble's law (cf. discussion relating to Fig. 9.3).

2. It must be an empty universe to have an age $t_0 = t_{\text{H}} = H_0^{-1}$. In a universe full of matter and energy we expect the gravitational attraction to slow down the expansion. This means that in the past the expansion rate was faster than H_0 leading to $t_0 < t_{\text{H}}$. From globular clusters, we deduced $t_{0\text{gc}} \gtrsim 12$ Gyr.

3. The "rotation curves" are the plots of matter's rotational speed as a function of radial distance to the center of the mass distribution (e.g. a galaxy). Gravitational theory would lead us to expect a rotational curve to drop as $v \sim r^{-1/2}$ outside the matter distribution. However, rotation curves are observed to stay flat, $v \sim r^0$, way beyond the luminous matter distribution.

4. A simple example: two masses $M \gg m$ illustrates the content of the virial theorem $\langle V \rangle = -2\langle T \rangle$ for a gravitational system: $G_{\text{N}} M \langle s^{-1} \rangle = \langle v^2 \rangle$.

5. Baryonic matter is the particle physics name for matter composed of ordinary (neutral or ionized) atoms. IGM can be detected by its electromagnetic absorption lines. It is not counted towards Ω_{DM} because we define dark matter as the one composed of particles having no electromagnetic interactions.

6. $\Omega_M \simeq 0.25$, $\Omega_{lum} \lesssim 0.005$, and $\Omega_B \simeq 0.04$. Thus $\Omega_{DM} = \Omega_M - \Omega_B \simeq 0.21$.

7. The cosmological principle: at any given instance of cosmic time, the universe appears the same at every point: space is homogeneous and isotropic. The comoving coordinates are a system where the time coordinate is chosen to be the proper time of each cosmic fluid element; the spatial coordinates are coordinate labels carried along by each fluid element (thus each fluid element has a fixed and unchanging comoving spatial coordinate).

8. See Eqs. (9.33), (9.37), and (9.38) for the Robertson–Walker metric in the spherical polar and cylindrical coordinate systems. The input used in the derivation is the cosmological principle.

9. $a(t)$ is the dimensionless scale which changes along with the cosmic time, and $k = \pm 1, 0$ is the curvature signature. The Hubble constant is related to the scale factor by $H(t) = \dot{a}(t)/a(t)$.

10. The scaling behavior of wavelength being $(\lambda_{rec}/\lambda_{em}) = a(t_{rec})/a(t_{em})$, the definition of redshift $z \equiv (\lambda_{rec} - \lambda_{em})/\lambda_{em}$ would lead to $a(t_0)/a(t_{em}) = 1 + z$.

11. Summarize the derivation of (9.40), (9.47), and (9.51): The proper distance measured at t_0 to a light source at d_p is the light ($ds^2 = 0$) path given by

$$d_p = \int_{t_{em}}^{t_0} \frac{c\,dt}{a(t)} = \int_a^1 \frac{c\,da}{a\dot{a}} \underset{(9.43)}{=} \int_a^1 \frac{c\,da}{a^2 H} \underset{(9.50)}{=} \int_0^z \frac{c\,dz}{H}.$$

12. Luminosity distance is defined through the observed flux in relation to the intrinsic luminosity of the source $d_L = \sqrt{\mathcal{L}/4\pi f}$, and is related to proper distance by $d_L = (1+z)d_p$.

Chapter 10: The expanding universe and thermal relics

1. The Friedmann equations are the Einstein equation subject to the cosmological principle, that is, the Robertson–Walker metric and ideal fluid $T_{\mu\nu}$. It has the Newtonian interpretation as being the energy balance equation, with E_{tot} being the sum of kinetic and potential energies. Newtonian interpretation is possible because of cosmological principle—large region behaves similarly to the small. Only quasi-Newtonian because we still need to supplement it with geometric concepts like curvature and scale factor, etc.

2. Both the critical density and escape velocity are used to compare the kinetic and potential terms to determine whether the total energy is positive (unbound system) or negative (bound system).

3. Radiation energy, being proportional to frequency (hence inverse wavelength), scales as a^{-1}. After dividing by the volume (a^3), the density scales as a^{-4}. Since matter density scales as a^{-3}, radiation dominates in the early universe $a \rightarrow 0$.

4. $p = w\rho c^2$ with $w_R = 1/3$ and $w_M = 0$. Flat RDU $a \sim t^{1/2}$ and $t_0 = \frac{1}{2}t_H$; MDU $a \sim t^{2/3}$ and $t_0 = \frac{2}{3}t_H$. Since radiation–matter equality time $t_{RM} \ll t_0$, MDU should be a good approximation for the whole period of t_0.

5. Figure 10.2.

6. Stefan–Boltzmann law: $\rho_R \sim T^4$ and radiation density scaling law $\rho_R \sim a^{-4}$, therefore $T \sim a^{-1}$. Blackbody radiation involves only scale invariant combinations of (volume)$\times E^2 dE$ (recall radiation energy $\sim a^{-1}$) and E/T.

7. Reaction rate faster than expansion rate. Cf. (10.45) and (10.46).

8. From big bang nucleosynthesis energy $E_{bbn} \approx$ MeV we have $T_{bbn} \approx 10^9$ K. Boltzmann distribution yields $n_n/n_p \simeq \exp{-\Delta m c^2 / k_b T_{bbn}}$ with $\Delta m = m_n - m_p$. Equation (10.55) leads to mass fraction of $\frac{1}{4}$, if $n_n/n_p \simeq \frac{1}{7}$.

9. Because the theoretical prediction of deuterium abundance by big bang nucleosynthesis is sensitive to Ω_B and number of neutrino flavors. The observed abundance can fix these quantities, cf. Fig. 10.3.

10. At t_γ the reversible reaction of $e + p \longleftrightarrow H + \gamma$ stopped proceeding from right to left. All charged particles turned into neutral atoms. The universe became transparent to photons. Average thermal energy O (eV) translates into $T(z_\gamma) \simeq 3000$ K. By the temperature scaling law

$$\frac{T(t_\gamma)}{T(t_0)} = \frac{a(t_0)}{a(t_\gamma)} = 1 + z_\gamma$$

leading to $T(t_0) \simeq 3$ K.

11. $t_{bbn} \simeq 10^2$ s, $t_{RM} \simeq 16\,000$ year, and $t_\gamma \simeq 360\,000$ year.

12. See discussion leading to (10.77).

13. Motion leads to frequency blueshift in one direction and redshift in the opposite direction. Frequency shift means energy change, hence temperature change.

14. Primordial density perturbation as amplified by gravity and resisted by radiation pressure set up acoustic waves in the photon–baryon fluid. Photons leaving denser regions would be gravitationally redshifted and thus bring about CMB temperature anisotropy.

15. Because cosmological theories predict only statistical distribution of hot and cold spots. To compare theory to experiment we need to average over an ensemble of identically prepared universes. Having only one universe, all we can do is to average samples from regions corresponding to different m moment number (for the same l). But for a given l, there are only $2l + 1$ values of m. For low l distributions there is large variance. Cf. (10.89).

Chapter 11: Inflation and the accelerating universe

1. Constant density means $dE = \varepsilon dV$ with ε being the constant energy density. The first law $dE = -pdV$ leads to $p = -\varepsilon$.

2. Cf. Section 11.1.1.

3. Cf. Eq. (11.16).

4. We see the same CMB temperature across patches of the sky $\gg 1°$ which is the horizon angular separation at t_γ. Since they could not have been causally connected and thermalized, how could they have the same property?

5. Cf. Fig. 11.3.

6. Repulsive expansion by the constant energy density is self-reinforcing: the more the volume increases the more gain in energy, leading in turn to more repulsive expansion, $\dot{a}(t) \propto a(t)$. Rapid expansion stretches out any curvature, solving the flatness problem. Also because it is possible to have $\dot{a}(t) > c$, one thermalized volume prior to inflation could be stretched out into such a large region with many horizon lengths across, resolving the horizon puzzle.

7. All came from the potential energy of inflation/Higgs field which turned into the false vacuum energy during the phase transition. Quantum fluctuation of the

Higgs field became the density fluctuations that seeded the cosmic structure. Cf. Section 11.2.3.

8. (I) The large angle region ($>1°$): we see the initial density perturbation. (II) The subdegree region: we see the signals of the acoustic waves of photon–baryon fluid, with gravity the driving force and radiation pressure the restoring force. (III) The small angle (less than arc-minute) region: photon decoupling shows up as exponential damping during this small finite interval.

9. The primary peak should correspond to the fundamental wave with a wavelength given by the sound horizon of the photon–baryon fluid. The corresponding angular separation is the sound horizon length at t_γ divided by the angular distance from t_γ to us at t_0. This distance would be affected by the curvature of space. The observation that the first peak at $l \approx 200$ agrees with the prediction of a flat universe.

10. An accelerating expansion means that expansion was slower in the past, hence a longer time interval for the cosmic recession to reach the present rate of H_0.

11. Dark energy has an equation of state parameter value $w < -1/3$. It corresponds to a negative pressure that can give rise to a repulsive gravitational force. The cosmological constant (unchanging vacuum energy) is one example of dark energy. Dark matter does not have strong or electromagnetic interactions. Examples of dark matter are neutrinos and WIMPs. Cf. introductory paragraph in Section 11.4. Furthermore, dark matter does not give rise to negative pressure.

12. Their intrinsic luminosities can be reliably calibrated, and they are extremely bright.

13. An accelerating expansion means that expansion was slower in the past. It would take a longer period, thus longer separation, before reaching a given redshift (recession velocity). The longer distance to the galaxy translates into a dimmer light.

14. If we live in an accelerating universe powered by the constant energy density of the cosmological constant, at earlier epochs the universe must be dominated by ordinary radiation and matter: instead of being a constant, ρ_R and ρ_M increases as a^{-4} and a^{-3} as $a \to 0$. Thus the accelerating phase must be preceded by deceleration.

15. Why should $\Omega_M \simeq \Omega_\Lambda$ now, or equivalently the matter-Λ equality time $t_{M\Lambda} \simeq t_0$? or why is $t_0 \simeq t_H$?

16. See bullet summary at the beginning of the chapter.

Chapter 12: Tensors in special relativity

1. The covariant components are the projections of the vector **A** onto the basis vectors, $A_\mu = \mathbf{A} \cdot \mathbf{e}_\mu$ and the contravariant components are the projections onto the inverse basis vectors, $A^\mu = \mathbf{A} \cdot \mathbf{e}^\mu$. These two kinds of vector (tensor) components are needed in order to construct invariants such as $A_\mu B^\mu$. One can raise or lower indices by using the metric tensor or inverse metric tensor as in (12.11).

2. The position 4-vector x^μ is naturally contravariant because its covariant components $x_\mu = (-ct, \vec{x})$ have an "unnatural" minus sign in the zeroth component; the gradient operator is naturally covariant because its contravariant components $\partial^\mu = (-c^{-1}\partial_t, \vec{\nabla})$ have the minus sign in the zeroth component.

3. $A'^\mu = \mathbf{L}^\mu_{\ \nu} A^\nu$ and $A'_\mu = \mathbf{L}^{-1\ \nu}_{\ \mu} A_\nu$.

4. $T'^\mu_{\ \nu} = \mathbf{L}^\mu_{\ \lambda} \mathbf{L}^{-1\ \rho}_{\ \nu} T^\lambda_{\ \rho}$.

5.

$$\begin{pmatrix} ct' \\ x' \end{pmatrix} = \begin{pmatrix} \gamma & -\beta\gamma \\ -\beta\gamma & \gamma \end{pmatrix} \begin{pmatrix} ct \\ x \end{pmatrix}$$

$$\begin{pmatrix} c^{-1}\partial'_t \\ \partial'_x \end{pmatrix} = \begin{pmatrix} \gamma & \beta\gamma \\ \beta\gamma & \gamma \end{pmatrix} \begin{pmatrix} c^{-1}\partial_t \\ \partial_x \end{pmatrix}.$$

6. Given $\partial_\mu F^{\mu\nu} + j^\nu/c = 0$, we can show that $\partial'_\mu F'^{\mu\nu} + j'^\nu/c = 0$:

$$\left(\partial' F' + j'/c\right) = \left[\mathbf{L}^{-1}\right] \partial \left[\mathbf{L}\right]\left[\mathbf{L}\right] F + \left[\mathbf{L}\right] j/c = \left[\mathbf{L}\right]\left(\partial F + j/c\right) = 0.$$

That is, because every term is a 4-vector, the form of the equation is unchanged under the Lorentz transformation. This equation includes the statement of electric charge conservation, $\partial_\mu j^\mu = 0$, because $\partial_\mu \partial_\nu F^{\mu\nu} = -\partial_\mu \partial_\nu F^{\mu\nu} = 0$ as $F^{\mu\nu} = -F^{\nu\mu}$ and $\partial_\mu \partial_\nu = +\partial_\nu \partial_\mu$.

7. $j^\mu = c\Delta q/\Delta S^\mu$, where ΔS^μ is the Minkowski volume with x^μ held fixed (i.e. it is a Minkowski 3-surface). For example, $\Delta S^0 = (\Delta x^1 \Delta x^2 \Delta x^3)$, etc., so that $j^0 = c \times$ charge density, and $\vec{j} =$ current density.

8. The energy-momentum tensor is the 4-current for the 4-momentum field: $T^{\mu\nu} = c\Delta p^\mu/\Delta S^\nu$ so that $T^{00} =$ energy density, $T^{0i} = T^{i0} =$ momentum density or energy current-density, and $T^{ij} =$ normal force per unit area (pressure) for $i = j$ and shear force per unit area for $i \neq j$.

9. See Eq. (12.72).

Chapter 13: Tensors in general relativity

1. Transformations in a curved space must necessarily be position dependent.
2. See Eqs. (13.7) and (13.12).
3. See Eqs. (13.17) and (13.20). Tensor equations are automatically relativistic.
4. $A^\mu = \mathbf{e}^\mu \cdot \mathbf{A}$ being coordinate dependent, its derivative will have extra term as in (13.19).
5. $D_\nu T^{\lambda\rho}_\mu = \partial_\nu T^{\lambda\rho}_\mu - \Gamma^\sigma_{\nu\mu} T^{\lambda\rho}_\sigma + \Gamma^\lambda_{\nu\sigma} T^{\sigma\rho}_\mu + \Gamma^\rho_{\nu\sigma} T^{\lambda\sigma}_\mu$.
6. See Eq. (13.37).
7. We can express the Riemann tensor as a commutator of covariant derivatives (13.67). Since every term other than $R^\mu_{\lambda\alpha\beta}$ is known to be a good tensor, by the quotient theorem, $R^\mu_{\lambda\alpha\beta}$ must also be a good tensor.
8. $G^{\mu\nu} = G^{\nu\mu}$ and $D_\mu G^{\mu\nu} = 0$.
9. At every point we can always find a coordinate system (LEF) in which $\partial_\mu g_{\nu\lambda} = 0$ and $\Gamma^\lambda_{\nu\sigma} = 0$. Therefore, in this LEF we have $D_\mu g_{\nu\lambda} = \partial_\mu g_{\nu\lambda} = 0$. Since $D_\mu g_{\nu\lambda} = 0$ is tensor equation, it must be valid in every frame.

Chapter 14: GR as a geometric theory of gravity – II

1. See first part of Section 14.1.
2. Just replace ordinary derivatives by covariant derivatives. This is required because the coordinate symmetry in GR is local. It involves position-dependent transformations. Covariant derivatives include the Christoffel symbols which, being the derivatives of the gravitational potential (i.e. the metric), constitute the gravitational field.
3. Following the procedure stated in Question 2, equation of motion in SR $dU^\mu/d\tau = 0$ leads to $DU^\mu/d\tau = 0$, which is the geodesic equation.
4. See Section 14.2.1.

5. See Eqs. (14.26) and (14.27).
6. Set the Ricci scalar for the 3D spatial metric for a spherically symmetric space to a constant, because homogeneous and isotropic space corresponds to a space having spherical symmetry with respect to every point. Namely, a homogeneous and isotropic space must be a space with a position-independent curvature scalar.
7. Friedmann equations are just the components of Einstein equation with Robertson–Walker metric and an energy-momentum tensor given by that of an ideal fluid.
8. $g^{\mu\nu} = g^{\nu\mu}$ and $D_\mu g^{\mu\nu} = 0$, see Question 8 in Chapter 13.
9. Moving the $-\Lambda g_{\mu\nu}$ term to the source side of the equation, we get

$$G_{\mu\nu} = \kappa(T_{\mu\nu} + \kappa^{-1}\Lambda g_{\mu\nu}) = \kappa(T_{\mu\nu} + T_{\mu\nu}^\Lambda).$$

Thus, even in the absence of matter/energy source $T_{\mu\nu} = 0$ (i.e. a vacuum), space can still be curved by the Λ term.

Chapter 15: Linearized theory and gravitational waves

1. Newton's is a static theory; it does not have nontrivial time dependence. Einstein's theory, being relativistic, treats space and time on an equal footing, hence has nontrivial time dependence.
2. This new window to the universe allows us to observe strong gravity regions which are often at the core of many interesting astrophysical phenomena.
3. Metric is slightly different from flat Minkowski metric $g_{\mu\nu} = \eta_{\mu\nu} + h_{\mu\nu}$. All GR equations are approximated only to the first order in $h_{\mu\nu}$. Propagation of a gravitational wave can be viewed a ripples of curvature moving in the flat Minkowski spacetime.
4. Both are long range forces. Their quanta are massless. EM wave (photon) has spin 1 and gravitational wave (graviton) spin 2. Leading EM radiation is dipole, gravitational radiation is quadrupole.
5. Coordinate transformations among the slightly curved coordinates. $\partial^\mu \bar{h}_{\mu\nu} = 0$ is the Lorentz gauge condition. Can make further gauge transformations as long as the vector gauge function satisfies the $\Box \chi_\mu = 0$ constraint.
6. Figure 15.1.
7. Equation (15.34).
8. The Poynting vector being the energy flux $S_{EM} = ct_{00}$ with the energy density for the EM wave $t_{00} \propto \vec{E}^2 = (\partial_t \vec{A})^2$ where $\vec{E}$ is the electric field and $\vec{A}$ the EM vector potential. We expect the gravitational wave energy flux to be proportional to $(\partial_t h)^2$ also. The dimensionful proportionality constant can be fixed by dimensional analysis to be c^3/G_N.
9. No monopole radiation according to Birkhoff's theorem. The amplitude must be the second derivatives of the mass moments. No dipole radiation because the dipole amplitude is just the rate of total momentum change which vanishes for an isolated system, see Eq. (15.55).
10. From the observed decrease in the rate of pulse arrival time, one can deduce that the binary orbit period is decreasing, which matches perfectly the GR prediction of energy loss due to gravitational wave emission by the circulating system.

Solutions to selected problems

2.2 **Inverse Lorentz transformation**   The Lorentz transformation (2.46), and its inverse, written out only for the nontrivial components are

$$\begin{pmatrix} ct' \\ x' \end{pmatrix} = \gamma \begin{pmatrix} 1 & -\beta \\ -\beta & 1 \end{pmatrix} \begin{pmatrix} ct \\ x \end{pmatrix}$$

$$\begin{pmatrix} ct \\ x \end{pmatrix} = \gamma \begin{pmatrix} 1 & \beta \\ \beta & 1 \end{pmatrix} \begin{pmatrix} ct' \\ x' \end{pmatrix}. \tag{1}$$

The inverse matrix relation is demonstrated, using $\gamma^2 = (1 - \beta^2)^{-1}$, by

$$\gamma^2 \begin{pmatrix} 1 & -\beta \\ -\beta & 1 \end{pmatrix} \begin{pmatrix} 1 & \beta \\ \beta & 1 \end{pmatrix} = \begin{pmatrix} 1 & 0 \\ 0 & 1 \end{pmatrix}.$$

2.3 **Lorentz transformation of derivative operators**

(a) Start with the chain rule,

$$\frac{\partial}{\partial x'} = \frac{\partial x}{\partial x'} \frac{\partial}{\partial x} + \frac{\partial t}{\partial x'} \frac{\partial}{\partial t} = \gamma \frac{\partial}{\partial x} + \gamma \beta \frac{\partial}{c \partial t}.$$

To reach the last equality, we used (1) showing (ct, x) as functions of (ct', x') to calculate $\partial x/\partial x' = \gamma$ and $\partial t/\partial x' = \gamma\beta/c$. Similarly, we have

$$\frac{\partial}{c \partial t'} = \gamma \frac{\partial}{c \partial t} + \gamma \beta \frac{\partial}{\partial x}.$$

(b) $\bar{\mathbf{L}}$ can be found by substituting into $\delta^\nu_\mu = \partial(x'_\nu)/\partial x'_\mu \equiv \partial'_\mu x'^\nu$ the respective Lorentz transformations $\mathbf{L}$ and $\bar{\mathbf{L}}$ for coordinates and coordinate derivatives Eqs. (2.47) and (2.48):

$$\delta^\nu_\mu = \partial'_\mu x'^\nu = \sum_{\lambda,\rho} \left(\bar{L}^\lambda_\mu \partial_\lambda \right) \left(L^\nu_\rho x^\rho \right)$$

$$= \sum_{\lambda,\rho} \bar{L}^\lambda_\mu L^\nu_\rho \delta^\rho_\lambda = \sum_\lambda \bar{L}^\lambda_\mu L^\nu_\lambda. \tag{2}$$

Namely, $\mathbf{1} = \bar{\mathbf{L}}\mathbf{L}$. Thus, the transformation for the coordinate derivative operators is just the inverse shown in (1)—as already indicated by the substitution of $v \to -v$.

2.4 **Lorentz covariance of Maxwell's equations**   Given (2.51), we show the validity of (2.50) by applying the Lorentz transformations for the fields and spacetime derivatives:

$$\nabla' \cdot \mathbf{B}' = \frac{\partial B'_x}{\partial x'} + \frac{\partial B'_y}{\partial y'} + \frac{\partial B'_z}{\partial z'}$$

$$= \gamma \left(\frac{\partial}{\partial x} + \beta \frac{\partial}{c \partial t} \right) B_x + \frac{\partial}{\partial y} \gamma \left(B_y + \beta E_z \right) + \frac{\partial}{\partial z} \gamma \left(B_z - \beta E_y \right)$$

$$= \gamma \underbrace{\left(\frac{\partial B_x}{\partial x} + \frac{\partial B_y}{\partial y} + \frac{\partial B_z}{\partial z} \right)}_{\nabla \cdot \mathbf{B} = 0} + \gamma \beta \underbrace{\left[\frac{\partial B_x}{c \partial t} + \left(\frac{\partial E_z}{\partial y} - \frac{\partial E_y}{\partial z} \right) \right]}_{\left(\nabla \times \mathbf{E} + \frac{1}{c} \frac{\partial \mathbf{B}}{\partial t} \right)_x = 0} \qquad (3)$$

where we have used Lorentz transformation of (2.49) and (2.16) to reach the second line. The x-component of Faraday's equation is singled out because we have assumed a Lorentz boost in the x direction.

2.5 **From Coulomb's to Ampere's law**   We illustrate the general approach by the example of a derivation of Faraday's law from the magnetic Gauss's law. We note that since the magnetic Gauss's law is valid in both frames, $\nabla \cdot \mathbf{B} = 0$ and $\nabla' \cdot \mathbf{B}' = 0$. Eq. (3) implies that the x component of $\nabla \times \mathbf{E} + \frac{1}{c} \frac{\partial \mathbf{B}}{\partial t}$ is zero. Hence, all three components of $\nabla \times \mathbf{E} + \frac{1}{c} \frac{\partial \mathbf{B}}{\partial t} = 0$ (all of the vector components are zero) as the y and z components can be similarly deduced by considering Lorentz boosts in the y and z directions.

2.6 **Length contraction and light-pulse clock**   In the rest frame of the clock, the total time $\Delta t'$ for a light pulse to go from one end to another and back is the sum $\Delta t' = \Delta t'_1 + \Delta t'_2$, where $\Delta t'_2$ is the time for the pulse to make the return trip. Clearly $\Delta t'_1 = \Delta t'_2 = L'/c$, where L' is the rest frame length of this clock. Now consider the clock in motion, moving with velocity v from left to right. The length the pulse must travel is lengthened when going from left to right, and shortened when going from right to left (on the return trip), due to the fact that the ends of the light clock are moving to the right:

$$c \, \Delta t_1 = L + v \, \Delta t_1, \quad c \, \Delta t_2 = L - v \, \Delta t_2,$$

where L and Δt are the length and time measured in the moving frame. We can solve the above equations for Δt_1 and Δt_2 to get the time it takes the light pulse to go from one end of the light clock to the the other in the moving frame:

$$\Delta t = \Delta t_1 + \Delta t_2 = \frac{L}{c - v} + \frac{L}{c + v} = \gamma^2 \frac{2L}{c}. \qquad (4)$$

Using the time-dilation formula (2.26), we can find Δt in terms of L'. By equating $\Delta t = \gamma \, \Delta t' = \gamma \, (2L/c)$ to the result of (4), which gives Δt in terms of L, we obtain the Lorentz length-contraction formula of $L = L'/\gamma$.

2.8 **Invariant spacetime interval and relativity of simultaneity**

(a) The invariant spacetime interval gives $-c^2 \Delta t'^2 + \Delta x'^2 = \Delta x^2$, or $c \Delta t' = \sqrt{\Delta x'^2 - \Delta x^2}$.

(b) The Lorentz transformation for the spatial coordinates, with $\Delta t = 0$, is $\Delta x' = \gamma \Delta x$. This implies that $\gamma = (1 - \beta^2)^{-1/2} = (\Delta x'/\Delta x)$ and $\gamma \beta = \sqrt{\gamma^2 - 1} = \sqrt{(\Delta x'/\Delta x)^2 - 1}$. The Lorentz transformation for the time coordinates then leads to the same result as in (a), $c \Delta t' = \gamma \beta \Delta x = \sqrt{\Delta x'^2 - \Delta x^2}$.

2.9   **More simultaneity calculations**

(a) Given the Lorentz transformation (2.34) and (2.36), as well as its inverse (2.37) and (2.38), it is clear that $\Delta t' = 0$ implies, through (2.36), $\Delta t = (\beta/c)\Delta x$, and through (2.38), $\Delta t = (\beta/c)\gamma\Delta x'$. These two equalities require the consistency condition $\Delta x = \gamma\Delta x'$, which is compatible with the Lorentz transformation (2.37) with $\Delta t' = 0$.

(b) Our derivation of length contraction in Section 2.2.3 would lead us to expect the result of $\Delta x' = \gamma^{-1}\Delta x$ because the key input of the two ends of an object being measured at the same time in the "moving frame" is satisfied by our $\Delta t' = 0$ condition.

(c) In Section 2.2.2, especially Eqs. (2.24) and (2.25), we have shown that the time intervals for the light signals to reach the back and front ends of the railcar as recorded by the platform observer are

$$t_1 = \frac{L}{2c}\frac{1}{1+\beta}, \quad t_2 = \frac{L}{2c}\frac{1}{1-\beta},$$

where L is the length of the moving railcar as seen by the platform observer. If we let the railcar length as seen by the railcar observer be $\Delta x'$, then the railcar length as seen by the platform observer should be $L = \gamma^{-1}\Delta x'$ due to length contraction. We can calculate the time difference, as in (2.25), to be

$$\Delta t = t_2 - t_1 = \frac{\beta}{c}\gamma^2 L = \frac{\beta}{c}\gamma\Delta x',$$

which agrees with $\Delta t = (\beta/c)\gamma\Delta x'$ obtained above. With respect to the O observer, the emission points are located at

$$x_1 = -ct_1 = -\frac{\Delta x'}{2\gamma}\frac{1}{1+\beta}, \quad x_2 = ct_2 = \frac{\Delta x'}{2\gamma}\frac{1}{1-\beta}.$$

Hence, according to the platform observer, the two emission events have a separation of

$$\Delta x = x_2 - x_1 = \frac{\Delta x'}{2\gamma}\left(\frac{1}{1-\beta} + \frac{1}{1+\beta}\right) = \gamma\Delta x'$$

which agrees with the result $\Delta x = \gamma\Delta x'$ gotten from Lorentz transformation above.

2.10   **Reciprocity of twin-paradox measurements**   In this reciprocal arrangement, the γ factors are exactly the same. Al's yearly flashes are received every 3 years by Bill at home during the outward-bound part (15 years) of Al's journey; thus, 15 flashes are seen by Bill during the 45 years between Al's departure and his turn-around. Thereafter, the flashes are received every 4 months; thus 15 flashes are seen by Bill in the last 5 years before Al's return. Therefore, Bill sees a total of 30 of Al's birthday fireworks over a period of 50 years of his time.

2.11   **Velocity addition in the twin paradox**   The speed of the rocketship before its turn-around is $\beta_1 = 4/5$ and after the turn-around it is $\beta_2 = -4/5$. Hence, the relative speed between the spaceship, before and after the turn-around, can be computed using the velocity addition rule of (2.22) $\beta_{12} = (\beta_1 - \beta_2)/(1 - \beta_1\beta_2) = 40/41$, which corresponds to a gamma factor of

$\gamma_{12} = (1 - \beta_{12}^2)^{-1/2} = 41/9$. Hence a reading of $t_1 = 9$ years by the clock on the rocketship before the turn-around will be seen by the clock after the turn-around, according to the SR time-dilation, to be $t_2 = \gamma_{12}t_1 = 41$ years.

3.2 Contraction and dummy indices After interchanging both pairs of indices, the symmetry properties of these two tensors yield $T_{\mu\nu}S^{\mu\nu} = -T_{\nu\mu}S^{\nu\mu}$. Since we can rename the dummy indices (e.g. rename μ as ν, and ν as μ) we have $T_{\mu\nu}S^{\mu\nu} = -T_{\nu\mu}S^{\nu\mu} = -T_{\mu\nu}S^{\mu\nu}$. Thus $T_{\mu\nu}S^{\mu\nu}$ equals to the negative of itself; it can only be zero.

3.4 Orthogonality fixes the rotation matrix The orthogonal condition

$$\begin{pmatrix} a & b \\ c & d \end{pmatrix} \begin{pmatrix} a & c \\ b & d \end{pmatrix} = \begin{pmatrix} 1 & 0 \\ 0 & 1 \end{pmatrix}$$

includes the diagonal conditions of $a^2 + b^2 = c^2 + d^2 = 1$, which can be solved by the parametrization of $a = \cos\phi$, $b = \sin\phi$ and $c = \sin\phi'$, $d = \cos\phi'$; while the off-diagonal condition of $ac + bd = \sin(\phi + \phi') = 0$ implies $\phi = -\phi'$. In terms of the actual rotation angle θ, we make the identification $\phi = \theta$.

$$\begin{pmatrix} a & b \\ c & d \end{pmatrix} = \begin{pmatrix} \cos\theta & \sin\theta \\ -\sin\theta & \cos\theta \end{pmatrix}.$$

3.5 Group property of Lorentz transformations We shall only display the group property of the boost transformation: Given the Lorentz boost (3.22), we have the combined transformation

$$[\mathbf{L}(\psi_1)][\mathbf{L}(\psi_2)] = \begin{pmatrix} c_1 & s_1 \\ s_1 & c_1 \end{pmatrix} \begin{pmatrix} c_2 & s_2 \\ s_2 & c_2 \end{pmatrix}$$

where $c_1 \equiv \cosh\psi_1$ and $s_1 \equiv \sinh\psi_1$. A straightforward matrix multiplication and the trigonometric identities, $c_{12} \equiv \cosh(\psi_1 + \psi_2)$ and $s_{12} \equiv \sinh(\psi_1 + \psi_2)$, of $c_{12} = c_1 c_2 + s_1 s_2$ and $s_{12} = s_1 c_2 + c_1 s_2$, lead us to

$$[\mathbf{L}(\psi_1)][\mathbf{L}(\psi_2)] = \begin{pmatrix} c_{12} & s_{12} \\ s_{12} & c_{12} \end{pmatrix} = [\mathbf{L}(\psi_1 + \psi_2)],$$

which is the stated result.

3.6 Group multiplication leads to velocity addition rule With the identification of (3.25) $\beta = -\tanh\psi$ so that $u/c = \beta_1 = -\tanh\psi_1$ and $-v/c = \beta_2 = -\tanh\psi_2$, and the group multiplication of (3.58), $u'/c = \beta_{12} = -\tanh\psi_{12} = -\tanh(\psi_1 + \psi_2)$, the velocity addition rule (2.22) follows from the trigonometric identity of

$$\tanh(\psi_1 \pm \psi_2) = \frac{\tanh\psi_1 \pm \tanh\psi_2}{1 \pm \tanh\psi_1 \tanh\psi_2}.$$

3.7 Lorentz transform and velocity addition rule Suppressing the transverse spatial components, the 4-velocities, according to (3.31), have components (in self-evident notations) $U^\mu = \gamma_u(c, u)$ and $U'^\mu = \gamma'_u(c, u')$, which are connected by Lorentz transformation $U'^\mu = L^\mu_\nu U^\nu$

$$\begin{pmatrix} \gamma'_u c \\ \gamma'_u u' \end{pmatrix} = \begin{pmatrix} \gamma_v & -\gamma_v\beta_v \\ -\gamma_v\beta_v & \gamma_v \end{pmatrix} \begin{pmatrix} \gamma_u c \\ \gamma_u u \end{pmatrix} = c\gamma_v\gamma_u \begin{pmatrix} 1 - \beta_v\beta_u \\ -\beta_v + \beta_u \end{pmatrix}.$$

Equating the first-row elements leads to $\gamma'_u = \gamma_v \gamma_u (1 - \beta_v \beta_u)$. When this is substituted into the equality of the second-row elements, we obtain the velocity addition rule of (2.22).

3.8   **Antiproton production threshold**   The minimum energy needed to produce the final state of three protons and one antiproton in the center-of-mass frame is $E_{final} = 4\,mc^2$. The square of the total 4-momentum of the final state, given the total 3-momentum being zero (because of CM frame), $p^{\mu}_{final} = (c^{-1}E_{final}, \vec{0})$ must then be $\eta_{\mu\nu} p^{\mu}_{final} p^{\nu}_{final} = -16\,m^2 c^2$. By energy–momentum conservation, this must also be the square of the total 4-momentum of the initial state of two protons (projectile and target): $p^{\mu}_{final} = p^{\mu}_{initial}$, where $p^{\mu}_{initial}$ is the initial total 4-momentum with total energy and total 3-momentum of the initial state, denoted by $(E, \vec{p})$, as its components. We then have, from (3.38),

$$-16m^2 c^4 = -E^2 + |\vec{p}|^2 c^2. \tag{5}$$

In the lab frame, in which the target proton is at rest, we have $E = E_1 + mc^2$, where E_1 is the energy of the projectile proton. The 3-momentum is given entirely by the projectile proton $\vec{p} = \vec{p}_1$, which is related to E_1 by the usual energy–momentum relation: $|\vec{p}|^2 c^2 = |\vec{p}_1|^2 c^2 = E_1^2 - m^2 c^4$. Substitute these two relations into (5) and solve for the projectile proton's lab energy to get $E_1 = 7\,mc^2$, which corresponds to a kinetic energy of the projectile $K_{lab} = E_1 - mc^2 = 6\,mc^2 = 5.6\,\text{GeV}$.

3.9   **More conventional derivation of Doppler effect**   Since the sender is at rest, we have $t = \tau$ and $d\phi = \omega dt$; on the other hand, for the moving receiver, we have $d\phi = \omega' dt'/\gamma = \sqrt{1 - \beta^2}\omega' dt'$. Thus invariance of the phase leads to $(\omega'/\omega) = (dt/dt')/\sqrt{1 - \beta^2}$. Now the two events (x, t) and (x', t') are connected by a light signal, we have $(x' - x) = c(t' - t)$ or $(dt/dt') = 1 - \beta$ (after using $dx'/dt' = v$). In this way we obtain

$$\frac{\omega'}{\omega} = \frac{1 - \beta}{\sqrt{1 - \beta^2}} = \sqrt{\frac{1 - \beta}{1 + \beta}}.$$

3.10   **Twin paradox measurements and Doppler effect**   We can view the sending and observing birthday fireworks as the sending and receiving of light signals. Thus the respective emission and receiving frequencies should obey the Doppler relation (3.47). The relative velocities for the outward and inward bound trips being $\beta = \pm 4/5$, the formula yields $\omega' = \omega/3$, and $\omega' = 3\omega$, respectively. This is just the frequency changes of birthday fireworks observed, for example, during the outward bound part $\beta = 4/5$, the emission frequency is red-shifted to $\omega' = \omega/3$ and Al sees Bill's annual firework every three years.

3.11   **Spacetime diagram for the twin paradox**   (see displayed diagram.)

3.12   **The twin paradox – the missing 32 years**

(a) The gamma factor between O and O'' frames being $\gamma = 5/3$, Al's inbound 15 years corresponds to the last 9 years ($= t_{QP''}$) of the total $t_Q = 50$ years. Hence $t_{P''} = t_Q - t_{QP''} = 41$ years.

(b) First we need to calculate the time dilation factor between O' and O'' frames. For this we need to work out the relative velocity $\bar{\beta}$ of these two frames. We can deduce it from the relative velocities $\beta_{1,2} = \pm 4/5$ of these two frames with respect to the O system by using the velocity addition rule of (2.22), $\bar{\beta} = (\beta_1 - \beta_2)/(1 - \beta_1 \beta_2) = 40/41$. This gives rise to a

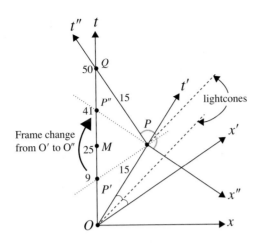

Problem 3.11 Three worldlines of the twin paradox: OQ is that for the stay-at-home Bill, OP that for the outward-bound part $(\beta = 4/5)$, PQ that for the inward-bound part $(\beta = -4/5)$ of the Al's journey. M is the midpoint between O and Q. These three lines define three inertial frames: O, O', and O'' systems. When Al changes from the O' to the O'' system at P the point that's simultaneous (with P) along Bill's worldline OQ jumps from point P' to P''. From the viewpoint of Bill, this is a leap of 32 years.

$\bar{\gamma} = 41/9$. We then find the turning point P as having an O'' frame time of $t''_{OP} = \bar{\gamma}t'_P = (41/9) \times 15 = (205/3)$ years. This moving clock time corresponds to the rest-frame time of $t_{P''} = (205/3)/(5/3) = 41$ years.

3.13 **Spacetime diagram for the pole-and-barn paradox (see displayed diagram.)**

4.1 **Inclined plane, pendulum and EP**

(a) **Inclined plane** The $F = ma$ equation along the inclined plane, is $m_{\mathrm{I}}a = m_{\mathrm{G}}g \sin\theta$, leading to a material-dependent acceleration: $a_{\mathrm{A}} = g\sin\theta \left(\frac{m_{\mathrm{G}}}{m_{\mathrm{I}}}\right)_{\mathrm{A}}$.

(b) **Pendulum** For the simple pendulum with a light string of length L, we have $m_{\mathrm{I}}L(d^2\theta/dt^2) = -m_{\mathrm{G}}g \sin\theta$. This has the form of a simple harmonic oscillator equation when approximated by $\sin\theta \approx \theta$, leading to a period of

$$T_{\mathrm{A}} = \frac{2\pi}{\omega} = 2\pi\sqrt{\frac{L}{g}\left(\frac{m_{\mathrm{I}}}{m_{\mathrm{G}}}\right)_{\mathrm{A}}}$$

for a blob made up of material A.

4.2 **Two EP brain-teasers**

(a) **Forward leaning balloon** According to EP the effective gravity is the vector sum $\mathbf{g}_{\mathrm{eff}} = \mathbf{g} + (-\mathbf{a})$, where $\mathbf{g}$ is the normal gravity (pointing vertically downward) while $\mathbf{a}$ is the acceleration of the vehicle. The buoyant force is always opposite to $\mathbf{g}_{\mathrm{eff}}$.

(b) **A toy for Einstein** Normally what is difficult to do is to have a net force pulling the ball back into the bowl. The net force is the combination of gravity and spring restoring force. But the task can be made easy by dropping the whole contraption—because EP informs us that gravity would disappear in this freely falling system. Without the interference of gravity, the spring will pull back the ball each time without any difficulty.

4.3 **The Global Position System**

(a) A satellite's centripetal acceleration is produced by earth's gravity: $v_s^2/r_s = G_{\mathrm{N}}M_\oplus/r_s$. The orbit period T_s is related to the radius and tangential velocity: $T_s = 2\pi r_s/v_s$. Knowing that $T_s = 12 \text{ h} = 4.32 \times 10^4$ s we can

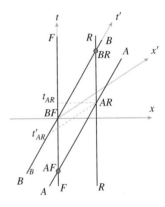

Problem 3.13 Spacetime diagram for the pole-and-barn paradox. Ground (barn) observer has (x, t) coordinates, while the runner (pole) rest frame has (x', t') coordinates. The heavy lines are the worldlines for the front-door (F), rear-door (R) of the barn, and front-end (A), back-end (B) of the pole. Note the order reversal: $t_{AR} > t_{BF}$ and $t'_{AR} < t'_{BF}$.

find r_s and v_s from these two equations: $r_s \simeq 2.7 \times 10^7$ m $\simeq 4.2R_\oplus$, $v_s \simeq$ 3.9 km/s, where $R_\oplus$ is earth's radius.

(b) The SR time dilation factor being $\gamma_s = (1 - \beta_s^2)^{-1/2} = 1 + \beta_s^2/2 + \cdots$ the fractional change is then $t/\tau - 1 = \gamma_s - 1 = \beta_s^2/2 = (v_s/c)^2/2 \simeq 0.85 \times 10^{-10}$. Here we have neglected the rotational speed of the clock on the ground—the corresponding β^2 value is a hundred times smaller even for the largest value on the equator.

(c) The gravitational time dilation effect is given by (4.31) with $\Phi = -G_N M/r$:

$$\frac{\Phi_\oplus - \Phi_s}{c^2} = -G_N \frac{M_\oplus}{c^2} \left(\frac{1}{R_\oplus} - \frac{1}{r_s} \right) \simeq -5.2 \times 10^{-10}.$$

Thus, the general relativity (GR) effect is about six times larger than the special relativity (SR) effect.

(d) In one minute duration $(\Delta t)_{GR} \simeq -30$ ns, and $(\Delta t)_{SR} \simeq 5$ ns. The gravitational effect makes the satellite clock go faster because it is at a higher gravitational potential. The SR dilation slows it down. The net effect is to make the clock in the satellite, when compared to the clock on the ground, run faster by about 25 ns for every passage of 1 min. This translates into a distance of $(2.5 \times 10^{-8}$ s$) \times (3 \times 10^8$ m/s$) = 7.5$ m.

Here is an example of the practical application in our daily life of this "pure science" of general relativity.

4.4 **Gravitational redshift directly from Doppler effect**   The receiver being in motion, moving with nonrelativistic velocity $\beta = \Delta u/c$, the SR Lorentz frequency transformation (3.47) becomes

$$\frac{\omega_{rec}}{\omega_{em}} = \sqrt{\frac{1 - \beta}{1 + \beta}} \simeq 1 - \beta.$$

Or, equivalently $\Delta\omega/\omega = -\Delta u/c$, which is just the expression shown in (4.22) leading to the final gravitational result of (4.24).

5.2 **Basis vectors on a spherical surface**   The respective basis vectors are

$$\mathbf{e}_\theta = R\hat{\mathbf{u}}_\theta, \quad \mathbf{e}_\phi = R \sin\theta \hat{\mathbf{u}}_\phi,$$

where $\hat{\mathbf{u}}_\theta$ is the unit vector in the polar angle direction, and $\hat{\mathbf{u}}_\phi$ is perpendicular to $\hat{\mathbf{u}}_\theta$ in the azimuthal direction. The resultant metric matrix, according to Eq. (5.7), is

$$g_{ab} = \begin{pmatrix} \mathbf{e}_\theta \cdot \mathbf{e}_\theta & \mathbf{e}_\theta \cdot \mathbf{e}_\phi \\ \mathbf{e}_\phi \cdot \mathbf{e}_\theta & \mathbf{e}_\phi \cdot \mathbf{e}_\phi \end{pmatrix} = \begin{pmatrix} R^2 & 0 \\ 0 & R^2 \sin^2\theta \end{pmatrix}.$$

5.3 **Coordinate transformation of the metric**   Given the transformation (5.18), we have the inverse matrix

$$\mathbf{R}^{-1} = \begin{pmatrix} (R \cos\theta)^{-1} & 0 \\ 0 & 1 \end{pmatrix}.$$

Using the metric transformation condition given in sidenote 11 of Chapter 3, we can obtain the metric for the cylindrical system from that of the polar system

$$\mathbf{R}^{-1T}\mathbf{gR}^{-1} = \mathbf{R}^{-1T} \begin{pmatrix} R^2 & 0 \\ 0 & R^2 \sin^2 \theta \end{pmatrix} \mathbf{R}^{-1}$$

$$= \begin{pmatrix} (\cos^2 \theta)^{-1} & 0 \\ 0 & R^2 \sin^2 \theta \end{pmatrix} = \begin{pmatrix} 1 - (\rho^2/R^2)^{-1} & 0 \\ 0 & \rho^2 \end{pmatrix}$$

$$= \mathbf{g}'.$$

5.4 Geodesics on simple surfaces

(a) **Flat plane** For this 2D space with Cartesian coordinates $(x^1, x^2) = (x, y)$, the metric $g_{ab} = \delta_{ab}$. The second term in the geodesic Eq. (5.30) vanishes, as well as the two components of the equation $d\dot{x}^\nu/d\lambda$ so that $\ddot{x} = 0$ and $\ddot{y} = 0$, which have respective solutions of $x = A + B\lambda$ and $y = C + D\lambda$. They can be combined as $y = \alpha + \beta x$, with (A, B, C, D) and (α, β) being constants. We recognize this as the equation for a straight line.

(b) **Spherical surface** For a 2-sphere, we choose the coordinates $(x^1, x^2) = (\theta, \phi)$ with a metric given by (5.13) For the θ component of the geodesic Eq. (5.30) is $\ddot{\theta} = \sin \theta \cos \theta \dot{\phi}^2$, the ϕ component equation, $2 \sin \theta \cos \theta \dot{\theta} \dot{\phi} + \sin^2 \theta \ddot{\phi} = 0$. Instead of working out the full parametric representation, we will just check that $\phi = $ constant and $\theta = \alpha + \beta \lambda$ solve these two equations. Clearly these solutions describe longitudinal great circles on the sphere.

5.5 Locally flat metric
The distance between two neighboring points can be rearranged by adding and subtracting a factor of $(g_{12}dx^2)^2/g_{11}$ so that

$$ds^2 = g_{11}(dx^1)^2 + 2g_{12}dx^1 dx^2 + g_{22}(dx^2)^2$$

$$= \left(\sqrt{g_{11}}dx^1 + \frac{g_{12}dx^2}{\sqrt{g_{11}}} \right)^2 + \left(g_{22} - \frac{g_{12}^2}{g_{11}} \right)(dx^2)^2.$$

The new coordinate system $(\bar{x}^1, \bar{x}^2)$ has the metric $\bar{g}_{ab} = \delta_{ab}$ because $ds^2 = (d\bar{x}^1)^2 + (d\bar{x}^2)^2$ where

$$d\bar{x}^1 = \sqrt{g_{11}}dx^1 + \frac{g_{12}dx^2}{\sqrt{g_{11}}}, \quad d\bar{x}^2 = \sqrt{g_{22} - \frac{g_{12}^2}{g_{11}}}dx^2.$$

On the other hand, if the original metric determinant is negative, $g_{11}g_{22} - g_{12}^2 < 0$, then $ds^2 = (d\bar{x}^1)^2 - (d\bar{x}^2)^2$ with

$$d\bar{x}^2 = \sqrt{\frac{g_{12}^2}{g_{11}} - g_{22}}dx^2.$$

5.7 3-sphere and 3-pseudosphere

(a) **3D flat space**

$$x = r \sin \theta \cos \phi, \quad y = r \sin \theta \sin \phi, \quad z = r \cos \theta.$$

The relation for the solid angle factor follows simply from the two expressions for the invariant separations in two coordinate systems:

$$ds^2 = dx^2 + dy^2 + dz^2 = dr^2 + r^2 d\Omega^2.$$

(b) **3-sphere** Given the metric for 3-sphere being

$$ds^2 = dr^2 + \left(R \sin \frac{r}{R}\right)^2 d\Omega^2, \tag{6}$$

the relation from part (a) $r^2 d\Omega^2 = dx^2 + dy^2 + dz^2 - dr^2$ suggests

$$\left(R \sin \frac{r}{R}\right)^2 d\Omega^2 = dX^2 + dY^2 + dZ^2 - \left[d\left(R \sin \frac{r}{R}\right)\right]^2.$$

Substituting this into (6), we have

$$ds^2 = dW^2 + dX^2 + dY^2 + dZ^2$$

where

$$dW^2 = dr^2 - \left[d\left(R \sin \frac{r}{R}\right)\right]^2 = \left[\sin \frac{r}{R} dr\right]^2$$

so we can identify $dW = \sin(r/R)dr$. This ds^2 invariant interval implies a Euclidian metric $g_{\mu\nu} = \mathrm{diag}(1, 1, 1, 1)$. Also, it suggests the embedding relation between (r, θ, ϕ) and (W, X, Y, Z) as

$$W = R \cos \frac{r}{R}, \quad X = \left(R \sin \frac{r}{R}\right) \sin \theta \cos \phi,$$

$$Y = \left(R \sin \frac{r}{R}\right) \sin \theta \sin \phi, \quad Z = \left(R \sin \frac{r}{R}\right) \cos \theta.$$

This set of relations lead immediately to the constraint $W^2 + X^2 + Y^2 + Z^2 = R^2$.

(c) **3-pseudosphere** With $W = R \cosh(r/R)$, the relations

$$W = R \cosh \frac{r}{R}, \quad X = \left(R \sinh \frac{r}{R}\right) \sin \theta \cos \phi,$$

$$Y = \left(R \sinh \frac{r}{R}\right) \sin \theta \sin \phi, \quad Z = \left(R \sinh \frac{r}{R}\right) \cos \theta,$$

lead, through the trigonometric relation $\cosh^2 \chi - \sinh^2 \chi = 1$ to

$$ds^2 = -dW^2 + dX^2 + dY^2 + dZ^2$$

thus a Minkowski metric $\eta_{\mu\nu} = \mathrm{diag}(-1, 1, 1, 1)$ and the condition

$$-W^2 + X^2 + Y^2 + Z^2 = -R^2.$$

5.8   **Volume of higher dimensional space**

$$dV = \sqrt{\det g} \prod_i dx^i. \tag{7}$$

(a) **3D flat space** For Cartesian coordinates $\sqrt{\det g} = 1$, (7) reduces to $dV = dx\,dy\,dz$, and for spherical coordinates $\sqrt{\det g} = r^2 \sin\theta$ and $dV = r^2 \sin\theta\,dr\,d\theta\,d\phi$.

(b) **3-sphere** From (7) we have $\sqrt{\det g} = R^2 \sin^2(r/R) \sin\theta$, thus the volume of a 3-sphere with radius R can be calculated:

$$R^2 \int_0^{\pi R} \sin^2 \frac{r}{R}\,dr \int_0^\pi \sin\theta\,d\theta \int_0^{2\pi} d\phi = 2\pi^2 R^3.$$

5.9 **Non-Euclidean relation between radius and circumference of a circle**

(a) **The case of a sphere** The radius of a circle being the displacement ds along a constant radial coordinate $(dr = 0)$, we have from either (5.48), (5.49) or (5.51), (5.52), $ds = R \sin(r/R)d\phi$. Thus, making a Taylor series expansion of the circumference $S = \int ds$, we have:

$$S = 2\pi R \sin\frac{r}{R} = 2\pi R \left(\frac{r}{R} - \frac{1}{3}\frac{r^3}{R^3} + \cdots \right)$$

$$= 2\pi r - \frac{1}{R^2}\frac{\pi r^3}{3} + \cdots$$

which is just the claimed result in (5.39) with $K = 1/R^2$.

(b) **The case of a pseudosphere:** For $k = -1$ surface, the displacement, according to either (5.48), (5.49) or (5.51), (5.52), is given by $ds = R \sinh(r/R)d\phi$, giving a circumference of $S = 2\pi R \sinh(r/R)$. Since the Taylor expansion of the hyperbolic sine differs from that for the sine function in the sign of the cubic term, again we obtain the result in agreement with (5.39) with $K = -1/R^2$. Thus, on a pseudospherical surface, the circumference of a circle with radius r is $S > 2\pi r$.

5.10 **Angular excess and polygon area** Any polygon is made up of triangles.

5.11 **Local Euclidean coordinates** From the chain rule of differentiation,

$$d\xi^1 = \frac{\partial \xi^1}{\partial x^1}dx^1 + \frac{\partial \xi^1}{\partial x^2}dx^2$$

$$d\xi^2 = \frac{\partial \xi^2}{\partial x^1}dx^1 + \frac{\partial \xi^2}{\partial x^2}dx^2.$$

Equate $ds^2 = (d\xi^1)^2 + (d\xi^2)^2 = g_{11}(dx^1)^2 + 2g_{12}(dx^1)(dx^2) + g_{22}(dx^2)^2$ we can immediately get the relation we are looking for:

$$g_{11} = \left(\frac{\partial \xi^1}{\partial x^1} \right)^2 + \left(\frac{\partial \xi^2}{\partial x^1} \right)^2$$

$$g_{12} = \left(\frac{\partial \xi^1}{\partial x^1} \right)\left(\frac{\partial \xi^1}{\partial x^2} \right) + \left(\frac{\partial \xi^2}{\partial x^1} \right)\left(\frac{\partial \xi^2}{\partial x^2} \right)$$

$$g_{22} = \left(\frac{\partial \xi^1}{\partial x^2} \right)^2 + \left(\frac{\partial \xi^2}{\partial x^2} \right)^2.$$

For the spherical polar coordinates $(x^1, x^2) = (\theta, \phi)$, we have $\xi^1 = R\theta$ and $\xi^2 = (R \sin\theta)\phi$ with their direction be along (θ, ϕ) so that $d\xi^1 = Rd\theta$ and $d\xi^2 = (R \sin\theta)d\phi$. In this way, we have

$$\left(\frac{\partial\xi^1}{\partial x^2}\right) = \left(\frac{\partial\xi^2}{\partial x^1}\right) = 0$$

to obtain the metric element $g_{12} = 0$ and

$$g_{11} = \left(\frac{\partial\xi^1}{\partial x^1}\right)^2 = R^2, \quad g_{22} = \left(\frac{\partial\xi^2}{\partial x^2}\right)^2 = R^2 \sin^2\theta.$$

6.2 Spatial distance and spacetime metric

The spacetime separation vanishes $(ds^2 = 0)$ for a light pulse:

$$g_{00}\left(dx^0\right)^2 + 2g_{0i}dx^i dx^0 + g_{ij}dx^i dx^j = 0.$$

Solving this quadratic equation for the coordinate time interval that takes the pulse going from A to B

$$dx^0_{AB} = -\frac{g_{0i}dx^i}{g_{00}} - \frac{\sqrt{(g_{0i}g_{0j} - g_{00}g_{ij})dx^i dx^j}}{g_{00}}$$

and the time for it to go from B to A (involving the change of $dx^i \to -dx^i$)

$$dx^0_{BA} = +\frac{g_{0i}dx^i}{g_{00}} - \frac{\sqrt{(g_{0i}g_{0j} - g_{00}g_{ij})dx^i dx^j}}{g_{00}}.$$

Therefore the total coordinate time

$$dx^0 = dx^0_{AB} + dx^0_{BA},$$

which is related to the proper time interval $d\tau_A$ (see Problem 6.1), hence the spatial distance dl,

$$dl \equiv \frac{cd\tau_A}{2} = \frac{\sqrt{-g_{00}}dx^0}{2} = \sqrt{\left(g_{ij} - \frac{g_{0i}g_{0j}}{g_{00}}\right)dx^i dx^j}.$$

Since $dl^2 = \gamma_{ij}dx^i dx^j$, we have $\gamma_{ij} = g_{ij} - (g_{0i}g_{0j}g_{00})$. Thus, $\gamma_{ij} \neq g_{ij}$ when $g_{0i} \neq 0$.

6.3 Non-Euclidean geometry of a rotating cylinder

Let us denote the spatial coordinates as follows:

$$(ct, r, \phi, z) \quad \text{lab observer,}$$

$$(ct, r_0, \phi_0, z) \quad \text{observer on the rotating disk.}$$

They are related by (see Fig. 6.1)

$$r = r_0, \quad \phi = \phi_0 + \omega t.$$

We shall ignore the vertical coordinate z below.

The line element written in terms coordinates at rest with respect to the observer on the rotating disk is, see (5.33)

$$ds^2 = -c^2 dt^2 + dr_0^2 + r_0^2 d\phi_0^2,$$

which can be written in terms of the lab coordinate (see Cook, 2004) by substituting in $d\phi_0 = d\phi - \omega dt$:

$$ds^2 = -\left[1 - \left(\frac{\omega r}{c}\right)^2\right] c^2 dt^2 + dr^2 + r^2 d\phi^2 - 2\omega r^2 dt d\phi.$$

The metric with respect to the (ct, r, ϕ) coordinates thus has elements

$$g_{00} = -\left[1 - \left(\frac{\omega r}{c}\right)^2\right], \quad g_{rr} = 1, \quad g_{\phi\phi} = r^2, \quad g_{0\phi} = -\frac{\omega r^2}{c}.$$

From Problem 6.2, we have the spatial distance

$$dl^2 = \left(g_{ij} - \frac{g_{0i} g_{0j}}{g_{00}}\right) dx^i dx^j = dr^2 + \frac{r^2 d\phi^2}{1 - (\omega r/c)^2}$$

showing clearly length contraction of the circumference, but not the radius.

6.5 **The geodesic equation and light deflection** The geodesic equation (6.9), after using $p^\mu = dx^\mu/d\lambda$, can be written as

$$\frac{d}{d\lambda} p^\mu + \Gamma^\mu_{\nu\sigma} p^\nu \frac{dx^\sigma}{d\lambda} = 0$$

or equivalently

$$dp^\mu = -\Gamma^\mu_{\nu\sigma} p^\nu dx^\sigma.$$

We are interested in the $\mu = 2$ component $p^2 \equiv p_y$:

$$dp_y = -\Gamma^2_{00} p^0 dx^0 - \Gamma^2_{11} p^1 dx^1 - \Gamma^2_{10} p^1 dx^0 - \Gamma^2_{01} p^0 dx^1$$
$$= -(\Gamma^2_{00} + \Gamma^2_{11} + 2\Gamma^2_{10}) p \, dx, \tag{8}$$

where we have used $dx^\mu = (dx, dx, 0, 0)$ and $p^\mu = (p, p, 0, 0)$. Christoffel symbols can be calculated by (6.10). Since we are working in the weak-field approximation, that is, the metric is very close to being the flat space Minkowski metric $\eta_{\mu\nu} = \text{diag}(-1, 1, 1, 1)$, and the Christoffel symbols (being the derivatives of the metric) must also be small. Thus the metric on the left-hand side (LHS) of (6.10) can be taken to be $\eta_{\mu\nu}$, which is diagonal. Consider the LHS component $\eta_{2\sigma} \Gamma^\sigma_{\mu\nu} = \Gamma^\sigma_{\mu\nu}$:

$$\Gamma^2_{\mu\nu} = \frac{1}{2} \left[\frac{\partial g_{\mu 2}}{\partial x^\nu} + \frac{\partial g_{\nu 2}}{\partial x^\mu} - \frac{\partial g_{\mu\nu}}{\partial x^2}\right].$$

If the only position-dependent metric element is $g_{00} = -1 - \Phi/c^2$ (as suggested by EP physics), then the only nonzero term on the RHS is the $\partial g_{00}/\partial x^2$.

That is,

$$\Gamma^2_{00} = \frac{-1}{c^2} \frac{\partial \Phi}{\partial y}$$

and Eq. (8) reduces to $dp_y = -\Gamma^2_{00} p \, dx$. This way we get

$$\delta\phi_{EP} = \int \frac{dp_y}{p} = \frac{1}{c^2} \int \frac{\partial \Phi}{\partial y} dx \qquad (9)$$

which is the result obtained by Huygens' principle in Eq. (4.44). For the argument that the GR value is twice that of the EP value, see Sections 7.2.1 and 7.3.2.

6.7   **The matrix for tidal forces is traceless**   We can take the trace of the tidal force matrix by contracting the two indices with the Kronecker delta:

$$\delta_{ij} \frac{\partial^2 \Phi}{\partial x^i \partial x^j} = \frac{\partial}{\partial x^i} \frac{\partial}{\partial x^i} \Phi = \nabla^2 \Phi.$$

Since the mass density vanishes ($\rho = 0$) at any field point away from the source, the Newtonian field equation (6.5) informs us that the gravitational potential satisfies the Laplace equation $\nabla^2 \Phi = 0$.

6.8   **G_N as a conversion factor**   One easily finds that this yields the dimension relation (curvature) = (length)$^{-2}$. This is consistent with the fact that curvature is the second derivative of the metric, which is dimensionless.

7.1   **Energy relation for a particle moving in the Schwarzschild spacetime**   Equation (7.44) with $r^* = 0$ is

$$-c^2 \left(\frac{dt}{d\tau}\right)^2 + \left(\frac{dr}{d\tau}\right)^2 + r^2 \left(\frac{d\phi}{d\tau}\right)^2 = -c^2,$$

where τ is the proper time $d/d\tau = \gamma d/dt$ with $\gamma = (1 - v^2/c^2)^{-\frac{1}{2}}$. Multiplying a factor of $-m^2 c^2$ on both sides, we obtain

$$\gamma^2 m^2 \left[c^4 - c^2 v^2\right] = m^2 c^4,$$

where $v^2 = (dr/dt)^2 + r^2 (d\phi/dt)^2$ is the velocity (squared) in the spherical coordinate system (r, θ, ϕ) when the polar angle θ is fixed. We recognize this is the energy–momentum relation $E^2 = p^2 c^2 + m^2 c^4$ after identifying the relativistic expression for energy $E = \gamma m c^2$ and momentum $p = \gamma m v$.

7.2   **Gravitational red shift via energy conservation**   The frequency ratio being

$$\frac{\omega_{em}}{\omega_{rec}} = \frac{\left(g_{00} p^0\right)_{em}}{\left(g_{00} p^0\right)_{rec}} \frac{U^0_{em}}{U^0_{rec}} = \frac{U^0_{em}}{U^0_{rec}},$$

where to reach the last equality we have used the energy conservation relation of $g_{\mu\nu} p^\mu K^\nu_{(t)em} = g_{\mu\nu} p^\mu K^\nu_{(t)rec}$ with the Killing vector $K^\mu_{(t)} = (1, 0, 0, 0)$. The invariance of the 4-velocity squared $(g_{00} U^0 U^0)_{em} = (g_{00} U^0 U^0)_{rec}$ leads to the desired result of

$$\frac{\omega_{em}}{\omega_{rec}} = \frac{U^0_{em}}{U^0_{rec}} = \sqrt{\frac{(g_{00})_{rec}}{(g_{00})_{em}}},$$

which can be translated into the standard expression of $(\Delta\omega)/\omega = -(\Delta\Phi)/c^2$ as done in Section 6.2.2.

7.3 **Light deflection via the geodesic equation**   The $L = ds^2/d\lambda^2 = 0$ equation in the form of (7.77) can be written slightly differently as

$$\left(\frac{dr}{d\lambda}\right)^2 + \left(1 - \frac{r^*}{r}\right)\frac{\lambda^2}{r^2} = \kappa^2.$$

After the usual change of variables $(r, \lambda) \to (u, \phi)$, we have

$$u'' + u - \epsilon u^2 = 0.$$

For a perturbative solution of $u = u_0 + \epsilon u_1$ with $\epsilon = O(r^*)$,

$$\left(u_0'' + u_0\right) + \epsilon\left(u_1'' + u_1 - u_0^2\right) + \cdots = 0.$$

The zeroth order, being a "simple harmonic oscillator" equation, has the solution $u_0 = r_{min}^{-1}\sin\phi$. To solve the first order equation

$$\frac{d^2u_1}{d\phi^2} + u_1 = \frac{1 - \cos 2\phi}{2r_{min}^2}$$

one tries $u_1 = \alpha + \beta\cos 2\phi$, and finds $\alpha = (2r_{min}^2)^{-1}$ and $\beta = (6r_{min}^2)^{-1}$. Putting the zeroth and first order terms together we get

$$\frac{1}{r} = \frac{\sin\phi}{r_{min}} + \frac{3 + \cos 2\phi}{4}\frac{r^*}{r_{min}^2}.$$

In the absence of gravity ($r^* = 0$), the asymptotes ($r = \mp\infty$) corresponds to $\phi_{-\infty} = \pi$ and $\phi_{+\infty} = 0$, and the trajectory is straight line (no deflection). When gravity is turned on, there is an angular deflection $\delta\phi = (\phi_{-\infty} - \phi_{+\infty} - \pi)$. Picking our coordinates so that $\phi_{-\infty} = \pi + \delta\phi/2$ and $\phi_{+\infty} = -\delta\phi/2$ and the trajectory equations yields (for either asymptote):

$$0 = -\sin\frac{\delta\phi}{2} + \frac{3 + \cos\delta\phi}{4}\frac{r^*}{r_{min}}.$$

For small deflection angle $\delta\phi$ we have $0 = -\delta\phi/2 + r^*/r_{min}$; we obtain the result of $\delta\phi_{GR} = 2r^*/r_{min}$.

7.5 **Total energy in curved spacetime**   We can check this claim by showing that starting with $\kappa/c = E/mc^2$ one can deduce $\mathcal{E} = E - mc^2$ in the NR limit. From the definition, we have

$$\mathcal{E} \equiv m\left(\kappa^2 - c^2\right)/2 = \frac{1}{2}mc^2\left[\left(\frac{E}{mc^2}\right)^2 - 1\right].$$

The NR total energy e_{NR} is defined in the NR limit by $E = mc^2 + e_{NR}$ with $mc^2 \gg e_{NR}$. Thus, the above equation does turn into $\mathcal{E} = e_{NR}$ in the NR limit.

7.6 Details for time-delay calculation

We substitute in the expansion of b given in (7.80) and expand

$$
c\frac{dt}{dr} = \left(1 - \frac{r^*}{r}\right)^{-1}\left[1 - \frac{b^2}{r^2}\left(1 - \frac{r^*}{r}\right)\right]^{-1/2}
$$

$$
\simeq \left(1 + \frac{r^*}{r}\right)\left[1 - \left(\frac{r_0^2}{r^2} + \frac{r_0 r^*}{r^2}\right)\left(1 - \frac{r^*}{r}\right)\right]^{-1/2}
$$

$$
\simeq \left(1 + \frac{r^*}{r}\right)\left[1 - \frac{r_0^2}{r^2} - \frac{r_0 r^*}{r^2} + \frac{r_0^2 r^*}{r^3}\right]^{-1/2}
$$

$$
\simeq \left(1 + \frac{r^*}{r}\right)\left(1 - \frac{r_0^2}{r^2}\right)^{-1/2}\left[1 + \frac{1}{2}\frac{\frac{r_0 r^*}{r^2}}{1 + \frac{r_0}{r}}\right]
$$

$$
\simeq \left(1 - \frac{r_0^2}{r^2}\right)^{-1/2}\left[1 + \frac{r^*}{r}\left(1 + \frac{\frac{1}{2}r_0}{r + r_0}\right)\right]
$$

$$
= \left(1 - \frac{r_0^2}{r^2}\right)^{-1/2}\left[1 + \frac{r^*}{r}\frac{r + \frac{3}{2}r_0}{r + r_0}\right].
$$

7.7 4-velocity of a particle in a circular orbit

(a) The orbit equation (7.59) for the variable $u \equiv 1/r$ has $u' = 0$ corresponding to a circular orbit case: $u^2 - (r^*c^2/\lambda^2)u - r^*u^3 = $ constant, with $\lambda = l/m = r^2 d\phi/d\tau$. If we differentiate this equation, we have $(r^*c^2/\lambda^2) = 2u - 3r^*u^2$. Putting in the specific value $u = 1/R$, it implies

$$
R^4\left(\frac{d\phi}{d\tau}\right)^2 = \lambda^2 = \frac{r^*c^2 R}{2}\left(1 - \frac{3r^*}{2R}\right)^{-1}
$$

or

$$
\left(U^\phi\right)^2 = \left(\frac{d\phi}{d\tau}\right)^2 = \frac{r^*c^2}{2R^3}\left(1 - \frac{3r^*}{2R}\right)^{-1}. \tag{10}
$$

As worked out in Problem (8.5), we can deduce from the energy balance equation (7.52) the conserved time-component of the 4-velocity

$$
\left(U^t\right)^2 = \left(\frac{dt}{d\tau}\right)^2 = \left(1 - \frac{3r^*}{2R}\right)^{-1}. \tag{11}
$$

(b) From these two equations (10) and (11), we can easily work out the orbital frequency

$$
\Omega^2 \equiv \left(\frac{d\phi}{dt}\right)^2 = \left(\frac{d\phi}{d\tau}\right)^2\left(\frac{d\tau}{dt}\right)^2
$$

$$
= \frac{r^*c^2}{2R^3} = \frac{G_N M}{R^3},
$$

which is Kepler's third law. We also note trivially that, for this simple kinematics, the proportionality constant for U^ϕ and U^t is just this orbital frequency: $U^\phi = \Omega U^t$.

(c) Finally we perform the consistency check for the invariant length of the 4-vector. For the particle in the circular orbit ($r = R$) in the equatorial plane ($\theta = \pi/2$), the other two components of the velocity vanihse $U^r = U^\theta = 0$, and at the trajectory the metric elements have values $g_{00} = -(1 - \frac{r^*}{R})$ and $g_{\phi\phi} = R^2$. We see that the 4-velocity indeed has the correct invariant length:

$$g_{\alpha\beta} U^\alpha U^\beta = g_{00} \left(U^0\right)^2 + g_{\phi\phi} \left(U^\phi\right)^2$$

$$= c^2 \left(-1 + \frac{r^*}{R} + \frac{r^*}{2R}\right) \left(1 - \frac{3r^*}{2R}\right)^{-1} = -c^2.$$

8.2 **The coordinate time across an event horizon** In the region outside the Schwarzschild surface, in order to find the full expression of the Schwarzschild coordinate time as a function of the radial distance, one can integrate (from r_0 to r) the equation in (8.5) to obtain

$$t = t_0 - \frac{2r^*}{3c} \left[\left(\frac{r}{r^*}\right)^{\frac{3}{2}} - \left(\frac{r_0}{r^*}\right)^{\frac{3}{2}}\right]$$

$$+ \frac{r^*}{c} \left\{\ln \left|\frac{\sqrt{r/r^*} + 1}{\sqrt{r/r^*} - 1} \cdot \frac{\sqrt{r_0/r^*} - 1}{\sqrt{r_0/r^*} + 1}\right| - 2 \left[\left(\frac{r}{r^*}\right)^{\frac{1}{2}} - \left(\frac{r_0}{r^*}\right)^{\frac{1}{2}}\right]\right\}. \quad (12)$$

In the limit of r and r_0 are much greater than r^*, the coordinate time of (12) approaches the proper time of (8.4) as it should. In order to study the limit of $r \to r^*$, we note that the above logarithmic term can be written as

$$\ln \left|\frac{\sqrt{r} + \sqrt{r^*}}{\sqrt{r} - \sqrt{r^*}} \cdot \frac{\sqrt{r_0} - \sqrt{r^*}}{\sqrt{r_0} + \sqrt{r^*}}\right| = \ln \left|\frac{\left(\sqrt{r} + \sqrt{r^*}\right)^2}{r - r^*} \cdot \frac{r_0 - r^*}{\left(\sqrt{r_0} + \sqrt{r^*}\right)^2}\right|.$$

When r is near r^*, we can drop all non-singular terms in (12) to recover the result shown in (8.4).

8.3 **Null 3-surface** We are discussing a $dr = 0$ surface. The light-like condition in the EF coordinates (*see* Table 8.1) is $ds^2 = (1 - r^*/r)dp^2 = 0$. The tangent discussed in the text holds for $d\theta = d\phi = 0$. Thus we can certainly pick the other two tangents one with $d\theta \neq 0$ and the other with $d\phi \neq 0$. With $t^\mu \equiv (t^p, t^r, t^\theta, t^\phi)$, the infinitesimal tangents $t_1^\mu = (dp, 0, 0, 0) = n^\mu$, $t_2^\mu = (dp, 0, d\theta, 0)$, $t_1^\mu = (dp, 0, 0, d\phi)$ are mutually orthogonal, with $\mathbf{t}_1 = \mathbf{n}$ being the normal to the surface as well: $g_{\mu\nu} t_1^\mu t_1^\nu = g_{\mu\nu} t_1^\mu t_2^\nu = g_{\mu\nu} t_1^\mu t_3^\nu = g_{\mu\nu} t_2^\mu t_3^\nu = 0$ because $g_{pp} = -(1 - r^*/r)$.

8.4 **Kruskal coordinates** Start from the definition $V = (p' + q')/2$, which in turn can be expressed in terms of $(\bar{t}, r)$ and $(\tilde{t}, r)$ through (8.25)

$$V = \frac{1}{2} \left[\exp\left(p/2r^*\right) - \exp\left(-q/2r^*\right)\right]$$

$$= \frac{1}{2} \left[\exp\left(\frac{\bar{t} + r}{2r^*}\right) - \exp\left(-\frac{\tilde{t} - r}{2r^*}\right)\right]. \quad (13)$$

From Table 18.1 we have

$$\frac{\bar{t}+r}{2r^*} = \frac{1}{2r^*}\left[ct + r^* \ln\left|\frac{r-r^*}{r^*}\right| + r\right],$$

$$-\frac{\bar{t}-r}{2r^*} = \frac{1}{2r^*}\left[-ct + r^* \ln\left|\frac{r-r^*}{r^*}\right| + r\right].$$

Since

$$\exp\left(\frac{1}{2}\ln\left|\frac{r-r^*}{r^*}\right|\right) = \left(\frac{r}{r^*} - 1\right)^{1/2} \quad \text{for} \quad r > r^* \tag{14}$$

it then follows from (13) that

$$V = \left(\frac{r}{r^*} - 1\right)^{1/2} e^{r/2r^*} \frac{e^{ct/2r^*} - e^{-ct/2r^*}}{2}$$

$$= \left(\frac{r}{r^*} - 1\right)^{1/2} e^{r/2r^*} \sinh\left(\frac{ct}{2r^*}\right)$$

which is the result quoted in Eq. (8.30). The expression for $U(t, r)$ can be similarly derived. As for the regime of $r < r^*$, it just changes the sign on the RHS of (14).

8.5   **Circular orbits**   For a circular orbit, the radial distance and the orbital angular momentum must satisfy a definite relation so that the effective potential (8.35)

$$\Phi_{\text{eff}} = -\frac{c^2 r^*}{2r} + \frac{l^2}{2m^2 r^2} - \frac{l^2 r^*}{2m^2 r^3}$$

is minimized (at this radial distance) $\partial\Phi_{\text{eff}}/\partial r = 0$, which fixes the angular momentum to be

$$l^2 = G_N M m^2 r \left(1 - \frac{3}{2}\frac{r^*}{r}\right)^{-1}. \tag{15}$$

Furthermore, $\dot{r} = 0$ for circular orbit, the total energy must equal the potential energy:

$$\mathcal{E} = m\Phi_{\text{eff}}$$

or, using the suggested form for Φ_{eff} and the relation $\mathcal{E}/m = (\kappa^2 - c^2)/2$ the total energy may be written as

$$\frac{\kappa^2 - c^2}{2} = \frac{c^2}{2}\left[\left(1 - \frac{r^*}{r}\right)\left(1 + \frac{l^2}{m^2 r^2 c^2}\right) - 1\right].$$

After plugging the result in (15), one finds

$$\kappa^2 = c^2 \left(1 - \frac{r^*}{r}\right)^2 \left(1 - \frac{3}{2}\frac{r^*}{r}\right)^{-1}. \tag{16}$$

From this we immediately see that at $r_0 = 3r^*$

$$[E(\infty)]_0 = mc\kappa_0 = \sqrt{\frac{8}{9}}\, mc^2.$$

8.6 **No stable circular orbit for light** Equation (7.77) can be written as $\kappa^2/2 = \dot{r}^2/2 + \Phi_{\text{eff}}$ with $\Phi_{\text{eff}}(r) = (\lambda^2/2r^2)(1 - (r^*/r))$. From this one can easily show an extremum is at $r_0 = 3r^*/2$ which is unstable as $\partial^2\Phi_{\text{eff}}/\partial r^2{}_{r_0} = -2r_0^4 < 0$.

8.7 **No counter-rotating light is possible in ergosphere** The Kerr metric for an extreme spinning black hole for $dr = 0$ at $\theta = \pi/2$ is

$$ds^2 = -\left(1 - \frac{r^*}{r}\right)c^2dt^2 - \frac{r^{*2}}{r}cdtd\phi + \bar{r}^2d\phi^2$$

where

$$\bar{r}^2 = r^2 + \frac{r^{*2}}{4} + \frac{r^{*3}}{4r}.$$

Thus the null interval $ds^2 = 0$ for a light ray obey is a quadratic equation for the angular velocity

$$\frac{\bar{r}^2}{c^2}\left(\frac{d\phi}{dt}\right)^2 - \frac{r^{*2}}{cr}\left(\frac{d\phi}{dt}\right) - \left(1 - \frac{r^*}{r}\right) = 0$$

with the solutions

$$\frac{d\phi}{dt} = \frac{cr^{*2}}{2r\bar{r}^2}\left[1 \pm \sqrt{1 + \left(\frac{2r\bar{r}^2}{cr^{*2}}\right)^2 \frac{c^2}{\bar{r}^2}\left(1 - \frac{r^*}{r}\right)}\right].$$

At the stationary limit surface (i.e. the outer boundary of the ergosphere) $r = r_S = r^*$ thus $\bar{r}^2 = (3/2)\,r^{*2}$,

$$\left[\frac{d\phi}{dt}\right]_S = \begin{cases} \frac{2c}{3r^*} & \text{co-rotating light} \\ 0 & \text{counter-rotating light.} \end{cases}$$

That is, because of frame dragging the angular velocity of counter-rotating light vanishes. At the horizon surface (i.e., the inner boundary of the ergosphere) $r = r_h = r^*/2$ thus $\bar{r} = r^*$,

$$\left[\frac{d\phi}{dt}\right]_h = \frac{c}{r^*}$$

for both co-rotating and counter-rotating light rays. Namely, inside the ergosphere, because of frame dragging, light can only rotate in the same direction as the source.

8.8   **Circulating light at the horizon**   On the equatorial plane $\theta = \pi/2$ (hence $\rho = r$), the event horizon $r = r_+$ of a Kerr black hole corresponds to the $\Delta = r_+^2 - r_+ r^* + a^2 = 0$, or

$$r_+^2 \left(1 - \frac{r^*}{r_+} \right) = -a^2.$$

Consequently, metric elements at r_+ have values of

$$g_{tt} = -\left(1 - \frac{r^*}{r_+} \right) c^2, \quad g_{\phi\phi} = r_+^2 + a^2 \left(1 + \frac{r^*}{r_+} \right), \quad g_{t\phi} = -\frac{r^*}{r_+} ac.$$

Thus,

$$g_{tt} g_{\phi\phi} = -r_+^2 \left(1 - \frac{r^*}{r_+} \right) - \left(1 - \frac{r^{*2}}{r_+^2} \right) a^2 c^2$$

$$= a^2 c^2 - (1 - \frac{r^{*2}}{r_+^2}) a^2 c^2 = g_{t\phi}^2.$$

With $g_{t\phi}^2 = g_{tt} g_{\phi\phi}$, it is clear that Eq. (8.45) has only one solution $d\phi/dt = \omega$ of (8.43) for both co-rotating and counter-rotating lights. For the extreme spinning case $r_+ = a = r^*/2$

$$\omega = -\frac{g_{t\phi}}{g_{\phi\phi}} = -\frac{-r^* c}{r^{*2}} = \frac{c}{r^*},$$

which agrees with the result shown in Problem 8.7.

8.9   **Binding energy of a particle in ISCO of a rotating black hole**   Given the effective potential $\left[\Phi_{\text{eff}}^{(K)} \right]$ as in (8.71) the radius can be found by the condition of $\partial \left[\Phi_{\text{eff}}^{(K)} \right] / \partial r = 0$ to yield $r^2 - 2Br + C = 0$ with

$$B = \frac{l^2 - a^2 m^2 (\kappa^2 - c^2)}{c^2 m^2 r^*}, \quad C = \frac{3 [l - am\kappa]^2}{c^2}.$$

The quadratic equation has the solutions $r = B \pm \sqrt{B^2 - C}$. The inner most stable orbit corresponds the case of vanishing square root, $B^2 = C$:

$$\frac{l_0^2 - a^2 m^2 (\kappa_0^2 - c^2)}{c^2 m^2 r^*} = \frac{\sqrt{3}}{c} \left[l_0 - am\kappa_0 \right] \tag{17}$$

and an orbit radius of

$$r_0 = \frac{\sqrt{3}}{mc} \left[l_0 - am\kappa_0 \right]. \tag{18}$$

For such orbit, the energy balance equation (as $\dot{r} = 0$) becomes

$$\frac{\kappa_0^2 - c^2}{2} = \left[\Phi_{\text{eff}}^{(K)} \right]_0 = -\frac{c^2 r^*}{2r_0} + \frac{l_0^2 - a^2 m^2 \left(\kappa_0^2 - c^2 \right)}{2m^2 r_0^2} - \frac{r^* (l_0 - am\kappa_0)^2}{2m^2 r_0^3}. \tag{19}$$

We thus have three equations (17), (18), and (19) for the three unknowns of (r_0, l_0, κ_0). It is straightforward exercise to check that $r_0 = r^*/2$, $l_0 = mcr^*/\sqrt{3}$, $\kappa_0 = c/\sqrt{3}$ satisfy these equations for an extreme spinning black hole $(a = r^*/2)$. This is the result of an extraordinary binding energy of $0.42 \, mc^2$.

9.2 **Luminosity distance to the nearest star**   The observed flux being $f = \mathfrak{L}/4\pi d^2$, we have

$$d_* = \left(\frac{f_\odot}{f_*}\right)^{1/2} \times \mathrm{AU} = 3 \times 10^5 \, \mathrm{AU} = 1.5 \, \mathrm{pc}.$$

9.3 **Gravitational frequency shift contribution to the Hubble redshift**   The gravitational redshift being given by (4.26), we can estimate to be

$$z_\mathrm{G} = \frac{M_\mathrm{G}}{M_\odot} \frac{R_\odot}{R_\mathrm{G}} z_\odot = O(10^{-7}),$$

Thus, the shift due to gravity is quite negligible.

9.4 **Energy content due to star light**   Let us denote the average stellar luminosity by $\mathfrak{L}_*$ and star number density by n. Their product is then the luminosity density as given by (9.19),

$$n\mathfrak{L}_* = 2 \times 10^8 \frac{\mathfrak{L}_\odot}{(\mathrm{Mpc})^3} = 2.6 \times 10^{-33} \, \mathrm{W \, m^{-3}},$$

which is the energy emitted per unit volume per unit time. Stars have been assumed to be emitting light at this luminosity during the entire $t_0 \simeq t_\mathrm{H} \simeq 13.6$ Gyr $= 4.3 \times 10^{17}$ s, leading to an energy density contribution at present of $\rho_* c^2 = n\mathfrak{L}_* t_\mathrm{H} \simeq 10^{-15} \, \mathrm{J \, m^{-3}}$ or, using (9.17), a density ratio $\Omega_* = (\rho_*/\rho_\mathrm{c}) \simeq 10^{-5}$.

9.5 **Night sky as bright as day**   Flux being in watts per unit area, the total flux due to all the starlights is, according to (9.2) and Problem 9.4,

$$f_* = (n\mathfrak{L}_*) \, ct_H \simeq \left(2 \times 10^8 \frac{\mathfrak{L}_\odot}{\mathrm{Mpc}^3}\right) ct_H$$

$$= 0.8 \times 10^{12} \frac{\mathfrak{L}_\odot}{\mathrm{Mpc}^2} = 2.5 \times 10^{-10} \frac{\mathfrak{L}_\odot}{4\pi (\mathrm{AU})^2}.$$

Thus, we need to lengthen the age by a factor of 4 billion before we can get a night sky as bright as day!

9.6 **The Virial theorem**   Time derivative of the virial yields (for notational simplicity we drop subscript n and the summation sign)

$$\frac{dG}{dt} \equiv \mathbf{p} \cdot \dot{\mathbf{r}} + \dot{\mathbf{p}} \cdot \mathbf{r} = mv^2 + \mathbf{F} \cdot \mathbf{r} = 2T - \frac{\partial V}{\partial r} r = 2T + V$$

where, to reach the last expression, we have used the r dependence of the gravitational potential $V = ar^{-1}$. We now time average $\langle .. \rangle = \tau^{-1} \int dt$ this equation; the LHS becomes

$$\left\langle \frac{dG}{dt} \right\rangle = \frac{1}{\tau} \int_0^\tau dG = \frac{1}{\tau} \left[G(\tau) - G(0) \right]$$

which vanishes for a periodic system. In this way we obtain the virial theorem of $2\langle T \rangle + \langle V \rangle = 0$.

9.8   **Wavelength in an expanding universe**   A radial light signal follows the null worldline in the RW geometry and its proper distance is given by (9.45). Consider two successive wavecrests with wavelength λ; the second one is emitted (and recieved) later by a time interval $\delta t = \lambda/c$. Both wavecrests travel the same distance $d_p(\xi, t_0)$:

$$\left(d_p = \right) \quad \int_{t_{em}}^{t_0} \frac{c\,dt}{a(t)} = \int_{t_{em}+\lambda_{em}/c}^{t_0+\lambda_0/c} \frac{c\,dt}{a(t)}.$$

After cancelling out the common interval from $t_{em} + \lambda_{em}/c$ to t_0 from both sides of the integral equality, we have

$$\int_{t_{em}}^{t_{em}+\lambda_{em}/c} \frac{c\,dt}{a(t)} = \int_{t_0}^{t_0+\lambda_0/c} \frac{c\,dt}{a(t)}.$$

Since the scale factor would not have changed much during the small time interval between these two crests

$$\frac{1}{a(t_{em})} \int_{t_{em}}^{t_{em}+\lambda_{em}/c} dt = \frac{1}{a(t_0)} \int_{t_0}^{t_0+\lambda_0/c} dt$$

which immediately leads to the expected result of $(\lambda_0/\lambda_{em}) = a(t_0)/a(t_{em})$.

9.9   **The deceleration parameter and Taylor expansion of the scale factor**

$$a(t) \simeq a(t_0) + (t - t_0)\dot{a}(t_0) + \frac{1}{2}(t - t_0)^2 \ddot{a}(t_0)$$

$$= 1 + (t - t_0)H_0 - \frac{1}{2}(t - t_0)^2 q_0 H_0^2 \tag{20}$$

and

$$\frac{1}{a(t)} \simeq 1 - (t - t_0)H_0 + (t - t_0)^2 \left(1 + \frac{q_0}{2}\right) H_0^2. \tag{21}$$

9.10   **The steady-state universe**

(a) "Perfect CP" means that the universe is not only homogeneous in space but also in time.

(b) From (9.43) we have $da/dt = H_0 a$, which has the solution $a(t) = \exp H_0(t - t_0)$. Thus $\dot{a} = H_0 a$ and $\ddot{a} = H_0^2 a$ so that $q_0 = -1$.

(c) According to (5.46), the curvature for the 3D space in the Steady-State Universe (SSU) is $K = kR^{-2}(t)$. Since the scale factor does depend on t, an unchanging K can come about only for the curvature signature $k = 0$. Namely, an SSU requires a 3D space with a flat geometry.

(d) For a constant density, the rate of mass increase must be proportional to that of volume increase

$$\frac{dM}{dt} = \frac{dM}{dV}\frac{dV}{dt} = \rho_M \frac{dV}{dt},$$

while the normalized volume increase rate can be directly related to the Hubble constant

$$\frac{\dot{V}}{V} = \frac{3\dot{a}}{a} = 3H.$$

The mass creation rate per unit volume can then be calculated

$$\frac{\dot{M}}{V} = \rho_M \frac{\dot{V}}{V} = 3H_0\rho_M \simeq 0.7 \times 10^{-24} \text{ g/year/km}^3.$$

Given that $m_p = 1.7 \times 10^{-24}$ g, this means the creation of one hydrogen atom, in a cubic kilometer volume, every 2–3 years.

9.11 z^2 **correction to the Hubble relation**

(a) From (9.45)

$$d_p(t_0) = a(t_0) \int_{t_{em}}^{t_0} \frac{c\,dt}{a(t)}$$

and the first two terms of the Taylor series (21) we have

$$d_p(t_0) = c(t_0 - t_{em}) + \frac{c}{2}H_0(t_0 - t_{em})^2. \tag{22}$$

The first term on the RHS is just the distance traversed by a light signal in a static environment; the second term represents the correction due to the expansion of the universe.

(b) $(t_0 - t_{em})$ can be related to the redshift z through (9.50) and (21):

$$z = -1 + \frac{1}{a(t_{em})} = (t_0 - t_{em})H_0\left[1 + (t_0 - t_{em})H_0\left(1 + \frac{q_0}{2}\right)\right]. \tag{23}$$

(c) Equation (23) can be inverted to yield

$$t_0 - t_{em} \simeq \frac{z}{H_0}\left[1 - (t_0 - t_{em})H_0\left(1 + \frac{q_0}{2}\right)\right]$$

$$\simeq \frac{z}{H_0}\left[1 - z\left(1 + \frac{q_0}{2}\right)\right]. \tag{24}$$

Plug this expression for the look-back time into (22), we have

$$D_p(t_0) \simeq \left[\frac{cz}{H_0} - \frac{cz^2}{H_0}\left(1 + \frac{q_0}{2}\right)\right] + \frac{cz^2}{2H_0}$$

$$= \frac{cz}{H_0}\left(1 - \frac{1 + q_0}{2}z\right).$$

10.2   **Newtonian interpretation of second Friedmann equation**   For the pressureless matter used for our Newtonian system, cf. Fig. 10.1, the gravitational attraction by the whole sphere being $-G_N M/r^2 = \ddot{r}$, or $-(4\pi/3)G_N a\rho = \ddot{a}$, which is just Eq. (10.2) without the pressure term.

10.4   **The empty universe**   The nontrivial solution to $\dot{a}^2 = -kc^2 R_0^{-2}$ is a negatively curved open universe $k = -1$ and $a = t/t_0$, with $t_0 = c^{-1}R_0$, which is just the straight-line $a(t)$ in Fig. 10.2. From (9.45) we can obtain the proper distance in terms of z.

$$d_p(t_0) = \int_{t_{em}}^{t_0} \frac{cdt}{a(t)} = ct_0 \int_{t_{em}}^{t_0} t^{-1} dt = ct_0 \ln\left(\frac{t_0}{t_{em}}\right) = ct_0 \ln(1+z),$$

where we have used $t_0/t_{em} = (a(t_{em}))^{-1} = (1+z)$. It is clear that for small redshift this equation reduces to the Hubble relation (9.5) with $H_0 = t_0^{-1}$. Namely, in an empty universe the age is given by the Hubble time $t_0 = t_H$, and the "radius" by the Hubble length $R_0 = l_H = ct_H$.

10.5 **Hubble plot in matter-dominated flat universe**   Since distance modulus is a simple logarithmic expression (9.62) of luminosity distance d_L, which in turn is related to our proper distance d_p by $d_L = (1+z)d_p$ of (9.57), all we need is to calculate the proper distance according to (9.47). This integration can be performed for this matter dominated flat universe, which has $a(t) = (t/t_0)^{2/3}$ as given in (10.30):

$$d_p(t_0) = \int_{t_{em}}^{t_0} \frac{cdt}{a(t)} = ct_0^{2/3} \int_{t_{em}}^{t_0} t^{-2/3} dt = 3ct_0 \left[1 - \left(\frac{t_{em}}{t_0}\right)^{1/3}\right]$$

$$= 3ct_0 \left(1 - [a(t_{em})]^{1/2}\right) = \frac{2c}{H_0}(1 - 1 + z^{-1/2}),$$

where we have used $a(t_{em}) = (t_{em}/t_0)^{2/3}$ one more time as well as the basic redshift relation of (9.50) and the age of a flat MDU $t_0 = \frac{2}{3}H_0^{-1}$ of (10.30). This way one finds the distance modulus to be

$$m - M = 5\log_{10}\frac{2cH_0^{-1}(1 + z - 1 + z^{1/2})}{10\ pc}.$$

10.7 **Distance to a light emitter at redshift z**   Plug (10.26), $a(t) = (t/t_0)^x$ into Eq. (9.47), we have

$$d_p(t_0) = \int_{t_{em}}^{t_0} \frac{cdt'}{a(t')} = \frac{ct_0}{1-x}\left[1 - \left(\frac{t_{em}}{t_0}\right)^{1-x}\right]. \tag{25}$$

On the other hand, the time of light emission from a receding galaxy with redshift z can be obtained through (9.50) and (10.26)

$$1 + z = \frac{a(t_0)}{a(t_{em})} = \left(\frac{t_0}{t_{em}}\right)^x$$

and thus $t_{em} = t_0(1+z)^{-1/x}$. Plugging into (25), we have

$$d_p(t_0) = \frac{ct_0}{1-x}\left[1 - \frac{1}{(1+z)^{(1-x)/x}}\right].$$

We note in particular for a matter-dominated flat universe $x = \frac{2}{3}$ we have

$$d_p(t_0) = 3ct_0\left[1 - \frac{1}{(1+z)^{1/2}}\right], \tag{26}$$

which for $t_0 = 2/(3H_0)$ agrees with the result obtained in Problem 10.5. For a radiation-dominated flat universe $x = \frac{1}{2}$, we have $d_p(t_0) = 2ct_0[1 - (1+z)^{-1}]$. NB: These simple relations between redshift and time hold only for a universe with a single-component on energy content; moreover, it does not apply to the situation when the equation-of-state parameter is negative ($w = -1$), even though the energy content is a single-component case.

10.8 **Scaling behavior of number density and Hubble's constant**

(a) For material particles the number density scales as the inverse volume factor, $n(t)/n_0 = a(t)^{-3}$. The basic relation (9.50) between scale factor and redshift leads to $n(t)/n_0 = (1+z)^3$. This scaling property also holds for radiation because $n \sim T^3 \sim a^{-3}$ as given in (10.35).

(b) We can obtain the scaling behavior of the Hubble parameter from Friedmann equation (10.1) for a flat universe: $\dot{a}^2/a^2 = 8\pi G_N \rho/3$, which can be written as $H^2/H_0^2 = \rho/\rho_{c,0}$. For an epoch when the density is dominated by radiation $\rho \simeq \rho_R = \rho_{R,0}a^{-4}$, the above expression for H becomes $H^2/H_0^2 = \Omega_{R,0}(1+z)^4$. Similarly, in a matter dominated epoch obeys $H^2/H_0^2 = \Omega_{M,0}(1+z)^3$.

10.9 **Radiation and matter equality time** Since the universe from t_{RM} to t_γ is matter-dominated, we have from (10.30), $a(t_{RM})/a(t_\gamma) = (t_{RM}/t_\gamma)^{2/3}$, or

$$t_{RM} = \left[\frac{a(t_{RM})}{a(t_\gamma)}\right]^{3/2} t_\gamma = \left[\frac{1+z_{RM}}{1+z_\gamma}\right]^{-3/2} t_\gamma \simeq \frac{t_\gamma}{8^{3/2}} \simeq 16\,000 \text{ years.}$$

10.10 **Density and deceleration parameter** Use the definition of w in (10.4), the second Friedmann Eq. (10.2) becomes

$$\frac{\ddot{a}(t)}{a(t)} = -\frac{4\pi G_N}{3} \sum_i \rho_i(1+3w_i).$$

In terms of the deceleration parameter (9.63) $q_0 \equiv -\ddot{a}(t_0)/a(t_0)H_0^2$ and the critical density (10.6) the second derivative equation leads to the claimed result

$$q_0 = \frac{1}{2}\sum_i \Omega_{i,0}(1+3w_i) = \Omega_{R,0} + \frac{1}{2}\Omega_{M,0} + \cdots$$

10.13 **Cosmological limit of neutrino mass** Even if we assume that all the non-baryonic dark matter is made of three species (flavors) of neutrinos $\rho_{DM} = \sum_{i=1}^3 \rho(\nu_i) = 3n_\nu\bar{m}$, where n_ν is the neutrino number density and $\bar{m}$ is the average neutrino mass. From the neutrino and photon temperature of (10.75) and density being the cubic power of temperature (10.35),

$$n_\nu = \left(\frac{T_\nu}{T_\gamma}\right)^3 n_\gamma \simeq (1.7)^{-3} \times 400 \simeq 150 \text{ cm}^{-3}.$$

The energy density ratio becomes $\Omega_{DM} = (3n_\nu\bar{m}c^2)/(\rho_c c^2) \simeq 0.21$. Using the critical energy density value of (9.17), we have the upper limit for the averaged neutrino mass of $\bar{m}c^2 \simeq (0.21 \times 5500)/(3 \times 150) \simeq 3$ eV.

10.14 **Temperature dipole anisotropy as Doppler effect** Recall that temperature scales as a^{-1}, that is, as inverse wavelength, or as frequency: $\delta T/T = \delta\omega/\omega$.

But the nonrelativistic Doppler effect (the small β limit of (3.46)) reads $\omega' = (1 - (v/c) \cos \theta)\omega$ or $(\delta\omega/\omega) = (v/c) \cos \theta$.

11.1 **Another form of the expansion equation** Consider the energy balance equation (10.11), $\dot{r}^2/2 - G_N M/r = $ const. leading to $\dot{a}^2 - (8\pi/3) G_N \rho a^2 = $ const'. which can also be obtained easily from (10.1). Dividing through by the second term and using the definition of critical density we have $\Omega^{-1} - 1 = $ const./(ρa^2).

11.2 **The epoch-dependent Hubble constant and** $a(t)$ Using (10.7) to replace the curvature parameter k in the Friedmann equation (10.1), we have

$$\frac{\dot{a}^2(t)}{a^2(t)} = \frac{8\pi G_N}{3}\rho + \frac{\dot{a}^2(t_0)}{a^2(t)}(1 - \Omega_0) = H_0^2 \left(\frac{\rho}{\rho_{c,0}} + \frac{1 - \Omega_0}{a^2(t)} \right). \quad (27)$$

Putting the time-dependence of the densities

$$\frac{\rho}{\rho_{c,0}} = \Omega(t) = \frac{\Omega_{R,0}}{a^4} + \frac{\Omega_{M,0}}{a^3} + \Omega_{\Lambda,0},$$

Eq. (27) becomes

$$\frac{H^2(t)}{H_0^2} = \frac{\Omega_{R,0}}{a^4} + \frac{\Omega_{M,0}}{a^3} + \Omega_{\Lambda,0} + \frac{1 - \Omega_0}{a^2}.$$

11.4 **Negative Λ and the "big crunch"** For the $\Omega_0 = 1$ flat universe with matter and dark energy, we have the Friedmann equation (11.38)

$$H(a) = H_0 \Omega_{M,0} a^{-3} + \Omega_\Lambda^{1/2}.$$

At $a = a_{max}$ the universe stops expanding and $H(a_{max}) = 0$, thus $a_{max} = (-\Omega_{M,0}/\Omega_\Lambda)^{1/3}$. The cosmic time for the big crunch being twice the time for the universe to go from a_{max} to $a = 0$, we calculate in a way similar to that shown in sidenote 28,

$$2t_H \int_0^{a_{max}} \frac{da}{\Omega_{M,0}a^{-1} + \Omega_\Lambda a^2{}^{1/2}} = \frac{4t_H}{3\sqrt{-\Omega_\Lambda}} \int_0^{a_{max}^{3/2}} \frac{dx}{a_{max}^3 - x^2{}^{1/2}}$$

$$= \frac{4t_H}{3\sqrt{-\Omega_\Lambda}} \left[\sin^{-1}\left(\frac{x}{a_{max}^{3/2}} \right) \right]_0^{a_{max}^{3/2}} = \frac{2\pi}{3\sqrt{-\Omega_\Lambda}} t_H = t_*.$$

11.5 **Estimate of matter and dark energy equality time** We define the matter and dark energy equality time $t_{M\Lambda}$ as $\rho_M(t_{M\Lambda}) = \rho_\Lambda(t_{M\Lambda})$. Using the scaling properties of these densities we have $\rho_{M,0}/a_{M\Lambda}^3 = \rho_{\Lambda,0}$ or $1 + z_{M\Lambda} = (a_{M\Lambda})^{-1} = (\Omega_\Lambda/\Omega_{M,0})^{1/3}$, which differs from the result in (11.48) by a factor of $2^{1/3} \approx 1.25$ to yield $a_{M\Lambda} = 0,7$ (and a redshift of $z_{M\Lambda} = 0.42$). Using the formula given in sidenote 25, we obtain the corresponding cosmic age $t_{M\Lambda} = t(a_{M\Lambda}) = 9.5$ Gyr.

12.1 **Basis and inverse basis vectors: a simple exercise**

(a) Given the basis vectors, the inverse basis vectors can be worked out: $\mathbf{e}^1 = \frac{1}{a}(1, -\cot\theta)$ and $\mathbf{e}^2 = \frac{1}{b}(0, \csc\theta)$. The condition $\mathbf{e}_1 \cdot \mathbf{e}^1 = \mathbf{e}_2 \cdot \mathbf{e}^2 = 1$ and $\mathbf{e}_1 \cdot \mathbf{e}^2 = \mathbf{e}_2 \cdot \mathbf{e}^1 = 0$ can be easily checked by explicit vector multiplication. For example, $\mathbf{e}_2 \cdot \mathbf{e}^1 = \frac{b}{a}(\cos\theta - \cos\theta) = 0$.

(b) Similarly by explicit vector multiplications, we have

$$g_{ij} = \mathbf{e}_i \cdot \mathbf{e}_j = \begin{pmatrix} a^2 & ab\cos\theta \\ ab\cos\theta & b^2 \end{pmatrix}$$

$$g^{ij} = \mathbf{e}^i \cdot \mathbf{e}^j = \begin{pmatrix} \dfrac{1}{a^2\sin^2\theta} & -\dfrac{\cos\theta}{ab\sin^2\theta} \\ -\dfrac{\cos\theta}{ab\sin^2\theta} & \dfrac{1}{b^2\sin^2\theta} \end{pmatrix}$$

so that $g_{ij}g^{jk} = \delta_{ik}$ can be checked by matrix multiplication.

(c) We can verify the completeness condition by calculating the direct-products of basis vectors,

$$\sum_i \mathbf{e}_i \otimes \mathbf{e}^i = \begin{pmatrix} 1 \\ 0 \end{pmatrix} (1 - \cot\theta) + \begin{pmatrix} \cos\theta \\ \sin\theta \end{pmatrix} (0 \ \csc\theta)$$

$$= \begin{pmatrix} 1 - \cot\theta & 0 \\ 0 & 0 \end{pmatrix} + \begin{pmatrix} 0 & \cot\theta \\ 0 & 1 \end{pmatrix} = \mathbf{1}.$$

12.4 **Transformation: coordinates vs. basis vectors** A^μ transform "oppositely" from the bases vectors $\mathbf{e}_\mu$

$$\mathbf{e}_\mu \longrightarrow \mathbf{e}'_\mu = \left[\mathbf{L}^{-1}\right]^\nu_\mu \mathbf{e}_\nu \tag{28}$$

because the vector itself $\mathbf{A} = A^\mu \mathbf{e}_\mu$ does not change under the coordinate transformations.

12.5 $g_{\mu\nu}$ **is a tensor**

(a) Plugging in the transformations of the basis vectors (28) in the metric definition $g'_{\mu\nu} = \mathbf{e}'_\mu \cdot \mathbf{e}'_\nu$ we immediately obtain that for the metric, (12.17).

(b) The invariance of the scalar product $\mathbf{A} \cdot \mathbf{B}$ can also expressed as

$$A_\mu B_\nu g^{\mu\nu} = A'_\lambda B'_\rho g'^{\lambda\rho} = A_\mu B_\nu \left[\mathbf{L}^{-1}\right]^\mu_\lambda \left[\mathbf{L}^{-1}\right]^\nu_\rho g'^{\lambda\rho},$$

or

$$g^{\mu\nu} = \left[\mathbf{L}^{-1}\right]^\mu_\lambda \left[\mathbf{L}^{-1}\right]^\nu_\rho g'^{\lambda\rho}.$$

We can invert this equation by multiplying two [**L**] factors on both sides to obtain

$$g'^{\mu\nu} = [\mathbf{L}]^\mu_\lambda [\mathbf{L}]^\nu_\rho g^{\lambda\rho}.$$

This shows, cf. Eq. (12.16), that the (inverse) metric is indeed a *bona fide* contravariant tensor.

12.6 **The quotient theorem** Given that the product $A^\mu B^\nu g_{\mu\nu}$ is a scalar, and vectors A^μ and B^ν are known to be tensors, their quotient $g_{\mu\nu}$ must also be a tensor.

12.7   **Transformation of the metric**   From (5.34) we can immediately deduce the coordinate transformation

$$\begin{pmatrix} dx \\ dy \end{pmatrix} = \begin{pmatrix} \cos\phi & -r\sin\phi \\ \sin\phi & r\cos\phi \end{pmatrix} \begin{pmatrix} dr \\ d\phi \end{pmatrix},$$

hence the transformation matrix

$$[\mathbf{L}] = \begin{pmatrix} \cos\phi & -r\sin\phi \\ \sin\phi & r\cos\phi \end{pmatrix} \text{ and } \left[\mathbf{L}^{-1}\right] = \begin{pmatrix} \cos\phi & \sin\phi \\ -r^{-1}\sin\phi & r^{-1}\cos\phi \end{pmatrix}.$$

Clearly the metric transformation checks out

$$\left[\mathbf{L}^{-1}\right]^{\mathsf{T}} \begin{pmatrix} 1 & 0 \\ 0 & r^2 \end{pmatrix} \left[\mathbf{L}^{-1}\right] = \begin{pmatrix} 1 & 0 \\ 0 & 1 \end{pmatrix}.$$

12.8   **Generalized orthogonality condition and the boost transformation**   We can work this out in a way that's entirely similar to Problem 3.4. Writing out the condition $\mathbf{L}\eta\mathbf{L}^{\mathsf{T}} = \eta$

$$\begin{pmatrix} a & b \\ c & d \end{pmatrix} \begin{pmatrix} -1 & 0 \\ 0 & 1 \end{pmatrix} \begin{pmatrix} a & c \\ b & d \end{pmatrix} = \begin{pmatrix} -1 & 0 \\ 0 & 1 \end{pmatrix}, \tag{29}$$

we have the conditions of $a^2 - b^2 = -c^2 + d^2 = 1$, which can be solved by the parametrization of $a = \cosh\psi$, $b = \sinh\psi$ and $c = \sinh\psi'$, $d = \cosh\psi'$, while the off-diagonal condition of $-ac + bd = -\cosh\psi\sinh\psi' + \sinh\psi\cosh\psi' = \sinh(\psi - \psi') = 0$ yields $\psi = \psi'$. The identification of $\tanh\psi = v/c$ was worked out in Box 3.2.

12.9   **Covariant Lorentz force law**

(a) This identification is justified because our relativistic force $\vec{F} = \gamma m\vec{a}$ becomes the usual $\vec{F} = m\vec{a}$ in the non-relativistic situation when $\gamma = 1$.

(b) For $\mu = i$,

$$K^i = \frac{q}{c} F^{i\nu} U_\nu = \frac{q}{c} \left(F^{i0} U_0 + F^{ij} U_j \right),$$

$$\gamma F_i = \frac{q}{c} \left[-E_i \left(-\gamma c \right) + \epsilon_{ijk} B_k \left(\gamma v_j \right) \right],$$

which, after the cancellation of the γ factor from both sides, is just the familiar Lorentz force law written in its components.

(c) For $\mu = 0$,

$$K^0 = \frac{q}{c} F^{0i} U_i = \gamma \frac{q}{c} \vec{E} \cdot \vec{v} \tag{30}$$

is indeed $\gamma \vec{F} \cdot \vec{v}/c$ because the dot product with the magnetic field term in Lorentz force vanishes.

12.11   **Homogeneous Maxwell's equations**   To show that $\partial_\mu F_{\nu\lambda} + \partial_\lambda F_{\mu\nu} + \partial_\nu F_{\lambda\mu} = 0$ follows from $\partial_\mu \tilde{F}^{\mu\nu} = 0$: From the definition of dual field tensor, we have $\partial_\mu F_{\lambda\rho} \epsilon^{\mu\nu\lambda\rho} = 0$, which is a trivial relation ($0 = 0$) if any pair of indices in (μ, λ, ρ) are equal. Thus, only when the indices are unequal do we get non-trivial relation: take the example of equation of $\partial_\mu \tilde{F}^{\mu 0} = \partial_\mu F_{\lambda\rho} \epsilon^{\mu 0 \lambda \rho} = 0$ we have

$$\partial_1 F_{23} + \partial_3 F_{12} + \partial_2 F_{31} = 0.$$

We can regard this as a relation in a particular coordinate frame with $\mu = 1$, $\nu = 2$, and $\lambda = 3$. Once written in the Lorentz covariant version, it must be valid in every frame. This is just the relation we set out to prove:

$$\partial_\mu F_{\nu\lambda} + \partial_\lambda F_{\mu\nu} + \partial_\nu F_{\lambda\mu} = 0.$$

To prove the converse statement, all we need to do is to contract $\epsilon^{\mu\nu\lambda\rho}$ onto the above equation.

12.15 $T^{\mu\nu}$ **for a system of EM field and charges**   We first calculate the divergence of $T^{\mu\nu}_{\text{charge}} = \rho'_{\text{mass}} U^\mu U^\nu$ to find that

$$\partial_\mu T^{\mu\nu}_{\text{charge}} = \rho'_{\text{mass}} \left(U^\mu \partial_\mu \right) U^\nu$$

where we have also used the mass conservation law of $\partial_\mu (\rho'_{\text{mass}} U^\mu) = 0$. The Lorentz invariant product $U^\mu \partial_\mu$ can be evaluated in any convenient reference frame; we choose the comoving frame $U^\mu = \gamma(c, \vec{0})$ to obtain $U^\mu \partial_\mu = \gamma \partial_t = \partial_\tau$, the differentiation with respect to the proper time τ. The term $\rho'_{\text{mass}} \partial_\tau U^\nu$ is the 4-force density (ie, mass replace by mass density). Use the formula (12.44) for the Lorentz force density (charge replaced by charge density), we then have

$$\partial_\mu T^{\mu\nu}_{\text{charge}} = \rho'_{\text{mass}} \partial_\tau U^\nu = \frac{\rho'_{\text{charge}}}{c} F^{\nu\lambda} U_\lambda = \frac{1}{c} F^{\nu\lambda} j_\lambda$$

where we have used the expression for the electromagnetic current for free charges $j_\lambda = \rho'_{\text{charge}} U_\lambda$.

We now calculate the divergence of $T^{\mu\nu}_{\text{field}}$ in (12.80) to find

$$\partial_\mu T^{\mu\nu}_{\text{field}} = \eta_{\alpha\beta} \left(\partial_\mu F^{\mu\alpha} \right) F^{\nu\beta}.$$

Here we have used the calculation performed in (12.82) and by noting the fact that, in the presence of charges, the inhomogeneous Maxwell's equation $\partial_\mu F^{\mu\alpha} = -\frac{1}{c} j^\alpha$ has a non-vanishing RHS

$$\partial_\mu T^{\mu\nu}_{\text{field}} = -\frac{1}{c} \eta_{\alpha\beta} j^\alpha F^{\nu\beta} = -\frac{1}{c} F^{\nu\lambda} j_\lambda.$$

This shows clearly that the sum $T^{\mu\nu} = T^{\mu\nu}_{\text{field}} + T^{\mu\nu}_{\text{charge}}$ has zero divergence.

12.16   **Radiation pressure and energy density**   The system of electromagnetic field can be viewed either as a system of field with energy–momentum tensor

$$T^{\mu\nu}_{\text{field}} = \eta_{\alpha\beta} F^{\mu\alpha} F^{\nu\beta} - \frac{1}{4} \eta^{\mu\nu} F^{\alpha\beta} F_{\alpha\beta},$$

or as a system of ideal fluid made up of photons with, cf. (12.72),

$$T^{\mu\nu}_{\gamma\,\text{fluid}} = \begin{pmatrix} \rho'c^2 & & & \\ & p & & \\ & & p & \\ & & & p \end{pmatrix}$$

with $\rho'c^2$ and p being the radiation energy density and pressure, respectively. Since these two representations both describe the same system we should expect $T^{\mu\nu}_{\gamma\,\text{fluid}} = T^{\mu\nu}_{\text{field}}$, in particular their traces should equal: $\eta_{\mu\nu}T^{\mu\nu}_{\gamma\,\text{fluid}} = \eta_{\mu\nu}T^{\mu\nu}_{\text{field}}$. But a simple inspection shows that $\eta_{\mu\nu}T^{\mu\nu}_{\text{field}} = 0$ because $\eta_{\mu\nu}\eta^{\mu\nu} = 4$. The vanishing trace $\eta_{\mu\nu}T^{\mu\nu}_{\gamma\,\text{fluid}} = 0$ leads to the result $p = \rho'c^2/3$. (That $T^{\mu\nu}$ is traceless is related to the scale invariance of the system.)

13.1 **Covariant derivative for covariant components**   Given that $A_\mu A^\mu$ is an invariant, in the notation of (13.43), we also have $\Delta(A_\mu A^\mu)_{\text{coord}} = 0$:

$$A_\mu \left[\Delta A^\mu\right]_{\text{coord}} + A^\mu \left[\Delta A_\mu\right]_{\text{coord}} = 0.$$

$\Delta A^\mu{}_{\text{coord}}$ being given by (13.45), we get

$$A^\mu \left[\Delta A_\mu\right]_{\text{coord}} = A_\mu \Gamma^\mu_{\nu\lambda} A^\nu dx^\lambda = A^\mu \left(\Gamma^\nu_{\mu\lambda} A_\nu dx^\lambda\right).$$

The last expression is reached by relabelling $\mu \leftrightarrow \nu$. The result of $\Delta A_{\mu\,\text{coord}} = +\Gamma^\nu_{\mu\lambda} A_\nu dx^\lambda$ implies that $D_\nu A_\mu = \partial_\nu A_\mu - \Gamma^\lambda_{\nu\mu} A_\lambda$.

13.2 **Christoffel symbols of polar coordinates for a flat plane**

(a) Explicitly differentiating the relation $\mathbf{r} = r\cos\theta\,\mathbf{i} + r\sin\theta\,\mathbf{j}$, we have

$$d\mathbf{r} \equiv dr\,\mathbf{e}_r + d\theta\,\mathbf{e}_\theta = dr\cos\theta\,\mathbf{i} - r\sin\theta d\theta\,\mathbf{i} + dr\sin\theta\,\mathbf{j} + r\cos\theta d\theta\,\mathbf{j}.$$

Collecting the dr and $d\theta$ terms,

$$\mathbf{e}_r = \cos\theta\,\mathbf{i} + \sin\theta\,\mathbf{j}, \quad \mathbf{e}_\theta = -r\sin\theta\,\mathbf{i} + r\cos\theta\,\mathbf{j}.$$

The inverse bases can be gotten by contracting with the inverse metric $g^{\mu\nu} = \text{diag}(1, r^{-2})$:

$$\mathbf{e}^r = \cos\theta\,\mathbf{i} + \sin\theta\,\mathbf{j}, \quad \mathbf{e}^\theta = -r^{-1}\sin\theta + r^{-1}\cos\theta\,\mathbf{j}.$$

(b) To calculate the Christoffel symbols through their definition of $\partial_\nu \mathbf{e}^\mu = -\Gamma^\mu_{\nu\lambda}\mathbf{e}^\lambda$ we first observe:

$$\frac{\partial\mathbf{e}^r}{\partial r} = 0, \quad \frac{\partial\mathbf{e}^\theta}{\partial r} = \frac{-1}{r^2}(-\sin\theta\,\mathbf{i} + \cos\theta\,\mathbf{j}) = \frac{-1}{r}\mathbf{e}^\theta.$$

Then the definitions

$$\frac{\partial\mathbf{e}^r}{\partial r} = \Gamma^r_{rr}\mathbf{e}^r + \Gamma^r_{r\theta}\mathbf{e}^\theta, \quad \frac{\partial\mathbf{e}^\theta}{\partial r} = \Gamma^\theta_{rr}\mathbf{e}^r + \Gamma^\theta_{r\theta}\mathbf{e}^\theta$$

allow us to read off the Christoffel symbols $\Gamma^r_{rr} = \Gamma^r_{r\theta} = \Gamma^\theta_{rr} = 0$ and $\Gamma^\theta_{r\theta} = r^{-1}$. Similarly, from

$$\frac{\partial \mathbf{e}^r}{\partial \theta} = -\sin\theta\, \mathbf{i} + \cos\theta\, \mathbf{j} = r\mathbf{e}^\theta,$$

$$\frac{\partial \mathbf{e}^\theta}{\partial \theta} = -r^{-1}\cos\theta\, \mathbf{i} - r^{-1}\sin\theta\, \mathbf{j} = -r^{-1}\mathbf{e}^r$$

we obtain $\Gamma^r_{\theta r} = \Gamma^\theta_{\theta\theta} = 0$, $\Gamma^r_{\theta\theta} = -r$ and $\Gamma^\theta_{\theta r} = r^{-1}$.

(c) Work out the components in

$$D_\mu A^\mu = \partial_\mu A^\mu + \Gamma^\mu_{\mu\nu} A^\nu$$

$$= \partial_r A^r + \partial_\theta A^\theta + \left(\Gamma^r_{rr} + \Gamma^\theta_{\theta r}\right) A^r + \left(\Gamma^r_{r\theta} + \Gamma^\theta_{\theta\theta}\right) A^\theta$$

$$= \partial_r A^r + \partial_\theta A^\theta + \frac{1}{r} A^r = \frac{1}{r}\frac{\partial}{\partial r}\left(r A^r\right) + \frac{\partial}{\partial\theta} A^\theta$$

$$= \left(\frac{1}{r}\frac{\partial}{\partial r} r \quad \frac{\partial}{\partial\theta}\right)\begin{pmatrix} A^r \\ A^\theta \end{pmatrix}.$$

(d) Because the scalar function $\Phi(x)$ is coordinate independent, $D_\mu\Phi = \partial_\mu\Phi$. To raise the index we must multiply it by the inverse metric $g^{\mu\nu}\partial_\mu\Phi$. Using the result obtained in (c) we have

$$D_\mu D^\mu \Phi(x) = D_\mu\left(g^{\mu\nu}\partial_\mu\Phi\right)$$

$$= \left(\frac{1}{r}\frac{\partial}{\partial r} r \quad \frac{\partial}{\partial\theta}\right)\begin{pmatrix} 1 & 0 \\ 0 & r^{-2} \end{pmatrix}\begin{pmatrix} \partial_r\Phi \\ \partial_\theta\Phi \end{pmatrix}$$

$$= \frac{1}{r}\frac{\partial}{\partial r}\left(r\frac{\partial\Phi}{\partial r}\right) + \frac{1}{r^2}\frac{\partial^2\Phi}{\partial\theta^2}.$$

(e) The metric in polar coordinates has only one nontrivial element $g_{\theta\theta} = r^2$. Checking the covariant differentiation with respect to the radial coordinate r, we get

$$D_r g_{\theta\theta} = \partial_r g_{\theta\theta} - 2\Gamma^\mu_{r\theta} g_{\mu\theta} = 2r - 2\frac{1}{r}r^2 = 0.$$

(f) Substituting $g_{\theta r} = 0$ and $g_{\theta\theta} = r^2$ into (13.37), we have

$$\Gamma^r_{\theta\theta} = \frac{1}{2}g^{r\mu}\left(\partial_\theta g_{\theta\mu} + \partial_\theta g_{\theta\mu} - \partial_\mu g_{\theta\theta}\right)$$

$$= \frac{1}{2}g^{rr}\left(2\partial_\theta g_{\theta r} - \partial_r g_{\theta\theta}\right) = -r$$

$$\Gamma^\theta_{\theta\theta} = \frac{1}{2}g^{\theta\theta}\partial_\theta g_{\theta\theta} = 0.$$

(g) We have a diagonal metric $g_{11} = g_{rr} = 1$ and $g_{22} = g_{\theta\theta} = r^2$ so that

$$R_{1212} = g_{1a}R^a_{212} = g_{11}\left(\partial_1\Gamma^1_{22} - \partial_2\Gamma^1_{21} + \Gamma^1_{b1}\Gamma^b_{22} - \Gamma^1_{b2}\Gamma^b_{21}\right)$$

From part (b), we have the only nonvanishing elements being $\Gamma^r_{\theta\theta} = -r$ and $\Gamma^\theta_{\theta r} = \Gamma^\theta_{r\theta} = r^{-1}$:

$$R_{1212} = \partial_r \Gamma^r_{\theta\theta} - \Gamma^r_{\theta\theta}\Gamma^\theta_{\theta r} = -1 + r/r = 0.$$

13.3 **Symmetry property of Christoffel symbols**   Because a scalar field $\Phi(x)$ is coordinate-independent, there is no difference between their covariant and ordinary derivatives, $D_\mu \Phi = \partial_\mu \Phi$. We then apply the result of Problem 13.1 to the torsion-free statement, after using $\partial_\nu \partial_\mu \Phi = \partial_\mu \partial_\nu \Phi$, to obtain

$$D_\nu D_\mu \Phi - D_\mu D_\nu \Phi = -\Gamma^\lambda_{\nu\mu}\partial_\lambda \Phi + \Gamma^\lambda_{\mu\nu}\partial_\lambda \Phi$$

$$= \left(-\Gamma^\lambda_{\nu\mu} + \Gamma^\lambda_{\mu\nu}\right)\partial_\lambda \Phi = 0.$$

13.4 **Metric is covariantly constant: by explicit calculation**   Take the covariant derivative of the metric tensor (with covariant indices) and then express the resulting Christoffel symbols in terms of derivatives of the metric

$$D_\mu g_{\nu\lambda} = \partial_\mu g_{\nu\lambda} - \Gamma^\rho_{\mu\nu}g_{\rho\lambda} - \Gamma^\rho_{\mu\lambda}g_{\rho\nu}$$

$$= \partial_\mu g_{\nu\lambda} - \frac{1}{2}g^{\rho\sigma}\left(\partial_\mu g_{\nu\sigma} + \partial_\nu g_{\mu\sigma} - \partial_\sigma g_{\mu\nu}\right)g_{\rho\lambda}$$

$$- \frac{1}{2}g^{\rho\sigma}\left(\partial_\mu g_{\lambda\sigma} + \partial_\lambda g_{\mu\sigma} - \partial_\sigma g_{\mu\lambda}\right)g_{\rho\nu}.$$

After summing over repeated indices, we find all terms cancel.

13.5 **$D_\nu V_\mu$ is a good tensor: another proof**   Start with (13.50) and use the fact that the σ -dependence is always through $x^\mu(\sigma)$

$$\frac{D}{d\sigma}\left(V_\mu \frac{dx^\mu}{d\sigma}\right) = \frac{D}{dx^\nu}\left(V_\mu \frac{dx^\mu}{d\sigma}\right)\frac{dx^\nu}{d\sigma} = 0.$$

We can use the geodesic equation in the form of $D(dx^\mu/d\sigma)/D\sigma = 0$ to obtain

$$\left(D_\nu V_\mu\right)\frac{dx^\mu}{d\sigma}\frac{dx^\nu}{d\sigma} = 0.$$

The quotient theorem then informs us that $D_\nu V_\mu$ is a good tensor, because it is contracted into a good tensor: $(dx^\mu/d\sigma)(dx^\nu/d\sigma)$.

13.6 **Parallel transport and the angular excess**   The triangle has three vertices (A, B, C) connected by geodesic curves with interior angles (α, β, γ). We now transport a vector around this triangle, along the three geodesic sides of the triangle. The key observation is that the angle subtended by the vector and the geodesic is unchanged (cf. the worked example in the text).

1. At vertex A, the vector makes an angle θ_1 with the tangent along AB.
2. At vertex B, the vector makes the same angle θ_1 with the tangent along AB, thus it makes $\theta_2 = \theta_1 + (\pi - \beta)$ along BC.
3. At vertex C, the vector makes $\theta_3 = \theta_2 + (\pi - \gamma)$ along CA.
4. Returning to A, the vector makes $\theta_4 = \theta_3 + (\pi - \alpha)$ along the original AB.

Plug in θ_i sequentially and take out a trivial factor of 2π, we obtain the directional change of the vector

$$\delta\theta = \theta_1 - \theta_4 = \alpha + \beta + \gamma - \pi,$$

which is just the angular excess ϵ.

13.7   **Riemann curvature tensor as the commutator of covariant derivatives**   Following the rule of (13.30), we have

$$D_\alpha D_\beta A_\mu = \partial_\alpha\left(D_\beta A_\mu\right) - \underbrace{\Gamma^\nu_{\alpha\beta}D_\nu A_\mu}_{\text{drop}} - \Gamma^\nu_{\alpha\mu}D_\beta A_\nu$$

$$= \underbrace{\frac{\partial_\alpha\partial_\beta A_\mu}{\text{drop}}} - \partial_\alpha\left(\Gamma^\nu_{\beta\mu}A_\nu\right) - \Gamma^\nu_{\alpha\mu}\partial_\beta A_\nu + \Gamma^\nu_{\alpha\mu}\Gamma^\lambda_{\beta\nu}A_\lambda$$

$$= -\left(\partial_\alpha\Gamma^\lambda_{\beta\mu}\right)A_\lambda \underbrace{\frac{-\Gamma^\nu_{\beta\mu}\partial_\alpha A_\nu - \Gamma^\nu_{\alpha\mu}\partial_\beta A_\nu}{\text{drop}}} + \Gamma^\nu_{\alpha\mu}\Gamma^\lambda_{\beta\nu}A_\lambda.$$

The underlined terms are symmetric in the indices (α, β) and will be cancelled when we include the $-D_\beta D_\alpha A_\mu$ calculation. From this we clearly get $D_\alpha, D_\beta A_\mu = -R^\lambda_{\mu\alpha\beta}A_\lambda$ with $R^\lambda_{\mu\alpha\beta}$ given by (13.58).

13.9   **Counting independent elements of Riemann tensor**   Write the curvature tensor as $R_{\{\mu\nu,\alpha\beta\}}$ to remind ourselves the symmetry properties of (13.69) to (13.71): antisymmetry of Eq. (13.69) as $\mu\nu$, that of (13.70) as $\alpha\beta$, and the symmetry of (13.71) as $\{\mu\nu, \alpha\beta\}$. An $n \times n$ matrix has $\frac{1}{2}n(n + 1)$ independent elements if it is symmetric, and $\frac{1}{2}n(n - 1)$ if antisymmetric. Hence, for the purpose of counting independent components, we can regard $R_{\{\mu\nu,\alpha\beta\}}$ as a $\frac{1}{2}n(n - 1)$ by $\frac{1}{2}n(n - 1)$ matrix, which is symmetric. This yields a count of

$$M_{(n)} = \frac{1}{2}\left[\frac{1}{2}n(n - 1)\right] \times \left[\frac{1}{2}n(n - 1) + 1\right]$$

$$= \frac{1}{8}n(n - 1)\left(n^2 - n + 2\right).$$

There are not as many independent elements as $M_{(n)}$ because we also need to factor-in the cyclic symmetry constraint of (13.72). Actually, (13.72) represents extra conditions that reduce the number of independent elements only if all four indices are different—because otherwise this cyclic condition reduces to the first three symmetry conditions. The number of additional constraint conditions as represented by (13.72) is given by:

$$C_{(n)} = \binom{n}{4} = n(n - 1)(n - 2)(n - 3)/4$$

Subtracting $C_{(n)}$ from $M_{(n)}$ leads to the the number of independent components of a curvature tensor in an n-dimensional space:

$$N_{(n)} = M_{(n)} - C_{(n)} = \frac{1}{12}n^2(n^2 - 1). \tag{31}$$

13.10   **Counting metric's independent second derivatives**

(a) Remembering that the number of independent elements of a symmetric $n \times n$ matrix is $n(n + 1)/2$, we see that the tensor $g_{\mu\nu}$ has 10 elements, and its first derivative $\partial_\alpha g_{\mu\nu}$ has 40, and its second derivative $\partial_\alpha\partial_\beta g_{\mu\nu}$

has 100 elements, when we used the fact that $\partial_\alpha \partial_\beta = \partial_\beta \partial_\alpha$. Namely,

	index sym	$A_{(4)}$
$g_{\mu\nu}$	$\{\mu\nu\}$	$(4 \times 5)/2 = 10$
$\partial_\alpha g_{\mu\nu}$	$\alpha \{\mu\nu\}$	$4 \times 10 = 40$
$\partial_\alpha \partial_\beta g_{\mu\nu}$	$\{\alpha\beta\} \{\mu\nu\}$	$10 \times 10 = 100$

In particular the number of components for the second derivative $\partial_\alpha \partial_\beta g_{\mu\nu}$ in an n-dimensional space is

$$A_{(n)} = \left[\frac{1}{2} n (n + 1) \right]^2 . \tag{32}$$

(b) Using the same notation as in (a), we find the number of parameters in the transformations for the four-dimensional space:

	index sym	$B_{(4)}$
$(\partial_\alpha x_\beta)$	$\alpha\beta$	$4 \times 4 = 16$
$\partial_\gamma (\partial_\alpha x_\beta)$	$\{\gamma\alpha\} \beta$	$10 \times 4 = 40$
$\partial_\gamma \partial_\delta (\partial_\alpha x_\beta)$	$\{\alpha\gamma\delta\} \beta$	$20 \times 4 = 80$

where, on the last line for the second derivative $\partial_\gamma \partial_\delta (\partial_\alpha x_\beta)$, we have used the fact that there are 20 possible totally symmetric combinations of three indices ($d = 3$) when each index can take on four possible values ($n = 4$). This is an example of the general result $N(d, n)$ being the number of symmetric combinations of d objects each can take on n possible values:

$$N(d, n) = \binom{d + n - 1}{d} = \frac{(n + d - 1)}{d (n - 1)} . \tag{33}$$

One can understand this result by thinking of the ways, for example, of placing d identical balls into n boxes, which is equivalent to the problem of permuting d identical balls together with the $n - 1$ partitions between the boxes.

(c) Of the results obtained in (a) and (b)

	$A_{(4)}$		$B_{(4)}$
$g_{\mu\nu}$	10	$(\partial_\alpha x_\beta)$	16
$\partial_\alpha g_{\mu\nu}$	40	$\partial_\gamma (\partial_\alpha x_\beta)$	40
$\partial_\alpha \partial_\beta g_{\mu\nu}$	100	$\partial_\gamma \partial_\delta (\partial_\alpha x_\beta)$	80

we note several features:

(i) **The $g_{\mu\nu}$ case**   Do we need the 16 parameters of $(\partial_\alpha x_\beta)$ to determine the 10 elements of $g_{\mu\nu}$? Yes, this count is correct, because the transformation includes the six parameter Lorentz transformations that leave the Euclidean metric $g_{\mu\nu} = \eta_{\mu\nu}$ invariant.

(ii) **The $\partial_\alpha g_{\mu\nu}$ case**   There are just the correct number (40) of parameters in $\partial_\gamma (\partial_\alpha x_\beta)$ to set all the 40 independent elements of $\partial_\alpha g_{\mu\nu}$ to zero. (Compare this to the flatness theorem.)

(iii) **The $\partial_\alpha \partial_\beta g_{\mu\nu}$ case**   We still have 20 yet undetermined elements in the second derivative $\partial_\alpha \partial_\beta g_{\mu\nu}$. This just corresponds to the number of independent elements in the four dimensional curvature tensor $N_{(4)} = 20$ as shown in Problem 13.9.

(d) For a general n dimensional space, the number of second derivatives of the transformation $\partial_\gamma \partial_\delta \partial_\alpha x_\beta$ as given by (33) for $d = 3$ (with a further multiplication of n for the β index) is

$$B_{(n)} = \frac{1}{6} n^2 (n + 2)(n + 1). \tag{34}$$

The number of independent elements of the second derivative must be the difference of (32) and (34): $N_{(n)} = A_{(n)} - B_{(n)} = n^2(n^2 - 1)/12$, which exactly matches the result of (31).

13.11 Reducing Riemann tensor to Gaussian curvature

(a) For a two-dimensional space with orthogonal coordinates, we have the metrics

$$g_{\mu\nu} = \begin{pmatrix} g_{11} & 0 \\ 0 & g_{22} \end{pmatrix}, \quad g^{\mu\nu} = \begin{pmatrix} g^{11} & 0 \\ 0 & g^{22} \end{pmatrix}$$

with $g^{11} = 1/g_{11}$ and $g^{22} = 1/g_{22}$ so that $g_{\mu\nu} g^{\nu\lambda} = \delta_\mu^\lambda$. The Christoffel symbols can be calculated from

$$\Gamma_{\mu\nu}^1 = \frac{1}{2} g^{11} \left(\partial_\mu g_{1\nu} + \partial_\nu g_{1\mu} - \partial_1 g_{\mu\nu} \right)$$

so that

$$\Gamma_{11}^1 = \frac{1}{2g_{11}} \partial_1 g_{11}, \quad \Gamma_{22}^1 = -\frac{1}{2g_{11}} \partial_1 g_{22}$$

$$\Gamma_{12}^1 = \Gamma_{21}^1 = \frac{1}{2g_{11}} \partial_2 g_{11}.$$

Similarly, we also have

$$\Gamma_{22}^2 = \frac{1}{2g_{221}} \partial_2 g_{22}, \quad \Gamma_{12}^2 = \Gamma_{21}^2 = \frac{1}{2g_{22}} \partial_1 g_{22}.$$

The only nontrivial (and independent) curvature element is

$$R_{1212} = g_{1\mu} R_{212}^\mu$$

$$= g_{11} \left(\partial_2 \Gamma_{21}^1 - \partial_1 \Gamma_{22}^1 + \Gamma_{21}^\nu \Gamma_{\nu 2}^1 - \Gamma_{22}^\nu \Gamma_{\nu 1}^1 \right)$$

$$= g_{11} \left(\partial_2 \Gamma_{21}^1 - \partial_1 \Gamma_{22}^1 + \Gamma_{21}^1 \Gamma_{12}^1 + \Gamma_{21}^2 \Gamma_{22}^1 - \Gamma_{22}^1 \Gamma_{1\nu 1}^1 - \Gamma_{22}^2 \Gamma_{21}^1 \right)$$

$$= \frac{1}{2} \left\{ \partial_2^2 g_{11} + \partial_1^2 g_{22} - \frac{1}{2g_{11}} \left[(\partial_1 g_{11})(\partial_1 g_{22}) + (\partial_2 g_{11})^2 \right] \right.$$

$$\left. - \frac{1}{2g_{22}} \left[(\partial_2 g_{11})(\partial_2 g_{22}) + (\partial_1 g_{22})^2 \right] \right\}$$

which, when divided by the metric determinant $g = g_{11} g_{22}$, the ratio $-R_{1212}/g$ is recognized as the Gaussian curvature of (5.35).

(b) The Ricci scalar is simply the twice contracted Riemann tensor $R = g^{\alpha\beta} g^{\mu\nu} R_{\alpha\mu\beta\nu} = 2g^{11} g^{22} R_{1212}$ because $R_{1212} = R_{2121}$. Since $g^{11} g^{22} = 1/g$, the result (a) leads to $R = -2K$.

(c) Equation (13.57) may be written as $dA^2 = R^2_{112}A^1\sigma$. Since the angular excess is related to the vector component change as $\epsilon = dA^2/A^1$, we can write this as

$$\epsilon = R^2_{112}\sigma = -g^{22}R_{1212}\sigma = g^{22}gK\sigma$$

$$= g^{22}(g_{11}g_{22})K\sigma = K\sigma,$$

where we have used $g^{22}g_{22} = 1$ and $g_{11} = 1$ as, e.g. in polar system (r, θ).

13.12   **Bianchi identities**   (1) The structure of the Bianchi identity (13.77) suggests that we consider the combination of double commutator of covariant derivatives, which manifestly vanishes (i.e. the Jacobi identity) when we expand out all the commutators:

$$\big[D_\lambda, [D_\mu, D_\nu]\big] + \big[D_\nu, [D_\lambda, D_\mu]\big] + \big[D_\mu, [D_\nu, D_\lambda]\big] \tag{35}$$

$$= D_\lambda D_\mu D_\nu - D_\lambda D_\nu D_\mu - D_\mu D_\nu D_\lambda + D_\nu D_\mu D_\lambda$$

$$+ D_\nu D_\lambda D_\mu - D_\nu D_\mu D_\lambda - D_\lambda D_\mu D_\nu + D_\mu D_\lambda D_\nu$$

$$+ D_\mu D_\nu D_\lambda - D_\mu D_\lambda D_\nu - D_\nu D_\lambda D_\mu + D_\lambda D_\nu D_\mu$$

$$= 0.$$

(2) We now express these double commutators as covariant derivatives of the Riemann curvature tensors. We will find that D_λ, D_μ, D_ν is essentially $D_\lambda R_{\mu\nu\alpha\beta}$ with the extra terms from the three double commutators mutually cancel. Using the expression of the Riemann tensor in terms of commutator of covariant derivatives as worked out in Problem 13.7 (because $D_\lambda A_\alpha$ is a good tensor), we have

$$\big[D_\lambda, [D_\mu, D_\nu]\big]A_\alpha = D_\lambda [D_\mu, D_\nu]A_\alpha - [D_\mu, D_\nu]D_\lambda A_\alpha$$

$$= -D_\lambda\big(R^\gamma_{\alpha\mu\nu}A_\gamma\big) + R^\gamma_{\alpha\mu\nu}D_\lambda A_\gamma + R^\gamma_{\lambda\mu\nu}D_\gamma A_\alpha$$

$$= -D_\lambda R^\gamma_{\alpha\mu\nu}A_\gamma - R^\gamma_{\alpha\mu\nu}D_\lambda A_\gamma + R^\gamma_{\alpha\mu\nu}D_\lambda A_\gamma$$

$$+ R^\gamma_{\lambda\mu\nu}D_\gamma A_\alpha$$

$$= -D_\lambda R^\gamma_{\alpha\mu\nu}A_\gamma + R^\gamma_{\lambda\mu\nu}D_\gamma A_\alpha, \tag{36}$$

as the two middle terms on the third line cancel. Applying this result to every double commutator, equation (35) can then be written as

$$0 = \big([D_\lambda, [D_\mu, D_\nu]] + [D_\nu, [D_\lambda, D_\mu]] + [D_\mu, [D_\nu, D_\lambda]]\big)A_\alpha$$

$$= -D_\lambda R^\gamma_{\alpha\mu\nu}A_\gamma - D_\nu R^\gamma_{\alpha\lambda\mu}A_\gamma - D_\mu R^\gamma_{\alpha\nu\lambda}A_\gamma$$

$$+ R^\gamma_{\lambda\mu\nu}D_\gamma A_\alpha + R^\gamma_{\nu\lambda\mu}D_\gamma A_\alpha + R^\gamma_{\mu\nu\lambda}D_\gamma A_\alpha$$

$$= -\big(D_\lambda R^\gamma_{\alpha\mu\nu} - D_\nu R^\gamma_{\alpha\lambda\mu} - D_\mu R^\gamma_{\alpha\nu\lambda}\big)A_\gamma$$

$$+ \big(R^\gamma_{\lambda\mu\nu} + R^\gamma_{\nu\lambda\mu} + R^\gamma_{\mu\nu\lambda}\big)D_\gamma A_\alpha.$$

The second term on the RHS vanishes because of the cyclic symmetry property of (13.72); the parentheses in the prior term must then vanish, leading to the Bianchi identities.

13.14 **Contraction of Christoffel symbols** The inverse matrix $g_{\mu\nu}{}^{-1}$ has elements $g^{\mu\nu}$, which are related to the determinant g of the matrix $g_{\mu\nu}$ and the cofactors $C^{\mu\nu}$ (associated with elements $g_{\mu\nu}$) as

$$g^{\mu\nu} = \frac{C^{\mu\nu}}{g}. \tag{37}$$

Also, the determinant g can be expanded as (for any fixed μ)

$$g = \sum_\nu g_{\mu\nu} C^{\mu\nu} \tag{38}$$

where we have displayed the summation sign to emphasize that there is no summation over the index μ. Because the determinant is a function of the matrix elements $g_{\mu\nu}$ which in turn are position dependent, we have

$$\frac{\partial g}{\partial x^\alpha} = \frac{\partial g}{\partial g_{\mu\nu}} \frac{\partial g_{\mu\nu}}{\partial x^\alpha} = C^{\mu\nu} \frac{\partial g_{\mu\nu}}{\partial x^\alpha} = g g^{\mu\nu} \partial_\alpha g_{\mu\nu} \tag{39}$$

where we have used (38) and (37) to reach the last two expressions. Knowing this identify, we proceed to make contraction of the Christoffel symbols

$$\Gamma^\mu_{\mu\alpha} = \frac{1}{2} g^{\mu\nu} \left[\partial_\alpha g_{\mu\nu} + \partial_\mu g_{\alpha\nu} - \partial_\nu g_{\mu\alpha} \right].$$

The last two terms cancel, $\partial^\nu g_{\alpha\nu} = \partial^\mu g_{\mu\alpha}$, so that the contraction can be rewritten by (39) as

$$\Gamma^\mu_{\mu\alpha} = \frac{1}{2} g^{\mu\nu} \partial_\alpha g_{\mu\nu} = \frac{1}{2g} \frac{\partial g}{\partial x^\alpha}$$

which is equivalent to the sought after result of

$$\Gamma^\mu_{\mu\alpha} = \frac{1}{\sqrt{-g}} \frac{\partial}{\partial x^\alpha} \sqrt{-g}.$$

13.15 **Contraction of Riemann tensor** Contracting the first two indices $R^\mu_{\mu\alpha\beta}$ (13.58): $\partial_\alpha \Gamma^\mu_{\mu\beta} - \partial_\beta \Gamma^\mu_{\mu\alpha} + \Gamma^\mu_{\nu\alpha} \Gamma^\nu_{\mu\beta} - \Gamma^\mu_{\nu\beta} \Gamma^\nu_{\mu\alpha}$. The dummy indices in the last two terms can be relabelled $\mu \leftrightarrow \nu$; we see that they cancel each other. A straightforward calculation of the first two terms by using the result obtained in Problem 13.14 shows that they cancel each other also.

14.4 **The equation of geodesic deviation** Following the procedure used in Box 6.3, let us consider two particles: one has the spacetime trajectory x^μ and another has $x^\mu + s^\mu$. These two particles, separated by the displacement vector s^μ, obey the respective equations of motion:

$$\frac{d^2 x^\mu}{d\tau^2} + \Gamma^\mu_{\alpha\beta}(x) \frac{dx^\alpha}{d\tau} \frac{dx^\beta}{d\tau} = 0$$

and

$$\left(\frac{d^2 x^\mu}{d\tau^2} + \frac{d^2 s^\mu}{d\tau^2} \right) + \Gamma^\mu_{\alpha\beta}(x+s) \left(\frac{dx^\alpha}{d\tau} + \frac{ds^\alpha}{d\tau} \right) \left(\frac{dx^\beta}{d\tau} + \frac{ds^\beta}{d\tau} \right) = 0.$$

When the separation distance s^μ is small, we can approximate the Christoffel symbols $\Gamma^\mu_{\alpha\beta}(x+s)$ by a Taylor expansion

$$\Gamma^\mu_{\alpha\beta}(x+s) = \Gamma^\mu_{\alpha\beta}(x) + \partial_\lambda \Gamma^\mu_{\alpha\beta} s^\lambda + \cdots .$$

From the difference of the two geodesic equations, we obtain, to first order in s^μ,

$$\frac{d^2 s^\mu}{d\tau^2} = -2\Gamma^\mu_{\alpha\beta} \frac{ds^\alpha}{d\tau} \frac{dx^\beta}{d\tau} - \partial_\lambda \Gamma^\mu_{\alpha\beta} s^\lambda \frac{dx^\alpha}{d\tau} \frac{dx^\beta}{d\tau}. \tag{40}$$

What we are seeking is the relative acceleration (the second derivative of the separation s^μ) along the worldline; thus, a double differentiation along the geodesic curve. From (13.47) we have the first derivative

$$\frac{Ds^\mu}{d\tau} = \frac{ds^\mu}{d\tau} + \Gamma^\mu_{\alpha\beta} s^\alpha \frac{dx^\beta}{d\tau}$$

and the second derivative

$$\frac{D^2 s^\mu}{d\tau^2} = \frac{D}{d\tau}\left(\frac{Ds^\mu}{d\tau}\right) = \frac{d}{d\tau}\left(\frac{Ds^\mu}{d\tau}\right) + \Gamma^\mu_{\alpha\beta}\left(\frac{Ds^\alpha}{d\tau}\right)\frac{dx^\beta}{d\tau}$$

$$= \frac{d}{d\tau}\left(\frac{ds^\mu}{d\tau} + \Gamma^\mu_{\alpha\beta} s^\alpha \frac{dx^\beta}{d\tau}\right) + \Gamma^\mu_{\alpha\beta}\left(\frac{ds^\alpha}{d\tau} + \Gamma^\alpha_{\lambda\rho} s^\lambda \frac{dx^\rho}{d\tau}\right)\frac{dx^\beta}{d\tau}$$

$$= \frac{d^2 s^\mu}{d\tau^2} + \partial_\lambda \Gamma^\mu_{\alpha\beta} \frac{dx^\lambda}{d\tau} s^\alpha \frac{dx^\beta}{d\tau} + \Gamma^\mu_{\alpha\beta} \frac{ds^\alpha}{d\tau}\frac{dx^\beta}{d\tau} + \Gamma^\mu_{\alpha\beta} s^\alpha \frac{d^2 x^\beta}{d\tau^2}$$

$$+ \Gamma^\mu_{\alpha\beta} \frac{ds^\alpha}{d\tau}\frac{dx^\beta}{d\tau} + \Gamma^\mu_{\alpha\beta} \Gamma^\alpha_{\lambda\rho} s^\lambda \frac{dx^\rho}{d\tau}\frac{dx^\beta}{d\tau}. \tag{41}$$

For the $d^2 s^\mu / d\tau^2$ term we use (40); for the $d^2 x^\beta / d\tau^2$ term we use the geodesic equation

$$\frac{d^2 x^\beta}{d\tau^2} = -\Gamma^\beta_{\lambda\rho} \frac{dx^\lambda}{d\tau}\frac{dx^\rho}{d\tau}.$$

This way one finds

$$\frac{D^2 s^\mu}{d\tau^2} = -2\Gamma^\mu_{\alpha\beta}\frac{ds^\alpha}{d\tau}\frac{dx^\beta}{d\tau} - \partial_\lambda\Gamma^\mu_{\alpha\beta}s^\lambda\frac{dx^\alpha}{d\tau}\frac{dx^\beta}{d\tau} + \partial_\lambda\Gamma^\mu_{\alpha\beta}\frac{dx^\lambda}{d\tau}s^\alpha\frac{dx^\beta}{d\tau}$$

$$+ 2\Gamma^\mu_{\alpha\beta}\frac{ds^\alpha}{d\tau}\frac{dx^\beta}{d\tau} - \Gamma^\mu_{\alpha\beta}s^\alpha\Gamma^\beta_{\lambda\rho}\frac{dx^\lambda}{d\tau}\frac{dx^\rho}{d\tau}$$

$$+ \Gamma^\mu_{\alpha\beta}\Gamma^\alpha_{\lambda\rho}s^\lambda\frac{dx^\rho}{d\tau}\frac{dx^\beta}{d\tau}.$$

After a cancellation of two terms and relabeling of several dummy indices, this becomes

$$\frac{D^2 s^\mu}{d\tau^2} = -\partial_\lambda\Gamma^\mu_{\alpha\beta}s^\lambda\frac{dx^\alpha}{d\tau}\frac{dx^\beta}{d\tau} + \partial_\alpha\Gamma^\mu_{\lambda\beta}\frac{dx^\alpha}{d\tau}s^\lambda\frac{dx^\beta}{d\tau}$$

$$- \Gamma^\mu_{\lambda\rho}s^\lambda\Gamma^\rho_{\alpha\beta}\frac{dx^\alpha}{d\tau}\frac{dx^\beta}{d\tau} + \Gamma^\mu_{\rho\beta}\Gamma^\rho_{\lambda\alpha}s^\lambda\frac{dx^\alpha}{d\tau}\frac{dx^\beta}{d\tau}$$

or

$$\frac{D^2 s^\mu}{d\tau^2} = -R^\mu_{\ \alpha\lambda\beta} s^\lambda \frac{dx^\alpha}{d\tau} \frac{dx^\beta}{d\tau},$$

where

$$R^\mu_{\ \alpha\lambda\beta} = \partial_\lambda \Gamma^\mu_{\alpha\beta} - \partial_\beta \Gamma^\mu_{\lambda\alpha} + \Gamma^\mu_{\lambda\rho}\Gamma^\rho_{\alpha\beta} - \Gamma^\mu_{\beta\rho}\Gamma^\rho_{\lambda\alpha}$$

in agreement with (13.58).

14.5 **From geodesic deviation to NR tidal forces** Besides slow moving particles, the Newtonian limit means a weak gravitational field: $g_{\mu\nu} = \eta_{\mu\nu} + h_{\mu\nu}$ with $h_{\mu\nu}$ being small. Thus (13.37) becomes

$$\Gamma^\mu_{\alpha\beta} = \frac{1}{2}\eta^{\mu\rho}\partial_\alpha h_{\beta\rho} + \partial_\beta h_{\alpha\rho} - \partial_\rho h_{\alpha\beta}.$$

Also, in this weak-field limit, we can drop the quadratic terms ($\Gamma\Gamma$) in the curvature so that there are only two terms, related by the interchange of (β, λ) indices

$$\begin{aligned} R^\mu_{\ \alpha\lambda\beta} &= \partial_\lambda \Gamma^\mu_{\alpha\beta} - \partial_\beta \Gamma^\mu_{\lambda\alpha} \\ &= \frac{1}{2}\eta^{\mu\rho}\partial_\lambda\partial_\alpha h_{\beta\rho} - \partial_\lambda\partial_\rho h_{\alpha\beta} - \partial_\beta\partial_\alpha h_{\lambda\rho} + \partial_\beta\partial_\rho h_{\alpha\lambda} \end{aligned}$$

after cancelling two terms. Thus

$$R^i_{\ 0j0} = \frac{1}{2}\partial_j\partial_0 h_{0i} - \partial_j\partial_i h_{00} - \partial_0\partial_0 h_{ji} + \partial_0\partial_i h_{0j} = -\frac{1}{2}\partial_i\partial_j h_{00}.$$

Because the Newtonian limit also has the static field condition, to reach the last line we have dropped all time derivatives. With $h_{00} = -2\Phi/c^2$ as given by (6.20), we have the sought-after relation of

$$R^i_{\ 0j0} = \frac{1}{c^2}\frac{\partial^2\Phi}{\partial x^i \partial x^j}.$$

14.6 **Relativistic spin precession** The Schwarzschild metric for a circular orbit (radius R) in the equatorial plane ($\theta = \pi/2$) has elements of

$$g_{tt} = -\left(1 - \frac{r^*}{R}\right), \quad g_{rr} = \left(1 - \frac{r^*}{R}\right)^{-1}, \quad g_{\theta\theta} = g_{\phi\phi} = R^2. \tag{42}$$

From this and the orthogonality condition $S^\mu U_\mu = 0$ we can immediately deduce the proportionality relation between S^t and S^ϕ:

$$g_{\alpha\beta}S^\alpha U^\beta = -\left(1 - \frac{r^*}{R}\right)S^t U^t + R^2 S^\phi \Omega U^t = 0$$

or

$$S^t = \left(1 - \frac{r^*}{R}\right)^{-1} R^2 \Omega S^\phi. \tag{43}$$

From (42) we can calculate the Christoffel symbols, which mostly vanish, with the nonzero ones being

$$\Gamma^r_{rt} = \frac{r^*}{2R^2}\left(1 - \frac{r^*}{R}\right), \quad \Gamma^r_{\phi\phi} = -R\left(1 - \frac{r^*}{R}\right),$$

$$\Gamma^t_{rt} = \frac{r^*}{2R^2}\left(1 - \frac{r^*}{R}\right)^{-1}, \quad \Gamma^\phi_{r\phi} = \frac{1}{R}. \tag{44}$$

The gyroscope equation $DS^\mu/d\tau = 0$, for the $\mu = \phi$ component, may then be written out as

$$\frac{dS^\phi}{d\tau} + \Gamma^\phi_{r\phi}S^r U^\phi = 0.$$

Substituting in the Christoffel symbol value and replacing the proper time derivative by the coordinate time derivative $d/d\tau = U^t d/dt$, we get

$$\frac{dS^\phi}{dt} + \frac{\Omega}{R}S^r = 0. \tag{45}$$

For the $\mu = r$ component, we have

$$\frac{dS^r}{d\tau} + \Gamma^r_{tt}S^t U^t + \Gamma^r_{\phi\phi}S^\phi U^\phi = 0$$

which, using the relation (43), becomes

$$\frac{dS^r}{dt} - R\Omega\left(1 - \frac{3r^*}{R}\right)S^\phi = 0. \tag{46}$$

For the two other components, $dS^\theta/d\tau = dS^\theta/dt = 0$ leads to $S^\theta(t) = S^\theta(0) = 0$, while the $\mu = t$ equation can be shown to be identical to (45). We can now solve for $S^r(t)$ by first time-differentiating Eq. (46)

$$\frac{d^2 S^r}{dt^2} - R\Omega\left(1 - \frac{3r^*}{R}\right)\frac{dS^\phi}{dt} = 0$$

and plug in the expression for dS^ϕ/dt from (45) to obtain

$$\frac{d^2 S^r}{dt^2} + \Omega'^2 S^r = 0 \tag{47}$$

with

$$\Omega' = \left(1 - \frac{3r^*}{R}\right)^{1/2}\Omega. \tag{48}$$

The simple harmonic oscillator equation (47), with the initial condition of (14.77), has the standard solution

$$S^r(t) = S^r_0 \cos \Omega' t.$$

The S^ϕ component can then be gotten by (46)

$$S^\phi = \frac{\Omega}{R\Omega'^2}\frac{dS^r}{dt} = -\frac{\Omega}{R\Omega'}S^r_0 \sin \Omega' t.$$

15.1 **Gauge transformations**

(a) Consider a coordinate (gauge) transformation as given in (15.12) so that, according to (15.17), $h'_{\alpha\beta} = h_{\alpha\beta} - \partial_\alpha \chi_\beta - \partial_\beta \chi_\alpha$. This implies (by contracting the indices on both sides) the transformation for the trace $h' = h - 2\partial^\beta \chi_\beta$. These two relations can be combined to yield the gauge transformation of $\bar{h}_{\alpha\beta}$,

$$h'_{\alpha\beta} - \frac{h'}{2}\eta_{\alpha\beta} = \bar{h}'_{\alpha\beta} = \bar{h}_{\alpha\beta} - \partial_\alpha \chi_\beta - \partial_\beta \chi_\alpha + \eta_{\alpha\beta}(\partial^\gamma \chi_\gamma). \quad (49)$$

(b) Taking the derivative on both sides of (49), $\partial^\alpha \bar{h}'_{\alpha\beta} = \partial^\alpha \bar{h}_{\alpha\beta} - \Box\chi_\beta$. The new metric perturbation field can be made to obey the Lorentz condition $\partial^\alpha \bar{h}'_{\alpha\beta} = 0$, if $\Box\chi_\beta = \partial^\alpha \bar{h}_{\alpha\beta}$.

(c) Plugging $\bar{h}_{\mu\nu} = \epsilon_{\mu\nu}e^{ikx}$ and $\chi_\nu = X_\nu e^{ikx}$ into the gauge transformation (49), we have

$$\epsilon'_{\mu\nu} = \epsilon_{\mu\nu} - ik_\mu X_\nu - ik_\nu X_\mu + i\eta_{\mu\nu}(k \cdot X) \quad (50)$$

which implies the trace relation $\epsilon'^\mu_\mu = \epsilon^\mu_\mu + 2ik^\mu X_\mu$. This means that if we start with a polarization tensor that is not traceless, it will be traceless $\epsilon'^\mu_\mu = 0$ in a new coordinate if the gauge vector function X_μ for the coordinate transformation is chosen to satisfy the condition $2ik \cdot X = -\epsilon^\mu_\mu$. Now we have used one of the four numbers in X_μ to fix the trace. How can we use the remaining three to obtain $\epsilon_{\mu0} = 0$ which would seem to represent four conditions? This is possible because we are working in the Lorentz gauge and k^μ is a null-vector. Here is the reason. Starting with $\epsilon_{\mu0} \neq 0$, new coordinate transformation leads to (50) with

$$\epsilon'_{\mu0} = \epsilon_{\mu0} - ik_\mu X_0 - ik_0 X_\mu + i\eta_{\mu0}(k \cdot X).$$

Formally $\epsilon'_{\mu0} = 0$ represents four conditions. But, because of $k^\mu \epsilon_{\mu0} = 0$ and $k^2 = 0$, these four equations must obey a constraint relation, obtained by a contraction with the vector k^μ:

$$k^\mu \epsilon_{\mu0} - ik^2 X_0 - ik_0(k \cdot X) + ik_0(k \cdot X) = 0.$$

That is, $k^\mu \epsilon'_{\mu0} = 0$. Thus, $\epsilon'_{\mu0} = 0$ actually stands for three independent relations.

(d) The polarization tensor being symmetric, $\epsilon_{\mu\nu} = \epsilon_{\nu\mu}$, it has 10 independent elements. The Lorentz gauge condition $k^\mu \epsilon_{\mu\nu} = 0$ represents four constraints, $\epsilon^\mu_\mu = 0$ is one, and $\epsilon_{\mu0} = 0$, as discussed above, is three. Thus there are only $10 - 4 - 1 - 3 = 2$ independent elements in the polarization tensor.

15.2 **The Schwarzschild solution** We work first in the Cartesian coordinates $x^\mu = (ct, x^i)$ with an approximate flat metric according to (15.1). The solution (15.74) must fulfill the Lorentz gauge condition (15.18)

$$\partial^\mu \bar{h}_{\mu\nu} = C_{\mu\nu}\partial^\mu \left(\frac{1}{r}\right) = C_{i\nu}\partial^i \left(\frac{1}{r}\right) = -\frac{x^i C_{i\nu}}{r} = 0.$$

for any x^i. This can only be satisfied by $C_{i\nu} = 0$. That is, every element of $\bar{h}_{\mu\nu}$ vanishes except $\bar{h}_{00} = C_{00}/r$. This also means that the trace $\bar{h} = \eta^{\mu\nu}\bar{h}_{\mu\nu} = -C_{00}/r$ and $h = \eta^{\mu\nu}h_{\mu\nu} = C_{00}/r$. From (15.19) we find the perturbation element $h_{00} = C_{00}/2r$. Because of spherical symmetry, we have $h_{11} = h_{22} = h_{33}$, and, to have the correct trace, each must equal to $C_{00}/2r$; namely, $h_{\mu\nu} = (C_{00}/2r)\delta_{\mu\nu}$. We can also fix the constant C_{00} by its Newtonian value as in (15.1) and (6.3)

$$g_{00} = -1 + \frac{C_{00}}{2r} = -1 - \frac{2\Phi}{c^2} = -1 + \frac{2G_N M}{c^2 r},$$

or $C_{00} = 4G_N M/c^2 = 2r^*$. In this way we find the approximate Schwarzschild metric in Cartesian coordinates as

$$ds^2 = -\left(1 - \frac{r^*}{r}\right)c^2 dt^2 + \left(1 + \frac{r^*}{r}\right)\left(dx_1^2 + dx_2^2 + dx_3^2\right).$$

In terms of the spherical coordinate we have

$$\left[dx_1^2 + dx_2^2 + dx_3^2\right] = \left[dr^2 + r^2\left(d\theta^2 + \sin^2\theta d\phi^2\right)\right].$$

In order to show our result in a form closer to the exact Schwarzschild metric of (14.54), we make the coordinate change of $\bar{r}^2 = (1 + (r^*/r))r^2$ so that, within the approximation of dropping $O(r^{*2})$ terms, we have

$$\frac{r^*}{r} \simeq \frac{r^*}{\bar{r}} \quad \text{and} \quad dr = d\bar{r}$$

and

$$ds^2 = -\left(1 - \frac{r^*}{\bar{r}}\right)c^2 dt^2 + \left(1 + \frac{r^*}{\bar{r}}\right)d\bar{r}^2 + \bar{r}^2\left(d\theta^2 + \sin^2\theta d\phi^2\right).$$

This is the expected result of (15.75).

15.3   **Wave effect via the deviation equation**   With a collection of nearby particles, we can consider velocity and separation fields, $U^\mu(x)$ and $S^\mu(x)$. The equation of geodesic deviation (Problem 14.4) may be written as

$$\frac{D^2}{d\tau^2} S^\mu = R^\mu_{\ \nu\lambda\rho} U^\nu U^\lambda S^\rho.$$

Since a slow moving particle $U^\mu = (c, 0, 0, 0) + O(h)$ and the Riemann tensor $R^\mu_{\ \nu\lambda\rho} = O(h)$, this equation has the structure

$$\frac{D^2}{d\tau^2} S^\mu = c^2 \eta^{\mu\sigma} R^{(1)}_{\sigma 00\rho} S^\rho + O(h^2).$$

The Christoffel symbols being of higher order, the covariant derivative may be replaced by ordinary differentiation; this equation at $O(h)$ is

$$\frac{d^2 S^\mu}{dt^2} = \frac{S^\rho}{2}\frac{d^2}{dt^2} h^\mu_{\ \rho}.$$

On the RHS we have used (15.6) and the TT gauge condition of $h_{00} = h_{0\mu} = 0$. The longitudinal component of the separation field S_z is not affected because $h_{3\rho} = 0$ in the TT gauge. For an incoming wave in the "plus" polarization state, the transverse components obey the equations

$$\frac{d^2 S_x}{dt^2} = \frac{S_x}{2} \frac{d^2}{dt^2}(h_+ e^{i(kx-\omega t)}), \quad \frac{d^2 S_y}{dt^2} = -\frac{S_y}{2} \frac{d^2}{dt^2}(h_+ e^{i(kx-\omega t)}).$$

These equations, to the lowest order, have solutions

$$S_x(x) = \left(1 + \frac{1}{2} h_+ e^{i(kx-\omega t)}\right) S_x(0),$$

$$S_y(x) = \left(1 - \frac{1}{2} h_+ e^{i(kx-\omega t)}\right) S_y(0)$$

in agreement with the result in (15.30) and (15.31).

15.4 $\Gamma^\mu_{\nu\lambda}$ and $R^{(2)}_{\mu\nu}$ in the TT gauge

(a) Christoffel symbols: we give samples of the calculation

$$\Gamma^1_{00} = \frac{1}{2} g^{11}(\partial_0 g_{10} + \partial_0 g_{01} - \partial_1 g_{00}) = 0$$

because $h_{10} = h_{01} = h_{00} = 0$ in the TT gauge:

$$\Gamma^1_{01} = \frac{1}{2}(1 - \tilde{h}_{11})(\partial_0 \tilde{h}_{11} + \partial_1 \tilde{h}_{01} - \partial_1 \tilde{h}_{10})$$

$$= \frac{1}{2}(\partial_0 \tilde{h}_+ - \tilde{h}_+ \partial_0 \tilde{h}_+).$$

(b) Ricci tensor: from what we know of Christoffel symbols having the non-vanishing elements of

$$\Gamma^1_{10} = \Gamma^1_{01} = \Gamma^0_{11} = \frac{1}{2} \partial_0 \tilde{h}_+,$$

$$\Gamma^1_{13} = \Gamma^1_{31} = -\Gamma^3_{11} = -\frac{1}{2} \partial_0 \tilde{h}_+$$

together with the same terms with the replacement of indices from 1 to 2, we can calculate the second-order Ricci tensor by

$$R^{(2)}_{\mu\nu} = \Gamma^\alpha_{\alpha\lambda} \Gamma^\lambda_{\mu\nu} - \Gamma^\alpha_{\mu\lambda} \Gamma^\lambda_{\alpha\nu}.$$

Thus

$$R^{(2)}_{00} = \Gamma^\alpha_{\alpha\lambda} \Gamma^\lambda_{00} - \Gamma^\alpha_{0\lambda} \Gamma^\lambda_{\alpha 0}$$

$$= 0 - 2\Gamma^1_{01} \Gamma^1_{10} = \frac{-1}{2}(\partial_0 \tilde{h}_+)^2 = R^{(2)}_{33},$$

$$R^{(2)}_{11} = \Gamma^\alpha_{\alpha\lambda} \Gamma^\lambda_{11} - \Gamma^\alpha_{1\lambda} \Gamma^\lambda_{\alpha 1}$$

$$= 2\Gamma^1_{1\lambda} \Gamma^\lambda_{11} - \Gamma^0_{1\lambda} \Gamma^\lambda_{01} - \Gamma^1_{1\lambda} \Gamma^\lambda_{11} - \Gamma^3_{1\lambda} \Gamma^\lambda_{31}$$

$$= 2\Gamma^1_{10}\Gamma^0_{11} + 2\Gamma^1_{13}\Gamma^3_{11} - \Gamma^0_{11}\Gamma^1_{01} - \Gamma^1_{10}\Gamma^0_{11} - \Gamma^1_{13}\Gamma^3_{11} - \Gamma^3_{11}\Gamma^1_{31}$$

$$= 0 = R^{(2)}_{22}.$$

15.5 **Trace calculation of** $\tilde{I}^{TT}_{ij}$ From the definition (15.58), we have

$$\tilde{I}^{TT}_{ij} = \Pi_{ik}\Pi_{jl}\tilde{I}_{kl} - \frac{1}{2}\Pi_{ij}\left(\Pi_{kl}\tilde{I}_{kl}\right).$$

To calculate its trace, we need to compute $\delta_{ij}\Pi_{ik}\Pi_{jl}$ and $\delta_{ij}\Pi_{ij}$:

$$\delta_{ij}\Pi_{ik}\Pi_{jl} = \delta_{ij}\left(\delta_{ik} - n_i n_k\right)\left(\delta_{jl} - n_j n_l\right)$$

$$= \left(\delta_{kl} - n_k n_l\right) = \Pi_{kl}.$$

Since $\delta_{ij}\Pi_{ij} = \delta_{ij}\left(\delta_{ij} - n_i n_j\right) = 3 - 1 = 2$, we have the trace of $\tilde{I}^{TT}_{ij}$ as

$$\delta_{ij}\tilde{I}^{TT}_{ij} = \left(\delta_{ij}\Pi_{ik}\Pi_{jl}\right)\tilde{I}_{kl} - \frac{1}{2}\left(\delta_{ij}\Pi_{ij}\right)\Pi_{kl}\tilde{I}_{kl}$$

$$= \Pi_{kl}\tilde{I}_{kl} - \Pi_{kl}\tilde{I}_{kl} = 0.$$

15.6 **Derive the relation (15.59)**   From the definition of (15.58) and the shorthand $\left(\Pi\tilde{I}\right) = \Pi_{kl}\tilde{I}_{kl}$, we have

$$\tilde{I}^{TT}_{ij}\tilde{I}^{TT}_{ij} = \left[\Pi_{ik}\Pi_{jl}\tilde{I}_{kl} - \frac{1}{2}\Pi_{ij}\left(\Pi\tilde{I}\right)\right]\left[\Pi_{im}\Pi_{jn}\tilde{I}_{mn} - \frac{1}{2}\Pi_{ij}\left(\Pi\tilde{I}\right)\right].$$

Using the result, $\Pi_{ik}\Pi_{il} = \Pi_{kl}$, obtained in Problem 15.5 we can carrying out the various multiplications:

$$\Pi_{ik}\Pi_{jl}\Pi_{im}\Pi_{jn}\tilde{I}_{kl}\tilde{I}_{mn} = \left(\Pi_{ik}\Pi_{im}\right)\left(\Pi_{jl}\Pi_{jn}\right)\tilde{I}_{kl}\tilde{I}_{mn}$$

$$= \Pi_{km}\Pi_{nl}\tilde{I}_{kl}\tilde{I}_{mn}$$

$$= \left(\delta_{km} - n_k n_m\right)\left(\delta_{nl} - n_n n_l\right)\tilde{I}_{kl}\tilde{I}_{mn}$$

$$= \tilde{I}_{ij}\tilde{I}_{ij} - 2n_k n_l\tilde{I}_{ki}\tilde{I}_{li} + n_k n_l n_m n_n\tilde{I}_{kl}\tilde{I}_{mn}$$

and

$$\Pi_{ij}\Pi_{ij}\left(\Pi\tilde{I}\right)^2 = \left(\delta_{ij} - n_i n_j\right)\left(\delta_{ij} - n_i n_j\right)\left(\Pi\tilde{I}\right)^2 = 2\left(\Pi\tilde{I}\right)^2,$$

and

$$\Pi_{ij}\Pi_{im}\Pi_{jn}\tilde{I}_{mn}\left(\Pi\tilde{I}\right) = \Pi_{jm}\Pi_{jn}\tilde{I}_{mn}\left(\Pi\tilde{I}\right) = \Pi_{mn}\tilde{I}_{mn}\left(\Pi\tilde{I}\right) = \left(\Pi\tilde{I}\right)^2$$

so we have

$$\tilde{I}^{TT}_{ij}\tilde{I}^{TT}_{ij} = \tilde{I}_{ij}\tilde{I}_{ij} - 2n_k n_l\tilde{I}_{ki}\tilde{I}_{li} + n_k n_l n_m n_n\tilde{I}_{kl}\tilde{I}_{mn} - \frac{1}{2}\left(\Pi\tilde{I}\right)^2$$

$$= \frac{1}{2}\left[2\tilde{I}_{ij}\tilde{I}_{ij} - 4\tilde{I}_{ik}\tilde{I}_{il}n_k n_l + \tilde{I}_{ij}\tilde{I}_{kl}n_i n_j n_k n_l\right]$$

because $\left(\Pi\tilde{I}\right) = -n_i n_j\tilde{I}_{ij}$ and $\left(\Pi\tilde{I}\right)^2 = \tilde{I}_{ij}\tilde{I}_{kl}n_i n_j n_k n_l$.

Glossary of symbols
and acronyms

Symbols

K^{μ}	3.2	covariant force 4-vector
$K^{\mu}_{(a)}$	7.3	Killing vector along a
l	7.3	angular momentum constant of motion (particle)
l	10.5	angular momentum number
l_{H}	9.1	Hubble length
l_{Pl}	8.5	Planck length
L	5.2	Lagrangian
$\mathfrak{L}$	9.1	luminosity
$\mathbf{L}^{\mu}_{\nu}$	3.1	element of a (Lorentz) coordinate transformation matrix
m	2.1	mass (test particle)
m	9.4	apparent magnitude of a light source
m_{G}	4.2	gravitational mass
m_{I}	4.2	inertial mass
M	2.1	mass (source)
M	9.4	absolute magnitude of a light source
$\mathcal{M}$	11.6	magnetization
M_{Pl}	8.5	Planck mass
n	4.3	index of refraction
n	10.3	number density
n_i	15.4	component of a unit radial vector
$O(x)$	1.2	of the order of x
p	3.2	magnitude of momentum 3-vector
p	8.1	time coordinate in Advanced EF system
p	10.2	pressure
p^{μ}	3.2	momentum 4-vector
p_{Pl}	8.5	Planck momentum (E_{Pl}/c)
P_{b}	1.1	binary orbit period
P_l	10.5	Legendre polynomial
q	8.1	time coordinate in Retarded EF system
q	11.4	cosmic deceleration parameter
r	5.1	radial coordinate in polar/spherical coordinates
r^*	7.1	Schwarzschild radius
$r_{\pm}$	8.4	event horizons of a Kerr black hole
$r_{S\pm}$	8.4	stationary limit (infinite redshift) surfaces of a Kerr black hole
R	5.1	radius of curvature
R	9.4	radius of the universe in Robertson–Walker geometry
R_0	9.4	radius of the universe in Robertson–Walker geometry now
R	13.3	Ricci curvature scalar
$\mathbf{R}_{ij}$	2.1	element of a rotational matrix
$R^{\mu}_{\lambda\alpha\beta}$	13.3	Riemann curvature tensor
$R_{\mu\nu}$	13.3	Ricci curvature tensor
s	2.3	invariant spacetime interval, proper length
t_0	9.1	cosmic time of the present epoch, age of the universe
t_{H}	9.1	Hubble time (=$1/H_0$)
t_{Pl}	8.5	Planck time
t_{RM}	10.5	radiation–matter equality time
t_{tr}	11.4	cosmic time of deceleration to acceleration transition
$T_{\mu\nu}$	11.1	energy-momentum tensor (also as $t_{\mu\nu}$)
U^{μ}	3.2	velocity 4-vector
w	10.1	cosmic equation of state parameter
Y^m_l	10.5	spherical harmonics
z	9.1	wavelength shift, redshift
β	2.2	v/c, velocity in unit of c

γ	2.2	the Lorentz factor $(1 - \beta^2)^{-1/2}$
Γ	10.3	reaction rate
$\Gamma^{\nu}_{\lambda\rho}$	6.2	Christoffel symbols
δ_{ij}	2.1	Kronecker delta
∂_{μ}	3.1	del operator (4D) with $\partial_0 = \frac{1}{c}\frac{\partial}{\partial t}$, etc.
$\triangle$	8.4	a Kerr metric parameter in Boyer–Lindquist coordinates
ε	4.2	gravity strength parameter $(G_N M r^{-1} c^{-2})$
ε_{ijk}	12.2	Levi-Civita symbol (3D)
$\varepsilon_{\mu\nu\lambda\rho}$	12.2	Levi-Civita symbols (4D)
ϵ	5.3	angular excess
$\epsilon_{\mu\nu}$	15.2	polarization tensor of a gravitational wave
$\eta_{\mu\nu}$	3.1	metric of (flat) Minkowski spacetime
θ	5.1	polar angle coordinate
κ	6.3	gravity strength $(-8\pi G_N/c^4)$
κ	7.3	energy constant of motion
λ	3.2	wavelength
λ	5.2	curve parameter as in $x^{\mu}(\lambda)$
Λ	11.1	cosmological constant
ξ	5.3	dimensionless cylindrical radial coordinate ρ/R
Π_{ij}	15.4	transverse projection operator
ρ	4.1	mass density
ρ	5.1	cylindrical radial coordinate
ρ	8.4	a Kerr metric parameter in Boyer–Lindquist coordinates
ρ_c	9.2	cosmic critical density
σ	15.3	fractional change of separation: "strain amplitude"
σ	5.3	area
σ_{ij}	13.3	area tensor
τ	2.3	proper time
Φ	4.1	gravitational potential
ϕ	5.1	azimuthal angle coordinate
$\phi(x)$	9.2	scalar field such as inflation/Higgs field
ϕ^{μ}	12.3	force density
χ	5.3	dimensionless radial coordinate r/R
ψ	3.1	rapidity parameter
ω	3.2	(angular) frequency
Ω	9.2	ratio of density to critical density (ρ/ρ_c)
Ω_B	9.2	ratio of baryonic matter density to critical density
Ω_B	1.1	relativistic (geodetic) spin precession in the double pulsar system
Ω_{lum}	9.2	ratio of luminous baryonic matter density to critical density
Ω_{gas}	9.2	ratio of nonluminous baryonic matter density to critical density
Ω_M	9.2	ratio of total mass density to critical density
Ω_{DM}	9.2	ratio of dark matter mass density to critical density
Ω_R	9.2	ratio of radiation density to critical density
Ω_{Λ}	11.1	ratio of dark energy density to critical density

Acronyms

2dF	9.1	Two-Degree Field Galaxy Redshift Survey
4D	3.1	four-dimensional
AGN	8.3	active galactic nucleus
B	9.2	baryon

CDM	9.2	cold dark matter
CMB	9.1	cosmic microwave background
COBE	10.5	Cosmic Background Explorer satellite observatory
CP	9.3	cosmological principle
DM	9.2	dark matter
EM	2.1	electromagnetic
EF	8.1	Eddington–Finkelstein
EP	4.1	equivalence principle
FLRW	11.1	Friedmann–Lemaître–Robertson–Walker
GR	1.1	general relativity
GRB	8.3	Gamma-ray burst
GUT	11.2	grand unified theory
IGM	9.1	intergalactic medium
ISCO	8.2	innermost stable circular orbit
LEF	5.2	local Euclidean frame
LHS	2.1	left-hand side
LIGO	15.3	Laser Interferometer Gravitational Observatory
LISA	15.3	Laser Interferometer Space Antenna
MACHO	7.2	MAssive Compact Halo Objects
MDU	10.1	matter-dominated universe
NR	3.2	non-relativistic
QFT	8.5	quantum field theory
RDU	10.1	radiation-dominated universe
RHS	2.1	right-hand side
RW	9.4	Robertson-Walker
SDSS	9.1	Sloan Digital Sky Survey
SNe	11.4	supernovae
SR	1.1	special relativity
SUSY	11.7	supersymmetry
TT	15.2	transverse-traceless gauge
WIMP	9.2	weakly interacting massive particle
WMAP	11.3	Wilkinson Microwave Anisotropy Probe satellite observatory

Some units

AU	7.2	Astronomical Unit (average distance between earth and sun)
d	15.4	day
GeV	8.5	giga electron volt
Gyr	9.1	billion years
kpc	9.1	kiloparsec
MeV	8.2	million electron volt
Mpc	9.1	megaparsec
pc	9.1	parsec

References and bibliography

Alcock, C. *et al.* (1997). The MACHO project: Large Magellanic Cloud microlensing results from the first two years and the nature of the galactic dark halo, *Astrophys. J.*, **486**, 697.

Alpher, R.A. and Herman, R. (1948). Evolution of the universe, *Nature*, **162**, 774.

Becker, K., Becker, M., and Schwarz, J.H. (2007). *String Theory and M-theory: A Modern Introduction*, Cambridge University Press, Cambridge.

Bennett, C.L. *et al.* (2003). First year WMAP observations: maps and basic results, *Astrophys. J.*, Suppl. ser., **143**, 1.

Bertone, G. Hooper, D., and Silk, J. (2005). Particle dark matter: Evidence, candidates and constraints, *Phys. Rept.*, **405**, 279.

Breton, R.P. *et al.* (2008). Relativistic spin precession in the double pulsar, *Science*, **321**, 104.

Burgay, M. *et al.* (2003). An increased estimate of the merger rate of double neutron stars from observation of a highly relativistic system, *Nature*, **426**, 531.

Burles, S. *et al.* (2001). Big-bang nucleosynthesis predictions for precision cosmology, *Astrophys. J. Lett.*, **552**, L1.

Cen, R. and Ostriker. J.P. (1999). Where are the baryons?. *Astrophys. J.*, **514**, 1.

Cheng, T.P. and Li, L.F. (1984). *Gauge Theory of Elementary Particle Physics*. Clarendon Press, Oxford.

Cheng, T.P. and Li, L.F. (1988). Resource letter: GI-1 gauge invariance, *Am. J. Phys.*, **56**, 596.

Cheng, T.P. and Li, L.F. (2000). *Gauge Theory of Elementary Particle Physics: Problems and Solutions*, Section 8.3, Clarendon Press, Oxford.

Clowe, D. *et al.* (2006). A direct empirical proof of the existence of dark matter. *Astrophys. J.*, **648**, L109.

Cook, R.J. (2004). Physical time and physical space in general relativity, *Am. J. Phys.*, **72**, 214.

Cornish, N.J. *et al.* (2004). Constraining the topology of the universe, *Phys. Rev. Lett.*, **92**, 201302.

Cram, T.R. *et al.* (1980). A complete, high-sensitivity 21-cm hydrogen line survey of M-31, *Astron. Astrophys.*, Suppl., **40**, 215.

Danforth, C.W. and Shull. J.M. (2008). The low-z intergalactic medium. III. H I and metal absorbers at $z < 0.4$, *Astrophys. J.*, **679**, 194.

Das, A. (1993). *Field Theory, A Path Integral Approach*. (Section 5.1), World Scientific, Singapore.

de Bernardis, P. *et al.* Boomerang collaboration (2000). A flat universe from high-resolution maps of the cosmic microwave background radiation, *Nature*, **404**, 955.

Dicke, R.H., Peebles, P.J.E., Roll, P.G., and Wilkinson, D.T. (1965). Cosmic blackbody radiation, *Astrophys. J.*, **412**, 414.

Doeleman, S.S. *et al.* (2008). Event-horizon-scale structure in the supermassive black hole candidate at the Galactic centre, *Nature*, **455**, 78.

Doroshkevich, A.G. and Novikov, I.D. (1964). Mean density of radiation in the metagalxy and certain problems in relativistic cosmology, *Soviet Phys. Doklady*, **9**, 111.

Einstein, A. (1989). *The Collected Papers of Albert Einstein*. Vols 2, 3, and 4, Princeton University Press, Princeton, NJ.

Einstein, A., Lorentz, H.A., Weyl, H., and Minkowski, H. (1952). *The Principle of Relativity—A Collection of Original Papers on the Special and General Theory of Relativity*. Dover, New York.

Ellis, G.F.R. and Williams, R.M. (1988). *Flat and Curved Space-Times*. Clarendon Press, Oxford.

Fixsen, D.J. *et al.* (1996). The cosmic microwave background spectrum from the full COBE FIRAS data set, *Astrophys. J.*, **473**, 576.

Freedman, W.L. and Turner, M.S. (2003). Colloquium: measuring and understanding the universe, *Rev. Mod. Phys.*, **75**, 1433.

Frieman, J.A., Turner, M.S., and Huterer, D. (2008). Dark energy and the accelerating universe, *Ann. Rev. Astron. Astrophys.*, **46**, 385.

Gamow, G. (1946). Expanding universe and the origin of elements, *Phys. Rev.*, **70**, 572.

Gamow, G. (1948). The evolution of the universe, *Nature*, **162**, 680.

Gamow, G. (1949). On relativistic cosmogony, *Rev. Mod. Phys.*, **21**, 367.

Gamow, G. (1970). *My World Line, An Informal Autobiography*. Viking, New York, p. 44.

Gott, J.R. *et al.* (2005). A map of the universe, *Astrophys. J.*, **624**, 463.

Guth, A.H. (1981). The inflationary universe: a possible solution to the horizon and flatness problems, *Phys. Rev. D*, **23**, 347.

Hafele, J.C. and Keating, R.E. (1972). Around-the-world atomic clocks: observed relativistic time gains, *Science*, **177**, 168.

Hamilton, A.J.S. (2002). Postulates of special relativity, http://casa.colorado.edu/ajsh/sr/postulate.html

Hanany, S. *et al.* (2000) Constraints on cosmological parameters from MAXIMA-1, *Astrophys. J. Lett.*, **545**, L5.

Hinshaw, G. *et al.* (2009) Five-year Wilkinson Microwave Anisotropy Probe (WMAP) observations: Data processing, sky maps, and basic results, *Astrophys. J. Supp. (Feb 2009)*, arXiv:0803.0732

Kibble, T.W.B. (1985). *Classical Mechanics*. 3rd edn, Longman Press, London.

Kiritsis, E. (2007). *String Theory in a Nutshell*, Princeton University. Press, Princeton, NJ.

Kramer, M. *et al.* (2006). Tests of general relativity from timing the double pulsar, *Science*, **314**, 97.

Krauss, L.M. and Chaboyer, B. (2003). Age estimates of globular clusters in the Milky Way: constraints on cosmology, *Science*, **299**, 65.

Logunov, A.A. (2001). *On the Articles by Henri Poincaré "On the Dynamics of the Electron,"* translated into English by G. Pontecorvo, 3rd edn, JINR, Dubna.

Luminet, J.-P. *et al.* (2003). Dodecahedral space topology as an explanation for weak wide-angle temperature correlations in the cosmic microwave background, *Nature*, **425**, 593.

Lyne, A.G. *et al.* (2004). A double-pulsar system: a rare laboratory for relativistic gravity and plasma physics, *Science*, **303**, 1153.

MacFadyen, A.I. and Woosley, S. (1999). Collapsars: Gamma-ray bursts and explosions in failed supernovae, *Astrophys. J.* **524**, 262.

Mather, J.C. *et al.* (1990). A preliminary measurement of cosmic microwave background spectrum by the Cosmic Background Explorer (COBE) satellite, *Astrophys. J. Lett.*, **354**, L37.

Miller, A.D. *et al.* TOCO collaboration (1999). A measurement of the angular power spectrum of the CMB from $l = 100$ to 400, *Astrophys. J. Lett.*, **524**, L1.

Okun, L.B., Selivanov, K.G., and Telegdi, V.L. (2000). On the interpretation of the redshift in a static gravitational field, *Am. J. Phys.*, **68**, 115.

Penzias, A.A. and Wilson, R.W. (1965). A measurement of excess antenna temperature at 4080 Mc/s, *Astrophys. J.*, **412**, 419.

Perlmutter, S. *et al.* Supernova Cosmology Project (1999). Measurements of omega and lambda from 42 high redshift supernovae, *Astrophys. J.*, **517**, 565.

Peters, P.C. and Mathews, J. (1963). Gravitational radiation from point masses in a Keplerian orbit, *Phys. Rev.*, **131**, 435.

Phillips, M.M. (1993). The absolute magnitudes of Type IA supernovae, *Astrophys. J.*, **413**, 105.

Pound, R.V. and Rebka, G.A. (1960). Apparent weight of photons, *Phys. Rev. Lett.*, **4**, 337.

Pound, R.V. and Snider, J.L. (1964). Effects of gravity on nuclear resonance, *Phys. Rev. Lett.*, **13**, 539.

Riess, A.G. *et al.* High-z Supernova Search Team (1998). Observational evidence from supernovae for an accelerating universe and a cosmological constant, *Astron. J.*, **116**, 1009.

Riess, A.G. (2000). The case for an accelerating universe from supernovae, *Publ. Astron. Soc. Pac.*, **112**, 1284.

Riess, A.G. *et al.* (2001). The farthest known supernova: support for an accelerating universe and a glimpse of the epoch of deceleration, *Astrophys. J.*, **560**, 49.

Riess, A.G. *et al.* (2004). Type Ia Supernova discoveries at $z > 1$ from the Hubble Space Telescope: evidence for past deceleration and constraints on dark energy evolution, *Astrophys. J.*, **607**, 665.

Sartori, L. (1996). The role of Lorentz and Poincaré in the birth of relativity, Section 4.9, in *Understanding Relativity*, University of California Press, Berkeley, CA.

Schwinger, J. (1986). *Einstein's Legacy—The Unity of Space and Time*, Chapter 4, Scientific American Books, New York.

Shapiro, I.I. (1964) Fourth test of general relativity, *Phys. Rev. Lett.*, **13**, 789.

Shapiro, I.I. *et al.* (1971) Fourth test of general relativity: New radar result, *Phys. Rev. Lett.*, **26**, 1132.

Smoot, G.F. *et al.* (1990). COBE Differential Microwave Radiometers: instrument design and implementation, *Astrophys. J.*, **360**, 685.

Smoot, G.F. *et al.* (1992). Structure in the COBE Differential Microwave Radiometer first year maps, *Astrophys. J.*, **396**, L1.

Tegmark, M. *et al.* (2006). Cosmological constraints from the SDSS luminous red galaxies, *Phys. Rev., D*, **74**, 123507.

Tolman, R.C. (1934). *Relativity, Thermodynamics and Cosmology.* Clarendon Press, Oxford.

Uhlenbeck, G. (1968). *Introduction to the General Theory of Relativity* (unpublished lecture notes, Rockefeller University).

Weisberg, J.M. and Taylor, J.H. (2003). The relativistic binary pulsar B1913+16, *Proceedings of Radio Pulsars*, Chania, Crete, 2002 (eds) M. Bailes, *et al.* (ASP. Conf. Series).

White, M. and Cohn, J.D. (2002). Resource letter: TACMB-1 the theory of anisotropies in the cosmic microwave background, *Am. J. Phys.*, **70**, 106.

Wilczek, F. (2004). Total relativity, *Physics Today*, **57** (No. 4), 10.

Zel'dovich, Y.B. (1968). The Cosmological constant and the theory of elementary particles, *Sov. Phys. Usp.*, **11**, 381.

Zwiebach, B. (2009). *A First Course in String Theory.* 2nd edn, Cambridge University Press.

Bibliography

This bibliography, by no means an exhaustive listing, contains titles that I have consulted while writing this book. They are arranged so that more recent publications and my personal favorites are placed at the top in each category.

1 Books at a level comparable to our presentation

    (a) *General relativity (including cosmology)*

        i Hartle, J.B., *Gravity: An Introduction to Einstein's General Relativity* (Addison-Wesley, San Francisco, 2003).

        i Hobson, M.P., Efstathiou, G. and Lasenby, A.N., *General Relativity: An Introduction for Physicists* (Cambridge University Press, 2006).

        ii Ohanian, H. and Ruffini, R., *Gravitation and Spacetime*, 2nd edn (Norton, New York, 1994).

        iii D'Inverno, R., *Introducing Einstein's Relativity* (Oxford University Press, 1992).

        iv Kenyon, I.R., *General Relativity* (Oxford University Press, 1990).

        v Schutz, B.F., *A First Course in General Relativity* (Cambridge University Press, 1985).

        vi Landau, L.D. and Lifshitz, E.M., *The Classical Theory of Fields* (Butterworth-Heinemann/Elsevier, Amsterdam, 1975).

    (b) *Cosmology*

        i. Ryden, B., *Introduction to Cosmology* (Addison-Wesley, San Francisco, 2003).

        ii. Raine, D.J. and Thomas, E.G., *An Introduction to the Science of Cosmology* (Institute of Physics, Bristol, 2001).

        iii. Rowan-Robinson, M., *Cosmology*, 4th edn (Oxford University Press, 2003).

    iv. Berry, M.V., *Principles of Cosmology and Gravitation* (Adam Hilger, Bristol, 1989).

    v. Harrison, E., *Cosmology: The Science of the Universe*, 2nd edn (Cambridge University Press, 2000).

    vi. Liddle, A. and Loveday, J., *The Oxford Companion to Cosmology* (Oxford University Press, 2008).

(c) *Special relativity*

    i. Sartori, L., *Understanding Relativity* (University of California Press, Berkeley, CA.,1996).

    ii. Kogut, J.B., *Introduction to Relativity* (Harcourt/Academic Press-Burlington, MA., 2001).

2 Books at a more advanced level

(a) *General relativity (including cosmology)*

    i Carroll, S.M., *Spacetime and Geometry* (Addison-Wesley, San Francisco, 2004).

    i Misner, C., Thorne, K., and Wheeler, J.A., *Gravitation* (W.H. Freeman, New York, 1970).

    ii Weinberg, S., *Gravitation and Cosmology* (Wiley, New York, 1972).

    iii Stephani, H., *General Relativity*, 2nd edn (Cambridge University Press, 1990).

    iv Wald, R.M., *General Relativity* (Chicago University Press, 1984).

(b) *Cosmology*

    i Weinberg, S., *Cosmology* (Oxford University Press, 2008).

    i Dodelson, S., *Modern Cosmology* (Academic Press, San Diego, CA. 2003).

    ii Peacock, J.A., *Cosmological Physics* (Cambridge University Press, 1999).

    iii Peebles, P.J.E., *Principles of Physical Cosmology* (Princeton University Press, 1993).

    iv Kolb, E.W. and Turner, M.S., *The Early Universe* (Addison-Wesley, San Francisco, 1990).

3 General interest and biographical books

(a) i. Thorne, K.S., *Black Holes & Time Warps: Einstein's Outrageous Legacy* (Norton, New York, 1994).

    ii. Pais, A., *Subtle is the Lord... The Science and Life of Albert Einstein* (Oxford University Press, 1982).

    iii. Weinberg, S., *The First Three Minutes* (Basic Books, New York, 1972).

    iv. Schwinger, J., *Einstein's Legacy—The Unity of Space and Time* (Scientific American Books, New York, 1986).

    v. Guth, A.H., *The Inflationary Universe* (Addison-Wesley, San Francisco, 1997).

    vi. Green, B., *The Elegant Universe* (Norton, New York, 1999).

    vii. Smolin, L., *Three Roads to Quantum Gravity* (Basic Books, New York, 2001).

viii. Zee, A., *Einstein's Universe: Gravity at Work and Play* (Oxford University Press, 2001).

ix. French, A. (ed.), *Einstein—A Centenary Volume* (Harvard University Press, 1979).

x. Howard, D. and J. Stachel (eds), *Einstein and the History of General Relativity* (Birkhäuser Boston, 1989).

Picture credits

Fig. 7.5 and book cover: Image from website (http://hubblesite.org/newscenter/newsdesk/archive/releases/2000/07/image/b). Credits: S. Baggett (STScI), A. Fruchter (NASA), R. Hook (ST-ECF), and Z. Levay (STScI).

Fig. 11.6:   Image from (de Bernardis *et al.*, 2000)

Fig. 15.3:   Courtesy of LIGO Hanford Observatory, funded by NSF. Image from website (http://www.ligo-wa.caltech.edu/).

Fig. 15.4:   Courtesy of the NAIC — Arecibo Observatory, a facility of the NSF. Image from website (http://www.ligo-wa.caltech.edu/).

Table of physical constants

1 Fundamental constants

Speed of light	$c = 3.00 \times 10^8 \, \mathrm{m\,s^{-1}}$
Gravitational constant	$G_N = 6.67 \times 10^{-11} \, \mathrm{m^3 \, kg^{-1} \, s^{-2}}$
Planck's constant	$\hbar = 1.05 \times 10^{-34} \, \mathrm{J\,s} = 6.58 \times 10^{-22} \, \mathrm{MeV\,s}$
Boltzmann's constant	$k_B = 1.38 \times 10^{-23} \, \mathrm{J\,K^{-1}} = 8.62 \times 10^{-11} \, \mathrm{MeV\,K^{-1}}$
Electron mass	$m_e = 0.911 \times 10^{-30} \, \mathrm{kg} = 0.511 \, \mathrm{MeV\,c^{-2}}$
Proton mass	$M_p = 1.67 \times 10^{-27} \, \mathrm{kg} = 938 \, \mathrm{MeV\,c^{-2}}$
Neutron mass	$M_n = M_p + 2.31 \times 10^{-30} \, \mathrm{kg} = M_p + 1.29 \, \mathrm{MeV\,c^{-2}}$

2 Conversion constants

Astronomical unit	$1 \, \mathrm{AU} = 1.50 \times 10^{11} \, \mathrm{m}$
Year (sidereal)	$1 \, \mathrm{yr} = 3.15 \times 10^7 \, \mathrm{s}$
Light-year	$1 \, \mathrm{ly} = 9.46 \times 10^{15} \, \mathrm{m}$
Parsec	$1 \, \mathrm{pc} = 3.09 \times 10^{16} \, \mathrm{m} = 3.26 \, \mathrm{ly}$
Arc second	$'' = 4.85 \times 10^{-6} \, \mathrm{radians}$
Electron volt	$1 \, \mathrm{eV} = 1.60 \times 10^{-19} \, \mathrm{J}$

3 Cosmological parameters

Hubble parameter	$h = 0.72 \pm 0.05$
Hubble constant	$H_0 \equiv h \times 100 \; \mathrm{(km/s)/\,Mpc}$
Hubble time	$t_H \equiv H_0^{-1} = 9.79 \, h^{-1} \, \mathrm{Gyr}$
Hubble length	$l_H \equiv c H_0^{-1} = 3000 \, h^{-1} \, \mathrm{Mpc}$
Critical density	$\rho_{c,0} \equiv 3H_0^2/8\pi G_N = 1.88 \times 10^{-26} \, h^2 \, \mathrm{kg\,m^{-3}}$

4 Astronomical constants

Mass of the sun	$M_\odot = 1.99 \times 10^{30}\,\text{kg}$
Schwarzschild radius of the sun	$r_\odot^* \equiv 2G_N M_\odot c^{-2} = 2.96\,\text{km}$
Radius of the sun	$R_\odot = 6.96 \times 10^5\,\text{km}$
Luminosity of the sun	$L_\odot = 3.85 \times 10^{26}\,\text{W}$
Mass of the earth	$M_\oplus = 5.98 \times 10^{24}\,\text{kg}$
Schwarzschild radius of the earth	$r_\oplus^* \equiv 2G_N M_\oplus c^{-2} = 0.886\,\text{cm}$
Radius of the earth (at equator)	$R_\oplus = 6.38 \times 10^3\,\text{km}$

Index